Nutrition in Chronic Conditions

Nutrition in Chronic Conditions

Editors

Omorogieva Ojo

Amanda Rodrigues Amorim Adegboye

MDPI • Basel • Beijing • Wuhan • Barcelona • Belgrade • Manchester • Tokyo • Cluj • Tianjin

Editors

Omorogieva Ojo
University of Greenwich
London
UK

Amanda Rodrigues
Amorim Adegboye
Coventry University
Coventry
UK

Editorial Office
MDPI
St. Alban-Anlage 66
4052 Basel, Switzerland

This is a reprint of articles from the Special Issue published online in the open access journal *Nutrients* (ISSN 2072-6643) (available at: https://www.mdpi.com/journal/nutrients/special_issues/nutrition_chronic_conditions).

For citation purposes, cite each article independently as indicated on the article page online and as indicated below:

LastName, A.A.; LastName, B.B.; LastName, C.C. Article Title. *Journal Name* **Year**, *Volume Number*, Page Range.

ISBN 978-3-0365-7132-4 (Hbk)
ISBN 978-3-0365-7133-1 (PDF)

Contents

About the Editors

Omorogieva Ojo

Prof. Omorogieva Ojo is a Professor in Nutrition and Diabetes at the University of Greenwich, London. He is a Registered Nutritionist with the Association for Nutrition, UK, and a Senior Fellow of the Higher Education Academy, UK. He is also Practice Lead in Centres for Chronic Illness and Ageing, and Exercise Activity and Rehabilitation, Institute for Lifecourse Development. The focus of Prof. Ojo's research is nutritional intervention in diabetes and other chronic conditions. He has widely published in many international journals, presented at many international conferences, and edited books. Prof. Ojo is an internationally acclaimed expert in nutrition and diabetes, and he is a recognised supervisor by the UK Council for Graduate Education. Prof. Ojo is on the editorial board of many international journals including Nutrients and International Journal of Environmental Research and Public Health. He has been Guest Editor of eight Special Issues in the areas of Nutrition and/or Diabetes. Prof. Ojo is a peer reviewer for many international journals and international grant awarding bodies.

Amanda Rodrigues Amorim Adegboye

Dr Amanda Rodrigues Amorim Adegboye is an Epidemiologist and registered Public Health Nutritionist with nearly 20 years of experience in research and academia. She is also a fellow of the Higher Education Academy (HEA). Currently, Dr Adegboye is an Associate Director of Research and Engagement at Coventry University. Dr Adegbye has vast experience in conducting primary research in Nutrition Science and Public Health, both writing and editing scientific manuscripts. Dr Adegboye is passionate about embedding Equity, Diversity and Inclusion (EDI) into all steps of the research process.

nutrients MDPI

Editorial

The Effects of Nutrition on Chronic Conditions

Omorogieva Ojo [1,*] and Amanda Rodrigues Amorim Adegboye [2,3]

1 School of Health Sciences, University of Greenwich, Avery Hill Campus, London SE9 2UG, UK
2 Centre for Agroecology, Water and Resilience, Coventry University, Coventry CV8 3LG, UK
3 Centre for Healthcare Research, Coventry University, Coventry CV1 5FB, UK
* Correspondence: o.ojo@greenwich.ac.uk

The effects of nutrition on chronic conditions, such as diabetes, obesity, heart disease, and stroke, continue to generate interest among researchers. This is based on the fact that diet is a modifiable risk factor [1]. The composition of diet, including the proportions and types of macronutrients and micronutrients, is a major contributor to chronic diseases [1].

The beneficial effects of nutritional interventions on chronic conditions have been well documented, although differences remain among researchers concerning their overall impact [1]. Evaluations of the role of nutrition in chronic conditions draw on diet's effects on body weight, body composition, glycemic and insulin excursions, and vascular remodelling. The effect of diet in modulating gut microbiota dysbiosis is also an evolving area of research [2].

This Special Issue, entitled "Nutrition in Chronic Conditions", aims to examine the effect of nutrition in the development, care, and management of chronic conditions. This Special Issue includes 11 original studies conducted in high- and middle-income countries, 3 systematic reviews with meta-analysis, and 3 literature reviews. This editorial provides an overview of the key findings of the papers published in this Special Issue. These papers are broadly divided into seven topics: the effects of diet on (1) insulin and glucose metabolism; (2) gut health; (3) brain and cognitive impairment; (4) infections, chronic conditions, malnutrition, and all-cause mortality; (5) obesity and dietary variables in postmenopausal women; (6) non-alcoholic fatty liver disease in mice, specifically the consumption of coffee; and (7) chronic conditions and COVID-19 infection.

1. Insulin and Glucose Metabolism

The literature on the use of diets with a low glycemic index (GI) and a low glycemic load (GL) in the management of diabetes in adult populations is vast. However, little is known if GI and GL peaks are related to glycemic control, particularly in young and healthy populations. Using a representative national school-based sample of students (12–17 years old) without diabetes, da Rocha et al. [3] investigated the association between dietary indicators of the quality of carbohydrate intake and markers of glycemic control. The authors found the GI of diet was better at predicting insulinemia, regardless of weight status, compared to the GL [3]. The authors argued that guidance on food consumption based on carbohydrate quality should be provided to adolescents as a measure of glycemic control, as higher GIs are highly associated with the intake of refined carbohydrates. Encouraging healthy lifestyle habits combined with a diet with low GI and low GL can also help control obesity and reduce the risk of developing type 2 diabetes [3].

Apart from the impact of low GI diets on metabolism, postprandial insulin, glucose, and triglyceride responses have been investigated. Louca et al. [4] found that individuals with hypertension had higher postprandial insulinemic and lipemic responses to two standardized test meals compared to the normotensive controls after adjustments for sex, age, and BMI. This effect was partially mediated by visceral fat mass. No significant difference was observed for postprandial glucose. These findings corroborate existing literature on the key role of visceral fat in metabolic syndrome [4].

Citation: Ojo, O.; Adegboye, A.R.A. The Effects of Nutrition on Chronic Conditions. *Nutrients* **2023**, *15*, 1066. https://doi.org/10.3390/nu15051066

Received: 1 February 2023
Accepted: 8 February 2023
Published: 21 February 2023

A drastic and prolonged reduction in carbohydrate intake leads to the exhaustion of glucose reserves in the body, shifting metabolism into ketogenesis and inducing hepatic oxidation of fatty acids. This process produces ketone bodies, an important alternative to glucose as the body's source of energy. Dilliraj et al. [5] conducted a review to explore current evidence regarding ketone bodies in relation to nutrition, metabolic pathways, signalling functions, and effects on clinical conditions. Based on the studies reviewed, ketone bodies, which are formed under normal metabolism in the absence of glucose, play a key role in controlling oxidative stress and inflammation, resulting in improved mitochondrial function and growth, energy rescue, and adaptative epigenetic control. However, clinical trials are needed to validate the results obtained from in vitro and in vivo studies as well as from animal models [5].

2. Gut Health

Ojo et al. [6] conducted a systematic review and meta-analysis of randomised controlled trials to evaluate the effects of almonds on gut microbiota, glycometabolism, and inflammatory markers in patients with type 2 diabetes. This review was conducted against the backdrop of rising global prevalence of type 2 diabetes and the recognition that nutritional interventions, including the use of almonds, which are rich sources of dietary fibre, essential minerals, protein, and monounsaturated fatty acids, may be effective in managing symptoms of type 2 diabetes.

Ojo et al. [6] found that an almond-based diet was effective in promoting the growth of short-chain fatty acid-producing bacteria and lowering glycated haemoglobin and body mass index in patients with type 2 diabetes. The nutrient composition of almond, such as high fibre content and low glycemic index, may be involved in the biological mechanism of its effect. However, almonds did not appear to have a significant effect ($p > 0.05$) on fasting blood glucose, postprandial blood glucose, inflammatory parameters, glucagon-like peptide 1, and Homeostatic Model Assessment of Insulin Resistance [6].

In a separate systematic review, Ojo et al. [7] carried out a network meta-analysis of randomised controlled trials. This review aimed to evaluate the effect of prebiotics and oral antidiabetic agents on the gut microbiome in patients with type 2 diabetes. Prebiotics are substrates (non-viable) that are resistant to gastric acid and intestinal absorption and are used selectively by host microorganisms, which leads to benefits [7]. Prebiotics may promote eubiosis of the gut microbiome and ensure glucose homeostasis in patients with type 2 diabetes. The network meta-analysis found that prebiotics significantly reduced ($p < 0.05$) glycated haemoglobin, compared to the control, in patients with type 2 diabetes. However, prebiotics and oral antidiabetic agents did not have a significant effect ($p > 0.05$) on the gut microbiome, body mass index, fasting blood glucose, and postprandial blood glucose [7].

3. Brain and Cognitive Impairment

Metabolic syndrome (MS) is a prevalent condition worldwide and is characterised by a cluster of conditions, including central obesity, hyperglycemia, insulin resistance, hypertension, and dyslipidemia. Insulin resistance, believed to be a key underlying mechanism responsible for MS, affects multiple tissues and organs, including the central nervous system, leading to cognitive impairment and Alzheimer's disease (AD). However, the inverse relationship between MS and cognitive impairment has not been fully explored. Rojas et al. [8] reviewed studies investigating a new hypothesis suggesting that cognitive impairment plays a role in the development of insulin resistance and, consequently, the appearance of MS. The authors concluded that a bidirectional relationship between MS and cognitive impairment seems to exist. However, large-scale longitudinal studies are still required to establish a causal relationship between these two factors [8].

In another study, Sochocka et al. [9] investigated the effect of *Ginkgo biloba* extract (Egb) as an alternative therapy on the mechanisms of innate immune response of peripheral blood leukocytes (PBLs) in patients with AD. The authors found that EGb has advantageous properties for health management in older adults and AD sufferers, especially in women with AD [9].

4. Infections, Chronic Conditions, Malnutrition, and All-Cause Mortality

Helicobacter pylori (H. pylori) infection is the most common cause of gastritis and other gastrointestinal disorders worldwide. Habbash et al. [10] investigated whether there is an association between dietary habits and *H. pylori* infections among 200 Bahraini adults. The authors found that among *H. pylori*-infected individuals, the consumption of coffee, green tea, and honey was significantly lower compared to non-infected individuals. They also found that vitamin D deficiency was a risk factor for *H. pylori* infection (OR = 1.1; 95% CI: 1.05, 1.18; $p < 0.001$). The authors suggested that coffee, green tea, and honey intake might be protective against *H. pylori* infection [10]. However, given the retrospective, cross-sectional study design, no causal relationship between dietary factors and *H. pylori* infection could be inferred.

Zupo et al. [11] conducted a systematic review and meta-analysis of the prevalence of zinc deficiency among patients suffering from inflammatory bowel disease (IBD). Zinc is essential for cell growth, tissue repair, and immune function. The authors included 17 studies and estimated an overall pooled prevalence of 50% (95% CI 0.48–0.52). However, the reviewed studies showed high heterogeneity, $I^2 = 96\%$ [11]. These studies were further divided into two groups: Crohn's disease (CD) ($n = 9$) and ulcerative colitis (UC) ($n = 8$). The prevalence of zinc deficiency was higher in patients with CD (54%) compared to those with UC (41%). The results point out that one in two patients with IBD has zinc deficiency, which can play a role in the severity of the disease. Therefore, clinicians should monitor zinc levels and other trace elements in patients with IBD.

Naber and Purohit [12] conducted a review to explore the role of diet in the management of chronic kidney disease (CKD). The authors focused on the Dietary Approaches to Stop Hypertension, the Mediterranean diet, and the whole-food, plant-based diet for their potential role in delaying CKD progression. They found strong evidence supporting the relevance of diets, which meet the daily nutritional requirement of patients, in the prevention and progression of CKD, particularly the whole-food, plant-based diet without the inclusion of animal products.

Malnutrition is prevalent among patients with chronic heart failure (CHF) due to the lack of appetite, unintentional weight loss, impaired intestinal function, catabolic metabolism, and other comorbidities. Schuetz et al. [13] investigated the cost-effectiveness of an individualised nutritional therapy in 645 hospitalised patients with CHF. The authors found that the overall incremental cost-effectiveness ratio for the individualised nutritional therapy vs. no nutritional therapy was 2625 Swiss Francs per life day gained. They concluded that the intervention increased life expectancy at an acceptable incremental cost-effectiveness ratio [13].

Malnutrition and loss of muscle mass are also prevalent among patients with cancer. In clinical assessments, handgrip strength (HGS) is used as a proxy of overall muscle strength. However, there are no population-specific values for HGS, particularly among oncology patients. Tribolet et al. [14] proposed sex-specific values for HGS stratified by age and tumour entity and tested their prognostic ability. The authors validated the prognostic value of HGS with respect to long-term mortality in hospitalised undernourished patients with cancer [14], which might aid clinical decisions.

Kwon et al. [15] examined the association between the intake of dietary fibres and CVD and all-cause mortality in the general population and among those with hypertension, diabetes, and dyslipidemia in a 10-year longitudinal study. After adjustments for confounders, the authors found that a higher intake of fibres reduced the risk of both all-cause mortality and CVD mortality [15].

5. Obesity and Dietary Variables in Postmenopausal Women

There are controversial results regarding the relationship between obesity and bone metabolism. López-Gómez et al. [16] investigated the differences in bone turnover among 250 postmenopausal women with and without obesity and compared their risk of fracture at five years of follow-up. The authors found that a bone formation marker (P1NP) was higher in women without obesity compared to women with obesity. However, postmenopausal women with obesity showed lower marker levels of bone formation, especially at younger ages. On the other hand, older women with obesity showed higher markers of bone resorption. This might be due to a decrease in vitamin D levels in women with obesity irrespective of age, which is associated with a high parathyroid hormone (PTH) level. However, no significant difference in the risk of fracture based on BMI was observed (OR = 0.90; 95% CI 0.30–2.72; $p = 0.85$). The authors concluded that the potential protective effect of obesity on bone mass and osteoporosis needs to be further investigated in other studies [16].

6. The Influence of Coffee Consumption on Non-Alcoholic Fatty Liver Disease in Mice

Di Mauro et al. [17] examined the effect of coffee consumption on non-alcoholic fatty liver disease in mice. In particular, this study aimed to establish if the intake of coffee might influence the expression of long non-coding ribonucleic acid (IncRNAs) in the liver. In this study, 24 four-week-old male mice were housed randomly in cages. Following one week of acclimation, the mice were randomly assigned to 1 of the 3 diets for 12 weeks, including a standard diet, a high-fat diet, and a high-fat diet plus decaffeinated coffee solution. This study found that decaffeinated coffee was effective in modulating the expression of IncRNAs, which are involved in the key pathways in the onset and progression of non-alcoholic fatty liver disease [17].

7. Chronic Conditions and COVID-19 Infection

In 2020, healthcare systems around the world were challenged by the COVID-19 pandemic. During the pandemic, it was observed that age and pre-existing health conditions (e.g., cancer, asthma, cancer, obesity, and diabetes) were risk factors for negative COVID-19 infection outcomes. Several micronutrient deficiencies were also associated with a higher risk for severe clinical symptoms. Voelkle et al. [18] found a heightened prevalence of micronutrient deficiencies (e.g., selenium, vitamin D, vitamin A, and zinc), particularly in older patients hospitalised for COVID-19. These deficiencies were also associated with more severe COVID-19 infection. The authors highlighted the need for further research regarding the effect of micronutrient supplementation on the treatment and prevention of COVID-19 infection [18].

The lockdown policies adopted by many countries to control the spread of the virus had a major impact on people's lifestyles. Arayess et al. [19] observed that a personalised lifestyle intervention in children with overweight and obesity was less successful in decreasing BMI z-score during the COVID-pandemic compared to the same intervention one year prior to the first lockdown in the Netherlands [19].

8. Conclusions

Based on the above research findings, it is clear that nutrition plays an important role in the development and severity of chronic conditions in children, adults, and older adults. Therefore, healthy dietary patterns should be promoted, and further research should be conducted to fully understand the biological pathways regarding how diet may influence chronic diseases.

Author Contributions: Conceptualization, O.O. and A.R.A.A.; methodology O.O. and A.R.A.A.; validation, O.O. and A.R.A.A.; formal analysis, O.O. and A.R.A.A.; writing—original draft preparation, O.O. and A.R.A.A.; writing—review and editing, O.O. and A.R.A.A. All authors have read and agreed to the published version of the manuscript.

Conflicts of Interest: The authors declare no conflict of interest.

References

1. Iriti, M.; Varoni, E.M.; Vitalini, S. Healthy Diets and Modifiable Risk Factors for Non-Communicable Diseases-The European Perspective. *Foods* **2020**, *9*, 940. [CrossRef] [PubMed]
2. Ojo, O.; Feng, Q.-Q.; Ojo, O.O.; Wang, X.-H. The Role of Dietary Fibre in Modulating Gut Microbiota Dysbiosis in Patients with Type 2 Diabetes: A Systematic Review and Meta-Analysis of Randomised Controlled Trials. *Nutrients* **2020**, *12*, 3239. [CrossRef] [PubMed]
3. da Rocha, C.M.M.; Gama, V.P.M.; de Moura Souza, A.; Massae Yokoo, E.; Verly Junior, E.; Bloch, K.V.; Sichieri, R. Comparison of Quality of Carbohydrate Metrics Related to Fasting Insulin, Glycosylated Hemoglobin and HOMA-IR in Brazilian Adolescents. *Nutrients* **2022**, *14*, 2544. [CrossRef] [PubMed]
4. Louca, P.; Berry, S.E.; Bermingham, K.; Franks, P.W.; Wolf, J.; Spector, T.D.; Valdes, A.M.; Chowienczyk, P.; Menni, C. Postprandial Responses to a Standardised Meal in Hypertension: The Mediatory Role of Visceral Fat Mass. *Nutrients* **2022**, *14*, 4499. [CrossRef] [PubMed]
5. Dilliraj, L.N.; Schiuma, G.; Lara, D.; Strazzabosco, G.; Clement, J.; Giovannini, P.; Trapella, C.; Narducci, M.; Rizzo, R. The Evolution of Ketosis: Potential Impact on Clinical Conditions. *Nutrients* **2022**, *14*, 3613. [CrossRef] [PubMed]
6. Ojo, O.; Wang, X.-H.; Ojo, O.O.; Adegboye, A.R.A. The Effects of Almonds on Gut Microbiota, Glycometabolism, and Inflammatory Markers in Patients with Type 2 Diabetes: A Systematic Review and Meta-Analysis of Randomised Controlled Trials. *Nutrients* **2021**, *13*, 3377. [CrossRef] [PubMed]
7. Ojo, O.; Wang, X.; Ojo, O.O.; Brooke, J.; Jiang, Y.; Dong, Q.; Thompson, T. The Effect of Prebiotics and Oral Anti-Diabetic Agents on Gut Microbiome in Patients with Type 2 Diabetes: A Systematic Review and Network Meta-Analysis of Randomised Controlled Trials. *Nutrients* **2022**, *14*, 5139. [CrossRef] [PubMed]
8. Rojas, M.; Chávez-Castillo, M.; Pirela, D.; Parra, H.; Nava, M.; Chacín, M.; Angarita, L.; Añez, R.; Salazar, J.; Ortiz, R.; et al. Metabolic Syndrome: Is It Time to Add the Central Nervous System? *Nutrients* **2021**, *13*, 2254. [CrossRef] [PubMed]
9. Sochocka, M.; Ochnik, M.; Sobczyński, M.; Gębura, K.; Zambrowicz, A.; Naporowski, P.; Leszek, J. Ginkgo Biloba Leaf Extract Improves an Innate Immune Response of Peripheral Blood Leukocytes of Alzheimer's Disease Patients. *Nutrients* **2022**, *14*, 2022. [CrossRef] [PubMed]
10. Habbash, F.; Alalwan, T.A.; Perna, S.; Ahmed, N.; Sharif, O.; Al Sayyad, A.; Gasparri, C.; Ferraris, C.; Rondanelli, M. Association between Dietary Habits and *Helicobacter pylori* Infection among Bahraini Adults. *Nutrients* **2022**, *14*, 4215. [CrossRef] [PubMed]
11. Zupo, R.; Sila, A.; Castellana, F.; Bringiotti, R.; Curlo, M.; De Pergola, G.; De Nucci, S.; Giannelli, G.; Mastronardi, M.; Sardone, R. Prevalence of Zinc Deficiency in Inflammatory Bowel Disease: A Systematic Review and Meta-Analysis. *Nutrients* **2022**, *14*, 4052. [CrossRef] [PubMed]
12. Naber, T.; Purohit, S. Chronic Kidney Disease: Role of Diet for a Reduction in the Severity of the Disease. *Nutrients* **2021**, *13*, 3277. [CrossRef] [PubMed]
13. Schuetz, P.; Sulo, S.; Walzer, S.; Krenberger, S.; Stagna, Z.; Gomes, F.; Mueller, B.; Brunton, C. Economic Evaluation of Individualized Nutritional Support for Hospitalized Patients with Chronic Heart Failure. *Nutrients* **2022**, *14*, 1703. [CrossRef] [PubMed]
14. Tribolet, P.; Kaegi-Braun, N.; Gressies, C.; Baumgartner, A.; Wagner, K.-H.; Stanga, Z.; Schuetz, P. Handgrip Strength Values Depend on Tumor Entity and Predict 180-Day Mortality in Malnourished Cancer Patients. *Nutrients* **2022**, *14*, 2173. [CrossRef] [PubMed]
15. Kwon, Y.-J.; Lee, H.-S.; Park, G.; Kim, H.-M.; Lee, J.-W. Association of Dietary Fiber Intake with All-Cause Mortality and Cardiovascular Disease Mortality: A 10-Year Prospective Cohort Study. *Nutrients* **2022**, *14*, 3089. [CrossRef] [PubMed]
16. López-Gómez, J.J.; Pérez-Castrillón, J.L.; García de Santos, I.; Pérez-Alonso, M.; Izaola-Jauregui, O.; Primo-Martín, D.; De Luis-Román, D.A. Influence of Obesity on Bone Turnover Markers and Fracture Risk in Postmenopausal Women. *Nutrients* **2022**, *14*, 1617. [CrossRef]
17. Mauro, S.; Salomone, F.; Scamporrino, A.; Filippello, A.; Morisco, F.; Guido, M.; Lembo, V.; Cossiga, V.; Pipitone, R.M.; Grimaudo, S.; et al. Coffee Restores Expression of lncRNAs Involved in Steatosis and Fibrosis in a Mouse Model of NAFLD. *Nutrients* **2021**, *13*, 2952. [CrossRef]

18. Voelkle, M.; Gregoriano, C.; Neyer, P.; Koch, D.; Kutz, A.; Bernasconi, L.; Conen, A.; Mueller, B.; Schuetz, P. Prevalence of Micronutrient Deficiencies in Patients Hospitalized with COVID-19: An Observational Cohort Study. *Nutrients* **2022**, *14*, 1862. [CrossRef]
19. Arayess, L.; Knockaert, N.; Winkens, B.; Lubrecht, J.W.; Verweij, M.; Vreugdenhil, A.C.E. The Side-Effects of the COVID-19 Pandemic: Increased BMI z-Score in Children with Overweight and Obesity in a Personalised Lifestyle Intervention One Year after the Start of the Pandemic in The Netherlands. *Nutrients* **2022**, *14*, 1942. [CrossRef]

nutrients

MDPI

Article

Comparison of Quality of Carbohydrate Metrics Related to Fasting Insulin, Glycosylated Hemoglobin and HOMA-IR in Brazilian Adolescents

Camilla Medeiros Macedo da Rocha [1,2,*], Vanessa Proêza Maciel Gama [3,4,5], Amanda de Moura Souza [6], Edna Massae Yokoo [4], Eliseu Verly Junior [2], Katia Vergetti Bloch [6] and Rosely Sichieri [2,*]

1 Instituto de Alimentação e Nutrição, Centro Mutidisciplinar UFRJ—Macaé, Universidade Federal do Rio de Janeiro, Av. Aluizio da Silva Gomes 50, Novo Cavaleiros, Macaé 27930-560, RJ, Brazil

2 Instituto de Medicina Social, Universidade do Estado do Rio de Janeiro, Rua São Francisco Xavier 524, Pavilhão João Lyra Filho, 7° Andar, Rio de Janeiro 20550-900, RJ, Brazil; eliseujunior@gmail.com

3 Instituto Federal de Educação, Ciência e Tecnologia Fluminense, Av. Souza Mota 350, Parque Fundão, Campos dos Goytacazes 28060-010, RJ, Brazil; vanessa.proeza@gmail.com

4 Departamento de Epidemiologia e Bioestatística, Universidade Federal Fluminense, Rua Marques de Paraná 303, 3° Andar, Centro, Niterói 24030-210, RJ, Brazil; eyokoo@gmail.com

5 Curso de Especialização em Nutrição Clínica, Instituto de Nutrição Josué de Castro, Universidade Federal do Rio de Janeiro, Av. Carlos Chagas Filho 373—Bloco J 2° Andar, Ilha do Fundão, Rio de Janeiro 21941-902, RJ, Brazil

6 Instituto de Estudos em Saúde Coletiva, Universidade Federal do Rio de Janeiro, Avenida Horácio Macedo s/n, Ilha do Fundão, Rio de Janeiro 21941-598, RJ, Brazil; amandamoura@iesc.ufrj.br (A.d.M.S.); kbloch@iesc.ufrj.br (K.V.B.)

* Correspondence: camillammrocha@gmail.com (C.M.M.d.R.); sichieri.rosely@gmail.com (R.S.)

Citation: da Rocha, C.M.M.; Gama, V.P.M.; de Moura Souza, A.; Massae Yokoo, E.; Verly Junior, E.; Bloch, K.V.; Sichieri, R. Comparison of Quality of Carbohydrate Metrics Related to Fasting Insulin, Glycosylated Hemoglobin and HOMA-IR in Brazilian Adolescents. *Nutrients* **2022**, *14*, 2544. https://doi.org/10.3390/nu14122544

Academic Editors: Omorogieva Ojo and Amanda R Amorim Adegboye

Received: 19 March 2022
Accepted: 13 June 2022
Published: 19 June 2022

Publisher's Note: MDPI stays neutral with regard to jurisdictional claims in published maps and institutional affiliations.

Abstract: Low glycemic index (GI) and glycemic load (GL) diets are effective for glycemic control (GC) associated with a carbohydrate-controlled meal plan. However, whether GI and GL peaks are related to GC is unknown. Objective: To compare the daily GI (DGI)/GL (DGL) and average GI (AvGI)/GL (AvGL) of meals (accounting for peaks) related to GC markers (GCM) in Brazilian adolescents. Methods: A representative national school-based (public/private) sample of students without diabetes, 12–17 years of age, was evaluated. Food intake was based on a 24 h recall. The models for complex cluster sampling were adjusted (sex, sexual maturation, age, and physical activity). Results: Of 35,737 students, 74% were from public schools, 60% girls, 17% overweight, and 8% obese. The minimum DGI and DGL were observed at lunch, with higher values at night. Fasting insulin was 1.5 times higher in overweight/obese (OW) girls, and 1.7 times higher in OW boys than in normal-weight (NW) girls. The same trend was observed for the homeostatic model assessment for insulin resistance (HOMA-IR) (OW = 2.82 vs. NW = 1.84 in girls; OW = 2.66 vs. NW = 1.54 in boys; $p < 0.05$). The daily and average metrics were greater for NW adolescents. Glycosylated hemoglobin was not associated with these metrics, except for AvGL. Insulin and HOMA-IR were associated with all metrics in NW adolescents, with greater coefficients associated with AvGL. Among overweight/obese adolescents, only GI metrics were associated ($\beta = 0.23$; AvGI and insulin) and appeared to have the best association with GCM. Conclusions: Among NW adolescents, GL is a better measure of carbohydrate quality, but for those with overweight/obesity, carbohydrate consumption is more associated with GC, probably because they eat/report small amounts of carbohydrates.

Keywords: glycemic control; intake; adolescents; glycemic index; glycemic load

1. Introduction

There is consistent evidence of the protective role of a low-glycemic index (GI) diet in diabetes. For adults with diabetes, studies have clearly indicated that diets with a low GI promote better glycemic control. Both glycated hemoglobin and fasting glucose levels are reduced compared to low- and high-GI diets, and a low-GI diet also improves insulin sensitivity [1,2]. GI refers to the increase in glycemia after the intake of 50 g of available carbohydrates and characteristics of carbohydrate chain, such as monosaccharides, type of starch with amylose being less readily digested compared to amylopectin, explain differences in GI. However, the actual overall glycemic effect of foods depends on the amount of ingested carbohydrates, which are measured by the glycemic load (GL) and other components of the meal, such as fat and protein content and organic acids [3]. For patients with diabetes, both the amount and quality of carbohydrates are usually controlled.

The role of carbohydrate metrics in the prevention of diabetes, cardiovascular disease and obesity is not well understood. A meta-analysis did not show a protective effect of low GI food intake [4]. A systematic review of 21 randomized controlled trials including 2538 participants indicated that there was no convincing evidence of the effect of a low-GI diet on blood pressure, serum lipids or cardiovascular events [5].

The lack of association between the quality of carbohydrate metrics and conditions other than diabetes may be related to the greater variability in the amount of carbohydrates consumed during the day.

A better understanding of the overall impact of carbohydrate intake is to evaluate the most commonly used markers of glycemic control in relation to glycemic control. Low-GI diets are commonly used to characterize the quality of carbohydrates in the diet and their impact on glycemic blood levels, but the quantity consumed is also important, as measured by the GL [3]. More recently, high peaks of glycemic load, more precisely breakfast, have also been associated with metabolic syndrome [6]. Although the quality of carbohydrate intake has been associated with non-infectious chronic diseases, the contribution of individual meals is not clear.

Adolescents are an interesting group for testing these possibilities. In addition, insulin levels increase from childhood until the development of type 2 diabetes in adulthood [7–10], preventing insulin resistance during adolescence.

This study aimed to evaluate the association of dietary indicators of the quality of carbohydrate intake with markers of glycemic control in a large representative survey of Brazilian adolescents. We measured the GI and GL of the whole diet, as well as the average GI and GL of the individual meals. Thus, the GI of the diet, the load and peaks of GI and GL were evaluated as indicators of the quality of carbohydrate intake. Insulin, HOMA-IR and glycosylated hemoglobin were used as indicators of glycemic control.

2. Materials and Methods

The present study is part of the Estudo de Risco Cardiovascular em Adolescentes (ERICA), a nationwide, cross-sectional, multicenter, school-based study conducted in 2013–2014 to estimate the prevalence of risk factors for cardiovascular diseases among adolescents enrolled in public and private schools in 273 Brazilian cities with over 100,000 inhabitants [11].

2.1. Subjects and Response Rates

The adolescents were selected by complex sampling with 32 strata (27 Federation Units and 5 sets of municipalities with more than 100,000 inhabitants representing each of the country's macroregions). The schools were selected in each geographic stratum with a probability of selection proportional to its size, and three classes were drawn in each school [12]. Adolescents aged 12–17 years were invited to participate in the study.

The sampling process selected 1247 schools that had a total of 114,162 students enrolled, but 10.4% of them were ineligible (215 pregnant girls, 364 with physical or cognitive disabilities and 11,256 were outside the age group studied), leaving 102,327 students. Owing to the need for fasting to analyze biochemical markers and because the blood collection was performed in schools, avoiding the displacement of adolescents to a laboratory, only students from the morning session participated in this stage. Almost three quarters (70.9%) were enrolled in the morning turn and could therefore have blood samples collected. However, only 40,732 students participated in the study. Among the reasons for the abstention of 43.8%, we can list refusals of blood collection, school absences on collection days and non-fasting.

The information contained in the ERICA questionnaire was organized into blocks, and during data processing to assess quality and consistency, each block of each adolescent was searched for certain key information that classified that block as complete or not. For the analysis of the current study, blocks named student questionnaire, 24 h recall (R24h), blood sample, anthropometry and blood pressure (data not included in this article) were used, which, combined, provided complete information on 36,956 adolescents.

Other information on the sampling design [12], the methodology adopted [11] in ERICA and the percentage of response and characterization of participants and those who refused to participate [13] can be found in previous publications.

The focus of this study was to identify directions for the prevention of type 2 diabetes; therefore, 1219 adolescents who reported having diabetes were excluded. A total of 35,737 students were analyzed, including 21,489 girls and 14,248 boys.

2.2. Variables and Missing Data

Fasting glucose, insulin and glycosylated hemoglobin levels were measured in fasting blood samples using the methods recommended by the Brazilian Society of Clinical Pathology, following the quality criteria for laboratory analysis [11]. The homeostatic model assessment for insulin resistance (HOMA-IR) was calculated considering the fasting glucose and fasting insulin levels. This is a recognized method for assessing glycemic control in children and adolescents [14–16]. Fasting insulin, glycosylated hemoglobin and HOMA-IR were used as markers of glycemic control in this study.

Some blood samples were lost during transport, storage or processing, resulting in missing data for glycated hemoglobin (female = 36; male = 16), fasting blood glucose (female = 91; male = 62), insulin (female = 84; male = 56) and HOMA-IR (female = 162; male = 114).

Food intake was assessed by one R24h using the United States Department of Agriculture Automated Multiple-Pass Method [17]. Information was collected in notebooks during face-to-face interviews conducted by trained field evaluators [11]. Ten percent of the sample answered the second R24h question.

The composition of macronutrients and fiber was calculated based on the compilation of information on the nutritional composition of food from the Family Budget Survey, 2008–2009 [18]. The amount of glycemic carbohydrates in each food was estimated by the difference between the total carbohydrates and total fibers, both measured in grams.

The GI of the majority of foods' GI were obtained from the database of the Boden Institute of Obesity, Nutrition, Exercise and Eating Disorders and Charles Perkins Centre at the University of Sydney [19] available at http://www.glycemicindex.com/, (accessed on 1 October 2018); for the 12 regional foods, data were searched in articles [20–25]; and for 30 foods without information, the glycemic indices of similar foods in composition were used. In addition, carbonated non-sugared beverages, classified as diet or light, were assigned a GI value of zero. The same was true for high-alcohol distilled spirits, such as cachaça and brandy. Beers and drinks are not included in this group because they contain a considerable amount of carbohydrates in their composition. Meat, offal and sausages have low or no carbohydrate concentrations and therefore have zero GI.

The daily glycemic index (DGI) and load (DGL) were calculated according to the recommendations of the Food and Agriculture Organization of the United Nations/World Health Organization of 1998 [26], with the amount in grams of glycemic carbohydrate consumed from a given food multiplied by its glycemic index and weighted by the total glycemic carbohydrate consumed on the day, defining the contribution of each food. The GI of each food on the day was summed to generate the DGI. DGL considers the amount of glycemic carbohydrate consumed from a given food multiplied by its glycemic index, summed for the whole day and divided by 100.

Similarly, GI and GL were calculated for each occasion of food intake reported at R24h and the average daily glycemic index (AvGI) and average daily glycemic load (AvGL) were calculated to synthesize the GI and GL peaks during the day for each of these intake moments.

Overweight and obesity are very important conditions in the development of glycemic control [7–10]. Therefore, in the present study, we chose to conduct an analysis stratified by weight status according to BMI. We considered normal-weight (NW) adolescents with a BMI z-score < 1, overweight adolescents with a BMI z-score $\geq$ 1 and obese adolescents with a BMI z-score $\geq$ 2. [27]. For the analysis, we stratified overweight and obese adolescents in the same stratum (i.e., OW).

Sex, age and sexual maturation were associated with insulin secretion [28], and the results were adjusted for them. In addition, physical activity level has an important influence on obesity and glycemic control and was included in the adjusted model [29].

The weekly time spent on physical activity was measured using a questionnaire adapted and validated for Brazilian adolescents, containing a list of 24 activities on which adolescents responded, when they were practiced, the weekly frequency, and the time spent in each session [30]. The estimate excluded low-intensity activities, such as walking the dog or caring for children, and commuting activities, such as walking as a means of transport to school, home, or work. Finally, reports of weekly physical activity greater than 2100 min/week were considered of low quality and were therefore disregarded in the analyses, resulting in 3169 missing data records (1091 girls and 1268 boys). To classify the physical activity levels, three categories were created: inactive (0 min/week), insufficiently active (<300 min/week), and active ($\geq$300 min/week).

2.3. Data Analysis

Descriptive analyses involved the calculation of means and their respective confidence intervals for continuous variables and frequencies for categorical variables. GI and GL have different dimensions, and for their effects to be comparable in linear regression, they were standardized to a distribution with a mean of zero and standard deviation. The association between GI and GL and each of the glycemic control indicators was evaluated using linear regression adjusted for the factors already described. All analyses were performed considering complex sampling using the proc survey command in SAS® OnDemand for Academics available at https://welcome.oda.sas.com/ (accessed on 15 January 2022).

3. Results

Among girls, 7.0% were classified as obese, and 17.6% were classified as overweight. Among boys, 9.6% were classified as obese, and 17.3% were classified as overweight. The majority of the students were from public schools and lived in the capital of the state. The mean adolescent age was 14 years (Table 1).

There were no differences in the DGI and AvGI according to sex or weight status, with a mean of 59. In contrast, the DGL was higher in boys, especially in those with normal weight (195). Both boys and girls with normal weight had greater DGL and AvGL values. (Table 1). Among normal weight adolescents the average of Insulin was 8.71 mU/L for girls and 7.10 mU/L for boys. Among overweight/obese individuals, these values were 1.5 times higher in females (13.20 mU/L) and 1.7 times higher in males (12.04 mU/L).

Glycosylated Hb levels were not predicted by the glycemic control markers. Only AvGL was associated with glycosylated Hb, but the magnitude of this association was small (β = 0.006). The glycemic index of the diet measured as daily and as average values (DGL and AvGL) were the best predictors of insulin and HOMA-IR, without significant differences for average or daily among NW adolescents, but greater values of the regression coefficient for the average GI among overweight/obese adolescents. In NW adolescents, AvGL was associated with insulin levels (Table 2).

Table 1. Frequencies, means of sociodemographic, anthropometric and dietary characteristics of adolescents, by sex and weight status. Brazil, 2013–2014.

Variable	Female *n* = (21,489)		Male (*n* = 14,248)	
	Normal Weight (*n* = 16,144)	Overweight/ Obese (*n* = 5345)	Normal Weight (*n* = 10,377)	Overweight/ Obese (*n* = 3871)
		Frequency (%) (95% CI)		
		School Type		
Public	75.00 (71.82; 78.19)	72.33 (68.81; 75.84)	75.76 (72.57; 78.96) [a]	65.15 (61.09; 69,21) [a]
Private	25.00 (21.81; 28.18)	27.67 (24.16; 31.19)	24.24 (21.04; 27.43) [b]	34.85 (30.79; 38.91) [b]
		Residence Area		
Capital	74.23 (72.85; 75.60)	72.4 (70.68; 74.20)	73.94 (72.35; 75.54)	73.62 (71.62; 75.62)
Countryside	25.77 (24.40; 27.15)	27.56 (25.80; 29.32)	26.06 (24.46; 27.65)	26.38 (24.38; 28.38)
		Physical activity level [1]		
Inactive	26.60 (25.80; 27.41) [c]	23.33 (22.10; 24.55) [c]	9.88 (9.23; 10.52)	10.22 (9.21; 11.23)
Insufficiently active	33.40 (32.58; 34.21)	32.11 (30.80; 33.42)	26.44 (25.42; 27.45)	28.52 (26.96; 30.07)
Active	40.00 (39.17; 40.84) [d]	44.56 (43.17; 45.95) [d]	63.69 (62.59; 64.81)	61.26 (59.64; 62.89)
		Mean (95% CI)		
Age (years)	14.74 (14.65–14.83)	14.47 (14.37–14.58)	14.72 (14.62–14.81) [e]	14.38 (14.27–14.48) [e]
Daily GI	59.35 (59.24–59.46)	59.14 (58.98–59.31)	59.53 (59.39–59.67)	59.21 (59.02–59.39)
Average GI	59.17 (59.05–59.28)	59.12 (58.95–59.29)	59.50 (59.35–59.65)	59.18 (58.98–59.37)
Daily GL	165.22 (163.40–167.04) [f]	140.61 (138.30–142.91) [f]	194.86 (192.52–197.19) [g]	164.94 (162.05–167.83) [g]
Average GL	37.92 (37.55–38.29) [h]	34.29 (33.75–34.83) [h]	45.27 (44.77–45.76) [i]	40.63 (39.97–41.30) [i]
Glucose (mg/dL) [2]	84.45 (84.25–84.66) [j]	85.26 (84.95–85.58) [j]	86.80 (86.57–87.03) [k]	88.04 (87.71–88.37) [k]
Glycosylated hemoglobin (%) [3]	5.33 (5.33–5.34) [l]	5.38 (5.37–5.39) [l]	5.40 (5.39–5.41)	5.42 (5.40–5.43)
Insulin (mU/L) [4]	8.71 (8.59–8.84) [m]	13.20 (12.91–13.48) [m]	7.10 (6.99–7.22) [n]	12.04 (11.74–12.34) [n]
HOMA-IR [5]	1.84 (1.81–1.87) [o]	2.82 (2.75–2.89) [o]	1.54 (1.52–1.57) [p]	2.66 (2.58–2.73) [p]

Missing data: [1] Physical activity level: female = 1091; male = 1268; [2] Fasting glucose: female = 91; male = 62. [3] Glycosylated hemoglobin: female = 36; male = 16. [4] Insulin: female = 84; male = 56. [5] HOMA-IR: female = 162; male = 114. [a–p] *p*-value < 0.05 (Equal letters identify the comparison group in which statistical relevance occurs).

Table 2. Regression coefficients of standardized values of glycemic index (GI) on measures of glycemic control, by weight status. Brazil, 2013–2014 [1].

Normal Weight						
	Glycosylated Hemoglobin		Insulin		HOMA-IR	
	ß	*p*-Value	ß	*p*-Value	ß	*p*-Value
Daily GI	0.000	0.883	0.091	0.002	0.021	0.002
Average GI	0.001	0.603	0.089	0.002	0.019	0.005
Daily GL	0.004	0.074	0.057	0.059	0.012	0.082
Average GL	0.006	0.011	0.124	<0.0001	0.029	<0.0001
Overweight/Obese						
	Glycosylated Hemoglobin		Insulin		HOMA-IR	
	ß	*p*-Value	ß	*p*-Value	ß	*p*-Value
Daily GI	−0.001	0.843	0.162	0.030	0.034	0.049
Average GI	0.003	0.3561	0.229	0.001	0.051	0.002
Daily GL	0.001	0.746	−0.084	0.308	−0.026	0.168
Average GL	−0.003	0.544	0.072	0.315	0.018	0.278

[1] linear regression adjusted for age, sex, self-evaluated sexual maturation and physical activity (inactive/insufficiently active/active).

The overall composition of the diet compared with the obesity status of participants was quite similar, with the greatest difference being the lower energy intake among those who were overweight/obese (Table 3).

Table 3. Energy intake and nutrients by weight status. Brazil, 2013–2014.

	Normal Weight (*n* = 26,521)		Overweight/Obese (*n* = 9216)	
	Mean (95% CI)	% Total Energy	Mean (95% CI)	% Total Energy
Energy (kcal)	2372 (2351–2393)	-	2059 (2034–2084) *	-
Total carbohydrate (g)	316 (313–319)	53	271 (267–274) *	52
Glycemic carbohydrate (g)	297 (295–300)	50	254 (251–258) *	49
Glycemic carbohydrate from food (g)	274 (272–277)	46	235 (232–238) *	45
Glycemic carbohydrate from added sugar (g)	23 (22–24)	4	20 (19–20) *	4
Fiber (g)	19 (19–19)	3	16 (16–17) *	3
Protein (g)	93 (92–94)	16	84 (83–85) *	16
Lipids (g)	83 (82–83)	31	72 (71–73) *	31

* *p*-value < 0.05.

Variations in GI and GL throughout the day are shown in Figure 1, with lower values of GI in occasions related to lunch, whereas intermediate intake associated with snacks had higher values without differences according to adolescents' weight status.

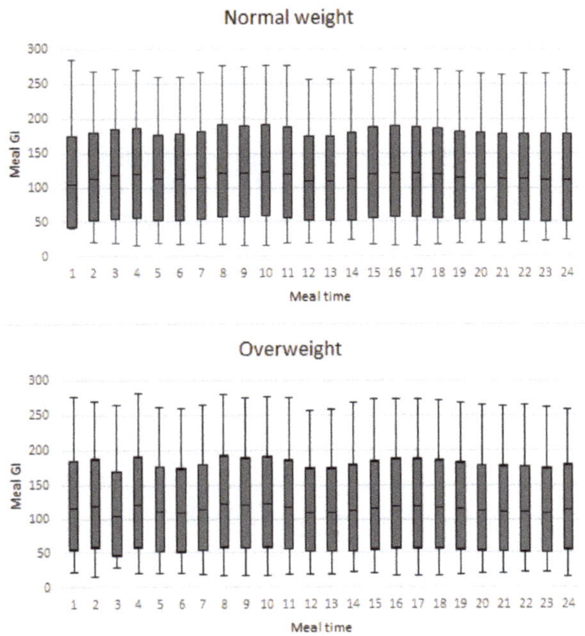

Figure 1. Boxplot of glycemic index of occasion meal time by weight status. Brazil 2013–2014.

4. Discussion

The present study provides an advance in the discussion of the effects of GI and GL on markers of glycemic control by comparing different methods of assessing these indices. The GI of chosen foods is associated with glycemic control in overweight/obese adolescents, who had a lower intake of carbohydrates, compared to NW adolescents. Differences in the weight status of adolescents and whether the index, load, average, or peak of carbohydrate intake were analyzed may explain the controversial findings in the literature.

Fasting glucose was not related to any of the carbohydrate metrics (results not shown). Although it seems controversial at first, it is necessary to consider that, in this study, we evaluated a sample of healthy adolescents. Zdravković [31] evaluated 87 obese adolescents and 17 normal-weight controls looking for symptoms of prediabetes in Belgrade. The oral glucose tolerance test showed isolated altered fasting glucose in only 13.9% of the obese group, indicating that it can be difficult to detect such an alteration.

In our study, average GI was the most important characteristic associated with insulin levels in both normal and overweight adolescents. GI meal values showed similar variation throughout the day (Figure 1) by weight status.

For mixed meals, GI peaks could be more important than the GI of individual foods. Chiavaroli [32] showed that the postprandial glycemic response of rice was greater than that of spaghetti. However, when tomato sauce and extra virgin olive oil are added to these foods, the maximum fasting glucose peak is reduced, mainly for rice. For the pesto sauce, the peaks were even smaller. The authors attributed this reduction to the addition of fat to the high-carbohydrate meal.

One study with 516 adolescents observed that GI peaks to daily mixed meals were not related, although the glycemic load of breakfast was a predictor of metabolic syndrome among girls; this finding is the only association found with only 17 cases of metabolic syndrome [6].

Cooper [33] showed that insulin increment was greater when female adolescents had a meal breakfast with a high GI compared to a low GI, with the same carbohydrate amount. Another study also showed an association between GL and metabolic syndrome in adolescence, with obesity being the most prevalent component [34]. Ojo [1] included six clinical trials that evaluated low GI diets as glycemic control markers in adults with type 2 diabetes. They concluded that a low-GI diet is more effective in controlling glycated hemoglobin and fasting blood than a high-GI diet.

According to Vega-Lopez [4], the association between GI and glycemic response with markers of glycemic control has been shown only in the strongest interventional studies, while findings from observational studies are weaker, and new diet quality metrics, such as average GI, need to be explored.

Although the GI of meals showed little variation, our findings indicate a regular pattern of intake related to the quality of carbohydrate intake. The small GI values for lunchtime are a consequence of the traditional intake of rice and beans in the Brazilian population, as shown in the last national survey [35–38].

The results of the variation in GI during the day indicate that intakes between the main meals had the greatest values, probably due to the intake of sweets, sodas and cookies as snacks. A significant intake of cookies and sugar-sweetened beverages was observed among adolescents in Brazil. The National School Health Survey (PeNSE 2012) found that 20% of 108,726 adolescents aged 14–16 years investigated regularly consume sweets and sodas [39].

A cross-sectional study of 351 children and adolescents (6–18 years of age), with and without overweight, analyzed the association between the insulinemic potential of the total diet and meals through the GI, GL, insulin index, insulin load and overweight risk. Dietary assessment was performed using a three-day food record. They found that higher insulin demand, especially at breakfast and dinner, was associated with being overweight, and night eating may be associated with eating compulsive behaviors related to obesity [40–42].

A limitation of the present study is that only one R24h was used, although it was based on the multiple passage method [43], which is considered a good way to stimulate memory for intake in all mealtimes [44]. In addition, measures of association using only one record are prone to sub-estimation, indicating that the effects may be greater than those observed. In addition, in large samples, it is possible to accept an isolated application when the objective is to estimate the energy and macronutrients [45].

Another possible limitation of our study is the lack of response to the blood drawn. However, Silva [13] showed no differences between participants and non-participants in relation to sex and age, but it was greater in public schools than in particular schools.

These findings suggest that carbohydrate quality metrics are associated with markers of glycemic control. The diet GI index was better at predicting insulinemia and, consequently, HOMA-IR, independent of weight status, than GL. In clinical practice, the findings show that the guidance of food consumption based on carbohydrate quality is a possibility for glycemic control, since higher GIs are highly associated with the intake of refined carbohydrates.

Encouraging healthy lifestyle habits combined with a low GI and low GL diet can also help control obesity, which is closely related to glycemic control, insulin resistance and the early development of increasing type 2 diabetes among children, adolescents and young adults. In addition, public policies that support the prevention of chronic diseases with the identification of individuals at risk of developing them, early diagnosis and individualized clinical follow-up have lower costs compared to the amount that has been spent on the treatment of type 2 diabetes and obesity worldwide.

Author Contributions: C.M.M.d.R. contributed to data analysis and interpretation and manuscript writing. V.P.M.G. contributed to data analysis and interpretation and manuscript writing. A.d.M.S., E.M.Y. and E.V.J. contributed to data interpretation and manuscript revision. K.V.B. and R.S. contributed to the study design and supervising, data analysis and interpretation and manuscript conception, writing and final revision. All authors have read and agreed to the published version of the manuscript.

Funding: This work was supported by the Brazilian Department of Science and Technology at the Secretariat of Science and Technology and Strategic Inputs of the Ministry of Health (Departamento de Ciência e Tecnologia da Secretaria de Ciência e Tecnologia e Insumos Estratégicos do Ministério da Saúde -Decit/SCTIE/MS).

Institutional Review Board Statement: ERICA was registered in Plataforma Brasil and was evaluated by the Research Ethics Committee of the Institute of Collective Health of the Federal University of Rio de Janeiro under case number 08/2008. Having fulfilled the prerogatives of resolution 196/96 of the National Health Council, approval was registered in 01/2009.

Informed Consent Statement: Informed consent was obtained from all subjects involved in the study.

Data Availability Statement: The data presented in this study are available on request from the corresponding author. The data are not publicly available due to the difficulty in identifying the 4 different databases and identifying the variables' names. All codebooks are in Portuguese.

Conflicts of Interest: The authors declare no conflict of interest.

References

1. Ojo, O.; Ojo, O.O.; Adebowale, F.; Wang, X.H. The effect of dietary glycaemic index on glycaemia in patients with type 2 diabetes: A systematic review and meta-analysis of randomized controlled trials. *Nutrients* **2018**, *10*, 3. [CrossRef] [PubMed]
2. Thomas, D.E.; Elliott, E.J. The use of low-glycaemic index diets in diabetes control. *Br. J. Nutr.* **2010**, *104*, 797–802. [CrossRef] [PubMed]
3. Venn, B.J.; Green, T.J. Glycemic index and glycemic load: Measurement issues and their effect on diet–disease relationships. *Eur. J. Clin. Nutr.* **2007**, *61*, S122–S131. [CrossRef] [PubMed]
4. Vega-López, S.; Venn, B.J.; Slavin, J.L. Relevance of the glycemic index and glycemic load for body weight, diabetes, and cardiovascular disease. *Nutrients* **2018**, *10*, 1361. [CrossRef]
5. Clar, C.; Al-Khudairy, L.; Loveman, E.; Kelly, S.A.M.; Hartley, L.; Flowers, N.; Germanò, R.; Frost, G.; Rees, K. Low glycaemic index diets for the prevention of cardiovascular disease. *Cochrane Database Syst. Rev.* **2017**, *2021*, CD004467. [CrossRef]
6. Nicholl, A.; du Heaume, M.; Mori, T.A.; Beilin, L.J.; Oddy, W.H.; Bremner, A.P.; O'Sullivan, T.A. Higher breakfast glycaemic load is associated with increased metabolic syndrome risk, including lower HDL-cholesterol concentrations and increased TAG concentrations, in adolescent girls. *Br. J. Nutr.* **2014**, *112*, 1974–1983. [CrossRef]
7. Liang, Y.; Hou, D.; Zhao, X.; Wang, L.; Hu, Y.; Liu, J.; Cheng, H.; Yang, P.; Shan, X.; Yan, Y.; et al. Childhood obesity affects adult metabolic syndrome and diabetes. *Endocrine* **2015**, *50*, 87–92. [CrossRef]
8. Murni, I.K.; Sulistyoningrum, D.C.; Susilowati, R.; Julia, M. Risk of metabolic syndrome and early vascular markers for atherosclerosis in obese Indonesian adolescents. *Paediatr. Int. Child Health* **2019**, *40*, 117–123. [CrossRef]
9. Weihe, P.; Weihrauch-Blüher, S. Metabolic syndrome in children and adolescents: Diagnostic criteria, therapeutic options and perspectives. *Curr. Obes. Rep.* **2019**, *8*, 472–479. [CrossRef]
10. Zimmet, P.; Alberti, K.G.M.; Kaufman, F.; Tajima, N.; Silink, M.; Arslanian, S.; Wong, G.; Bennett, P.; Shaw, J.; Caprio, S.; et al. The metabolic syndrome in children and adolescents? An IDF consensus report. *Pediatr. Diabetes* **2007**, *8*, 299–306. [CrossRef]
11. Bloch, K.V.; Szklo, M.; Kuschnir, M.C.C.; de Azevedo Abreu, G.; Barufaldi, L.A.; Klein, C.H.; de Vasconcelos, M.T.; da Veiga, G.V.; Figueiredo, V.C.; Dias, A.; et al. The study of cardiovascular risk in adolescents—ERICA: Rationale, design and sample characteristics of a national survey examining cardiovascular risk actor profile in Brazilian adolescents. *BMC Public Health* **2015**, *15*, 94.
12. De Vasconcellos, M.T.L.; Silva, P.L.D.N.; Szklo, M.; Kuschnir, M.C.C.; Klein, C.H.; Abreu, G.D.A.; Barufaldi, L.A.; Bloch, K.V. Sampling design for the Study of Cardiovascular Risks in Adolescents (ERICA). *Cad. Saude Publica* **2015**, *31*, 921–930. [CrossRef] [PubMed]
13. Silva, T.L.N.; Klein, C.H.; Souza, A.D.M.; Barufaldi, L.A.; Abreu, G.D.A.; Kuschnir, M.C.C.; de Vasconcellos, M.T.L.; Bloch, K.V. Response rate in the study of cardiovascular risks in adolescents—ERICA. *Rev. Saude Publica* **2016**, *50*, 1s–13s. [CrossRef] [PubMed]
14. Conwell, L.S.; Trost, S.G.; Brown, W.J.; Lote, J.A. Indexes of insulin resistance and secretion in obese children and adolescents: A validation study. *Diabetes Care* **2004**, *27*, 314–319. [CrossRef] [PubMed]
15. Hoffman, R.P.; Vicini, P.; Cobelli, C. Pubertal changes in HOMA and QUICKI: Relationship to hepatic and peripheral insulin sensitivity. *Pediatr. Diabetes* **2004**, *5*, 122–125. [CrossRef]

16. Keskin, M.; Kurtoglu, S.; Kendirci, M.; Atabek, M.E.; Yazici, C. Homeostasis model assessment is more reliable than the fasting glucose/insulin ratio and quantitative insulin sensitivity check index for assessing insulin resistance among obese children and adolescents. *Pediatrics* **2005**, *115*, e500–e503. [CrossRef]
17. Moshfegh, A.J.; Rhodes, D.G.; Baer, D.J.; Murayi, T.; Clementes, J.C.; Rumpler, W.V.; Paulo, D.R.; Sebastião, R.S.; Kuczynski, K.J.; Ingwersen, L.A.; et al. The US Department of Agriculture Automated Multiple-Pass Method reduces bias in the collection of energy intakes. *Am. J. Clin. Nutr.* **2008**, *88*, 324–332. [CrossRef]
18. Instituto Brasileiro de Geografia e Estatística (IBGE). *Pesquisa de Orçamentos Familiares—POF 2008–2010: Tabelas de Composição Nutricional dos Alimentos Consumidos no Brasil*; Instituto Brasileiro de Geografia e Estatística (IBGE): Rio de Janeiro, Brazil, 2011. (In Portuguese)
19. The University of Sydney. Glycemic Index. Available online: https://www.glycemicindex.com/index.php (accessed on 1 October 2018).
20. Balisteiro, D.M. Efeito dos Compostos Fenólicos de Frutas Nativas Brasileiras na Glicemia pós Prandial. Master's Thesis, Universidade de São Paulo, São Paulo, Brazil, 2013.
21. Cardoso, A.M.C. Índice glicêmico de alimentos típicos da Amazônia. *Rev. Bras. Nutr. Clínica* **2003**, *18*, 190–192.
22. Cordenunsi, B.R.; de Menezes, E.W.; Genovese, M.I.; Colli, C.; de Souza, A.G.; Lajolo, F.M. Chemical composition and glycemic index of Brazilian pine (*Araucaria angustifolia*) seeds. *J. Agric. Food Chem.* **2004**, *52*, 3412–3416. [CrossRef]
23. Foster-Powell, K.; Holt, S.H.; Brand-Miller, J.C. International table of glycemic index and glycemic load values: 2002. *Am. J. Clin. Nutr.* **2002**, *76*, 5–56. [CrossRef]
24. Oriondo, R.; Valdivieso, R.; Oré, R.; Arnao, I.; Palomino, M.; Estrada, E. Evaluación de la capacidad antioxidante y el índice glicémico de los frutos promisorios amazónicos del Perú. *An. Fac. Med.* **2013**, *73*, 19. [CrossRef]
25. Passos, T.U.; Sampaio, H.A.D.C.; Sabry, M.O.D.; De Melo, M.L.P.; Coelho, M.A.M.; Lima, J.W.D.O. Glycemic index and glycemic load of tropical fruits and the potential risk for chronic diseases. *Food Sci. Technol.* **2015**, *35*, 66–73. [CrossRef]
26. Food and Agriculture Organization of the United Nations; World Health Organization (Eds.) *Carbohydrates in Human Nutrition: Report of a Joint FAO/WHO Expert Consultation, Rome, 14–18 April 1997*; World Health Organization: Geneva, Switzerland, 1998.
27. De Onis, M.; Onyango, A.W.; Borghi, E.; Siyam, A.; Nishida, C.; Siekmann, J. Development of a WHO growth reference for school-aged children and adolescents. *Bull. World Health Organ.* **2007**, *85*, 660–667. [CrossRef] [PubMed]
28. DeBoer, M.D. Assessing and managing the metabolic syndrome in children and adolescents. *Nutrients* **2019**, *11*, 1788. [CrossRef]
29. Marson, E.C.; Delevatti, R.S.; Prado, A.K.G.; Netto, N.; Kruel, L.F.M. Effects of aerobic, resistance, and combined exercise training on insulin resistance markers in overweight or obese children and adolescents: A systematic review and meta-analysis. *Prev. Med.* **2016**, *93*, 211–218. [CrossRef]
30. Sallis, J.F.; Strikmiller, P.K.; Harsha, D.W.; Feldman, H.A.; Ehlinger, S.; Stone, E.J.; Williston, J.; Woods, S. Validation of interviewer- and self-administered physical activity checklists for fifth grade students. *Med. Sci. Sports Exerc.* **1996**, *28*, 840–851. [CrossRef]
31. Zdravković, V.; Sajić, S.; Mitrović, J.; Stefanović, I.; Pavićević, P.; Nikolić, D.; Dimić, J.; Lalić, N.M. The diagnosis of prediabetes in adolescents. *J. Med. Biochem.* **2015**, *34*, 38–45. [CrossRef]
32. Chiavaroli, L.; Di Pede, G.; Dall'Asta, M.; Cossu, M.; Francinelli, V.; Goldoni, M.; Scazzina, F.; Brighenti, F. The importance of glycemic index on post-prandial glycaemia in the context of mixed meals: A randomized controlled trial on pasta and rice. *Nutr. Metab. Cardiovasc. Dis.* **2021**, *31*, 615–625. [CrossRef]
33. Cooper, S.B.; Dring, K.J.; Morris, J.G.; Cousins, B.E.W.; Nute, M.L.; Nevill, M. Sex differences in adolescents' glycaemic and insulinaemic responses to high and low glycaemic index breakfasts: A randomised control trial. *Br. J. Nutr.* **2017**, *117*, 541–547. [CrossRef]
34. Cornejo-Monthedoro, A.; Negreiros-Sánchez, I.; Del Águila, C.; Ysla-Marquillo, M.; Mayta-Tristán, P. Association between dietary glycemic load and metabolic syndrome in obese children and adolescents. *Arch. Argent. Pediatr.* **2017**, *115*, 323–330. [CrossRef]
35. De Almeida Alves, M.; de Moura Souza, A.; Barufaldi, L.A.; Tavares, B.M.; Bloch, K.V.; de Vasconcelos, F.D.A.G. Padrões alimentares de adolescentes brasileiros por regiões geográficas: Análise do Estudo de Riscos Cardiovasculares em Adolescentes (ERICA). *Cad. Saúde Pública* **2019**, *35*, e00153818. [CrossRef] [PubMed]
36. Massarani, F.A.; Cunha, D.B.; Muraro, A.P.; da Silva Nalin de Souza, B.; Sichieri, R.; Yokoo, E.M. Agregação familiar e padrões alimentares na população brasileira. *Cad. Saúde Pública* **2015**, *31*, 2535–2545. [CrossRef] [PubMed]
37. Souza, A.M.; Pereira, R.A.; Yokoo, E.M.; Levy, R.B.; Sichieri, R. Alimentos mais consumidos no Brasil: Inquérito Nacional de Alimentação 2008–2009. *Rev. Saude Publica* **2013**, *47*, 190S–199S. [CrossRef] [PubMed]
38. Souza, A.M.; Barufaldi, L.A.; Abreu, G.D.A.; Giannini, D.T.; De Oliveira, C.L.; dos Santos, M.M.; Leal, V.S.; de Vasconcelos, F.D.A.G. ERICA: Intake of macro and micronutrients of Brazilian adolescents. *Rev. Saúde Pública* **2016**, *50*, 5S. [CrossRef]
39. Ferreira, N.L.; Claro, R.M.; Lopes, A.C.S. Consumption of sugar-rich food products among Brazilian students: National School Health Survey (PeNSE 2012). *Cad. Saude Publica* **2015**, *31*, 2493–2504. [CrossRef]
40. Caferoglu, Z.; Erdal, B.; Akin, L.; Kurtoglu, S. Breakfast and dinner insulin index and insulin load in relation to overweight in children and adolescents. *Eur. J. Nutr.* **2021**, *60*, 2819–2829. [CrossRef]
41. Hernández, E.; Kim, M.; Kim, W.G.; Yoon, J. Nutritional aspects of night eating and its association with weight status among Korean adolescents. *Nutr. Res. Pract.* **2016**, *10*, 448–455. [CrossRef]
42. Striegel-Moore, R.H.; Rosselli, F.; Wilson, G.T.; Perrin, N.; Harvey, K.; DeBar, L. Nocturnal eating: Association with binge eating, obesity, and psychological distress. *Int. J. Eat. Disord.* **2010**, *43*, 520–526. [CrossRef]

43. Conway, J.M.; Ingwersen, L.A.; Vinyard, B.T.; Moshfegh, A.J. Effectiveness of the US Department of Agriculture 5-step multiple-pass method in assessing food intake in obese and nonobese women. *Am. J. Clin. Nutr.* **2003**, *77*, 1171–1178. [CrossRef]
44. Rutishauser, I.H.E. Dietary intake measurements. *Public Health Nutr.* **2005**, *8*, 1100–1107. [CrossRef]
45. Bingham, S.A.; Nelson, M. Assessment of food consumption and nutrient intake. In *Design Concepts in Nutritional Epidemiology*; Margetts, B.M., Nelson, M., Eds.; Oxford University Press: Oxford, UK, 1997.

nutrients

MDPI

Article

Postprandial Responses to a Standardised Meal in Hypertension: The Mediatory Role of Visceral Fat Mass

Panayiotis Louca [1], Sarah E. Berry [2], Kate Bermingham [1,2], Paul W. Franks [3], Jonathan Wolf [4], Tim D. Spector [1], Ana M. Valdes [5], Phil Chowienczyk [6,†] and Cristina Menni [2,*,†]

1 Department of Twin Research, King's College London, St Thomas' Hospital Campus, London SE1 7EH, UK
2 Department of Nutritional Sciences, King's College London, Franklin Wilkins Building, London SE1 9NH, UK
3 Genetic & Molecular Epidemiology Unit, Department of Clinical Sciences, Lund University, SE-20502 Malmo, Sweden
4 Zoe, London SE1 7RW, UK
5 Nottingham NIHR Biomedical Research Centre, School of Medicine, University of Nottingham, Nottingham NG5 1PB, UK
6 Vascular Risk & Surgery, King's College London, St Thomas' Hospital Campus, London SE1 7EH, UK
* Correspondence: cristina.menni@kcl.ac.uk; Tel.: +44-(0)-207-188-7188 (ext. 52594)
† These authors contributed equally to this work.

Abstract: Postprandial insulinaemia, triglyceridaemia and measures of inflammation are thought to be more closely associated with cardiovascular risk than fasting measures. Although hypertension is associated with altered fasting metabolism, it is unknown as to what extent postprandial lipaemic and inflammatory metabolic responses differ between hypertensive and normotensive individuals. Linear models adjusting for age, sex, body mass index (BMI), visceral fat mass (VFM) and multiple testing (false discovery rate), were used to investigate whether hypertensive cases and normotensive controls had different fasting and postprandial (in response to two standardised test meal challenges) lipaemic, glycaemic, insulinaemic, and inflammatory (glycoprotein acetylation (GlycA)) responses in 989 participants from the ZOE PREDICT-1 nutritional intervention study. Compared to normotensive controls, hypertensive individuals had significantly higher fasting and postprandial insulin, triglycerides, and markers of inflammation after adjusting for age, sex, and BMI (effect size: Beta (Standard Error) ranging from 0.17 (0.08), $p = 0.04$ for peak insulin to 0.29 (0.08), $p = 4.4 \times 10^{-4}$ for peak GlycA). No difference was seen for postprandial glucose. When further adjusting for VFM effects were attenuated. Causal mediation analysis suggests that 36% of the variance in postprandial insulin response and 33.8% of variance in postprandial triglyceride response were mediated by VFM. Hypertensive individuals have different postprandial insulinaemic and lipaemic responses compared to normotensive controls and this is partially mediated by visceral fat mass. Consequently, reducing VFM should be a key focus of health interventions in hypertension. **Trial registration:** The ClinicalTrials.gov registration identifier is NCT03479866.

Keywords: postprandial; hypertension; insulinaemia; triglyceridaemia; inflammation

Citation: Louca, P.; Berry, S.E.; Bermingham, K.; Franks, P.W.; Wolf, J.; Spector, T.D.; Valdes, A.M.; Chowienczyk, P.; Menni, C. Postprandial Responses to a Standardised Meal in Hypertension: The Mediatory Role of Visceral Fat Mass. *Nutrients* **2022**, *14*, 4499. https://doi.org/10.3390/nu14214499

Academic Editors: Omorogieva Ojo and Amanda R. Amorim Adegboye

Received: 18 September 2022
Accepted: 18 October 2022
Published: 26 October 2022

1. Introduction

Hypertension is the most prevalent modifiable risk factor for cardiovascular morbidity and mortality affecting over 1.3 billion people around the world [1]. Studies have shown that hypertension clusters with metabolic factors including glucose intolerance, hyperinsulinaemia, and dyslipidaemia [2]. Indeed, results from the prospective follow-up study, Pressioni Arteriose Monitorate E Loro Associazioni (PAMELA), suggest that elevated blood pressure (BP) is the most common component of the metabolic syndrome (MetS), with 95.4% of participants with MetS having elevated BP, and up to 80% of individuals with MetS being hypertensive [3,4]. Moreover, hypertensive individuals that fulfil the criteria for MetS have up to a 73% increased age and risk factor-adjusted risk for cardiovascular events [5]. Hypertension may also be linked with the onset of new MetS [6]; it is strongly associated with insulin resistance, a component of MetS, independently of other risk factors, including obesity [7]. Mechanisms behind this may centre around the hormonal actions of insulin, which can regulate renal sodium clearance [8], a key mechanism involved in BP regulation. The increased cardiovascular risk associated with MetS and hypertension may also be linked to endothelial dysfunction and atherosclerosis [2].

Although previous research has explored fasting metabolism in hypertensive individuals, the majority of the population spend most of their waking hours in a postprandial state [9,10] and postprandial glycaemia, insulinaemia, lipaemia, and inflammation are thought to be more closely associated with cardiovascular risk than fasting levels [11], it is therefore of utmost importance to understand postprandial metabolic responses in hypertensive individuals. Hypertensive individuals have been found to have higher postprandial triglyceride levels [12], and postprandial hypertriglyceridaemia also correlates with levels of visceral adiposity [13], and causal links have been shown in murine models [14,15]. Additionally, postprandial glucose disposal in the presence of insulin resistance may promote hypertension through various atherogenic processes [2].

However, a comprehensive exploration of the fasting and postprandial differences in metabolic markers (triglycerides, insulin, glucose, and inflammation), between hypertensive and normotensive individuals, when challenged by a standardised mixed-nutrient meal, is lacking. Here, we investigate whether individuals with hypertension have a different postprandial response compared to normotensive controls. We further explore whether visceral fat mass (VFM), thought to be a key marker of glucose homeostasis and lipid metabolism, is a mediator of associations between hypertension and postprandial insulin and triglyceride response in the ZOE UK PREDICT 1 study [9]—a single-arm, randomized cross-over trial of standardized meal interventions designed to quantify and predict individual variations in postprandial responses (NCT03479866).

2. Materials and Methods

A consort diagram with the study design is presented in Figure 1.

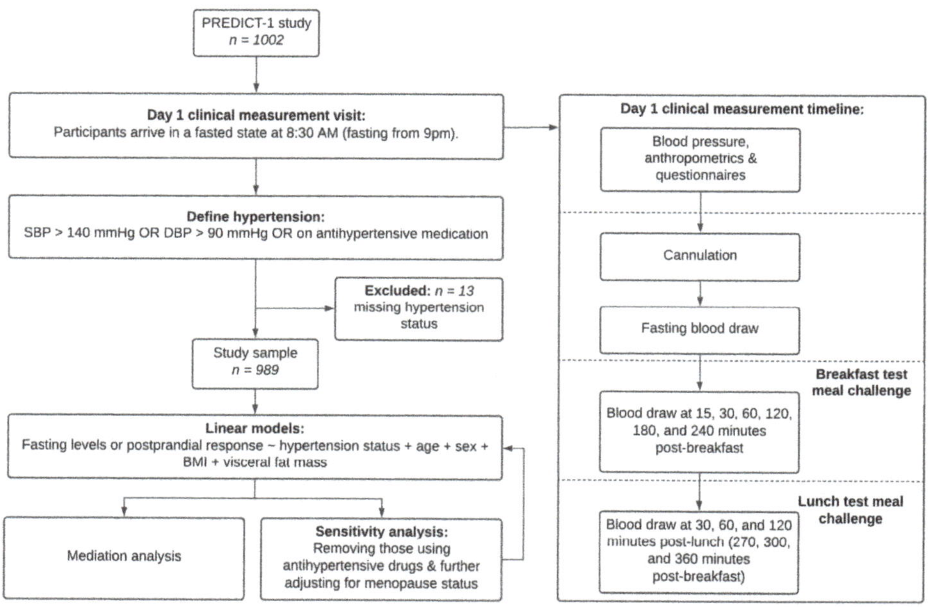

Figure 1. Flow chart of study analytical pipeline and clinical visit timeline. Study population and design.

We included 989 individuals from the UK based ZOE PREDICT-1 study. The ZOE PREDICT-1 study [9] was a single-arm nutritional intervention conducted between June 2018 and May 2019. Study participants were apparently healthy individuals but included those with risk factors such as hypertension. Participants were aged between 18–65 years recruited from the TwinsUK registry [16], and the general population using online advertising. Participants attended a full day clinical visit consisting of test meal challenges followed by a 13-day home-based phase, as previously described [9].

Data relevant to this analysis pertain to the day 1 baseline clinical measurement visit at St. Thomas' Hospital. As shown in Figure 1, during their visit, participants arrived at 8:30 am in a fasted state (fasting from 9 pm the previous night). On arrival, participants provided baseline characteristics, including age, sex, anthropometric measurements (including adiposity as described below) and BP was recorded. Participants were cannulated and a fasting blood sample was taken. Within a tightly controlled clinical setting, participants consumed meal 1: breakfast muffins and a milkshake (890 kcal, 85.5 g carbohydrate, 52.7 g fat, and 16.1 g protein at the 0-h timepoint, following baseline blood draw, BP, and anthropometrics). Venous blood samples were collected at 15, 30, 60, 120, 180, 240, 300, 360 min post meal 1. Meal 2: lunch muffins (502 kcal, 71.2 g carbohydrate, 22.2 g fat, and 9.6 g protein) was consumed at the 240-min timepoint (after the 240-min blood sample). Participants were permitted to sip water throughout (Figure 1). Outcome variables from blood sampling were blood triglyceride, glucose, insulin, and glycoprotein acetylation (GlycA) (as a marker of inflammation) levels [9]. GlycA is a particular proton nuclear magnetic resonance spectroscopy signal that reflects the methyl groups bound to N-acetylglucosamine residues attached to circulating plasma proteins and is recognised and validated as a biomarker of systemic inflammation [17]. GlycA moderately correlates with several other biomarkers of inflammation but has greater analytical precision and lower-intra-individual variability [18]. Moreover, GlycA levels have also been shown to associate with both acute and chronic inflammation, severity of inflammatory disorders, and cardiovascular events independent of other inflammatory markers [19,20]. For each of these variables, we considered (i) the baseline fasting measures; (ii) the peak (over the 6-h (360 min) visit for triglycerides and GlycA, and 2-h (240 min) for insulin and glucose) [9] and (iii) the magnitude of increase (delta increase = peak − baseline). Postprandial peaks were previously identified using line trajectories as detailed in Berry et al. 2018 [9] and the specific timepoints used here are based on these previous reports.

2.1. Blood Pressure

Prior to the breakfast test meal challenge, BP was measured by a trained nurse with the patient in a seated position for 3 min. The cuff was placed on the subject's arm so that it was approximately 2–3 cm above the elbow joint of the inner arm, with the air tube lying over the brachial artery. The subject's arm was placed on the table or supported with the palm facing upwards, so that the tab of the cuff was placed at the same level of the heart. Triplicate measurements were taken with an interval of approximately 1 min between each reading. The first reading was discarded and the mean of the second and third measurements recorded.

Participants were classified into hypertensive cases if their systolic BP $\geq$ 140 mmHg OR their diastolic BP $\geq$ 90 mmHg OR they were using antihypertensive medication, otherwise they were considered as non-hypertensive controls.

2.2. Adiposity

Body fat distribution was determined by whole body dual-energy X-ray absorptiometry (DXA) using a fan beam X-ray bone densitometer (QDR-4500 W, Hologic, Inc., Marlborough, MA, USA) with the participant in the supine position and analysed with QDR Systems software version 12.6 (Hologic, Inc., MA, USA), as previously described [21]. DXA fat mass from the abdominal region was recorded as VFM in a similar manner to Bertin et al. [22].

2.3. Statistical Analysis

Statistical analysis was performed using R version 4.0.2. Circos plot was generated using the R package 'ggplot2'.

Continuous variables were standardised using z-scores. Linear models were used to investigate whether hypertensive cases and non-hypertensive controls had different fasting levels and postprandial responses after adjusting for age, sex, body mass index (BMI), and multiple testing (Benjamini-Hochberg false discovery rate (FDR < 0.05)). For each of these comparisons, we report effect size (Betas) and standard error (SE). To explore links between metabolic factors and visceral adiposity we conducted a sensitivity analysis by additionally adjusting for VFM.

To investigate the mediatory effects of VFM (indirect effect) in the relationship between hypertension and postprandial insulin and triglyceride levels (direct effects), we constructed a causal mediation analysis using the R package "mediation" [23]. The variance accounted for (VAF) score depicts the ratio of indirect-to-total effect and determines the proportion of the variance that can be explained by the mediator, in this instance, VFM.

We conducted additional sensitivity analyses by (i) removing any individuals on antihypertensive medication; (ii) adjusting for menopausal status; and (iii) stratifying by sex. We also investigated associations between hypertension status and insulin resistance using the Homeostasis Model Assessment for Insulin Resistance (HOMA-IR) formulated as fasting insulin (μU/mL) × fasting glucose (mmol/L)/22.5 [24].

3. Results

We analysed data from 989 participants who attended a full day (6-h) clinical visit consisting of two test meal challenges and had BP measurements. We included 203 hypertensive cases and 786 normotensive controls, aged 45.57 (mean, SD = 11.94) years, mainly females (72.7%), and were on average slightly overweight (BMI = 25.61, SD = 5.07) kg/m^2 (Table 1) with an average waist-to-hip ratio of 0.85.

Table 1. Demographic characteristics of the study population overall and by hypertension status.

	Overall (n = 989)		Hypertensive Cases (n = 203)		Normotensive Controls (n = 786)	
	n	%	n	%	n	%
Antihypertensive drug use	56	5.7	56	27.6	0	0
Females	719	72.7	135	66.5	584	74.3
Peri-menopausal	54	8.8	15	12.5	39	7.9
Post-menopausal	201	32.7	72	60	129	26.1
	Mean	*Sd*	*Mean*	*Sd*	*Mean*	*Sd*
Age (years)	45.6	11.9	52	10.2	43.9	11.8
BMI (kg/m^2)	25.6	5.1	27.4	5.6	25.2	4.8
Waist to hip ratio	0.85	0.08	0.88	0.09	0.84	0.08
VFM (g)	527.2	311.6	689.8	358.8	485.1	283.7
HOMA-IR	1.4	1.1	1.9	1.8	1.3	0.9

Abbreviations: BMI, body mass index; VFM, visceral fat mass; HOMA-IR, Homeostasis Model Assessment for Insulin Resistance.

3.1. Fasting Levels

We first investigated differences in the fasting (baseline) states between hypertensive cases and normotensive controls. As depicted in Figure 2, after adjusting for multiple testing we found that individuals with hypertension had significantly higher fasting glucose (Beta (SE) = 0.18 (0.08), $p = 0.02$), insulin (Beta (SE) = 0.34 (0.07), $p = 8.61 \times 10^{-7}$), triglycerides (Beta (SE) = 0.38 (0.08), $p = 2.6 \times 10^{-6}$), and GlycA (Beta (SE) = 0.26 (0.08), $p = 1.2 \times 10^{-3}$) (Figure 2), in line with the literature [7]. We also found that hypertensive patients have higher HOMA-IR (Beta (SE) = 0.36 (0.07), $p = 3.6 \times 10^{-7}$) compared to normotensive controls.

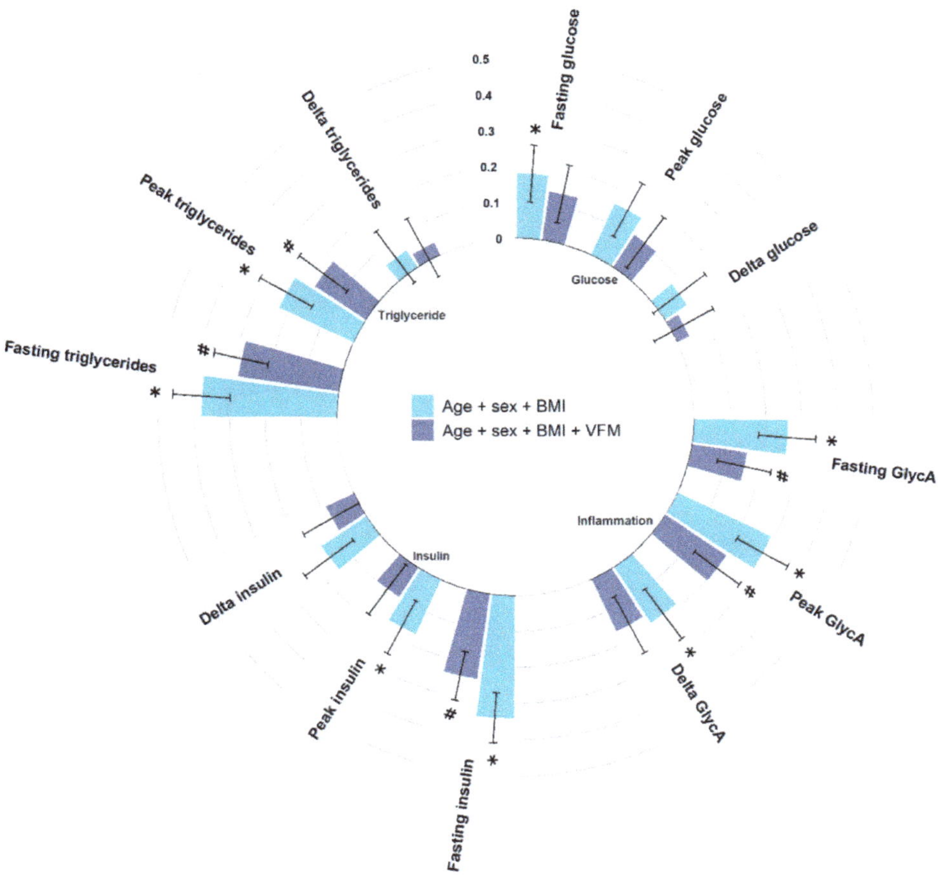

Figure 2. Circos bar plot with bars representing standardised coefficients of linear models between metrics and hypertension status with error bars representing standard error. Bars are colour coded based on covariates. Light blue bars indicate adjustment for age, sex, and BMI, while navy bars indicated adjustment for age, sex, BMI, and VFM. * FDR < 0.05 (for age, sex, and BMI adjusted model), # nominal significant (for age, sex, BMI, and VFM adjusted model). Abbreviations: BMI, body mass index; VFM, visceral fat mass.

3.2. Peak Levels

We observed significantly higher postprandial peaks in insulin (2-h insulin peak, Beta (SE) = 0.17 (0.08), $p = 4.4 \times 10^{-2}$), triglycerides (6-h peak triglyceride, Beta (SE) = 0.23 (0.08), $p = 5.6 \times 10^{-3}$) and GlycA (6-h peak GlycA, Beta (SE) = 0.29 (0.08), $p = 4.4 \times 10^{-4}$) in hypertensive cases, after adjusting for age, sex, BMI and multiple testing (Figure 2). However, when additionally adjusting for VFM, a marker of adipose tissue strongly related to metabolic disturbances and hypertension [25], effects were attenuated (peak triglycerides, Beta (SE) = 0.17 (0.08), $p = 0.04$; peak insulin, Beta (SE) = 0.1 (0.09), $p = 0.25$); (peak GlycA, Beta (SE) = 0.2 (0.08), $p = 0.01$). Given the links between VFM and, cardiometabolic risk factors, including insulin resistance and triglycerides, we conducted a causal mediation analysis using bootstrapping and 1000 simulations to determine the indirect effect of VFM on the relationship between hypertension and postprandial triglyceride, and insulin responses. This suggested that VFM was fully mediating the positive association between hypertension and 2-h peak insulin (VAF = 36% (0.02, 0.1) $p = 0.002$), and partially mediating the positive association between hypertension and 6-h triglyceride peak (variance accounted for (VAF) = 33.8% (0.03, 0.16) $p = 0.004$) (Figure 3).

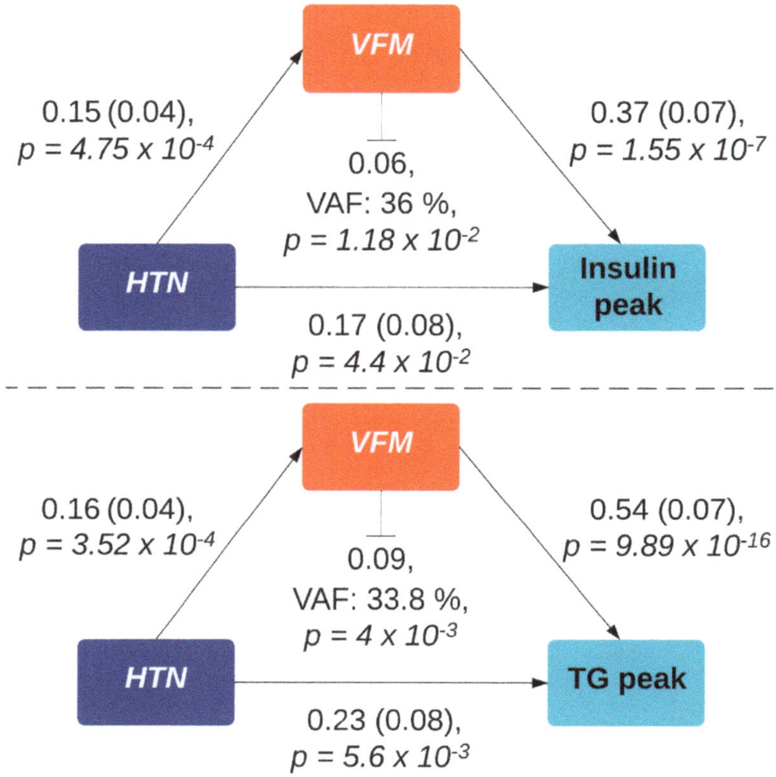

Figure 3. Mediation analysis of the association between hypertension and peak postprandial insulin and triglyceride response via visceral fat. Path coefficients are illustrated beside each path and indirect effect and variance accounted for score is denoted below the mediator. Abbreviations: TG, triglycerides; HTN, hypertension; VAF, variance accounted for.

3.3. Change from Fasting Levels

Results were consistent when we rerun the analyses investigating the correlation between hypertension status and postprandial metabolic changes measured as delta. Beta coefficients were in the same direction, although p values were attenuated (Figure 2).

3.4. Sensitivity Analysis

To account for potential confounding, we conducted sensitivity analyses by (i) excluding those on antihypertensive drugs, (ii) adjusting for menopausal status (pre-, peri- and post-menopausal, as determined by a health and lifestyle questionnaire), (iii) stratifying by sex. Results/effect sizes remained consistent. See Supplementary Table S1.

4. Discussion

In the largest study of its kind to look at differential postprandial responses in hypertensive individuals compared to normotensive subjects, we find that in addition to a disrupted fasting metabolic state, individuals with hypertension have higher postprandial insulin, triglycerides, and inflammatory responses after adjusting for traditional risk factors. Causal mediation analysis further suggests that VFM, a key risk factor in metabolic syndrome [26], fully mediates the associations between hypertension and insulin responses and partially mediates that with postprandial triglyceride response. Moreover, we find an increased level of insulin resistance in hypertensive participants compared to normotensive controls.

Postprandial lipaemia is a greater predictor of cardiovascular events, in contrast to fasting triglyceride levels [11,27]. Here, we report higher spikes in postprandial triglycerides in hypertensive cases, after consumption of a mixed-nutrient challenge. In support of our findings, Hwu et al. [12] found higher postprandial triglycerides in hypertensive participants 4-h after a 1000 kcal, high fat meal (65.9% fat, 18.9%, carbohydrates, 15.2% protein). Moreover, when compared to normotensive participants, Kolovou and colleagues [28] also report higher postprandial triglycerides in 25 individuals with essential hypertension following a meal dense in fat (83.5% fat, 14% carbohydrates, 2.5% protein). Although, in contrast to our study, the hypertensive participants were found to have normal fasting triglycerides [28]. Moreover, Kolovou et al. show significant positive correlations between BMI and maximal postprandial triglyceride concentration within the hypertensive participants. Extending this link between elevated postprandial triglycerides, hypertension, and body fat. Here, we find that VFM partially mediates the relationship between hypertension status and postprandial lipaemia, suggesting that despite the link with cardiovascular events [9], the lipaemia is likely to have a smaller effect on blood pressure than previously thought. Rather, this data suggests that the relevance of triglyceride metabolism in hypertension lies mainly in higher adipose fat. A possible explanation, is the variety of vasoactive factors (both vasodilators such as leptin, adiponectin, apelin and omectin) and vasoconstrictors like resistin, chemerin, and visfatin released by adipocytes [25] and that levels of visceral fat are directly involved in blood pressure regulation or influence blood pressure through activation of sympathetic nervous system activity [25].

Approximately, half of all patients with essential hypertension are thought to be insulin-resistant [2,29]. Indeed, here we report that hypertensive individuals have higher levels of insulin resistance, as determined by the HOMA-IR index. As expected, we also find hypertensive individuals to have postprandial hyperinsulinaemia. The links between hyperinsulinaemia and hypertension is thought to be driven via a few key mechanisms, (i) a decrease in insulin sensitivity, (ii) insulin mediated glucose disposal [2], both of which are thought to promote hypertension and atherogenesis. (iii) increased plasma aldosterone levels [30], and (iv) upregulated angiotensin II receptors [31], two components of the renin-angiotensin-aldosterone system, a critical regulator of BP. Visceral fat, particularly that deposited around the liver has been linked with impaired insulin clearance or hepatic insulin action [32], which would further exacerbate these actions, and may explain the mediation effects we observe.

Additionally, evidence of a causal role of VFM in insulin and triglyceride actions has also been found in murine models, where the removal of VFM restored insulin action and improved lipid profiles [14,15], which supports our reports of a strong mediatory role of VFM in postprandial response.

We also observe higher fasting glucose concentrations in hypertensive cases. However, despite our findings on insulin resistance, fasting glucose, and fasting/postprandial hyperinsulinaemia, we did not find significant differences in glycaemic responses. In contrast, in a cross-sectional, longitudinal analysis of 3437 individuals, of which 497 developed hypertension, fasting and postprandial glucose were independent predictors of incident hypertension [33]. This may be explained by an early stage of insulin resistance in the hypertensive cases (HOMA-IR = 1.9, Table 1), whereas hyperglycaemia is thought to become prevalent at more advanced stages [2].

These results suggest that hypertensive individuals may be more prone to cardiovascular events as a result of exacerbated metabolic responses. Regardless of fasting levels, an exacerbated postprandial increase in insulin, triglycerides, glucose and inflammation have detrimental effects on vascular health [26]. Postprandial hyperlipaemia has been linked with impaired lipid metabolism, endothelial dysfunction, hypercoagulability, all of which are key factors involved in atherogenesis [34]. Detrimental effects of postprandial hyperinsulinaemia relate to the hormonal action of insulin, which has the capacity to stimulate numerous cellular responses and has been shown to promote protein synthesis, de novo lipogenesis, and cellular proliferation while inhibiting autophagy, and lipolysis, necessary actions for cellular turnover. Likewise systemic inflammation is independently linked with atherogenesis and coronary heart disease events [12].

Our findings also suggest that BMI is unable to capture the true effects brought about by adiposity. Although BMI is easy to measure in contrast to VFM, the utility of BMI to distinguish between fat and muscle has long been questioned [35]. The present results suggest visceral fat mass should be more routinely measured and used as an actionable target with dietary efforts seeking to reduce visceral fat in hypertensive patients, and in turn mitigate adverse postprandial responses.

The finding that VFM is a causal mediator between hypertension and an atherogenic postprandial triglyceride response has a number of implications for management of hypertension. Firstly, it emphasises the importance of VFM and suggests that VFM should be measured and actioned as a target for treatment. Many lifestyle interventions that are advocated for hypertension, for example, weight loss and reductions in alcohol intake are expected to reduce VFM as well as hypertension but in the context of hypertension they tend to be evaluated according to the reduction in blood pressure achieved. The present results suggest that normalisation of both blood pressure and VFM is likely to achieve optimal risk prevention. Interventions such as bariatric surgery may be very effective in reducing both blood pressure and VFM [36] and their benefit in terms of reduction in VFM may strengthen this indication. The relative benefits and risk of such an intervention can, however, only be rigorously assessed by randomised clinical trials. Secondly, the importance of VFM as contributing to atherogenic risk in hypertension raises the possibility that it could be incorporated in a risk score guiding the indication for statin therapy to offset the atherogenic risk. Again, this would need to be guided by prospective studies evaluating cardiovascular risk and would require more widely available measures of VFM.

Although our study is strengthened by numerous factors, including the tightly controlled nature of the study, there are important limitations. These include (i) the use of office blood pressure measured on a single day to define hypertension, which is prone to measurement error, and white coat effect, i.e., BP increases due to physiological changes when in the presence of a clinician [37], which may have resulted in misclassification bias. These limitations can be overcome by using other means of BP measurement such as ambulatory BP monitoring. However, ambulatory BP was not available for the full sample due to the associated costs and participant burden. (ii) the predominantly female (72.5%) sex of our sample; further research may be required to accurately elucidate any differences between the sexes. (iii) Our sensitivity analysis with menopause status was based on a self-reported questionnaire rather than hormone profiles and may lack accuracy in identifying those pre- and post-menopausal. (iv) While fasting metabolic levels have been widely explored in hypertensive individuals, there is a lacuna of research administering meal challenges in individuals with hypertension and measuring postprandial metabolic responses. Accordingly, there is relative novelty in our study and a lack of studies to compare our findings to.

5. Conclusions

Our findings further the clinical perspective of hypertension as a metabolic disorder and suggest that visceral adiposity is a key factor exacerbating postprandial hypertriglyceridaemia and hyperinsulinaemia. Consequently, reducing VFM should be a key focus of health interventions in hypertension.

Supplementary Materials: The following supporting information can be downloaded at: https://www.mdpi.com/article/10.3390/nu14214499/s1, Supplementary Table S1: Sensitivity analysis of fasting, and postprandial metabolic responses between hypertensive cases and controls, including overall results adjusted for age, sex, and BMI; removing those using antihypertensives; further adjusting for menopause; and when stratifying by sex.

Author Contributions: C.M. and P.C. conceived and designed the experiment; P.L. ran the analysis; P.L., S.E.B. and C.M., wrote the original manuscript. A.M.V., J.W., P.W.F., T.D.S. and K.B. contributed methods/materials/analysis tools. All authors have read and agreed to the published version of the manuscript.

Funding: This work was supported by the Chronic Disease Research Foundation, ZOE, and in part by the Wellcome Trust [Grant number: 212904/Z/18/Z]. For the purpose of open access, the authors have applied a CC BY public copyright licence to any Author Accepted Manuscript version arising from this submission. TwinsUK receives funding from the Wellcome Trust, the European Commission H2020 grants SYSCID (contract #733100); the National Institute for Health Research (NIHR) Clinical Research Facility and the Biomedical Research Centre based at Guy's and St Thomas' NHS Foundation Trust in partnership with King's College London, the Chronic Disease Research foundation, the UKRI Medical Research Council (MRC)/British Heart Foundation Ancestry and Biological Informative Markers for Stratification of Hypertension (AIM-HY; MR/M016560/1), and Zoe. C.M. is funded by the Chronic Disease Research Foundation and by the UKRI Medical Research Council (MRC) AIM-HY project grant (MR/M016560/1). P.L. is supported by the Chronic Disease Research Foundation (CDRF–15/2018). A.M.V. is supported by the National Institute for Health Research Nottingham Biomedical Research Centre.

Institutional Review Board Statement: In accordance with the declaration of Helsinki, all participants provided informed written consent and the study was approved by the Research Ethics Committee and Integrated Research Application System (IRAS 236407). The trial was registered on ClinicalTrials.gov (registration number: NCT03479866).

Informed Consent Statement: Informed consent was obtained from all subjects involved in the study.

Data Availability Statement: The data used in this study are held by the department of Twin Research at King's College London. The data can be released to bona fide researchers using our normal procedures overseen by the Wellcome Trust and its guidelines as part of our core funding (https://twinsuk.ac.uk/resources-for-researchers/access-our-data/).

Acknowledgments: We express our thanks to all the participants of the ZOE PREDICT 1 study for contributing their time and effort and supporting our research.

Conflicts of Interest: T.D.S. is a co-founder and shareholder of ZOE, A.M.V. and P.W.F. are consultants for and J.W. is an employee of ZOE. All other authors declare no competing financial interests.

References

1. Mills, K.T.; Stefanescu, A.; He, J. The Global Epidemiology of Hypertension. *Nat. Rev. Nephrol.* **2020**, *16*, 223–237. [CrossRef] [PubMed]
2. Mulè, G.; Calcaterra, I.; Nardi, E.; Cerasola, G.; Cottone, S. Metabolic Syndrome in Hypertensive Patients: An Unholy Alliance. *World J. Cardiol.* **2014**, *6*, 890–907. [CrossRef]
3. Katsimardou, A.; Imprialos, K.; Stavropoulos, K.; Sachinidis, A.; Doumas, M.; Athyros, V. Hypertension in Metabolic Syndrome: Novel Insights. *Curr. Hypertens. Rev.* **2020**, *16*, 12–18. [CrossRef] [PubMed]
4. Mancia, G.; Bombelli, M.; Corrao, G.; Facchetti, R.; Madotto, F.; Giannattasio, C.; Trevano, F.Q.; Grassi, G.; Zanchetti, A.; Sega, R. Metabolic Syndrome in the Pressioni Arteriose Monitorate E Loro Associazioni (PAMELA) Study. *Hypertension* **2007**, *49*, 40–47. [CrossRef] [PubMed]
5. Schillaci, G.; Pirro, M.; Vaudo, G.; Gemelli, F.; Marchesi, S.; Porcellati, C.; Mannarino, E. Prognostic Value of the Metabolic Syndrome in Essential Hypertension. *J. Am. Coll. Cardiol.* **2004**, *43*, 1817–1822. [CrossRef]
6. Cuspidi, C.; Facchetti, R.; Bombelli, M.; Sala, C.; Tadic, M.; Grassi, G.; Mancia, G. Risk of New-Onset Metabolic Syndrome Associated with White-Coat and Masked Hypertension: Data from a General Population. *J. Hypertens.* **2018**, *36*, 1833–1839. [CrossRef]
7. Ferrannini, E.; Natali, A.; Capaldo, B.; Lehtovirta, M.; Jacob, S.; H. Yki-Järvinen for the European Group for the Study of Insulin Resistance. Insulin Resistance, Hyperinsulinemia, and Blood Pressure. *Hypertension* **1997**, *30*, 1144–1149. [CrossRef]

8. Horita, S.; Seki, G.; Yamada, H.; Suzuki, M.; Koike, K.; Fujita, T. Insulin Resistance, Obesity, Hypertension, and Renal Sodium Transport. *Int. J. Hypertens.* **2011**, *2011*, 391762. [CrossRef]
9. Berry, S.E.; Valdes, A.M.; Drew, D.A.; Asnicar, F.; Mazidi, M.; Wolf, J.; Capdevila, J.; Hadjigeorgiou, G.; Davies, R.; Al Khatib, H.; et al. Human Postprandial Responses to Food and Potential for Precision Nutrition. *Nat. Med.* **2020**, *26*, 964–973. [CrossRef]
10. Lopez-Miranda, J.; Marin, C. Frontiers in neuroscience dietary, physiological, and genetic impacts on postprandial lipid metabolism. In *Fat Detection: Taste, Texture, and Post Ingestive Effects*; Montmayeur, J.P., le Coutre, J., Eds.; Taylor and Francis: Abingdon, UK, 2010.
11. Bansal, S.; Buring, J.E.; Rifai, N.; Mora, S.; Sacks, F.M.; Ridker, P.M. Fasting Compared with Nonfasting Triglycerides and Risk of Cardiovascular Events in Women. *JAMA* **2007**, *298*, 309–316. [CrossRef]
12. Hwu, C.M.; Kwok, C.F.; Kuo, C.S.; Hsiao, L.C.; Lee, Y.S.; Wei, M.J.; Kao, W.Y.; Lee, S.H.; Ho, L.T. Exacerbation of Insulin Resistance and Postprandial Triglyceride Response in Newly Diagnosed Hypertensive Patients with Hypertriglyceridaemia. *J. Hum. Hypertens.* **2002**, *16*, 487–493. [CrossRef] [PubMed]
13. Elffers, T.W.; de Mutsert, R.; Lamb, H.J.; de Roos, A.; Willems van Dijk, K.; Rosendaal, F.R.; Jukema, J.W.; Trompet, S. Body Fat Distribution, in Particular Visceral Fat, is Associated with Cardiometabolic Risk Factors in Obese Women. *PLoS ONE* **2017**, *12*, e0185403. [CrossRef] [PubMed]
14. Franczyk, M.P.; He, M.; Yoshino, J. Removal of Epididymal Visceral Adipose Tissue Prevents Obesity-Induced Multi-Organ Insulin Resistance in Male Mice. *J. Endocr. Soc.* **2021**, *5*, bvab024. [CrossRef] [PubMed]
15. Gabriely, I.; Ma, X.H.; Yang, X.M.; Atzmon, G.; Rajala, M.W.; Berg, A.H.; Scherer, P.; Rossetti, L.; Barzilai, N. Removal of Visceral Fat Prevents Insulin Resistance and Glucose Intolerance of Aging: An Adipokine-Mediated Process? *Diabetes* **2002**, *51*, 2951–2958. [CrossRef]
16. Verdi, S.; Abbasian, G.; Bowyer, R.C.E.; Lachance, G.; Yarand, D.; Christofidou, P.; Mangino, M.; Menni, C.; Bell, J.T.; Falchi, M.; et al. TwinsUK: The UK Adult Twin Registry Update. *Twin Res. Hum. Genet.* **2019**, *22*, 523–529. [CrossRef]
17. Ballout, R.A.; Remaley, A.T. GlycA: A New Biomarker for Systemic Inflammation and Cardiovascular Disease (CVD) Risk Assessment. *J. Lab. Precis. Med.* **2020**, *5*, 17. [CrossRef]
18. Otvos, J.D.; Shalaurova, I.; Wolak-Dinsmore, J.; Connelly, M.A.; Mackey, R.H.; Stein, J.H.; Tracy, R.P. GlycA: A Composite Nuclear Magnetic Resonance Biomarker of Systemic Inflammation. *Clin. Chem.* **2015**, *61*, 714–723. [CrossRef]
19. Akinkuolie, A.O.; Buring, J.E.; Ridker, P.M.; Mora, S. A Novel Protein Glycan Biomarker and Future Cardiovascular Disease Events. *J. Am. Heart Assoc.* **2014**, *3*, e001221. [CrossRef]
20. Connelly, M.A.; Otvos, J.D.; Shalaurova, I.; Playford, M.P.; Mehta, N.N. GlycA, a Novel Biomarker of Systemic Inflammation and Cardiovascular Disease Risk. *J. Transl. Med.* **2017**, *15*, 219. [CrossRef]
21. Direk, K.; Cecelja, M.; Astle, W.; Chowienczyk, P.; Spector, T.D.; Falchi, M.; Andrew, T. The Relationship between DXA-Based and Anthropometric Measures of Visceral Fat and Morbidity in Women. *BMC Cardiovasc. Disord.* **2013**, *13*, 25. [CrossRef]
22. Bertin, E.; Marcus, C.; Ruiz, J.C.; Eschard, J.P.; Leutenegger, M. Measurement of Visceral Adipose Tissue by DXA Combined with Anthropometry in Obese Humans. *Int. J. Obes.* **2000**, *24*, 263–270. [CrossRef] [PubMed]
23. Tingley, D.; Yamamoto, T.; Hirose, K.; Keele, L.; Imai, K. Mediation: R Package for Causal Mediation Analysis. *J. Stat. Softw.* **2014**, *59*, 38. [CrossRef]
24. Qu, H.-Q.; Li, Q.; Rentfro, A.R.; Fisher-Hoch, S.P.; McCormick, J.B. The Definition of Insulin Resistance Using HOMA-IR for Americans of Mexican Descent Using Machine Learning. *PLoS ONE* **2011**, *6*, e21041. [CrossRef] [PubMed]
25. Saxton, S.N.; Clark, B.J.; Withers, S.B.; Eringa, E.C.; Heagerty, A.M. Mechanistic Links between Obesity, Diabetes, and Blood Pressure: Role of Perivascular Adipose Tissue. *Physiol. Rev.* **2019**, *99*, 1701–1763. [CrossRef]
26. Kolb, H.; Kempf, K.; Röhling, M.; Martin, S. Insulin: Too Much of a Good Thing is Bad. *BMC Med.* **2020**, *18*, 224. [CrossRef]
27. Jackson, K.G.; Poppitt, S.D.; Minihane, A.M. Postprandial Lipemia and Cardiovascular Disease Risk: Interrelationships Between Dietary, Physiological and Genetic Determinants. *Atherosclerosis* **2012**, *220*, 22–33. [CrossRef]
28. Kolovou, G.D.; Daskalova, D.; Iraklianou, S.A.; Adamopoulou, E.N.; Pilatis, N.D.; Hatzigeorgiou, G.C.; Cokkinos, D.V. Postprandial Lipemia in Hypertension. *J. Am. Coll. Nutr.* **2003**, *22*, 80–87. [CrossRef]
29. Mancusi, C.; Izzo, R.; di Gioia, G.; Losi, M.A.; Barbato, E.; Morisco, C. Insulin Resistance the Hinge Between Hypertension and Type 2 Diabetes. *High Blood Press Cardiovasc. Prev.* **2020**, *27*, 515–526. [CrossRef]
30. Chaudhary, K.; Buddineni, J.P.; Nistala, R.; Whaley-Connell, A. Resistant Hypertension in the High-Risk Metabolic Patient. *Curr. Diab. Rep.* **2011**, *11*, 41–46. [CrossRef] [PubMed]
31. Nickenig, G.; Röling, J.; Strehlow, K.; Schnabel, P.; Böhm, M. Insulin Induces Upregulation of Vascular AT_1 Receptor Gene Expression by Posttranscriptional Mechanisms. *Circulation* **1998**, *98*, 2453–2460. [CrossRef]
32. Najjar, S.M.; Perdomo, G. Hepatic Insulin Clearance: Mechanism and Physiology. *Physiology* **2019**, *34*, 198–215. [CrossRef] [PubMed]
33. Si, Y.; Wang, A.; Yang, Y.; Liu, H.; Gu, S.; Mu, Y.; Lyu, Z. Fasting Blood Glucose and 2-h Postprandial Blood Glucose Predict Hypertension: A Report from the Reaction Study. *Diabetes* **2021**, *12*, 1117–1128. [CrossRef] [PubMed]
34. Zhao, Y.; Liu, L.; Yang, S.; Liu, G.; Pan, L.; Gu, C.; Wang, Y.; Li, D.; Zhao, R.; Wu, M. Mechanisms of Atherosclerosis Induced by Postprandial Lipemia. *Front. Cardiovasc. Med.* **2021**, *8*, 636947. [CrossRef] [PubMed]
35. Bouchard, C. BMI, Fat Mass, Abdominal Adiposity and Visceral Fat: Where is the 'Beef'? *Int. J. Obes.* **2007**, *31*, 1552–1553. [CrossRef] [PubMed]

36. Schiavon, C.A.; Bersch-Ferreira, A.C.; Santucci, E.V.; Oliveira, J.D.; Torreglosa, C.R.; Bueno, P.T.; Frayha, J.C.; Santos, R.N.; Damiani, L.P.; Noujaim, P.M.; et al. Effects of Bariatric Surgery in Obese Patients with Hypertension: The Gateway Randomized Trial (Gastric Bypass to Treat Obese Patients with Steady Hypertension). *Circulation* **2018**, *137*, 1132–1142. [CrossRef]
37. Pickering, T.G. Measurement of Blood Pressure in and Out of the Office. *J. Clin. Hypertens.* **2005**, *7*, 123–129. [CrossRef]

nutrients

MDPI

Review

The Evolution of Ketosis: Potential Impact on Clinical Conditions

Latha Nagamani Dilliraj [1,†], Giovanna Schiuma [1,†], Djidjell Lara [1], Giovanni Strazzabosco [1], James Clement [1,2], PierPaolo Giovannini [1], Claudio Trapella [1], Marco Narducci [1,‡] and Roberta Rizzo [1,*,‡]

1 Department of Chemical, Pharmaceutical and Agricultural Sciences, University of Ferrara, 44121 Ferrara, Italy
2 BetterHumans, Inc., 3653 NE 77th Avenue, Gainesville, FL 32609, USA
* Correspondence: rbr@unife.it
† These authors contributed equally to this work.
‡ These authors contributed equally to this work.

Abstract: Ketone bodies are small compounds derived from fatty acids that behave as an alternative mitochondrial energy source when insulin levels are low, such as during fasting or strenuous exercise. In addition to the metabolic function of ketone bodies, they also have several signaling functions separate from energy production. In this perspective, we review the main current data referring to ketone bodies in correlation with nutrition and metabolic pathways as well as to the signaling functions and the potential impact on clinical conditions. Data were selected following eligibility criteria accordingly to the reviewed topic. We used a set of electronic databases (Medline/PubMed, Scopus, Web of Sciences (WOS), Cochrane Library) for a systematic search until July 2022 using MeSH keywords/terms (i.e., ketone bodies, BHB, acetoacetate, inflammation, antioxidant, etc.). The literature data reported in this review need confirmation with consistent clinical trials that might validate the results obtained in in vitro and in vivo in animal models. However, the data on exogenous ketone consumption and the effect on the ketone bodies' brain uptake and metabolism might spur the research to define the acute and chronic effects of ketone bodies in humans and pursue the possible implication in the prevention and treatment of human diseases. Therefore, additional studies are required to examine the potential systemic and metabolic consequences of ketone bodies.

Keywords: beta-hydroxybutyrate; evolution; ketogenesis; anti-inflammatory

Citation: Dilliraj, L.N.; Schiuma, G.; Lara, D.; Strazzabosco, G.; Clement, J.; Giovannini, P.; Trapella, C.; Narducci, M.; Rizzo, R. The Evolution of Ketosis: Potential Impact on Clinical Conditions. *Nutrients* **2022**, *14*, 3613. https://doi.org/10.3390/nu14173613

Academic Editors: Omorogieva Ojo and Amanda R Amorim Adegboye

Received: 7 August 2022
Accepted: 30 August 2022
Published: 1 September 2022

1. Introduction

Ketone bodies are small compounds created from fatty acids that serve as an alternative mitochondrial energy source when insulin levels are low, such as during fasting or strenuous exercise [1].

The most important ketone bodies in humans are acetoacetate (AcAc) and β-hydroxybutyrate (BHB), in particular, the R-enantiomer of BHB. Ketone bodies are believed to be adaptive molecules secreted by the liver and quickly distributed to vital organs as a part of an integrated survival mechanism evolved and conserved to provide bioenergetic and signaling advantages when humans face life-threatening conditions or risk factors that could increase the likelihood of premature death [2]. During times of scarce glucose, for example, during fasting or strenuous exercise, BHB is the currency by which energy stored in adipose tissue is turned into fuel that serves the cells to maintain their functions. BHB derives from fatty acids mobilized from adipose tissue and transported to the liver. BHB circulates in the blood to all tissues. After being absorbed into a cell, BHB is broken down in the mitochondria to generate acetyl-CoA, which is further metabolized into ATP. This is the canonical "energy currency" function of BHB.

By reducing carbohydrate ingestion, there is an exhaustion of the body's glucose reserve, shifting the metabolism into ketogenesis, inducing hepatic oxidation of fatty acids, and producing ketones as an important alternative to glucose as the body's energy source [3].

Ketosis is a physiological metabolic state characterized by an increased serum ketone body level from ~0.2 mM to above 3.0 mM [4], caused by exercise [5], fasting/starvation [6], or diabetes [7,8]. In peculiar conditions, the ketosis might develop into overt ketoacidosis, with a decreased serum bicarbonate level and pH, causing serious illness and hospitalization. Ketoacidosis is mainly associated with alcoholism and diabetes mellitus type I, starvation, particularly during malnutrition, and poor dietary intake in people following low-carbohydrate and/or low-caloric diets [9].

In addition to the metabolic function of ketone bodies, they also have several signaling functions separate from energy production. Ketone bodies are involved in epigenetic changes [10–12], controlling cellular signaling metabolites [13], gut microbiota, and butyrogenesis [14]. The epigenetic changes regulate cellular gene expression and metabolism, with an implication for physiological and pathological conditions [15,16]

In this review, we discuss not only the metabolic effect of ketone bodies but also their implication in clinical conditions, suggesting possible future research fields in the use of these molecules as a clinical approach.

2. Methods

In this perspective, we have reviewed the main current data referring to ketone bodies in correlation with nutrition and metabolic pathways as well as to signaling functions, and the potential impact on clinical conditions. Data were selected following eligibility criteria according to the reviewed topic. We used a set of electronic databases (Medline/PubMed, Scopus, Web of Sciences (WOS), Cochrane Library) for a systematic search until July 2022 using MeSH keywords/terms, such as "ketone bodies", "BHB", "acetoacetate", "inflammation", "anti-oxidant", "critical illness". We applied no date or language restrictions. We followed the Preferred Reporting Items for the Systematic Review and Meta-Analysis (PRISMA) statement [17]. Two independent reviewers performed title-abstract screening on all selected studies, then the full texts of the selected articles were reviewed. In cases of duplicate information, the data were checked and combined. Studies reporting ketone bodies as well as BHB and acetoacetate were selected. Publications were selected using specific keywords (i.e., ketone bodies, BHB, acetoacetate, inflammation, antioxidant, etc.) also according to the date of publication (not older than 1990) and for the fulfillment of the topic of this review. Studies that were just case reports and commentaries were excluded. The extraction of the data from included studies was performed by two reviewers separately, considering key characteristics including publication year, author, type of study, country, sample size, and laboratory findings. The funnel plot and Egger's regression test were used to assess publication bias [18].

3. Metabolism of Ketones: Ketogenesis and Ketolysis

The human body produces energy as ATP generated by the mitochondria, for survival. The main energy sources are carbohydrates, fats, amino acids (predominantly glutamine), lactate, and ketone bodies (Figure 1).

Ketone bodies, that are produced during ketogenesis are of three types: acetone, acetoacetate (AcAc), and 3-hydroxybutyrate (BHB). The mainly produced ketone body molecule during ketogenesis is BHB, mainly the R-BHB enantiomer. BHB levels increase in plasma much faster than acetoacetate or acetone, for example, in prolonged fasting [19]. During the conditions such as fasting or vigorous exercise, the consumption of blood glucose lowers insulin levels, turning on ketogenesis, and triglycerides are catabolized to fatty acids that are converted into ketone bodies in the liver [2]. Ketone bodies arrive in metabolically active tissues (muscle, brain) via the blood stream to be metabolized into acetyl-CoA and eventually ATP (Figure 1). BHB is converted, in extrahepatic tissues, by the enzyme 3-hydroxybutyrate dehydrogenase (BDH1) to AcAc [2], which generates Acetyl-CoA [20,21] by exchanging the CoA-fraction from succinyl-CoA [22] by succinyl CoA-oxoacid transferase (SCOT) [23]. Acetyl-CoA enters the TCA cycle and produces

22 ATP per molecule post oxidative phosphorylation. Acetoacetate can be converted also to 3-hydroxybutyrate by BDH or to acetone via a non-enzymatic decarboxylation [20].

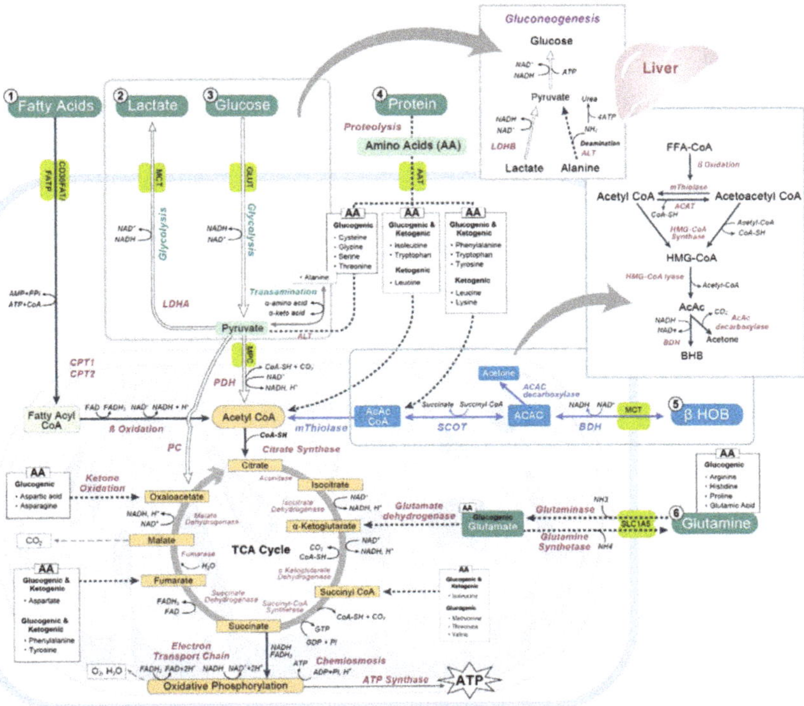

Figure 1. Schematic Diagram of the metabolic pathways of key energy sources in the human body. NAD: nicotinamide adenine dinucleotide; FADH: reduced flavin adenine dinucleotide; NADH: nicotinamide adenine dinucleotide (NAD) + hydrogen (H); PDH: pyruvate dehydrogenase; LDH: lactate dehydrogenase; CPT: carnitine palmitoyl transferase; ß HOB: beta hydroxybutyrate; BDH: D-3-hydroxybutyrate dehydrogenase; SCOT: succinyl-CoA acetoacetate transferase; AcAc: acetoacetate; CoA: coenzyme A; CoA-SH: coenzyme A with sulfhydryl functional group; CO2: carbon dioxide; H2O: water; ADP: adenosine diphosphate; ATP: adenosine triphosphate; Pi: phosphorylated forms of phosphatidylinositol; NH3: ammonia; NH4: ammonium; H+: hydrogen ion; HMG-CoA: ß-hydroxy-ß-methylglutaryl-CoA; MCT: monocarboxylate transporters; GLUT: glucose transporter; AAT: amino acid transporter; ALT: alanine amino transferase; PC: pyruvate carboxylase; MPC: mitochondrial pyruvate carrier.

The short-chain fatty acids (SCFAs), butyrate, acetate, and propionate (in a molar ratio of 3:1:1), are produced as microbial fermentative end-products of undigested/unabsorbed dietary carbohydrates [24,25]. Butyrate is produced by acetate and/or lactate-utilizing butyrate-producing bacteria through the butyryl-CoA: acetate CoA-transferase pathway [24,26,27]. This pathway is typically present in Firmicutes, within Lachnospiraceae (*Eubacterium hallii, Eubacterium rectale, Coprococcuscatus, Roseburia intestinalis*), Ruminococaceae (*Faecalibacterium prausnitzii*), and *Clostridium* spp. in which butyrate and acetyl-CoA are formed from butyryl-CoA and the transformation of the CoA moiety to the external acetate molecule [25–29], while *Bifidobacterium* and *Lactobacillus* spp. use lactate to produce SCFAs. These molecules, similarly, and in combination with ketone bodies, have an anti-inflammatory role via an epigenetic mechanism such as butyrate-associated HDAC inhibition [30].

4. Endogenous Sources of Ketone Bodies

Ketone bodies might derive from endogenous or exogenous sources (Figure 2): endogenous ketones are normally present in our bloodstream and are produced mostly by the liver and by certain species of gut bacteria.

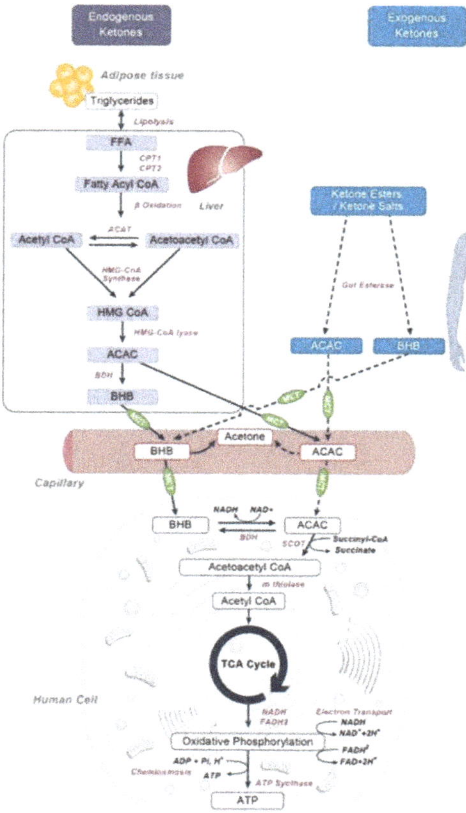

Figure 2. Endogenous and exogenous metabolic pathways. NAD: nicotinamide adenine dinucleotide; FADH: reduced flavin adenine dinucleotide; NADH: nicotinamide adenine dinucleotide (NAD) + hydrogen (H); CPT: carnitine palmitoyl transferase; ß HOB: beta hydroxybutyrate; BDH: D-3-hydroxybutyrate dehydrogenase; SCOT: succinyl-CoA acetoacetate transferase; AcAc: acetoacetate; CoA: coenzyme A; ADP: adenosine diphosphate; ATP: adenosine triphosphate; Pi: phosphorylated forms of phosphatidylinositol; H+: hydrogen ion; HMG-CoA: ß-hydroxy-ß-methylglutaryl-CoA; MCT: monocarboxylate transporters.

Microbiota-derived SCFAs, primarily butyrate, acetate, and propionate are metabolites produced by gut microbiota via dietary non-digestible carbohydrate fermentation [24,25]. SCFAs play a significant role in CHO and lipid metabolism. Butyrate and acetate are used as precursors for lipid synthesis (cholesterol, long-chain fatty acids), whereas propionate is used as a precursor for hepatic gluconeogenesis [24,31]. SCFAs are influenced by the diet (fiber, fats, plant-based proteins) and are important during pregnancy and lactation, controlling the formation of infant gut microbiota. The difference in SCFA-producing bacteria in gut microbiota leads to a different anti-inflammatory state with an impact on inflammatory conditions (e.g., obesity, asthma) [11,12].

In physiology, ketosis can be achieved by fasting, via exogenous supplementation, or the consumption of a ketogenic diet. Fasting from 12–16 h up to 48 h increases serum ketone body concentrations but may elicit inconsistent effects on performance in humans and animal models [32].

5. Exogenous Sources of Ketone Bodies

Exogenous ketone bodies are able to obtain ketosis [1]. Exogenous ketone bodies can be acquired from diet or supplements, such as medium-chain triglyceride (MTC) oils, ketone salts, or esters. MTCs are usually sold as oils or as lyophilized powders and are used to provide the mitochondria with non-carbohydrate energy. They are composed of 8–10 carbon fatty acid chains and they are capable of inducing ketosis thanks to the excess of acetyl CoA produced by the liver while it metabolizes MTC. Ketone salts can be a good alternative to ketogenic diets thanks to their application versatility. They are easy to take and can quickly raise the blood level of ketones. These compounds have a low risk of health issues in humans, and the only concern happens because of the sodium salts as they could increase blood-free sodium levels [33]. Another issue could be the racemic mixture that is present in this product; usually, BHB salts are a combination of D-BHB and L-BHB. It is known that the L form cannot be metabolized by the liver and remains in the blood until it is eliminated through urine or feces. Ketone esters are produced by synthesis, linking one molecule of BHB. Ketone esters are more capable of increasing and maintaining the state of ketosis compared with ketone salts. Some small clinical trials have compared the two different forms of the same molecule, concluding that the ketone esters were able to increase free BHB levels 50% higher than ketone salts [34]. Recently, supplementation with ketone body esters has shown an improvement in exercise performance [35,36].

A ketone diet is characterized by the supply of approximately 80–90% of calories from fat, 10–15% of calories from protein, and <5% calories from carbohydrates [37], stimulating fat oxidation and promoting fat loss, which are important in obesity conditions [38]. The ketone diet has demonstrated successful results in the treatment of epilepsy and other neurological conditions [39]. However, prolonged ketone diets seem to have an impact on liver steatosis [40,41], glucose homeostasis [40–42], and dyslipidemia [43].

6. Metabolic Functions of Ketone Bodies in Vital Organs

During low-carbohydrate conditions, ketone bodies are used in proportion to their plasma concentrations, and consequently, liver production.

It has been demonstrated that the acetoacetate generated from an oral intake of BHB esters reaches first the heart, followed by the kidneys, brain, skeletal muscles, and intestine. Normally, heart tissue can oxidize ketones which enter cardiomyocyte cells, thanks to a specific carrier called MCT1 (monocarboxylate transporter) [44]. In the mitochondria of these cardiomyocytes, ketones are converted into AcAc and then into aceto-acetyl CoA, thanks to the activity of the enzyme SCOT. After this conversion, a thiolase transforms the last product into two molecules of acetyl CoA, which enter the TCA cycle for the production of ATP. In stressful situations, such as heart failure and aortic stenosis-induced left ventricular hypertrophy, the circulating levels of ketones and their oxidation states are heightened, supporting the role of ketones in providing the increased energy demanded by the heart [45].

In the ketosis physiological state, BHB can easily enter proximal tubular cells because there is no saturation mechanism [10]. There, BHB is oxidized through the TCA cycle that produces acetyl CoA, which is then transformed via oxidative phosphorylation into ATP.

An adequate and consistent energy supply is necessary to maintain brain cell functions since low glycogen storage is present inside the brain [46]. This is evidenced in pathological conditions with defects in the brain (e.g., glucose transporter type 1 (GLUT-1 deficiency), with impaired cerebral glucose uptake and consequent seizures, movement disorders, and cognitive impairments [47]. The uptake of ketone bodies across the blood–brain barrier (BBB) is possible via monocarboxylate transporter (MCT) [48], with MCT1-isoform expressed by endothelial cells and astroglia [9]. The brain uses ketone bodies, which can give more than 50% of its energy requirements.

Ketone bodies are also important energy substrates for skeletal muscle, with a robust anticatabolic response, reducing phenylalanine efflux from muscle [49].

Ketone bodies derived from short-chain fatty acids are employed by colonic epithelial cells as respiratory fuels where they predominantly use butyrate sequentially followed by acetoacetate, glutamine, and glucose, notwithstanding the interaction of substrates [50]. Ketone body signaling facilitates Lgr5+ intestinal stem cell (ISC) homeostasis aiding in post-injury intestinal regeneration, restoring intestinal regeneration, and resuming ISC function [51].

7. Biological Properties of Ketone Bodies

The increase in ketone body blood levels derives from fatty acid breakdown during low carbohydrate availability, resulting in a danger signal of starvation and a physiological response for improving survival during starvation. The energy production from ketone bodies is associated with increased radical oxygen species (ROS) release in the mitochondria, an increase in NAD+ levels, and a lower AMP/ATP ratio. Ketone bodies and the ketogenic diet act in upregulating anti-oxidant and anti-inflammatory mechanisms (Figure 3).

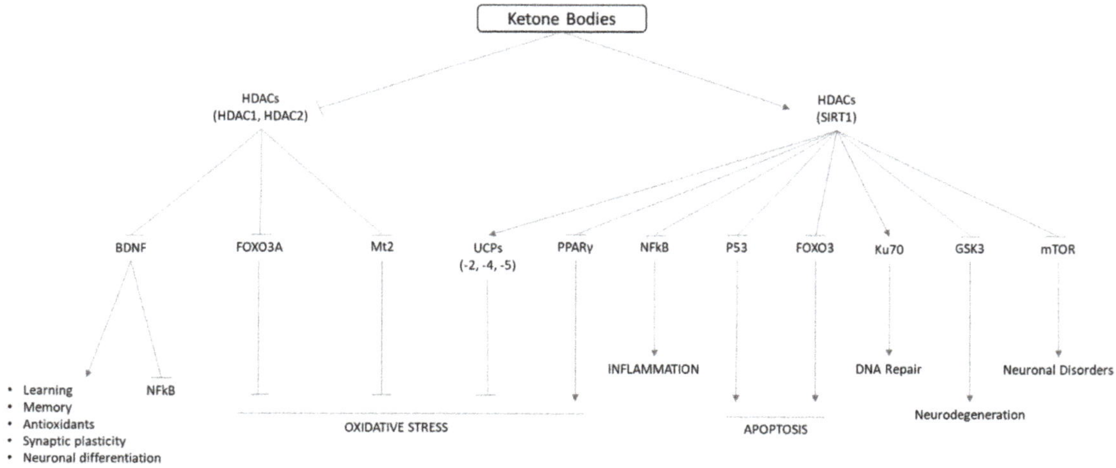

Figure 3. Molecular pathways involved in ketone bodies effect on oxidative stress, inflammation, and epigenetic control. HDAC: histone deacetylase; BDNF: brain-derived neurotrophic factor; NFkB: nuclear factor kappa-light-chain-enhancer of activated B cells; FOXO3A: forkhead box O3; Mt2: mammalian metallothionein-2; UCP: uncoupling protein; PPARγ: peroxisome proliferator-activated receptor γ; p53: protein 53; Ku70: DNA repair subunit protein; GSK3: serine/threonine protein kinase; mTOR: mammalian target of rapamycin.

Consistently, ketone bodies preserve mitochondria and their role in cellular energy homeostasis [52]. An increased uptake of Ca2+ into mitochondria enhances ROS production and blocks the synthesis of ATP, inducing cytochrome c release and mitochondrial membrane potential [53]. These modifications cause mitochondrial swelling and apoptosis [54], with consequently impaired energy homeostasis [55]. BHB is able to maintain mitochondrial density and function [56] by controlling the mitophagy of damaged mitochondria and inducing the renewal of the mitochondria population [57]. The mitochondrial biogenesis is enhanced via Nrf2 activation, which induces the transcription of PGC-1 (peroxisome proliferator-activated receptor gamma coactivator-1) [58], which controls the transcription of TFAM (mitochondrial transcription factor A), resulting in the replication of mitochondrial DNA and the mitochondrial biogenesis [59].

The block of glutathione peroxidase (GSH-Px), a key rate-limiting enzyme in ROS formation [60] by BHB, reduces the production of ROS, with a consequent improvement in mitochondrial respiration and homeostasis [61–65], ATP production, activation of adenosine receptors that lower oxidative stress [66], upregulation of antioxidant genes and activation of antioxidant enzymes that control lipid peroxidation and protein oxidation [67].

Ketones' anti-inflammatory effects are related to the inhibition of the NLRP3 (NOD-like receptor pyrin-domain containing-3) inflammasome, which activates caspase-1 and the release of the pro-inflammatory cytokines (IL-1β and IL-18) [68]. Ketone bodies block the K+ efflux, which activates the NLPR3 inflammasome [69]. During brain injury induced by the middle cerebral artery occlusion (MCAO) model, ketone bodies inhibited NLRP3 and inflammation [70]. In a randomized, controlled dietary intervention trial with 40 overweight subjects aged 18–55 years fed with a diet very low in carbohydrates or an isocaloric diet low in fat for 12 weeks, the subjects following the ketogenic diet showed lower inflammation, with the reduction of interleukin-8 (IL-8), TNF-alpha, plasminogen activator inhibitor-1 (PAI-1), monocyte chemoattractant protein (MCP-1), E-selectin, and intercellular adhesion molecule-1 (ICAM-1) in the presence of mild inflammation [71].

The immune response is controlled by ketones by the binding of BHB to HCAR2 (hydroxycarboxylic acid receptor 2), which results in the induction of prostaglandin D2 (PGD2) production by cyclooxygenase 1 (COX1) [72] and the inhibition of NF-KB (nuclear factor kappa-light-chain-enhancer of activated B cells) mediated inflammation through the blockage of IKB kinase (IKK), by a metabolite of PGD2.

BHB is involved in controlling cellular function via epigenomic regulation. Ketone bodies control histone post-translational modifications, including histone methylation (Kme), histone/lysine acetylation (Kac), and β-hydroxybutyrylation (Kbhb), which regulate chromatin architecture and gene expression. BHB is able to inhibit the histone deacetylase (Class 1 and Class IIa HDACs) leading access to transcription factors of genes encoding oxidative stress resistance factors like FOXO3 (forkhead box O3) and Mt2 (Mammalian metallothionein-2) [73]. The induction via BHB of histone Kbhb levels with site-specific lysine residues (H3K4, H4K8, H3K9, H4K12, H3K56) is increased in human embryonic kidney 293 (HEK293) cells during prolonged fasting, supporting a direct role in chromatin structure and functions regulation [74]. In HEK293 cells transiently transfect with ORM (yeast)-Like protein isoform 3 (ORMDL3) mRNA expression, an asthma susceptibility gene [75], BHB controlled inflammation inhibiting ER stress response pathway proteins and enhancing both Foxp3 and manganese superoxide dismutase (MnSOD) transcription via AMP-activated protein kinase (AMPK) activation, leading to a decrease in cellular oxidative stress [76].

BHB inhibition of HDACs leads to increase expression of brain-derived neurotrophic factor (BDNF), which exerts neuroprotective effects against various insults to the central nervous system, as the functional recovery after traumatic brain injury (TBI) in mice [77].

BHB can act as an inducer of transcription factor Nrf2 (nuclear factor-erythroid factor 2-related factor 2). Nrf2 is a transcription factor that regulates the cellular defense against toxic and oxidative insults through the upregulation of the expression of genes involved in the oxidative stress response and drug detoxification [78]. Human microvascular endothelial cells (HMEC-1) exposed to ketone bodies increased NRF2 expression with a clear translocation to the nucleus and induction of antioxidant proteins [79].

8. Role of Ketone Bodies in Pathophysiological Ailments

Critical illness has demonstrated various disruptions in metabolism and mitochondrial function. Whether it arises from organ failure or microbial infections, the metabolic response to these ailments requires maintenance of metabolic balance, nutrient utilization for different activities, and functional recovery. On the other hand, a metabolic response can create clinical consequences such as catabolic processes that can impair physiological stability. Furthermore, a long period of metabolic imbalance could produce mitochondrial dysfunction, including energy crisis and high free-radical production, resulting in a compromised immune system, tissue and organ failure, and death [80–83].

Critical-illness-affected patients require an energy source to support physiological stress responses and to give robust protection to critical organs such as the heart, brain, liver, and kidney [81,83] (Figure S1). Additionally, increasing evidence across animal and human studies has shown ketones are a beneficial alternative substrate due to reloading acetyl-coenzyme-A through an independent pathway irrespective of glucose levels [84], and are useful to maintain the cytosolic NAD+ (nicotinamide adenine dinucleotide) pool which is pivotal for cellular survival, antioxidants, and pro-survival pathways [85]. As serum ketone body concentration varies during physiological and pathological conditions and acts as potent endogenous signaling molecules, they may act in cellular protection and repair, mitochondrial biogenesis, antioxidant defenses, and enhanced autophagy [80,82,86–88]. With the known and possible mechanistic properties of ketones, their use as an individual or adjunct therapy for different conditions in critical illness has been explored.

8.1. Heart Failure

During heart failure, the heart undergoes a metabolic switch favoring ketone metabolism in cardiomyocytes, which are more efficiently used than in the normal heart. Moreover, ketone body oxidation is a more efficient energy substrate than terminal fatty acid oxidation [88]. Significant improvement in the cardiac output of about 24% was observed after ketone infusion in both chronic heart failure patients and animal models with heart failure [84,89]. These cardioprotective properties can be attributed to their increased utilization in the heart or upregulation of crucial oxidative phosphorylation mediators [44]. It has also been demonstrated that ketones improve blood lipid profile in obese adults by lowering LDL (low-density lipoprotein) cholesterol and raising HDL (high-density lipoprotein) cholesterol, reducing adipocyte cell volume, and lowering serum lipolytic products [84]. Ketones' mechanistic action in inhibiting NLRP3 inflammasome and activation of the GPCR109A (G-protein-coupled receptor) has been shown to rescue heart failure by reducing mitochondrial hyperacetylation, resulting in lower inflammation and oxidative stress and preventing atherosclerosis [84]. A study in mice also showed that BHB reduces sympathetic outflow and lowers heart rate and total energy expenditure by antagonizing GPR41, a G-protein—coupled receptor for short-chain fatty acids [88].

8.2. Kidneys and Liver Diseases

During the development of kidney diseases, impaired lipolysis and mTORC1 (mammalian target of rapamycin complex 1) hyperactivation are observed; with this pathological condition, ketone supplementation might be an alternative energy source for mitochondrial respiration. Moreover, ketones' reno-protective roles are mainly via the endogenous inhibition of HDAC (histone deacetylase) and NLRP3, both increased expression of protective genes; consequently, it inhibits mTORC1, inflammation, oxidative stress, and tissue fibrosis [84,89]. Additionally, one mice study involving high-fat liver injury showed that oxidation of the ketone body acetoacetate by liver-resident macrophage-like Kupffer cells lowers fibrosis [89].

8.3. Brain Injury and Neuronal Diseases

Substantial evidence has shown ketone bodies' pleiotropic neuroprotection properties due to their pivotal role in cerebral energy homeostasis and an active signaling molecule. BHB directly regulates inflammation and neurotrophic factors by inhibiting the activation of innate immune sensor NLRP3 and inhibiting HDAC, which upregulates BDNF (brain-derived neurotrophic factor) [90]. BDNF is crucial in the maintenance, restoration, and improvement of neural networks and brain functions after a brain injury. Thus, BDNF production in the brain plays an essential role in the prolonged maintenance of neuroplasticity [90]. Ketones have been known to have anti-seizure effects, which can be achieved through their action in altering synaptic neurotransmission via increasing GABA (gamma-aminobutyric acid) synthesis and decreasing glutamate synthesis [91]. Similarly, the anti-epileptic effects of ketones via activation of the KATP channels and GABA signaling lead to lower neuronal firing [92]. Multiple in vitro studies have demonstrated that ketones can increase the survival rate of cultured neocortical neurons and isolated cortical mitochondria from exposure to hydrogen peroxide both with and without glucose addition [93]. It also reduced apoptosis after hypoxia in rat hippocampal neuron cultures from various insults, including hypoglycemia, hypoxia, and N-methyl-D-aspartate-induced excitotoxicity [93]. Furthermore, ketones demonstrated strong neuroprotective properties in various animal models of brain injury, reducing neuronal apoptosis and brain edema and enhancing sensory-motor and cognitive performance [89,93]. It has also reduced neuronal loss and infarct size in animal models of stroke and reduced the glutamate release and seizure severity in a mouse model of epilepsy [89,93]. In spinal cord injury models in rats, ketosis reduced spinal lesions enhanced the GLUT 1 and MCT1 vascular transporters, and forelimb motor function improvement [94]. Preclinical studies in adult rats with moderate and severe traumatic brain injury strongly demonstrated therapeutic actions, where it showed a significant reduction of infarct and penumbral volume in MRI, decreased tissue death and edema, and improved neurological scores [91]. The administration of ketones and hypertonic saline (HTS) showed beneficial effects in managing intracranial hypertonic pressure with enhanced cerebral metabolism [95]. Thus, the administration of exogenous ketones in patients with different stages of brain injury might benefit clinical outcomes due to the suggested neuroprotective property in maintaining mitochondrial function and decreasing inflammation, oxidative stress, and seizure problems [91]. Ketogenic interventions might facilitate the brain's utilization of ketones as an essential energy source and as a signaling molecule that may slow down the disease progression and delay or even prevent the disease onset if started earlier [96]. In vitro models of Parkinson's disease in mouse neuronal cultures demonstrated increased cell survival, enhanced mitochondrial membrane potential, and lower cytochrome c release, while a 60% increase in cell survival was also observed in human neuroblastoma cell culture [97]. Motor function was restored and prevented losing dopaminergic neurons after several infusions of sodium BHB (1.6 mmol/kg/day) [44]. Similarly, in the in vitro model of Alzheimer's disease, the cultured mouse hippocampal neurons were protected from amyloid beta 1 to 42 toxicity after 4 mmol/L ß-OHB was administered [44]. Ketone supplementation provided cognitive protection for several months [91] and even better performance in paragraph recall and Alzheimer's assessment scale tests in humans subjected to a ketogenic diet [97]. While in animal models of amyotrophic lateral sclerosis, the clinical and biological manifestations of the disease were beneficially altered after exposure to hyperketonemia, which had more remarkable motor neuron survival and enhanced motor function [98].

8.4. Muscle Weakness

Post-recovery weakness involves more than 50 percent of patients recovering from critical illness. This condition is characterized by muscle dysfunction, atrophy, and damage, with consequent immobility, inflammation, and catabolism. The maintenance of mobility is the basis of critical care management [44]. Ketone bodies seem to have a direct effect on protein turnover, as their increase is associated with a decrease in proteins and amino acid efflux from skeletal muscle [81,99]. Additionally, ketone supplementation reduces muscle atrophy, and increased cholesterol myofiber, which is associated with muscle force [100] and stimulates muscle regeneration [81,84].

9. Neuroprotective Actions of Ketone Bodies

It has been observed that ketone bodies have multiple key roles in the brain, which are exerted not only during fasting but also in the newborn period. The development and function of a healthy neonatal brain appear to be related to locally produced ketones, which are the preferred precursors for fatty acids and cholesterol for the creation of dry matter in the brain [22,101,102]. Human newborns are characterized by extensive subcutaneous adipose stores [103], which provide fatty acids and ketones. The medium-chain fatty acids (MCFAs) in breast milk, synthesized de novo from glucose within the epithelium of the milk duct, promote the production of ketone bodies in the infant liver [104] and gut [105]. Some of the MCFAs in breast milk create adipose stores in the infant, that prolongs mild ketonemia after lactation ends. The energetic support of the development of the brain in infants is mainly supported by ketone uptake and oxidation, with a molar utilization of BHB that is around 50% greater than that of glucose [106,107]. The newborn rats start ketosis from the beginning of the suckling period, thanks to the MCFAs in the dam's milk [28,108]. The levels of BHB and AcAc are 3- to 4-fold higher at the blood–brain barrier in newborns in comparison to adults [109,110].

Ketone Bodies Influence Neurotransmitters

The brain's major excitatory neurotransmitter is glutamate [111], which is not transported from blood, but is synthesized in the brain and delivered to neurons upon depolarization. Ketone bodies control neurotransmitters metabolism as acetyl-CoA production by ketone bodies decreases oxaloacetate, increasing glutamate levels and inducing GABA synthesis [112,113] (Figure 4). In the presence of ketone bodies, it metabolizes to acetyl-CoA and oxaloacetate follows citrate synthesis [110,114,115], reducing the activity of glutamic oxaloacetic transaminase (GOT) and preserving glutamate for the glutamate decarboxylase reaction to yield GABA. Ketone bodies DL-β-hydroxybutyrate were shown also to control GABA activity during the developmental, resulting in a switch from being predominantly depolarizing–excitatory to predominantly hyperpolarizing–inhibitory [116].

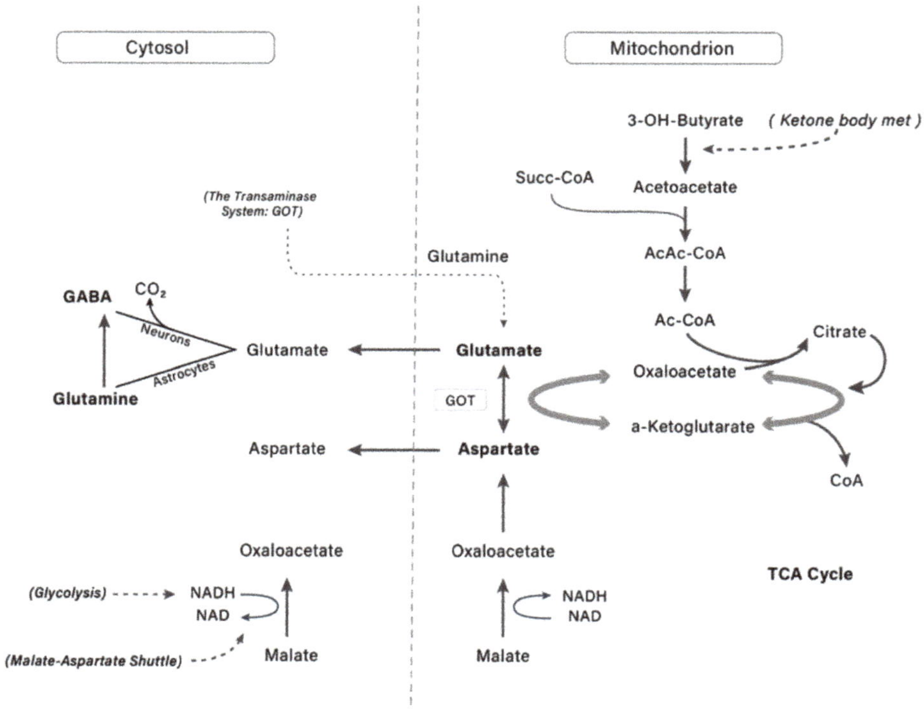

Figure 4. Scheme illustrating the relationship between brain metabolism of ketone bodies and that of glutamate and GABA, where the metabolism of ketone bodies of acetyl-CoA induces the increase of glutamate and GABA. 3-OH_Butyrate: β-hydroxybutyrate; Succ-CoA: succinyl-CoA; AcAc-CoA: acetoacetyl-CoA; Ac- CoA: acetyl-CoA; CoA: coenzyme A; NADH/NAD: nicotinamide adenine dinucleotide; GOT: glutamate-oxaloacetate transaminase; GABA: gamma-aminobutyric acid.

10. Discussion

This review has highlighted the impact of ketone bodies on several human physiological and pathological conditions. Ketone bodies are an alternative energy source in hypoglycemia conditions, such as when fasting or during strenuous exercise, they have several signaling functions inside human cells, affecting cell genome acetylation and consequently gene expression, controlling adipose tissue metabolism, changing sympathetic nervous system activation and the whole-body metabolic rate, inhibition, and inflammasome activation. These ketone bodies' effects on human cells might suggest their implication in the control of human pathological conditions.

The psychophysiological and metabolic milieu that triggers the secretion of ketone bodies includes (i) starvation; (ii) severe injuries; (iii) acute infections or viral illnesses [89] (iv) physical exhaustion, and (v) in the presence of harsh ecological stressors. In these contexts, ketone bodies galvanize and modulate the body's survival factors during these unfavorable conditions by offsetting physiological dyshomeostasis and psychophysical functionality. Therefore, the advantage of ketone metabolism is the conservation of precious glycogen reserves and the immediate supply of a potent and effective fuel for the brain. The most apparent sign that ketone bodies are a well-preserved and highly adaptive trait of evolution is the fact that even infants and embryos utilize ketone bodies as a critical bioenergetic buffer to sustain the tremendous growth of the neonatal brain. Evolutionary forces selected ketone bodies to ensure self-preservation during the most critical time for any specie evolution. Ketone bodies enter the TCA cycle with fewer steps than glucose and produce more ATP per mole than pyruvate with a lower oxygen requirement to produce more ATP per mole than glucose, preventing the depletion of NAD+ and endogenous antioxidants while increasing cellular bioenergetic efficiency [101]. Cotter et al. [117] demonstrated that postnatal mice without ketone bodies oxidation present a lethal metabolic state, even in the presence of alternative metabolic fuels supplied through milk. A similar condition seems to be present in humans, where the sudden infant death syndrome (SIDS) has been attributed to SCOT deficiency [117]. In the light of this evolutionary perspective, nutritional ketosis when induced by exogenously or endogenously seems to enhance survival during hemorrhagic shocks, severe hypoxia, cerebrovascular ischemia, heart attacks, deep wounds, traumatic brain injuries, sepsis, poisoning, and severe intoxications in in vivo animal models. Nutritional ketosis might affect the biophysical state with a possible role in controlling central fatigue, anxiety, aggression, clinical depression, sense of hunger, or perceived pain while increasing focus and mental performance. In line with seminal studies emerging from calorie restriction, more recent evidence shows that being in nutritional ketosis might control degenerative conditions including recalcitrant metabolic diseases that manifest dysfunctional homeostatic adaptations and deteriorations.

The ketone metabolism is a constitutive feature of organ functions, mainly in the brain [118]. These findings suggest the need for clinical studies to evaluate the possible effect of the administration of exogenous ketone bodies to enhance general brain health.

Currently, there is considerable interest in ketone body supplementation, such as drinks containing ketone esters and ketone salts, which can increase ketone bodies' blood concentration without dietary changes [119], positively affecting the ketone body's brain uptake and metabolism. For example, the oral ingestion of exogenous BHB is able to obtain rapid and significant ketosis (i.e., above 6 mmol/L) in humans. The oral BHB administration (3 mg KE/g of body weight) in non-fasted mice, increased acetyl-CoA and citric cycle intermediates in the brain [120], with a preferential distribution in the neocortex. Acetate supplementation increased plasma acetate and brain acetyl-CoA levels in rats [121], with no modification in brain ATP, ADP, NAD, GTP levels, or the energy charge ratio, glycogen, and mitochondrial biogenesis when compared to controls [121]. The literature data suggest that ketone bodies had a major impact on the evolution of our brains over 2 million years, when the life of hominid monkeys was characterized by intermittent starving and fat intake, optimal for the generation of ketone bodies, supporting the "ketone-brain expansion" hypothesis [122].

11. Conclusions

In conclusion, ketone bodies showed a significant role in controlling oxidative stress and inflammation, which result in improved mitochondrial function and growth, energy rescue, and adaptative epigenetic control (Figure 5). In this context, ketolysis is an adaptive response of the human body to resist acute and chronic diseases, acting as an alternative fuel during periods of deficient food supply, with the reduction of oxidative stress in the mitochondria, and the protection of cell functions. The absence of consistent clinical trials partially dampens the interesting results obtained in vitro and in vivo in animal models. However, the data on exogenous ketones consumption and its effect on the ketone bodies' brain uptake and metabolism might spur the research to define the acute and chronic effects of ketone bodies in humans and pursue the possible implication in the prevention and treatment of human diseases. Therefore, additional studies are required to examine the potential systemic and metabolic consequences of ketone bodies.

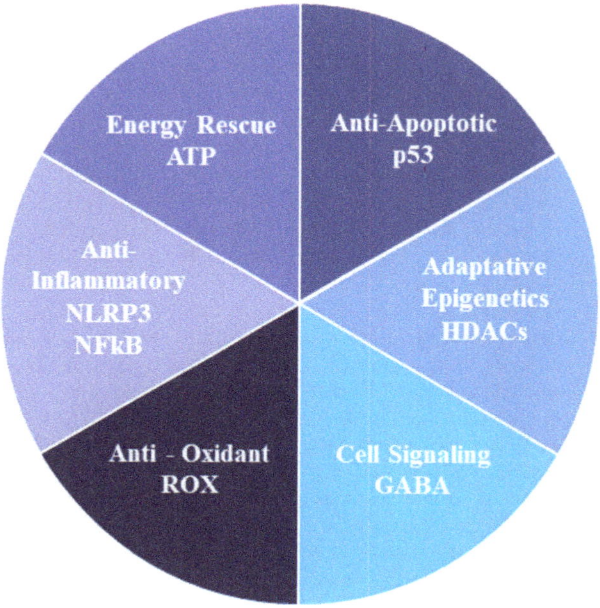

Figure 5. The six hallmarks of ketonic action. ATP: Adenosine triphosphate; NLRP3: NOD-, LRR- and pyrin domain-containing protein 3; NFkB: nuclear factor kappa-light-chain-enhancer of activated B cells; ROX: chemical reduction oxidation; GABA: gamma-aminobutyric acid; HDAC: histone deacetylase.

Supplementary Materials: The following supporting information can be downloaded at: https://www.mdpi.com/article/10.3390/nu14173613/s1. Figure S1: Target organs and pathophysiological conditions of ketone administration; possible mechanism of action as alternative fuel and signaling molecule.

Author Contributions: Conceptualization, R.R., M.N., L.N.D. and G.S. (Giovanna Schiuma); data curation, L.N.D., G.S. (Giovanna Schiuma), D.L., G.S. (Giovanni Strazzabosco), J.C., P.G. and C.T.; writing—original draft preparation, R.R., M.N., L.N.D., G.S. (Giovanna Schiuma) and J.C. All authors have read and agreed to the published version of the manuscript.

Funding: This research received no external funding.

Institutional Review Board Statement: Not applicable.

Informed Consent Statement: Not applicable.

Data Availability Statement: Not applicable.

Conflicts of Interest: The authors declare no conflict of interest.

References

1. Evans, M.; Cogan, K.E.; Egan, B. Metabolism of ketone bodies during exercise and training: A physiological basis for exogenous supplementation. *J. Physiol.* **2016**, *595*, 2857–2871. [CrossRef]
2. Rui, L. Energy Metabolism in the Liver. *Compr. Physiol.* **2014**, *4*, 177–197. [CrossRef] [PubMed]
3. Gomez-Arbelaez, D.; Crujeiras, A.B.; Castro, A.I.; Goday, A.; Mas-Lorenzo, A.; Bellon, A.; Tejera, C.; Bellido, D.; Galban, C.; Sajoux, I.; et al. Acid–base safety during the course of a very low-calorie-ketogenic diet. *Endocrine* **2017**, *58*, 81–90. [CrossRef] [PubMed]
4. Harvey, K.L.; Holcomb, L.E.; Kolwicz, S.C., Jr. Ketogenic Diets and Exercise Performance. *Nutrients* **2019**, *11*, 2296. [CrossRef] [PubMed]
5. Rowe, P.; O'Neill, C.; DeWitt, E.; Kolwicz, S.C. Endurance Exercise Capacity and Substrate Metabolism in Male and Female Mice. *FASEB J.* **2019**, *33*, 698.1. [CrossRef]
6. Wentz, A.; D'Avignon, D.A.; Weber, M.L.; Cotter, D.G.; Doherty, J.M.; Kerns, R.; Nagarajan, R.; Reddy, N.; Sambandam, N.; Crawford, P.A. Adaptation of Myocardial Substrate Metabolism to a Ketogenic Nutrient Environment. *J. Biol. Chem.* **2010**, *285*, 24447–24456. [CrossRef] [PubMed]
7. Avogaro, A.; Crepaldi, C.; Miola, M.; Maran, A.; Pengo, V.; Tiengo, A.; Del Prato, S. High blood ketone body concentration in Type 2 non-insulin dependent diabetic patients. *J. Endocrinol. Investig.* **1996**, *19*, 99–105. [CrossRef]
8. Guerci, B.; Benichou, M.; Floriot, M.; Bohme, P.; Fougnot, S.; Franck, P.; Drouin, P. Accuracy of an Electrochemical Sensor for Measuring Capillary Blood Ketones by Fingerstick Samples During Metabolic Deterioration After Continuous Subcutaneous Insulin Infusion Interruption in Type 1 Diabetic Patients. *Diabetes Care* **2003**, *26*, 1137–1141. [CrossRef]
9. Aksu, N.; Akcora, Z.; Ilhan, B.; Bayar, O.; Akkas, M. Ketacidosis due to starvation. *Eur. J. Emerg. Med.* **2018**, *17*, 39–40.
10. Cabrera-Mulero, A.; Tinahones, A.; Bandera, B.; Moreno-Indias, I.; Macías-González, M.; Tinahones, F.J. Keto microbiota: A powerful contributor to host disease recovery. *Rev. Endocr. Metab. Disord.* **2019**, *20*, 415–425. [CrossRef]
11. Alsharairi, N.A. The Role of Short-Chain Fatty Acids in the Interplay between a Very Low-Calorie Ketogenic Diet and the Infant Gut Microbiota and Its Therapeutic Implications for Reducing Asthma. *Int. J. Mol. Sci.* **2020**, *21*, 9580. [CrossRef] [PubMed]
12. Alsharairi, N.A. The Role of Short-Chain Fatty Acids in Mediating Very Low-Calorie Ketogenic Diet-Infant Gut Microbiota Relationships and Its Therapeutic Potential in Obesity. *Nutrients* **2021**, *13*, 3702. [CrossRef] [PubMed]
13. Newman, J.C.; Verdin, E. β-hydroxybutyrate: Much more than a metabolite. *Diabetes Res. Clin. Pract.* **2014**, *106*, 173–181. [CrossRef] [PubMed]
14. Sasaki, K.; Sasaki, D.; Hannya, A. In vitro human colonic microbiota utilises D–hydroxybutyrate to increase butyro-genesis. *Sci. Rep.* **2020**, *10*, 8516. [CrossRef]
15. Tzika, E.; Dreker, T.; Imhof, A. Epigenetics and Metabolism in Health and Disease. *Front. Genet.* **2018**, *9*, 361. [CrossRef]
16. Prescott, S.; Saffery, R. The role of epigenetic dysregulation in the epidemic of allergic disease. *Clin. Epigenet.* **2011**, *2*, 223–232. [CrossRef]
17. Moher, D.; Shamseer, L.; Clarke, M.; Ghersi, D.; Liberati, A.; Petticrew, M.; Shekelle, P.; Stewart, L.A.; Group, P.-P. Preferred reporting items for systematic review and meta-analysis protocols (prisma-p) 2015 statement. *Syst. Rev.* **2015**, *4*, 1. [CrossRef]
18. Van Enst, W.A.; Ochodo, E.; Scholten, R.J.P.M.; Hooft, L.; Leeflang, M.M. Investigation of publication bias in meta-analyses of diagnostic test accuracy: A meta-epidemiological study. *BMC Med. Res. Methodol.* **2014**, *14*, 70. [CrossRef]
19. Desrochers, S.; David, F.; Garneau, M.; Jetté, M.; Brunengraber, H. Metabolism of *R*- and *S*-1,3-butanediol in perfused livers from meal-fed and starved rats. *Biochem. J.* **1992**, *285*, 647–653. [CrossRef]
20. Najac, C.; Radoul, M.; Le Page, L.M.; Batsios, G.; Subramani, E.; Viswanath, P.; Gillespie, A.M.; Ronen, S.M. In vivo investigation of hyperpolarized [1,3-13C2]acetoacetate as a metabolic probe in normal brain and in glioma. *Sci. Rep.* **2019**, *9*, 3402. [CrossRef]
21. Newman, J.C.; Verdin, E. β-Hydroxybutyrate: A Signaling Metabolite. *Annu. Rev. Nutr.* **2017**, *37*, 51–76. [CrossRef]
22. Puchalska, P.; Crawford, P.A. Multi-dimensional Roles of Ketone Bodies in Fuel Metabolism, Signaling, and Therapeutics. *Cell Metab.* **2017**, *25*, 262–284. [CrossRef] [PubMed]
23. Fukao, T.; Song, X.-Q.; A Mitchell, G.; Yamaguchi, S.; Sukegawa, K.; Or, T.; Kondo, N. Enzymes of Ketone Body Utilization in Human Tissues: Protein and Messenger RNA Levels of Succinyl-Coenzyme A (CoA):3-Ketoacid CoA Transferase and Mitochondrial and Cytosolic Acetoacetyl-CoA Thiolases. *Pediatr. Res.* **1997**, *42*, 498–502. [CrossRef] [PubMed]
24. Ríos-Covián, D.; Ruas-Madiedo, P.; Margolles, A.; Gueimonde, M.; De Los Reyes-Gavilán, C.G.; Salazar, N. Intestinal Short Chain Fatty Acids and their Link with Diet and Human Health. *Front. Microbiol.* **2016**, *7*, 185. [CrossRef]
25. Louis, P.; Flint, H.J. Formation of propionate and butyrate by the human colonic microbiota. *Environ. Microbiol.* **2017**, *19*, 29–41. [CrossRef] [PubMed]
26. Gerhart, D.Z.; Enerson, B.E.; Zhdankina, O.Y.; Leino, R.L.; Drewes, L.R. Expression of monocarboxylate transporter MCT1 by brain endothelium and glia in adult and suckling rats. *Am. J. Physiol. Metab.* **1997**, *273*, E207–E213. [CrossRef] [PubMed]
27. Vacca, M.; Celano, G.; Calabrese, F.M.; Portincasa, P.; Gobbetti, M.; De Angelis, M. The Controversial Role of Human Gut Lachnospiraceae. *Microorganisms* **2020**, *8*, 573. [CrossRef]
28. Chowdhury, G.M.; Jiang, L.; Rothman, D.L.; Behar, K.L. The Contribution of Ketone Bodies to Basal and Activity-Dependent Neuronal Oxidation in Vivo. *J. Cereb. Blood Flow Metab.* **2014**, *34*, 1233–1242. [CrossRef]
29. Oliphant, K.; Allen-Vercoe, E. Macronutrient metabolism by the human gut microbiome: Major fermentation by-products and their impact on host health. *Microbiome* **2019**, *7*, 91. [CrossRef]

30. Indrio, F.; Martini, S.; Francavilla, R.; Corvaglia, L.; Cristofori, F.; Mastrolia, S.A.; Neu, J.; Rautava, S.; Spena, G.R.; Raimondi, F.; et al. Epigenetic Matters: The Link between Early Nutrition, Microbiome, and Long-term Health Development. *Front. Pediatr.* **2017**, *5*, 178. [CrossRef]
31. Bridgman, S.L.; Azad, M.B.; Field, C.; Haqq, A.M.; Becker, A.B.; Mandhane, P.J.; Subbarao, P.; Turvey, S.; Sears, M.R.; Scott, J.A.; et al. Fecal Short-Chain Fatty Acid Variations by Breastfeeding Status in Infants at 4 Months: Differences in Relative versus Absolute Concentrations. *Front. Nutr.* **2017**, *4*, 11. [CrossRef] [PubMed]
32. Zouhal, H.; Saeidi, A.; Salhi, A.; Li, H.; Essop, M.F.; Laher, I.; Rhibi, F.; Amani-Shalamzari, S.; Ben Abderrahman, A. Exercise Training and Fasting: Current Insights. *Open Access J. Sports Med.* **2020**, *11*, 224919. [CrossRef] [PubMed]
33. Kesl, S.L.; Poff, A.M.; Ward, N.P.; Fiorelli, T.N.; Ari, C.; Van Putten, A.J.; Sherwood, J.W.; Arnold, P.; D'Agostino, D.P. Effects of exogenous ketone supplementation on blood ketone, glucose, triglyceride, and lipoprotein levels in Sprague–Dawley rats. *Nutr. Metab.* **2016**, *13*, 9. [CrossRef]
34. White, H.; Heffernan, A.J.; Worrall, S.; Grunsfeld, A.; Thomas, M. A Systematic Review of Intravenous β-Hydroxybutyrate Use in Humans–A Promising Future Therapy? *Front. Med.* **2021**, *8*, 740374. [CrossRef] [PubMed]
35. Cox, P.J.; Kirk, T.; Ashmore, T.; Willerton, K.; Evans, R.; Smith, A.; Murray, A.J.; Stubbs, B.; West, J.; McLure, S.W.; et al. Nutritional Ketosis Alters Fuel Preference and Thereby Endurance Performance in Athletes. *Cell Metab.* **2016**, *24*, 256–268. [CrossRef]
36. Murray, A.J.; Knight, N.S.; Cole, M.A.; Cochlin, L.E.; Carter, E.; Tchabanenko, K.; Pichulik, T.; Gulston, M.K.; Atherton, H.J.; Schroeder, M.A.; et al. Novel ketone diet enhances physical and cognitive performance. *FASEB J.* **2016**, *30*, 4021–4032. [CrossRef]
37. Veech, R.L. The therapeutic implications of ketone bodies: The effects of ketone bodies in pathological conditions: Ketosis, ketogenic diet, redox states, insulin resistance, and mitochondrial metabolism. *Prostaglandins Leukot. Essent. Fatty Acids* **2004**, *70*, 309–319. [CrossRef]
38. Paoli, A. Ketogenic Diet for Obesity: Friend or Foe? *Int. J. Environ. Res. Public Health* **2014**, *11*, 2092–2107. [CrossRef]
39. Wheless, J.W. History of the ketogenic diet. *Epilepsia* **2008**, *49* (Suppl. 8), 3–5. [CrossRef]
40. Browning, J.D.; A Baker, J.; Rogers, T.; Davis, J.; Satapati, S.; Burgess, S.C. Short-term weight loss and hepatic triglyceride reduction: Evidence of a metabolic advantage with dietary carbohydrate restriction. *Am. J. Clin. Nutr.* **2011**, *93*, 1048–1052. [CrossRef]
41. Garbow, J.R.; Doherty, J.M.; Schugar, R.C.; Travers, S.; Weber, M.L.; Wentz, A.; Ezenwajiaku, N.; Cotter, D.G.; Brunt, E.M.; Crawford, P.A. Hepatic steatosis, inflammation, and ER stress in mice maintained long term on a very low-carbohydrate ketogenic diet. *Am. J. Physiol. Gastrointest. Liver Physiol.* **2011**, *300*, G956–G967. [CrossRef] [PubMed]
42. De Koning, L.; Fung, T.T.; Liao, X.; Chiuve, S.E.; Rimm, E.B.; Willett, W.C.; Spiegelman, D.; Hu, F.B. Low-carbohydrate diet scores and risk of type 2 diabetes in men. *Am. J. Clin. Nutr.* **2011**, *93*, 844–850. [CrossRef] [PubMed]
43. Dashti, H.M.; Al-Zaid, N.S.; Mathew, T.C.; Al-Mousawi, M.; Talib, H.; Asfar, S.K.; Behbahani, A.I. Long Term Effects of Ketogenic Diet in Obese Subjects with High Cholesterol Level. *Mol. Cell. Biochem.* **2006**, *286*, 19641727. [CrossRef] [PubMed]
44. Cotter, D.G.; Schugar, R.C.; Crawford, P.A. Ketone body metabolism and cardiovascular disease. *Am. J. Physiol. Heart Circ. Physiol.* **2013**, *304*, H1060–H1076. [CrossRef]
45. Kruszynska, Y.T.; McCormack, J.G.; McIntyre, N. Effects of glycogen stores and non-esterified fatty acid availability on insulin-stimulated glucose metabolism and tissue pyruvate dehydrogenase activity in the rat. *Diabetologia* **1991**, *34*, 205–211. [CrossRef] [PubMed]
46. Bak, L.K.; Walls, A.B.; Schousboe, A.; Waagepetersen, H.S. Astrocytic glycogen metabolism in the healthy and diseased brain. *J. Biol. Chem.* **2018**, *293*, 7108–7116. [CrossRef]
47. Pearson, T.S.; Akman, C.; Hinton, V.J.; Engelstad, K.; De Vivo, D.C. Phenotypic Spectrum of Glucose Transporter Type 1 Deficiency Syndrome (Glut1 DS). *Curr. Neurol. Neurosci. Rep.* **2013**, *13*, 342. [CrossRef]
48. Pierre, K.; Pellerin, L. Monocarboxylate transporters in the central nervous system: Distribution, regulation and function. *J. Neurochem.* **2005**, *94*, 15953344. [CrossRef]
49. Koutnik, A.P.; D'Agostino, D.P.; Egan, B. Anticatabolic Effects of Ketone Bodies in Skeletal Muscle. *Trends Endocrinol. Metab.* **2019**, *30*, 227–229. [CrossRef]
50. Roediger, W.E.W. The starved colon—Diminished mucosal nutrition, diminished absorption, and colitis. *Dis. Colon Rectum* **1990**, *33*, 858–862. [CrossRef]
51. Cheng, C.-W.; Biton, M.; Haber, A.L.; Gunduz, N.; Eng, G.; Gaynor, L.T.; Tripathi, S.; Calibasi-Kocal, G.; Rickelt, S.; Butty, V.L.; et al. Ketone Body Signaling Mediates Intestinal Stem Cell Homeostasis and Adaptation to Diet. *Cell* **2019**, *178*, 1115–1131.e15. [CrossRef] [PubMed]
52. Kann, O.; Kovács, R. Mitochondria and neuronal activity. *Am. J. Physiol. Cell Physiol.* **2007**, *292*, C641–C657. [CrossRef] [PubMed]
53. Brustovetsky, N.; Brustovetsky, T.; Jemmerson, R.; Dubinsky, J.M. Calcium-induced Cytochrome c release from CNS mitochondria is associated with the permeability transition and rupture of the outer membrane. *J. Neurochem.* **2002**, *80*, 207–218. [CrossRef]
54. Hansson, M.J.; Mansson, R.; Mattiasson, G.; Ohlsson, J.; Karlsson, J.; Keep, M.F.; Elmer, E. Brain-derived respiring mitochondria exhibit homogeneous, complete and cyclosporin-sensitive permeability transition. *J. Neurochem.* **2004**, *89*, 715–729. [CrossRef]
55. Calabrese, V.; Scapagnini, G.; Stella, A.M.G.; Bates, T.; Clark, J.B. Mitochondrial Involvement in Brain Function and Dysfunction: Relevance to Aging, Neurodegenerative Disorders and Longevity. *Neurochem. Res.* **2001**, *26*, 739–764. [CrossRef]
56. Stafstrom, C.E.; Rho, J.M. The Ketogenic Diet as a Treatment Paradigm for Diverse Neurological Disorders. *Front. Pharmacol.* **2012**, *3*, 59. [CrossRef] [PubMed]

57. Peterson, C.M.; Johannsen, D.L.; Ravussin, E. Skeletal Muscle Mitochondria and Aging: A Review. *J. Aging Res.* **2012**, *2012*, 194821. [CrossRef]
58. Olmos, Y.; Sanchez-Gomez, F.J.; Wild, B.; Garcia-Quintans, N.; Cabezudo, S.; Lamas, S.; Monsalve, M. SirT1 Regulation of Antioxidant Genes Is Dependent on the Formation of a FoxO3a/PGC-1α Complex. *Antioxidants Redox Signal.* **2013**, *19*, 1507–1521. [CrossRef]
59. Aquilano, K.; Baldelli, S.; Pagliei, B.; Ciriolo, M.R. Extranuclear localization of SIRT1 and PGC-1α: An insight into possible roles in diseases associated with mitochondrial dysfunction. *Curr. Mol. Med.* **2013**, *13*, 140–154. [CrossRef]
60. Milder, J.; Patel, M. Modulation of oxidative stress and mitochondrial function by the ketogenic diet. *Epilepsy Res.* **2011**, *100*, 295–303. [CrossRef]
61. Tieu, K.; Perier, C.; Caspersen, C.; Teismann, P.; Wu, D.C.; Yan, S.D.; Naini, A.; Vila, M.; Jackson-Lewis, V.; Ramasamy, R.; et al. D-beta-hydroxybutyrate rescues mitochondrial respiration and mitigates features of Parkinson disease. *J. Clin. Investig.* **2003**, *112*, 892–901. [CrossRef]
62. Achanta, L.B.; Rae, C.D. β-Hydroxybutyrate in the Brain: One Molecule, Multiple Mechanisms. *Neurochem. Res.* **2016**, *42*, 35–49. [CrossRef]
63. Haces, M.L.; Hernández-Fonseca, K.; Medina-Campos, O.N.; Montiel, T.; Pedraza-Chaverri, J.; Massieu, L. Antioxidant capacity contributes to protection of ketone bodies against oxidative damage induced during hypoglycemic conditions. *Exp. Neurol.* **2008**, *211*, 85–96. [CrossRef]
64. Vergati, M.; Krasniqi, E.; Monte, G.D.; Riondino, S.; Vallone, D.; Guadagni, F.; Ferroni, P.; Roselli, M. Ketogenic Diet and Other Dietary Intervention Strategies in the Treatment of Cancer. *Curr. Med. Chem.* **2017**, *24*, 1170–1185. [CrossRef]
65. Veech, R.L. Ketone ester effects on metabolism and transcription. *J. Lipid Res.* **2014**, *55*, 2004–2006. [CrossRef]
66. Gough, S.M.; Casella, A.; Ortega, K.J.; Hackam, A.S. Neuroprotection by the Ketogenic Diet: Evidence and Controversies. *Front. Nutr.* **2021**, *8*, 782657. [CrossRef]
67. Greco, T.; Glenn, T.; A Hovda, D.; Prins, M.L. Ketogenic diet decreases oxidative stress and improves mitochondrial respiratory complex activity. *J. Cereb. Blood Flow Metab.* **2016**, *36*, 1603–1613. [CrossRef]
68. Swanson, K.V.; Deng, M.; Ting, J.P.-Y. The NLRP3 inflammasome: Molecular activation and regulation to therapeutics. *Nat. Rev. Immunol.* **2019**, *19*, 477–489. [CrossRef]
69. Youm, Y.-H.; Nguyen, K.Y.; Grant, R.W.; Goldberg, E.L.; Bodogai, M.; Kim, D.; D'Agostino, D.; Planavsky, N.; Lupfer, C.; Kanneganti, T.-D.; et al. The ketone metabolite β-hydroxybutyrate blocks NLRP3 inflammasome–mediated inflammatory disease. *Nat. Med.* **2015**, *21*, 263–269. [CrossRef]
70. Guo, M.; Wang, X.; Zhao, Y.; Yang, Q.; Ding, H.; Dong, Q.; Chen, X.; Cui, M. Ketogenic Diet Improves Brain Ischemic Tolerance and Inhibits NLRP3 Inflammasome Activation by Preventing Drp1-Mediated Mitochondrial Fission and Endoplasmic Reticulum Stress. *Front. Mol. Neurosci.* **2018**, *11*, 86. [CrossRef]
71. Forsythe, C.E.; Phinney, S.D.; Fernandez, M.L.; Quann, E.E.; Wood, R.J.; Bibus, D.M.; Kraemer, W.J.; Feinman, R.D.; Volek, J.S. Comparison of Low Fat and Low Carbohydrate Diets on Circulating Fatty Acid Composition and Markers of Inflammation. *Lipids* **2007**, *43*, 65–77. [CrossRef]
72. Offermanns, S. Hydroxy-Carboxylic Acid Receptor Actions in Metabolism. *Trends Endocrinol. Metab.* **2017**, *28*, 227–236. [CrossRef] [PubMed]
73. Hartman, A.L.; Rho, J.M. The New Ketone Alphabet Soup: BHB, HCA, and HDAC. *Epilepsy Curr.* **2014**, *14*, 355–357. [CrossRef] [PubMed]
74. Xie, Z.; Zhang, D.; Chung, D.; Tang, Z.; Huang, H.; Dai, L.; Qi, S.; Li, J.; Colak, G.; Chen, Y.; et al. Metabolic Regulation of Gene Expression by Histone Lysine β-Hydroxybutyrylation. *Mol. Cell* **2016**, *62*, 194–206. [CrossRef] [PubMed]
75. Ono, J.G.; Worgall, T.S.; Worgall, S. 17q21 locus and ORMDL3: An increased risk for childhood asthma. *Pediatr. Res.* **2013**, *75*, 165–170. [CrossRef]
76. Bae, H.R.; Kim, D.H.; Park, M.H.; Lee, B.; Kim, M.J.; Lee, E.K.; Chung, K.W.; Kim, S.M.; Im, D.S.; Chung, H.Y. β-Hydroxybutyrate suppresses inflammasome formation by ameliorating endoplasmic reticulum stress *via* AMPK activation. *Oncotarget* **2016**, *7*, 66444–66454. [CrossRef]
77. Sada, N.; Fujita, Y.; Mizuta, N.; Ueno, M.; Furukawa, T.; Yamashita, T. Inhibition of HDAC increases BDNF expression and promotes neuronal rewiring and functional recovery after brain injury. *Cell Death Dis.* **2020**, *11*, 655. [CrossRef]
78. He, F.; Ru, X.; Wen, T. NRF2, a Transcription Factor for Stress Response and Beyond. *Int. J. Mol. Sci.* **2020**, *21*, 4777. [CrossRef]
79. Meroni, E.; Papini, N.; Criscuoli, F.; Casiraghi, M.C.; Massaccesi, L.; Basilico, N.; Erba, D. Metabolic Responses in Endothelial Cells Following Exposure to Ketone Bodies. *Nutrients* **2018**, *10*, 250. [CrossRef]
80. Gunst, J. Recovery from critical illness-induced organ failure: The role of autophagy. *Crit. Care* **2017**, *21*, 209. [CrossRef]
81. Flower, L.; Page, A.; Puthucheary, Z. Should nutritional therapy be modified to account for mitochondrial dysfunction in critical illness? *J. Parenter. Enter. Nutr.* **2021**, *45*, 60–65. [CrossRef] [PubMed]
82. Rojas-Morales, P.; Pedraza-Chaverri, J.; Tapia, E. Ketone bodies, stress response, and redox homeostasis. *Redox Biol.* **2019**, *29*, 101395. [CrossRef] [PubMed]

83. McClave, S.A.; Taylor, B.E.; Martindale, R.G.; Warren, M.M.; Johnson, D.R.; Braunschweig, C.; McCarthy, M.S.; Davanos, E.; Rice, T.W.; Cresci, G.A.; et al. Guidelines for the Provision and Assessment of Nutrition Support Therapy in the Adult Critically Ill Patient: Society of Critical Care Medicine (SCCM) and American Society for Parenteral and Enteral Nutrition (ASPEN). *JPEN J. Parenter. Enter. Nutr.* **2016**, *40*, 159–211. [CrossRef] [PubMed]
84. Yao, A.; Li, Z.; Lyu, J.; Yu, L.; Wei, S.; Xue, L.; Wang, H.; Chen, G.-Q. On the nutritional and therapeutic effects of ketone body d-β-hydroxybutyrate. *Appl. Microbiol. Biotechnol.* **2021**, *105*, 6229–6243. [CrossRef]
85. Amjad, S.; Nisar, S.; Bhat, A.A.; Shah, A.R.; Frenneaux, M.P.; Fakhro, K.; Haris, M.; Reddy, R.; Patay, Z.; Baur, J.; et al. Role of NAD+ in regulating cellular and metabolic signaling pathways. *Mol. Metab.* **2021**, *49*, 101195. [CrossRef]
86. Waldman, H.S.; McAllister, M.J. Exogenous Ketones as Therapeutic Signaling Molecules in High-Stress Occupations: Implications for Mitigating Oxidative Stress and Mitochondrial Dysfunction in Future Research. *Nutr. Metab. Insights* **2020**, *13*, 1178638820979029. [CrossRef]
87. Bradshaw, P.C.; Seeds, W.A.; Miller, A.C.; Mahajan, V.R.; Curtis, W.M. COVID-19: Proposing a Ketone-Based Metabolic Therapy as a Treatment to Blunt the Cytokine Storm. *Oxidative Med. Cell. Longev.* **2020**, *2020*, 6401341. [CrossRef]
88. Lopaschuk, G.D.; Karwi, Q.G.; Ho, K.L.; Pherwani, S.; Ketema, E.B. Ketone metabolism in the failing heart. *Biochim. et Biophys. Acta (BBA) Mol. Cell Biol. Lipids* **2020**, *1865*, 158813. [CrossRef]
89. Stubbs, B.J.; Koutnik, A.P.; Goldberg, E.L.; Upadhyay, V.; Turnbaugh, P.J.; Verdin, E.; Newman, J.C. Investigating Ketone Bodies as Immunometabolic Countermeasures against Respiratory Viral Infections. *Med* **2020**, *1*, 43–65. [CrossRef]
90. Wood, T.R.; Stubbs, B.J.; Juul, S.E. Exogenous Ketone Bodies as Promising Neuroprotective Agents for Developmental Brain Injury. *Dev. Neurosci.* **2018**, *40*, 451–462. [CrossRef]
91. Simeone, T.A.; Simeone, K.A.; Rho, J.M. Ketone Bodies as Anti-Seizure Agents. *Neurochem. Res.* **2017**, *42*, 2011–2018. [CrossRef] [PubMed]
92. Ma, W.; Berg, J.; Yellen, G. Ketogenic Diet Metabolites Reduce Firing in Central Neurons by Opening KATP Channels. *J. Neurosci.* **2007**, *27*, 3618–3625. [CrossRef]
93. Newman, J.C.; Verdin, E. Ketone bodies as signaling metabolites. *Trends Endocrinol. Metab.* **2013**, *25*, 42–52. [CrossRef] [PubMed]
94. Prins, M.L.; Matsumoto, J.H. The collective therapeutic potential of cerebral ketone metabolism in traumatic brain injury. *J. Lipid Res.* **2014**, *55*, 2450–2457. [CrossRef]
95. White, H.; Venkatesh, B. Clinical review: Ketones and brain injury. *Crit. Care* **2011**, *15*, 219. [CrossRef] [PubMed]
96. Altayyar, M.; Nasser, J.A.; Thomopoulos, D.; Bruneau, M. The Implication of Physiological Ketosis on The Cognitive Brain: A Narrative Review. *Nutrients* **2022**, *14*, 513. [CrossRef]
97. Kashiwaya, Y.; Takeshima, T.; Mori, N.; Nakashima, K.; Clarke, K.; Veech, R.L. d-β-Hydroxybutyrate protects neurons in models of Alzheimer's and Parkinson's disease. *Proc. Natl. Acad. Sci. USA* **2000**, *97*, 5440–5444. [CrossRef]
98. Zhao, Z.; Lange, D.J.; Voustianiouk, A.; MacGrogan, D.; Ho, L.; Suh, J.; Humala, N.; Thiyagarajan, M.; Wang, J.; Pasinetti, G.M. A ketogenic diet as a potential novel therapeutic intervention in amyotrophic lateral sclerosis. *BMC Neurosci.* **2006**, *7*, 29. [CrossRef]
99. Koutnik, A.P.; Poff, A.M.; Ward, N.; DeBlasi, J.; Soliven, M.A.; Romero, M.A.; Roberson, P.A.; Fox, C.; Roberts, M.D.; D'Agostino, D.P. Ketone Bodies Attenuate Wasting in Models of Atrophy. *J. Cachex Sarcopenia Muscle* **2020**, *11*, 973–996. [CrossRef]
100. Goossens, C.; Weckx, R.; Derde, S.; Perre, S.V.; Derese, I.; Van Veldhoven, P.P.; Ghesquière, B.; Berghe, G.V.D.; Langouche, L. Altered cholesterol homeostasis in critical illness-induced muscle weakness: Effect of exogenous 3-hydroxybutyrate. *Crit. Care* **2021**, *25*, 252. [CrossRef]
101. Jensen, N.J.; Wodschow, H.Z.; Nilsson, M.; Rungby, J. Effects of Ketone Bodies on Brain Metabolism and Function in Neurodegenerative Diseases. *Int. J. Mol. Sci.* **2020**, *21*, 8767. [CrossRef] [PubMed]
102. Tracey, T.J.; Steyn, F.J.; Wolvetang, E.J.; Ngo, S.T. Neuronal Lipid Metabolism: Multiple Pathways Driving Functional Outcomes in Health and Disease. *Front. Mol. Neurosci.* **2018**, *11*, 10. [CrossRef] [PubMed]
103. Cunnane, S.C.; Crawford, M.A. Survival of the fattest: Fat babies were the key to evolution of the large human brain. *Comp. Biochem. Physiol. Part A Mol. Integr. Physiol.* **2003**, *136*, 17–26. [CrossRef]
104. Ahmed, T.B.; Eggesbø, M.; Criswell, R.; Uhl, O.; Demmelmair, H.; Koletzko, B. Total Fatty Acid and Polar Lipid Species Composition of Human Milk. *Nutrients* **2021**, *14*, 158. [CrossRef]
105. Békési, A.; Williamson, D.H. An Explanation for Ketogenesis by the Intestine of the Suckling Rat: The Presence of an Active Hydroxymethylglutaryl-Coenzyme A Pathway. *Neonatology* **1990**, *58*, 160–165. [CrossRef]
106. Cunnane, S.C.; Crawford, M.A. Energetic and nutritional constraints on infant brain development: Implications for brain expansion during human evolution. *J. Hum. Evol.* **2014**, *77*, 88–98. [CrossRef]
107. Vannucci, S.J.; Simpson, I.A. Developmental switch in brain nutrient transporter expression in the rat. *Am. J. Physiol. Endocrinol. Metab.* **2003**, *285*, E1127–E1134. [CrossRef]
108. Larqué, E.; Zamora, S.; Gil, A. Dietary trans Fatty Acids Affect the Essential Fatty-Acid Concentration of Rat Milk. *J. Nutr.* **2000**, *130*, 847–851. [CrossRef]
109. Steiner, P. Brain Fuel Utilization in the Developing Brain. *Ann. Nutr. Metab.* **2019**, *75* (Suppl. 1), 8–18. [CrossRef]
110. Tildon, J.T.; McKenna, M.C.; Stevenson, J.H. Transport of 3-hydroxybutyrate by cultured rat brain astrocytes. *Neurochem. Res.* **1994**, *19*, 1237–1242. [CrossRef] [PubMed]
111. Meldrum, B.S. Glutamate as a Neurotransmitter in the Brain: Review of Physiology and Pathology. *J. Nutr.* **2000**, *130*, 1007S–1015S. [CrossRef] [PubMed]

112. Yudkoff, M.; Daikhin, Y.; Nissim, I.; Grunstein, R.; Nissim, I. Effects of Ketone Bodies on Astrocyte Amino Acid Metabolism. *J. Neurochem.* **2002**, *69*, 682–692. [CrossRef] [PubMed]
113. Yudkoff, M.; Nelson, D.; Daikhin, Y.; Erecińska, M. Tricarboxylic acid cycle in rat brain synaptosomes. Fluxes and interactions with aspartate aminotransferase and malate/aspartate shuttle. *J. Biol. Chem.* **1994**, *269*, 27414–27420. [CrossRef]
114. Hasselbalch, S.G.; Knudsen, G.M.; Jakobsen, J.; Hageman, L.P.; Holm, S.; Paulson, O.B. Blood-brain barrier permeability of glucose and ketone bodies during short-term starvation in humans. *Am. J. Physiol. Metab.* **1995**, *268*, E1161–E1166. [CrossRef] [PubMed]
115. Kuzawa, C.W.; Chugani, H.T.; Grossman, L.I.; Lipovich, L.; Muzik, O.; Hof, P.R.; Wildman, D.E.; Sherwood, C.C.; Leonard, W.R.; Lange, N. Metabolic costs and evolutionary implications of human brain development. *Proc. Natl. Acad. Sci. USA* **2014**, *111*, 13010–13015. [CrossRef]
116. Kirmse, K.; Witte, O.W.; Holthoff, K. GABA Depolarizes Immature Neocortical Neurons in the Presence of the Ketone Body -Hydroxybutyrate. *J. Neurosci.* **2010**, *30*, 16002–16007. [CrossRef]
117. Cotter, D.G.; D'Avignon, D.A.; Wentz, A.E.; Weber, M.L.; Crawford, P.A. Obligate Role for Ketone Body Oxidation in Neonatal Metabolic Homeostasis. *J. Biol. Chem.* **2011**, *286*, 6902–6910. [CrossRef]
118. Cunnane, S.; Nugent, S.; Roy, M.; Courchesne-Loyer, A.; Croteau, E.; Tremblay, S.; Castellano, A.; Pifferi, F.; Bocti, C.; Paquet, N.; et al. Brain fuel metabolism, aging, and Alzheimer's disease. *Nutrition* **2011**, *27*, 3–20. [CrossRef]
119. Stubbs, B.J.; Cox, P.J.; Evans, R.D.; Santer, P.; Miller, J.J.; Faull, O.K.; Magor-Elliott, S.; Hiyama, S.; Stirling, M.; Clarke, K. On the Metabolism of Exogenous Ketones in Humans. *Front. Physiol.* **2017**, *8*, 848. [CrossRef]
120. Suissa, L.; Kotchetkov, P.; Guigonis, J.-M.; Doche, E.; Osman, O.; Pourcher, T.; Lindenthal, S. Ingested Ketone Ester Leads to a Rapid Rise of Acetyl-CoA and Competes with Glucose Metabolism in the Brain of Non-Fasted Mice. *Int. J. Mol. Sci.* **2021**, *22*, 524. [CrossRef]
121. Bhatt, D.P.; Houdek, H.M.; Watt, J.A.; Rosenberger, T.A. Acetate supplementation increases brain phosphocreatine and reduces AMP levels with no effect on mitochondrial biogenesis. *Neurochem. Int.* **2013**, *62*, 296–305. [CrossRef] [PubMed]
122. García-Rodríguez, D.; Giménez-Cassina, A. Ketone Bodies in the Brain Beyond Fuel Metabolism: From Excitability to Gene Expression and Cell Signaling. *Front. Mol. Neurosci.* **2021**, *14*, 732120. [CrossRef] [PubMed]

nutrients

MDPI

Review

The Effects of Almonds on Gut Microbiota, Glycometabolism, and Inflammatory Markers in Patients with Type 2 Diabetes: A Systematic Review and Meta-Analysis of Randomised Controlled Trials

Omorogieva Ojo [1,*], Xiao-Hua Wang [2], Osarhumwese Osaretin Ojo [3] and Amanda Rodrigues Amorim Adegboye [4]

1 Faculty of Education, Health and Human Sciences, School of Health Sciences, University of Greenwich, Avery Hill Campus, Avery Hill Road, London SE9 2UG, UK
2 The School of Nursing, Soochow University, Suzhou 215006, China; wangxiaohua@suda.edu.cn
3 South London and Maudsley NHS Foundation Trust, University Hospital, Lewisham High Street, London SE13 6LH, UK; osarhumwese.ojo@slam.nhs.uk
4 Faculty of Health and Life Sciences, School of Nursing, Midwifery and Health, Coventry University, Priory Street, Coventry CV1 5FB, UK; ad6287@coventry.ac.uk
* Correspondence: o.ojo@greenwich.ac.uk

Abstract: The use of nutritional interventions for managing diabetes is one of the effective strategies aimed at reducing the global prevalence of the condition, which is on the rise. Almonds are the most consumed tree nut and they are known to be rich sources of protein, monounsaturated fatty acids, essential minerals, and dietary fibre. Therefore, the aim of this review was to evaluate the effects of almonds on gut microbiota, glycometabolism, and inflammatory parameters in patients with type 2 diabetes. Methods: This systematic review and meta-analysis was carried out according to the preferred reporting items for systematic review and meta-analysis (PRISMA). EBSCOhost, which encompasses the Health Sciences Research Databases; Google Scholar; EMBASE; and the reference lists of articles were searched based on population, intervention, control, outcome, and study (PICOS) framework. Searches were carried out from database inception until 1 August 2021 based on medical subject headings (MesH) and synonyms. The meta-analysis was carried out with the Review Manager (RevMan) 5.3 software. Results: Nine randomised studies were included in the systematic review and eight were used for the meta-analysis. The results would suggest that almond-based diets have significant effects in promoting the growth of short-chain fatty acid (SCFA)-producing gut microbiota. Furthermore, the meta-analysis showed that almond-based diets were effective in significantly lowering ($p < 0.05$) glycated haemoglobin (HbA1c) levels and body mass index (BMI) in patients with type 2 diabetes. However, it was also found that the effects of almonds were not significant ($p > 0.05$) in relation to fasting blood glucose, 2 h postprandial blood glucose, inflammatory markers (C-reactive protein and Tumour necrosis factor α, TNF-α), glucagon-like peptide-1 (GLP-1), homeostatic model assessment of insulin resistance (HOMA–IR), and fasting insulin. The biological mechanisms responsible for the outcomes observed in this review in relation to reduction in HbA1c and BMI may be based on the nutrient composition of almonds and the biological effects, including the high fibre content and the low glycaemic index profile. Conclusion: The findings of this systematic review and meta-analysis have shown that almond-based diets may be effective in promoting short-chain fatty acid-producing bacteria and lowering glycated haemoglobin and body mass index in patients with type 2 diabetes compared with control. However, the effects of almonds were not significant ($p > 0.05$) with respect to fasting blood glucose, 2 h postprandial blood glucose, inflammatory markers (C-reactive protein and TNF-α), GLP-1, HOMA–IR, and fasting insulin.

Keywords: type 2 diabetes; almonds; tree nuts; glycated haemoglobin; gut microbiota; body mass index

Citation: Ojo, O.; Wang, X.-H.; Ojo, O.O.; Adegboye, A.R.A. The Effects of Almonds on Gut Microbiota, Glycometabolism, and Inflammatory Markers in Patients with Type 2 Diabetes: A Systematic Review and Meta-Analysis of Randomised Controlled Trials. *Nutrients* **2021**, *13*, 3377. https://doi.org/10.3390/nu13103377

Academic Editor: Roberto Cangemi

Received: 1 September 2021
Accepted: 23 September 2021
Published: 26 September 2021

1. Introduction

The use of nutritional interventions is one of the strategies for managing diabetes, which is on the increase worldwide. It is projected that the global prevalence of diabetes could reach 700 million by 2045, up by 51% from 463 million who were living with the condition in 2019 [1]. Over 90% of people with diabetes have type 2 diabetes, which is linked to lifestyle factors [2], and this has implications in terms of morbidity and mortality. Poor diabetes control increases the costs of healthcare as a result of potentially avoidable hospital treatment and drug prescription and in the UK, the total annual spending on patients with type 2 diabetes is expected to rise to about £2.2 billion by 2040–2050 [3,4]. Therefore, nutritional interventions, which are effective in terms of clinical outcomes, are often recommended for diabetes management [5]. In this regard, the use of nuts, including tree nuts, such as almond, walnut, hazelnut, cashew nuts and Brazil nuts, and groundnuts (mainly peanuts), which are high in unsaturated fatty acids and are rich sources of bioactive nutrients that have significant metabolic and cardiovascular health benefits, have been suggested [6,7].

Almonds are the most consumed tree nut and they are known to be rich sources of protein, monounsaturated fatty acids, essential minerals, and dietary fibre [6,8]. The role of dietary fibre in modulating gut microbiota dysbiosis and in the regulation of glycaemic parameters have been demonstrated in previous systematic reviews and meta-analyses [9,10] and in randomised controlled trials [11,12].

1.1. Description of the Intervention

Nuts have been part of the human diet for centuries. Nuts are included in different recipes and, more recently, nuts, particularly almonds, have been consumed as a healthy snack [13]. However, the level of consumption of nuts may vary globally, across different populations. Almonds are tree nuts that have a low glycaemic index, are rich in dietary fibre and unsaturated fatty acids, and have low carbohydrate content [6]. The macro- and micronutrient components of almonds, including monounsaturated fatty acids, polyunsaturated fatty acids, fibre, vitamins, minerals, phytosterols, and polyphenols, have been associated with health benefits including anti-inflammatory and lipid-lowering properties [6,8]. Almonds also have antioxidant properties [8]. The polyphenols and fibre content of almonds may be used as substrates for gut microbial growth and regulation of gut microbiota [8]. It has been suggested that there is an inverse relationship between the consumption of nuts and the risk of developing type 2 diabetes [6].

1.2. How This Intervention Might Work

It has been reported that almond consumption increases satiety, decreases postprandial glycaemia, and regulates oxidative stress [6]. Almond consumption may also decrease the rate of nutrient digestion, reduce glucose response, and stimulate incretin and the production of glucagon-like peptide- 1 (GLP-1) [6,14]. The fermentation of the dietary fibre component of almonds may lead to improvement in the composition and metabolic products of gut microbiota, such as an increase in the prevalence of health-promoting bacteria and short-chain fatty acid production, including propionic, butyric, and acetic acid [10,15,16]. The short-chain fatty acids produced during this process have been shown to improve glycometabolism in patients with diabetes [10,15,17]. An almond-based low-calorie diet has also been found to be effective in reducing weight [18], which is useful in promoting insulin sensitivity and regulating glycaemic control.

1.3. Why It Is Important to Do This Review

Incorporating almonds in well-balanced healthy diets have been shown to confer beneficial effects on glycaemic control in patients with type 2 diabetes [6,14,19]. However, it would appear that previous systematic reviews and meta-analyses in this area of research have either focused on the effects of tree nuts in general [20–22], on blood pressure [23], or on fasting blood lipids [24]. For example, Mohammadifard et al. [23], conducted a systematic review and meta-analysis on the effect of tree nuts, peanuts, and soy nuts on blood pressure, while Blanco-Mejia et al.'s [20] review focused on the effects of tree nuts on metabolic syndrome. Muley et al. [21], on the other hand, evaluated the effects of tree nuts on glycaemic control in adults with type 2 diabetes, while Musa–Veloso et al. [24] examined the effects of almond consumption on fasting blood lipids. Viguinliouk et al.'s [22] review examined the effect of tree nuts on glycaemic control in patients with diabetes.

However, the present systematic review and meta-analysis will complement the existing literature by providing evidence that focuses on the role of almonds on gut microbiota, glycaemic control, and inflammatory markers. There are indications that increased markers of inflammation and disequilibrium of the gut microbial community are associated with the dysregulation of glycaemic control and type 2 diabetes [10,25,26].

1.4. Aim

To evaluate the effects of almonds on gut microbiota, glycometabolism, and inflammatory parameters in patients with type 2 diabetes.

2. Methods

This systematic review and meta-analysis was carried out according to the preferred reporting items for systematic review and meta-analysis (PRISMA) [27].

2.1. Types of Studies

Only randomised controlled trials (RCTs) including crossover and parallel designs were included in this review.

2.2. Types of Participants

Adult participants with type 2 diabetes regardless of the existence of co-morbidities (e.g., obesity) were selected for the review.

2.3. Types of Interventions

We included RCTs comparing the provision of almonds or advice to increase almond consumption with a control group (no intervention or habitual diet or other types of nuts) also with type 2 diabetes. There was no restriction regarding the minimum and maximum amount of almonds consumed. RCT including multiple interventions (diet and exercise) were not considered. There was no restriction regarding the duration of the interventions.

2.4. The Inclusion Criteria

Randomised controlled trials involving participants with type 2 diabetes and aged 18 years and over were included in this review. Studies with outcomes of interest involving gut microbiota, glycometabolism, anthropometric measurements, and inflammatory parameters were also included in this review.

2.5. The Exclusion Criteria

Studies excluded were those with prediabetes; involving other tree nuts other than almonds, such as walnuts; involving patients with gestational diabetes, type 1 diabetes, and only healthy populations. Cluster randomised trials were not eligible for inclusion. Furthermore, studies with participants aged below 18 years and animal studies were excluded from this review. Pregnant and lactating women were not included.

2.6. Types of Outcome Measures

The following were the primary outcome measures of interest:

- Gut microbiota;
- Blood glucose parameters: glycated haemoglobin (HbA1c, %);
- Inflammatory markers: tumour necrosis factor α (TNF-α); high-sensitivity C-reactive protein (hsCRP);

- Body mass index (BMI) (Kg/m^2).
- Secondary outcome measures of interest:
- Fasting blood glucose (FBG, mmol/L);
- Postprandial blood glucose (2 h PBG, mmol/L);
- Homeostatic model assessment of insulin resistance (HOMA–IR);
- Glucagon-like peptide-1 (GLP-1);
- Fasting insulin.

2.7. Search Methods for Identification of Studies

EBSCO-host (which encompasses the Health Sciences Research Databases, including MEDLINE, Academic Search Premier, APA PsycInfo, Psychology and Behavioural Sciences Collection, APA PsycArticles databases, and CINAHL Plus with Full Text), Google Scholar, and EMBASE were the databases searched for relevant articles. In addition, the reference lists of articles were also searched based on the population, intervention, control, outcome, and study (PICOS) framework (Table 1). Searches were carried out from database inception until 1st August 2021. Search terms were drawn from medical subject headings (MesH) and synonyms and were combined using Boolean operators (OR/AND). Two members of the research team (O.O. and O.O.O.) conducted the searches independently and these were cross-checked by the other two members of the team (X.W. and A.R.A.A). Resolution of differences was through discussion and consensus. Search results from databases were exported to EndNote (Analytics, Philadelphia, PA, USA) and de-duplicated.

Table 1. Search terms and search strategy.

Patient/Population	Intervention	Comparator	Outcome (Primary)	Study Designs	Combining Search Terms
Patients with diabetes	Almonds	Control	Glycometabolism	Randomised controlled trial	
Patients with diabetes OR type 2 diabetes OR Diabetes OR Diabetes complications OR diabetes mellitus, type 2 OR diabetes mellitus	Almond OR Tree, Almond OR Almond Tree OR Sweet Almond OR Almond Trees OR Tree Nuts OR Almond, Sweet			1. Randomised controlled trial OR controlled clinical trial OR randomized OR placebo OR drug therapy OR randomly OR trial OR groups 2. "Animals" NOT "Humans" 3. 1 NOT 2	Column 1 AND Column 2 AND Column 3

3. Data Collection and Analysis

3.1. Selection of Studies

The PRISMA flow chart (Figure 1) was based on a set of inclusion and exclusion criteria that were used to select the studies included.

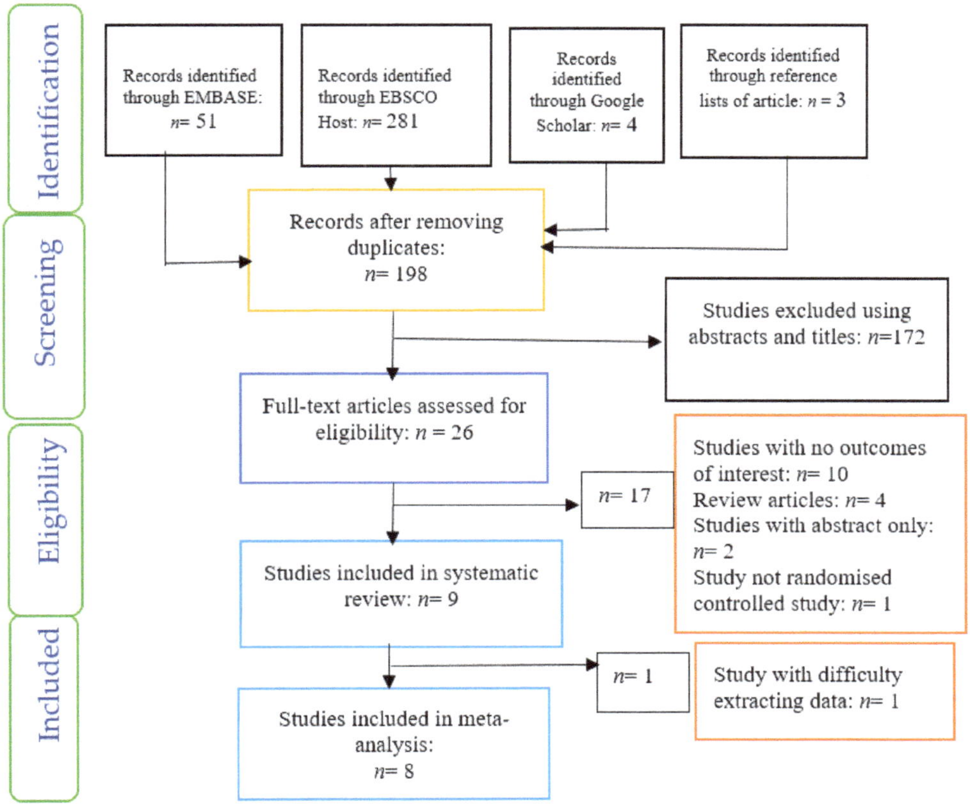

Figure 1. PRISMA flow chart on selection and inclusion of studies.

3.2. Data Extraction and Management

The data were extracted in a pre-piloted and standardised form. We extracted the following information: the country where the study was conducted, characteristics of the study population (e.g., mean age), sample size, outcome data, intervention details (duration) (Table 2).

Where the findings of more than one study were reported in one article, only the data from the study pertaining to patients with diabetes were included in the analysis.

The data was extracted by one researcher (O.O.) from the articles included and the three other members of the research team (O.O.O., X.W., A.R.A.A) cross-checked the information. Final values and changes from baseline were used to compare the intervention group with the control group. The units of measurements for some of the parameters were converted to ensure the same unit of measurements for all the studies for that parameter. In studies reporting values in median and 1st and 3rd quartile values, these were converted to means and standard deviations.

3.3. Assessment of Risk of Bias in Included Studies

Two members of the research team (O.O. and O.O.O.) evaluated the risk of bias of the included studies using the domain-based risk assessment tool [28]. The results were cross-checked by the other two members of the team (X.W. and A.R.A.A). Allocation concealment, the random sequence generation, blinding of outcome assessment, blinding of participants and personnel, selective reporting, incomplete outcome data, and other biases were the domains evaluated [29].

Table 2. General characteristics of included studies.

Citation/Country of Study	Type of Study	Sample Details and Duration of Study	Mean Age (Years)	Aim	Interventions	Results
Bodnaruc et al. [30] Canada	A randomised cross-over study	7 men with type 2 diabetes. Data were collected during two experimental sessions separated by a ≥7day washout period.	63.9 ± 2.5	To evaluate the effects of almonds on postprandial glucose response.	Participants completed 2 experimental visits and control (white bread, butter, cheese) and test (white bread, almonds) diets were ingested.	The test meal was associated with lower postprandial glycemia and insulinemia.
Chen et al. [31] Taiwan	A randomised cross-over controlled study	40 patients with type 2 diabetes. 12-weeks duration.	54.9 ± 10.5	To examine the effect of almonds on glycaemia	Approximately 60 g/day almonds added to a National Cholesterol Education Programme Step II diet (NCEP II) compared to NCEP II diet alone as control	Both almond-based and control diets did not significantly affect body weight and BMI or change HbA1c, fasting serum glucose, insulin, or HOMA-IR values.
Cohen et al. [32] USA	A randomised parallel study	13 participants diagnosed with type 2 diabetes Almond-based diets (n = 6) Control (n = 7) 12-weeks duration.	Almond group: 66 ± 3.3 Control group: 66 ± 3.3	To examine the impact of chronic almond ingestion on glycaemic control in patients with type 2 diabetes.	Participants were randomised to almond group (1 oz of almonds, 5 days/week) or cheese group (2 cheese sticks, 5 days/week)	HbA1c was the only blood marker that changed significantly between the treatment groups (p = 0.045). Chronic almond ingestion resulted in a 4% reduction in BMI compared with control (p = 0.047).
Hou et al. [6] China	A randomised controlled study	Almond group (n = 14) Peanut group (n = 11) 12-weeks duration.	Almond group: 70.86 ± 8.21 Peanut group: 68 ± 5.80	To compare the effects of peanuts and almonds incorporated into a low-carbohydrate diet on cardiometabolic and inflammatory parameters in patients with type 2 diabetes	Peanuts or almonds were incorporated into a low-carbohydrate diet and both groups were compared after a 3-month intervention.	Almonds and peanuts have similar effect on improving fasting and postprandial blood glucose among patients with type 2 diabetes when incorporated into a low-carbohydrate diet.

Table 2. *Cont.*

Citation/Country of Study	Type of Study	Sample Details and Duration of Study	Mean Age (Years)	Aim	Interventions	Results
Li et al. [33] Taiwan	Randomised cross-over clinical trial	20 Chinese patients with type 2 diabetes. 12-weeks duration.	58 ± 2	To evaluate the effect of almond consumption on glycaemia in Chinese patients with type 2 diabetes	Incorporation of almonds into National Cholesterol Education Programme Step II diet (NCEP II) to replace 20% of total daily calorie intake compared with NCEP II diet alone as control.	Adding almonds into a healthy diet has beneficial effects on adiposity and glycaemic control.
Liu et al. [34] Taiwan	Randomised cross-over controlled feeding trial	20 Chinese patients with type 2 diabetes. 12-weeks duration.	58 ± 2	To examine the effect of almond consumption on inflammation and oxidative stress in patients with type 2 diabetes	Addition of almonds (approximately 56 g/day) into National Cholesterol Education Programme Step II diet (NCEP II) to replace 20% of total daily calorie intake compared with NCEP II diet alone as control.	Adding almonds into a healthy diet could ameliorate inflammation and oxidative stress in patients with type 2 diabetes.
Lovejoy et al. [35] USA	Randomized double-blind crossover design	30 participants with type 2 diabetes. 16-weeks duration.	53.8 ± 1.9	To assess the effects of almond-enriched diets on insulin sensitivity in patients with diabetes	The 4 diets were as follows: (1) high-fat, high-almond (HFA; 37% total fat, 10% from almonds); (2) low-fat, high-almond (LFA; 25% total fat, 10% from almonds); (3) high-fat control (HFC; 37% total fat, 10% from the MUFAs from olive or canola oil); and (4) low-fat control (LFC; 25% total fat, 10% from olive or canola oil). The almond-containing diets provided 57–113 g almonds/d depending on the total energy level	Almond-enriched diets did not influence glycaemia in patients with diabetes.

Table 2. *Cont.*

Citation/Country of Study	Type of Study	Sample Details and Duration of Study	Mean Age (Years)	Aim	Interventions	Results
Ren et al. [14] China	Randomised controlled trial	45 participants with type 2 diabetes. 12-weeks duration.	LCD group: 73.55 ± 4.99 LFD group: 70.48 ± 5.91	To determine the effect of almond-based low-carbohydrate diet on glycometabolism, gut microbiota, and GLP-1 in patients with type 2 diabetes.	The intervention group consumed a low-carbohydrate diet, which included 56 g/day almonds that replaced 150 g/day staple food, while the control group adopted a low-fat diet education programme.	Almond-based LCD may be effective in regulating glycometabolism in patients with diabetes by stimulating the growth of SCFA-producing bacteria, increasing SCFA production and promoting GLP-1 secretion. The almond-based LCD significantly increased the SCFA-producing bacteria *Roseburia*, *Ruminococcus*, and *Eubacterium*.
Sweazea et al. [36] USA	Randomised controlled study	21 participants with type 2 diabetes. 12-weeks duration.	Almond group: 57.8 ± 5.6 Control group: 54.7 ± 8.9	To evaluate if almond supplementation without further dietary advice improves glycaemic control compared with control.	The almond group consumed 43 g almonds 5–7 times per week and to maintain their usual diet and activity pattern while the control group maintained their usual diet and activity pattern.	Daily almond consumption in the absence of other dietary or physical activity activities is useful in reducing inflammation in patients with type 2 diabetes.

Abbreviations: LCD, low-carbohydrate diet; LFD, low-fat diet; GLP-1, glucagon-like peptide-1; MUFAs, monounsaturated fats; SCFAs, short-chain fatty acid.

The risk assessment was conducted using the Review Manager 5.3 software (Copenhagen, Denmark) [28].

3.4. Data Analysis

Whenever there were enough trial reporting data on the same outcome, we performed a meta-analysis. Continuous data were analysed as mean difference (MD) with 95% confidence intervals (CIs), except for the fasting insulin due to differences in the units of measurements of the studies included and, thus, the standardised mean difference (SMD) was used for the meta-analysis. Forest plots were used to depict the results of the meta-analysis and in respect of statistical significance of the overall effect of the intervention, this was set at $p < 0.05$.

Sensitivity analysis was also conducted by removing studies one by one from the meta-analysis to assess the level of consistency of the results. The I^2 statistic expressed as percentage was used to measure the degree of heterogeneity of studies included [29] in the review. A fixed-effects model was used for the meta-analysis for all the parameters of interest except for the fasting insulin due to differences in the units of measurements of the studies included and the standardised mean difference was used for the meta-analysis. Whenever a substantial heterogeneity ($\geq$50%) was observed and there were enough studies included in the outcome, subgroup analysis was conducted. In addition, final values and changes from baseline were used to compare the intervention group with the control group [29]. If 10 or more studies were included, we would have performed a funnel plot to assess the presence of publication bias and small study effect. The meta-analysis was carried out in Review Manager (RevMan) 5.3 software [28].

4. Results

Nine studies were included in the systematic review and eight were used for the meta-analysis (Figure 1). The description and characteristics of eligible studies, including the type of study, details of sample, mean age, the aim of study, interventions, and results are outlined in Table 2. While one study was conducted in Canada [30], three each were conducted in Taiwan [31,33,34] and the USA [32,35,36], and two in China [6,14].

4.1. Evaluation of the Risk of Bias of Included Studies

The risk of bias of included studies is shown in Figure 2a,b. All studies showed a low risk of bias in relation to the random sequence generation (selection bias), incomplete outcome data (attrition bias), and selective reporting (reporting bias). However, unclear risk of bias was found in relation to allocation concealment, blinding of participants and personnel, and blinding of outcome assessments in some of the studies [31–34,36].

The presentation of the results of the systematic review and meta-analysis were divided into.

Gut microbiota, glycaemic control, inflammatory parameters, body mass index, homeostatic model assessment of insulin resistance (HOMA-IR), glucagon-like peptide-1 (GLP-1), and fasting insulin.

4.2. Gut Microbiota

Only one study [14] examined the effects of almonds on gut microbiota. Ren et al. [14] found that the almond-based low-carbohydrate diet (LCD) significantly increased the short-chain fatty acid (SCFA)-producing bacteria *Roseburia, Ruminococcus,* and *Eubacterium*. In particular, the LCD group had a significantly higher population of *Roseburia* ($p < 0.01$) at the genus level compared with the low-fat diet (LFD) group by the third month, and compared to the baseline, *Eubacterium* ($p < 0.01$) and *Roseburia* increased significantly ($p < 0.05$) and *Bacteroides* ($p < 0.05$) significantly decreased in the almond-based LCD group.

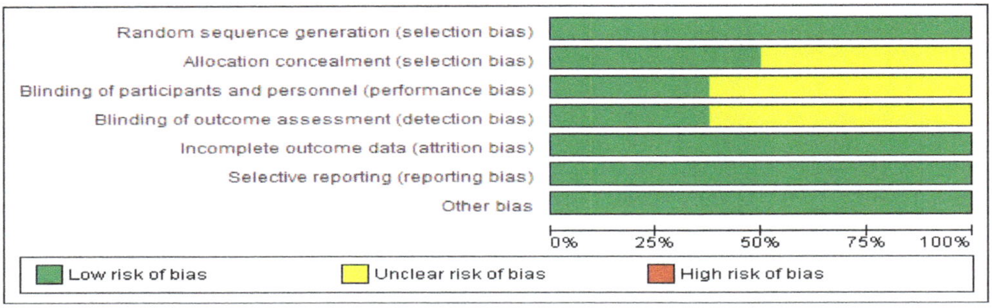

(a)

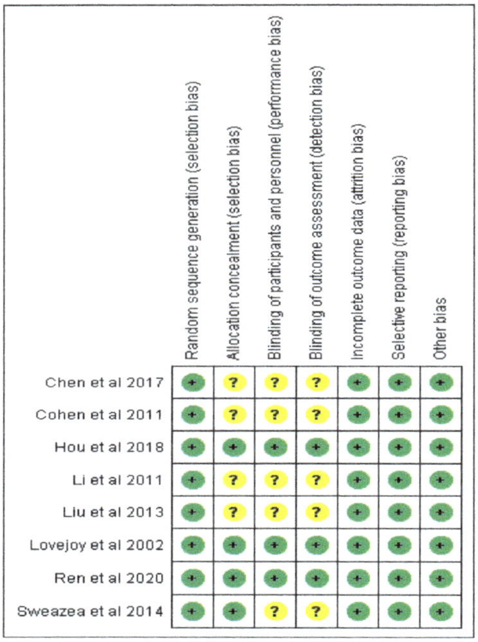

(b)

Figure 2. Shows (**a**) risk of bias graph and (**b**) risk of bias summary of studies included.

4.3. Glycaemic Control

Chen et al. [31] did not find any significant effect with respect to change in glycated haemoglobin (HbA1c) and fasting serum glucose values in the almond-based and control diets. However, in the study by Cohen et al. [32], there was a significant reduction ($p = 0.045$) in HbA1c in the almond-based diet group compared with control group. Ren et al. [14] also showed that almond-based LCD may be effective in modulating glycometabolism in patients with diabetes.

Bodnaruc et al. [30] noted that the almond-based meal was associated with lower postprandial glucose. According to Hou et al. [6], while the almond-based diet did improve fasting, postprandial blood glucose, and glycated haemoglobin in patients with type 2 diabetes, these were not significantly different from the control group. Li et al. [33] observed that including almonds in a healthy diet led to significant improvement ($p < 0.05$) in glycaemic control, while Lovejoy et al. [35] showed that an almond-enriched diet had no significant effect ($p > 0.05$) on glycaemia in patients with diabetes.

With respect to the meta-analysis, the results of the effects of almonds on glucose control are shown in Figure 3a–d. Six studies each contributed data for the HbA1c analysis (almond group (gp), n = 115; control gp, n = 114) (Figure 3a; sub-group analysis, Figure 3b) and fasting blood glucose analysis (almond gp, n = 113; control gp, n = 111) (Figure 3c). The almond-based diet group experienced a significant reduction ($p < 0.001$) in HbA1c levels compared to the control group with a mean difference of −0.52 (95% CI: −0.58, −0.46). Regarding the 2-hour postprandial blood glucose levels, two studies contributed to the data analysis (almond gp, n = 44; control gp, n = 41) (Figure 3d). The levels of fasting blood glucose and 2-hour postprandial blood glucose were lower in the almond group compared to the control group, although the differences were not significant ($p > 0.05$). The mean differences were −0.03 (95% CI: −0.18, 0.11) for fasting blood glucose and −0.15 (95% CI: −0.44, 0.13) for postprandial blood glucose.

The sensitivity analysis conducted by removing studies one by one from the meta-analysis did not change the results in relation to HbA1c ($p < 0.05$), fasting blood glucose ($p > 0.05$), and 2 h postprandial blood glucose ($p > 0.05$). The sub-group analysis for HbA1c showed that, although there was a significant difference ($p < 0.001$) between the almond group and control with respect to the meta-analysis of the randomised parallel studies, the differences were not significant ($p = 0.25$) in relation to the cross-over studies (Figure 3b).

4.4. *Inflammatory Markers*

The study by Liu et al. [34] observed that the addition of almonds into a healthy diet could ameliorate inflammation and oxidative stress in patients with type 2 diabetes. Similarly, Sweazea et al. [36] noted that the daily consumption of almonds in the absence of other dietary or physical activity activities could be an effective approach in reducing inflammation in patients with type 2 diabetes.

The meta-analyses of the effects of almonds on inflammatory markers are shown in Figure 4a,b. Three studies contributed data for the C-reactive protein analysis (almond gp, n = 63; control gp, n = 63) (Figure 4a) and two studies for tumour necrosis factor- α (TNF- α) analysis (almond gp, n = 30; control gp, n = 31) (Figure 4b). The levels of C-reactive protein and TNF- α were lower in the almond group compared to the control group. However, the differences between the two groups were not significant ($p > 0.05$), with mean differences of −0.54 (95% CI: −1.61, 0.53) for C-reactive protein and −16.67 (95% CI: −53.25, 19.91) for TNF- α respectively. The results did not change between the almond group and control group ($p > 0.05$) with respect to C-reactive protein and TNF- α following sensitivity tests.

4.5. *Body Mass Index (BMI) (Kg/m^2)*

Chen et al. [31] did not find any significant effect of the almond-based diet with respect to body weight and BMI. In contrast, Cohen et al. [32] found that chronic almond ingestion resulted in a 4% reduction in BMI compared with control ($p = 0.047$). Six studies contributed to the results of the analysis for BMI (almond gp, n = 105; control gp, n = 105).

The meta-analysis showed that the BMI was significantly lower ($p < 0.05$) in the almond group compared with the control group (Figure 5), with a mean difference of −0.36 (95% CI: −0.52, −0.19). The results of the sensitivity analysis showed consistency in terms of the significant difference between the almond group and the control group, except when the Hou et al. [6] study was removed.

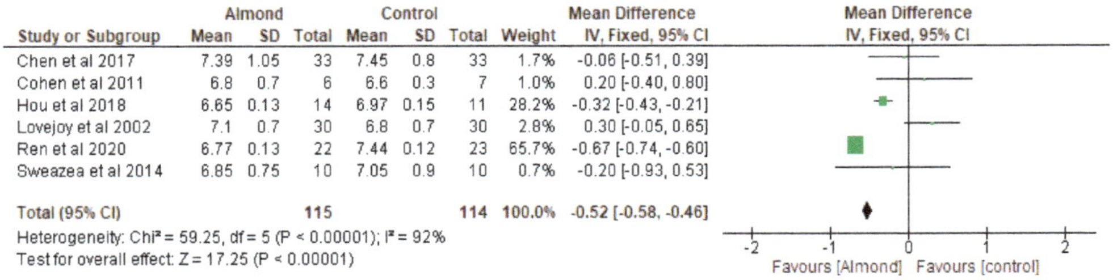

(a)

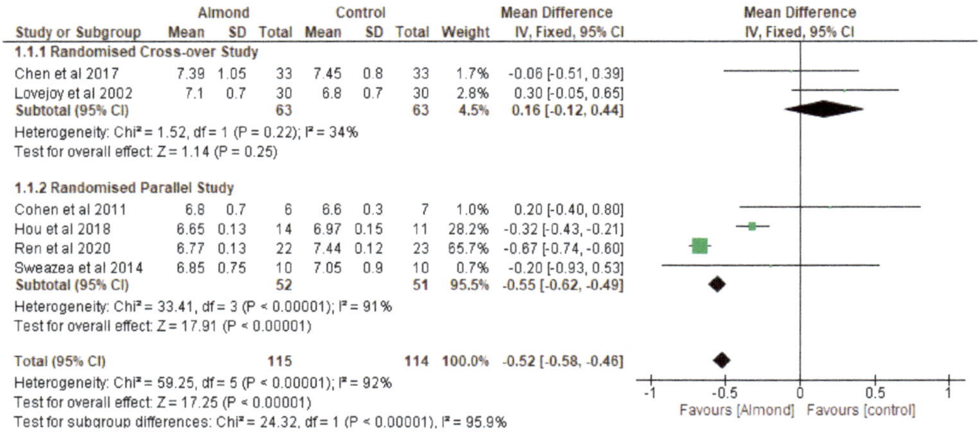

(b)

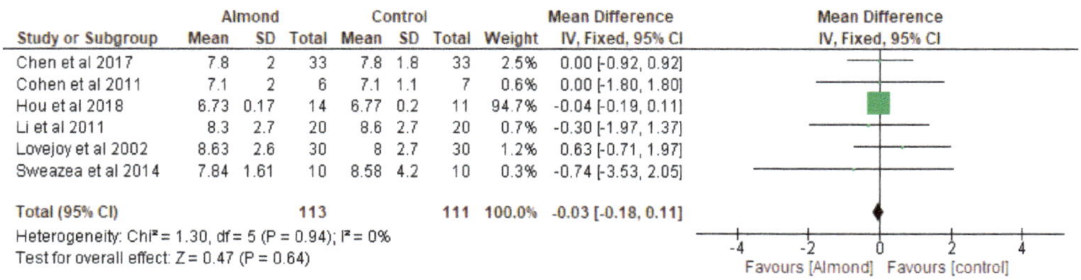

(c)

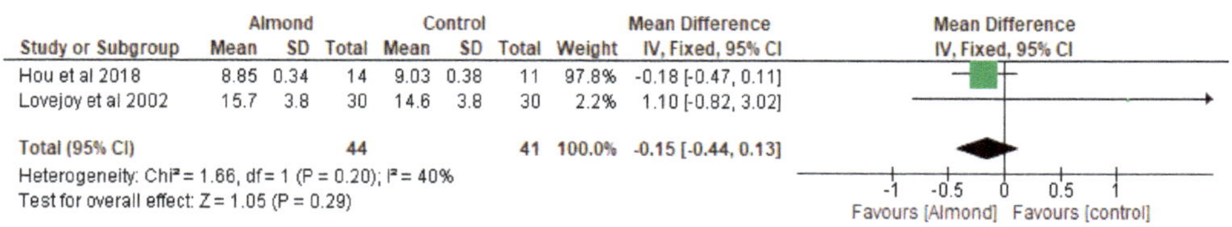

(d)

Figure 3. The effect of almonds on (**a**) glycated haemoglobin (Hba1c, %), (**b**) Hba1c (%)—subgroup analysis; (**c**) fasting blood glucose (mmol/L); (**d**) 2 h postprandial blood glucose (mmol/L).

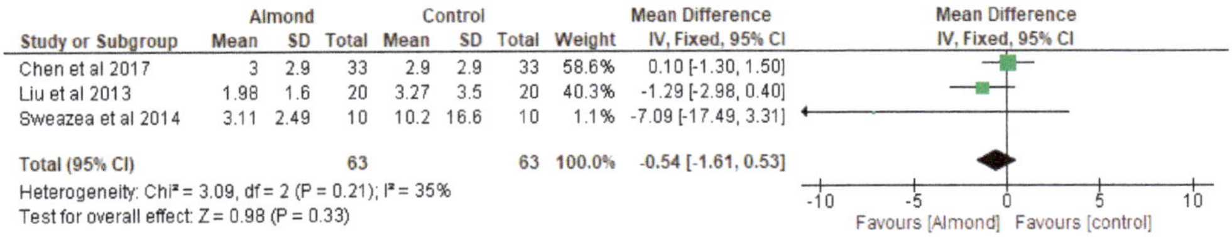

(a)

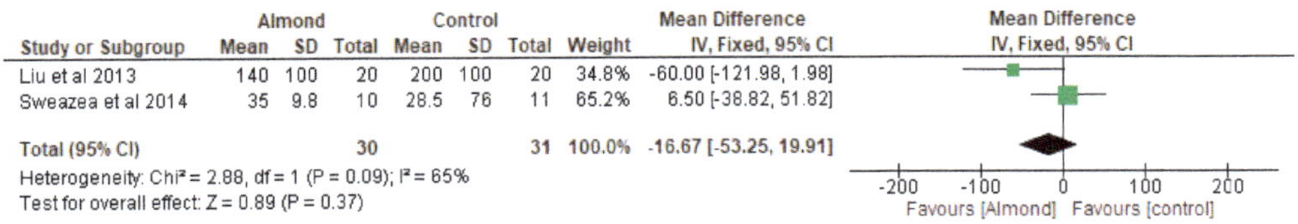

(b)

Figure 4. The effect of almonds on (**a**) C-reactive protein (mg/L) and (**b**) tumour necrosis factor–α (pg/mL).

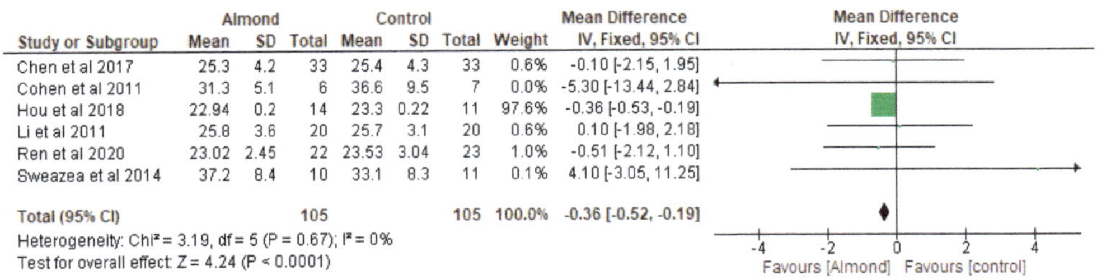

Figure 5. The effect of almonds on body mass index (Kg/m^2).

4.6. Homeostatic Model Assessment of Insulin Resistance (HOMA–IR)

According to Chen et al. [31], the almond-based diet did not show a significant effect with respect to HOMA-IR compared with control.

Three studies contributed data for HOMA-IR meta-analysis (almond gp, $n = 63$; control gp, $n = 63$) and the difference between the almond and control groups was not significant ($p > 0.05$) (Figure 6). The mean difference was −0.41 (95% CI: −1.32, 0.50). The sensitivity analysis did not change the results between the almond group and the control group ($p > 0.05$) in respect of HOMA–IR.

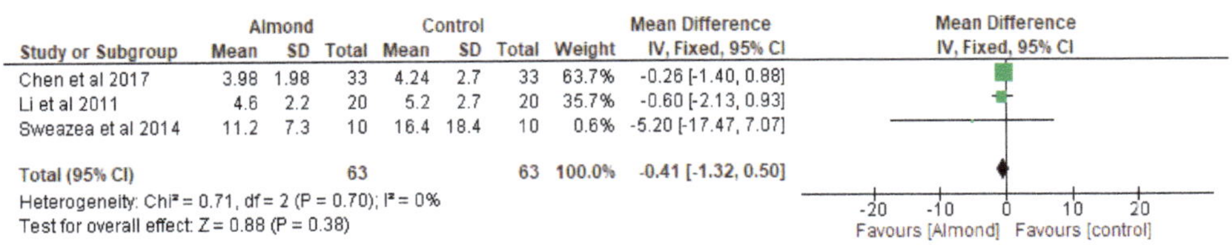

Figure 6. The effect of almond on HOMAR–IR.

4.7. Glucagon-Like Peptide-1 (GLP-1)

There was significant difference ($p < 0.05$) between the almond-based diet group and the control group with respect to the GLP-1 in the study by Ren et al. [14], although Cohen et al. [32] did not find any significant differences ($p > 0.05$) between the two groups.

Regarding the GLP-1 meta-analysis, two studies contributed data (almond gp, $n = 28$; control gp, $n = 30$) (Figure 7). GLP-1 was higher in the almond-based diet group compared with control, although the difference was not statistically significant (mean difference: 0.65; 95% CI: −0.16, 1.47; p-value = 0.12). The sensitivity analysis did not change the results between the almond group and the control group ($p > 0.05$) with to respect to GLP-1.

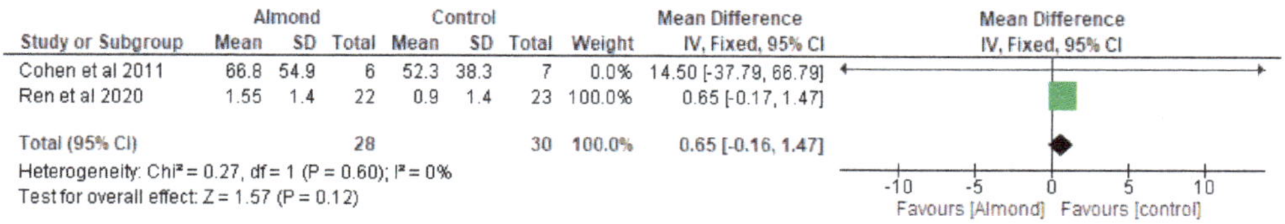

Figure 7. The effect of almond on GLP-1 (pmol/L).

4.8. Fasting Insulin

Bodnaruc et al. [30] found that an almond-based diet was associated with lower insulinemia, while Chen et al. [31] did not find any significant effect with respect to insulin levels in the almond-based and control diets. Five studies contributed data for this outcome (almond gp, $n = 99$; control gp, $n = 100$) (Figure 8). It was observed that there was no significant difference between the almond-based group compared to the control group in relation to insulin (standardised mean difference: −0.12; 95% CI: −0.40, 0.16; p-value = 0.39). There was also no significant difference ($p > 0.05$) between the almond group and the control group following the sensitivity analysis in regards to fasting insulin.

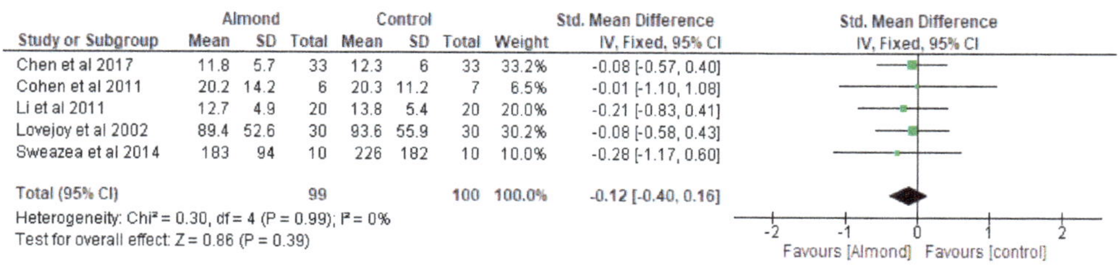

Figure 8. The effect of almond on fasting insulin (standardised mean difference).

5. Discussion

The results of the systematic review suggest that almond-based diets can promote the growth of short-chain fatty acid (SCFAs)-producing gut microbiota. Furthermore, the meta-analysis showed that almond-based diets were effective in significantly lowering ($p < 0.05$) glycated haemoglobin and body mass index (BMI) in patients with type 2 diabetes. However, it was also found that the effects of almond-based diets were not significant ($p > 0.05$) in relation to fasting blood glucose, 2 h postprandial blood glucose, inflammatory markers (C-reactive protein and TNF-α), GLP-1, HOMA–IR, and fasting insulin.

Our findings of the beneficial effects of almond-based diets on glycated haemoglobin are consistent with a previous study on almond supplementation in patients with type 2 diabetes [19] and an earlier review on the effect of tree nuts on glycaemic control in patients with diabetes [22]. Similarly, our results in relation to BMI are consistent with the findings of a previous study on the effect of almond consumption in the general population [18] and an earlier review of the effect of almonds on BMI [37].

The biological mechanisms responsible for the outcomes observed in this review in relation to reduction in glycated haemoglobin and BMI may be based on the nutrient composition of almond and its biological effects [37]. When compared to other nuts, it has been reported that almonds have the highest levels of fibre, monounsaturated and polyunsaturated fats, flavonoids, phytosterols, and phenolic acids [5,37]. Almonds also have a low glycaemic index [5] and almond-based diets have been shown to modulate gut microbiota dysbiosis and promote the production of GLP-1 in patients with type 2 diabetes [14].

The glycaemic index (GI) of food is an important measure of the quality of the food and it is a reflection of the digestibility of the available carbohydrates in the food compared with the reference food, often glucose [38]. It is a measure that ranks food based on the blood glucose response that they produce when ingested compared with the response to glucose or white wheat bread, which are reference foods [39]. Therefore, foods with low GI, such as almonds, usually breakdown slowly during digestion, and are slowly assimilated and, thus, have a slower impact on blood glucose levels and insulin response [40–42]. In a previous systematic review and meta-analysis, Ojo et al. [40], found that diets with low GI were more effective in improving glycated haemoglobin and fasting blood glucose compared with high-GI diets in patients with type 2 diabetes. In contrast, diets with high GI have been associated with type 2 diabetes and cardiovascular diseases due to their effect on blood glucose and insulin levels [38].

Due to the gradual entrance of glucose into the blood leading to reduced and more sustained insulin release, low-GI diets are more effective in controlling glycaemia compared with high GI diets [41]. In addition, low GI diets may be effective in increasing insulin sensitivity by reducing fluctuations in blood glucose levels and minimising insulin secretion over the day [41]. Based on the effectiveness of low-GI diets in controlling glycemia in patients with diabetes, the FAO [42] has recommended the use of a glycaemic index of foods along with the information about food composition in clinical applications in patients with diabetes.

Apart from the potential to improve glycaemic control, it has been suggested that diets with low GI may be useful in reducing weight because they produce a low insulin response [43]. This view is based on the lipogenic effect of hyperinsulinaemia [43]. On the other hand, high-GI diets may elicit a higher postprandial insulin response and this may lead to quicker hunger response and overeating through the reduction in metabolic fuels in the body [43]. Increased satiety and reduced voluntary food intake has been proposed as another mechanism through which foods with low GI can reduce weight [43].

Nuts, including almonds, are rich in energy density and high in fat, therefore, the greater fat availability could lead to reduced gastric emptying and increased satiety [5,14].

Another area of interest is the high soluble fibre and unsaturated fatty acid content of almonds [6]. According to Huo et al. [6], unsaturated fatty acids have been reported to promote the movement of glucose receptors to the cell surface and this could enhance insulin sensitivity. The role of polyunsaturated fatty acids on insulin sensitivity may be based on the fatty acid composition of the cell membrane, which relies on the fatty acid composition of the diet and regulates insulin action [44]. Kien et al. [45] suggested that a possible mechanism of dietary fatty acids in reducing the risk of insulin resistance may be due to the presence of a high level of unsaturated fatty acids in the cell membrane that could influence the physical properties, including plasticity, which promotes the movement of glucose receptors to the cell surface. It has also been shown that skeletal muscle insulin resistance due to obesity or dietary fatty acids may result from defective mitochondrial oxidation of fatty acids, which could lead to the accumulation of ceramides that may inhibit insulin signalling [45]. In addition, a high saturated fatty acid level of the membrane phospholipids increases insulin resistance [44].

Haag and Dippenaar [46] noted that the high saturated fat content of the cell membrane may lead to rigid and unresponsive membranes, while membranes that are high in unsaturated fatty acids promote fluidity and responsiveness. Therefore, the polyunsaturated content and omega-3/omega-6 ratio in the muscle and fat membranes are of significant importance in the aetiology of insulin resistance [46]. Furthermore, fatty acid-derived entities such as long chain acyl-CoA (coenzymes) may impact negatively on insulin mediated glucose transport and disrupt the insulin signalling cascade [46]. These findings were confirmed in randomised controlled trials in overweight individuals conducted by Kahleova et al. [47], who found that fat quantity and quality were related to body weight and body composition, insulin secretion, and insulin resistance.

Unsaturated fatty acids can also promote the efficiency of β-cell function through their action in stimulating GLP-1 secretion [6]. The findings of this review did reveal that an almond-based diet was effective in promoting the secretion of GLP-1, although this was not significant compared to the control. GLP-1 is a 30-amino-acid agent, which regulates glucose by stimulating insulin after ingesting a meal [32].

High dietary fibre in almonds can also increase gastric distension, viscosity of gastrointestinal tract, and slower absorption of macronutrients, including slowing the absorption of carbohydrates and the level of postprandial blood glucose [6]. High dietary fibre has been reported to promote the growth of SCFAs producing bacteria, increasing the production of SCFAs and promoting GLP-1 secretion [14].

In the study by Zhao et al. [48], it was found that the presence of greater diversity and abundance of fibre-promoting SCFA producers improved glycated haemoglobin levels in patients with type 2 diabetes through the production of glucagon-like peptide-1. The dietary fibre undergoes fermentation by colonic microbiota to produce SCFAs, including propionic, acetic, and butyric acid, which have significant effects on host physiology [49]. The SCFAs are useful in regulating the metabolic and immune system of the host as well as in cell proliferation [50]. However, low dietary fibre intake can cause microbiota dysbiosis, reduction in SCFAs production, and lead to the utilisation of less favourable substrates, such as proteins and fat [50,51]. The lipopolysaccharides resulting from the use of a high-fat diet can elicit an inflammatory response and contribute to the development of insulin resistance and type 2 diabetes [51].

Limitations

One of the studies included [32] was a pilot study with a small sample size. Furthermore, the number of studies included in the meta-analyses was eight or smaller in the different parameters. These could affect the broader application of the findings of the review. Therefore, more studies are required to further explore the role of almonds in patients with type 2 diabetes.

6. Conclusions

The findings of this systematic review and meta-analysis have shown that almond-based diets may be effective in promoting short-chain fatty acid-producing bacteria, and lowering glycated haemoglobin and body mass index in patients with type 2 diabetes compared with control. However, the effects of almonds were not significant ($p > 0.05$) with respect to fasting blood glucose, 2 h postprandial blood glucose, inflammatory markers (C-reactive protein and TNF-α), GLP-1, HOMA–IR, and fasting insulin.

Author Contributions: Conceptualization, O.O. and O.O.O.; methodology, O.O., X.-H.W. and O.O.O.; A.R.A.A.; validation, O.O., X.-H.W., O.O.O. and A.R.A.A.; formal analysis, O.O.; writing—original draft preparation, O.O.; writing—review and editing, O.O., X.-H.W., O.O.O. and A.R.A.A. All authors have read and agreed to the published version of the manuscript.

Funding: This research received no external funding.

Institutional Review Board Statement: Not applicable.

Informed Consent Statement: Not applicable.

Data Availability Statement: Not applicable.

Conflicts of Interest: The authors declare no conflict of interest.

References

1. International Diabetes Federation. Promoting Diabetes Care, Prevention and a Cure Worldwide. 2019. Available online: https://sites.pitt.edu/~{}super1/Metabolic/IDF5.pdf (accessed on 23 August 2021).
2. National Collaborating Centre for Chronic Conditions (NCCCC). *Type 2 Diabetes: National Clinical Guideline for Management in Primary and Secondary Care (Update)*; Royal College of Physicians: London, UK, 2008.
3. Bagust, A.; Hopkinson, P.K.; Maslove, L.; Currie, C.J. The projected health care burden of Type 2 diabetes in the UK from 2000 to 2060. *Diabet. Med.* **2002**, *19*, 1–5. [CrossRef] [PubMed]
4. Diabetes UK (2012) NHS Spending on Diabetes 'to Reach £16.9 Billion by 2035'. Available online: https://www.diabetes.org.uk/about_us/news_landing_page/nhs-spending-on-diabetes-to-reach-169-billion-by-2035 (accessed on 29 June 2021).
5. Mori, A.M.; Considine, R.V.; Mattes, R.D. Acute and second-meal effects of almond form in impaired glucose tolerant adults: A randomized crossover trial. *Nutr. Metab.* **2011**, *8*, 6–13. [CrossRef] [PubMed]
6. Hou, Y.-Y.; Ojo, O.; Wang, L.-L.; Wang, Q.; Jiang, Q.; Shao, X.-Y.; Wang, X.-H. A Randomized Controlled Trial to Compare the Effect of Peanuts and Almonds on the Cardio-Metabolic and Inflammatory Parameters in Patients with Type 2 Diabetes Mellitus. *Nutrients* **2018**, *10*, 1565. [CrossRef]
7. Jenkins, D.J.A.; Kendall, C.W.C.; Lamarche, B.; Banach, M.S.; Srichaikul, K.; Vidgen, E.; Mitchell, S.; Parker, T.; Nishi, S.; Bashyam, B.; et al. Nuts as a replacement for carbohydrates in the diabetic diet: A reanalysis of a randomised controlled trial. *Diabetologia* **2018**, *61*, 1734–1747. [CrossRef] [PubMed]
8. Barreca, D.; Nabavi, S.M.; Sureda, A.; Rasekhian, M.; Raciti, R.; Silva, A.S.; Annunziata, G.; Arnone, A.; Tenore, G.C.; Süntar, İ.; et al. Almonds (Prunus Dulcis Mill. D. A. Webb): A Source of Nutrients and Health-Promoting Compounds. *Nutrients* **2020**, *12*, 672. [CrossRef]
9. Ojo, O.; Feng, Q.-Q.; Ojo, O.O.; Wang, X.-H. The Role of Dietary Fibre in Modulating Gut Microbiota Dysbiosis in Patients with Type 2 Diabetes: A Systematic Review and Meta-Analysis of Randomised Controlled Trials. *Nutrients* **2020**, *12*, 3239. [CrossRef]
10. Ojo, O.; Ojo, O.O.; Zand, N.; Wang, X. The Effect of Dietary Fibre on Gut Microbiota, Lipid Profile, and Inflammatory Markers in Patients with Type 2 Diabetes: A Systematic Review and Meta-Analysis of Randomised Controlled Trials. *Nutrients* **2021**, *13*, 1805. [CrossRef]
11. Reimer, R.A.; Wharton, S.; Green, T.J.; Manjoo, P.; Ramay, H.R.; Lyon, M.R.; Gahler, R.J.; Wood, S. Effect of a functional fibre supplement on glycemic control when added to a year-long medically supervised weight management program in adults with type 2 diabetes. *Eur. J. Nutr.* **2021**, *60*, 1237–1251. [CrossRef]
12. Birkeland, E.; Gharagozlian, S.; Birkeland, K.I.; Valeur, J.; Måge, I.; Rud, I.; Aas, A.-M. Prebiotic effect of inulin-type fructans on faecal microbiota and short-chain fatty acids in type 2 diabetes: A randomised controlled trial. *Eur. J. Nutr.* **2020**, *59*, 3325–3338. [CrossRef]
13. King, J.C.; Blumberg, J.; Ingwersen, L.; Jenab, M.; Tucker, K.L. Tree nuts and peanuts as components of a healthy diet. *J. Nutr.* **2008**, *138*, 1736S–1740S. [CrossRef]
14. Ren, M.; Zhang, H.; Qi, J.; Hu, A.; Jiang, Q.; Hou, Y.; Feng, Q.; Ojo, O.; Wang, X. An Almond-Based Low Carbohydrate Diet Improves Depression and Glycometabolism in Patients with Type 2 Diabetes through Modulating Gut Microbiota and GLP-1: A Randomized Controlled Trial. *Nutrients* **2020**, *12*, 3036. [CrossRef]
15. Fang, Q.; Hu, J.; Nie, Q.; Nie, S. Effects of polysaccharides on glycometabolism based on gut microbiota alteration. *Trends Food Sci. Technol.* **2019**, *92*, 65–70. [CrossRef]
16. Kaoutari, A.E.; Armougom, F.; Gordon, J.I.; Raoult, D.; Henrissat, B. The abundance and variety of carbohydrate-active enzymes in the human gut microbiota. *Nat. Rev. Microbiol.* **2013**, *11*, 497–504. [CrossRef]
17. Nie, Q.; Chen, H.; Hu, J.; Fan, S.; Nie, S. Dietary compounds and traditional Chinese medicine ameliorate type 2 diabetes by modulating gut microbiota. *Crit. Rev. Food Sci. Nutr.* **2019**, *59*, 848–863. [CrossRef]
18. Wien, M.A.; Sabate, J.M.; Ikle, D.N.; Cole, S.E.; Kandeel, F.R. Almonds vs complex carbohydrates in a weight reduction program. *Int. J. Obes. Relat. Metab. Disord.* **2003**, *27*, 1365. [CrossRef] [PubMed]
19. Gulati, S.; Misra, A.; Pandey, R. Effect of Almond Supplementation on Glycemia and Cardiovascular Risk Factors in Asian Indians in North India with Type 2 Diabetes Mellitus: A 24-Week Study. *Metab. Syndr. Relat. Disord.* **2017**, *15*, 98–105. [CrossRef] [PubMed]
20. Blanco Mejia, S.; Kendall, C.W.C.; Viguiliouk, E.; Augustin, L.S.; Ha, V.; Cozma, A.I.; Mirrahimi, A.; Maroleanu, A.; Chiavaroli, L.; Leiter, L.A.; et al. Effect of tree nuts on metabolic syndrome criteria: A systematic review and meta-analysis of randomised controlled trials. *BMJ Open* **2014**, *4*, e004660. [CrossRef] [PubMed]
21. Muley, A.; Fernandez, R.; Ellwood, L.; Muley, P.; Shah, M. Effect of tree nuts on glycemic outcomes in adults with type 2 diabetes mellitus: A systematic review. *JBI Evid. Synth.* **2021**, *19*, 966–1002. [CrossRef] [PubMed]
22. Viguiliouk, E.; Kendall, C.W.C.; Blanco Mejia, S.; Cozma, A.I.; Ha, V.; Mirrahimi, A.; Jayalath, V.H.; Augustin, L.S.A.; Chiavaroli, L.; Leiter, L.A.; et al. Effect of Tree Nuts on Glycemic Control in Diabetes: A Systematic Review and Meta-Analysis of Randomized Controlled Dietary Trials. *PLoS ONE* **2014**, *9*, e0103376. [CrossRef] [PubMed]
23. Mohammadifard, N.; Salehi-Abargouei, A.; Salas-Salvadó, J.; Guasch-Ferré, M.; Humphries, K.; Sarrafzadegan, N. The effect of tree nut, peanut, and soy nut consumption on blood pressure: A systematic review and meta-analysis of randomized controlled clinical trials. *Am. J. Clin. Nutr.* **2015**, *101*, 966–982. [CrossRef]

24. Musa-Veloso, K.; Paulionis, L.; Poon, T.; Lee, H.Y. The effects of almond consumption on fasting blood lipid levels: A systematic review and meta-analysis of randomised controlled trials. *J. Nutr. Sci.* **2016**, *5*, e34. [CrossRef] [PubMed]
25. Tsai, Y.-L.; Lin, T.-L.; Chang, C.-J.; Wu, T.-R.; Lai, W.-F.; Lu, C.-C.; Lai, H.-C. Probiotics, prebiotics and amelioration of diseases. *J. Biomed. Sci.* **2019**, *26*, 3. [CrossRef] [PubMed]
26. Carrera-Quintanar, L.; López Roa, R.I.; Quintero-Fabián, S.; Sánchez-Sánchez, M.A.; Vizmanos, B.; Ortuño-Sahagún, D. Phytochemicals that influence gut microbiota as prophylactics and for the treatment of obesity and inflammatory diseases. *Mediat. Inflamm.* **2018**, *2018*, 9734845. [CrossRef] [PubMed]
27. Moher, D.; Liberati, A.; Tetzlaff, J.; Altman, D.G. The PRISMA Group. Preferred reporting items for systematic reviews and meta-analyses: The PRISMA statement. *Ann. Intern. Med.* **2009**, *151*, 264–269. [CrossRef]
28. The Nordic Cochrane Centre. *Review Manager, Version 5.3*; The Nordic Cochrane Centre; The Cochrane Collaboration: Copenhagen, Denmark, 2014.
29. Higgins, J.P.T.; Green, S. *Cochrane Handbook for Systematic Reviews of Interventions*; Wiley-Blackwell: Hoboken, NJ, USA, 2009.
30. Bodnaruc, A.M.; Prud'homme, D.; Giroux, I. Acute effects of an isocaloric macronutrient-matched breakfast meal containing almonds on glycemic, hormonal, and appetite responses in men with type 2 diabetes: A randomized crossover study. *Appl. Physiol. Nutr. Metab.* **2020**, *45*, 520–529. [CrossRef]
31. Chen, C.-M.; Liu, J.-F.; Li, S.-C.; Huang, C.-L.; Hsirh, A.-T.; Weng, S.-F.; Chang, M.-L.; Li, H.-T.; Mohn, E.; Oliver Chen, C.-Y. Almonds ameliorate glycemic control in Chinese patients with better controlled type 2 diabetes: A randomized, crossover, controlled feeding trial. *Nutr. Metab.* **2017**, *14*, 1–12. [CrossRef]
32. Cohen, A.E.; Johnston, C.S. Almond ingestion at mealtime reduces postprandial glycemia and chronic ingestion reduces hemoglobin A(1c) in individuals with well-controlled type 2 diabetes mellitus. *Metab. Clin. Exp.* **2011**, *60*, 1312–1317. [CrossRef]
33. Li, S.-C.; Liu, Y.-H.; Liu, J.-F.; Chang, W.-H.; Chen, C.-M.; Chen, C.-Y.O. Almond consumption improved glycemic control and lipid profiles in patients with type 2 diabetes mellitus. *Metab. Clin. Exp.* **2011**, *60*, 474–479. [CrossRef]
34. Liu, J.-F.; Liu, Y.-H.; Chen, C.-M.; Chang, W.-H.; Chen, C.-Y. The effect of almonds on inflammation and oxidative stress in Chinese patients with type 2 diabetes mellitus: A randomized crossover controlled feeding trial. *Eur. J. Nutr.* **2013**, *52*, 927–935. [CrossRef]
35. Lovejoy, J.C.; Most, M.M.; Lefevre, M.; Greenway, F.L.; Rood, J.C. Effect of diets enriched in almonds on insulin action and serum lipids in adults with normal glucose tolerance or type 2 diabetes. *Am. J. Clin. Nutr.* **2002**, *76*, 1000–1006. [CrossRef]
36. Sweazea, K.; Johnston, C.; Ricklefs, K.D.; Petersen, K.N. Almond supplementation in the absence of dietary advice significantly reduces C-reactive protein in subjects with type 2 diabetes. *J. Funct. Foods* **2014**, *10*, 252–259. [CrossRef]
37. Dreher, M.L. A Comprehensive Review of Almond Clinical Trials on Weight Measures, Metabolic Health Biomarkers and Outcomes, and the Gut Microbiota. *Nutrients* **2021**, *13*, 1968. [CrossRef]
38. Mohan, V.; Anjana, R.M.; Gayathri, R.; Ramya Bai, M.; Lakshmipriya, N.; Ruchi, V.; Sudha, V. Glycemic Index of a Novel High-Fiber White Rice Variety Developed in India—A Randomized Control Trial Study. *Diabetes Technol. Ther.* **2016**, *18*, 164–170. [CrossRef]
39. Similä, M.E.; Valsta, L.M.; Kontto, J.P.; Albanes, D.; Virtamo, J. Low-, medium- and high-glycaemic index carbohydrates and risk of type 2 diabetes in men. *Br. J. Nutr.* **2011**, *105*, 1258–1264. [CrossRef]
40. Ojo, O.; Ojo, O.O.; Adebowale, F.; Wang, X.-H. The Effect of Dietary Glycaemic Index on Glycaemia in Patients with Type 2 Diabetes: A Systematic Review and Meta-Analysis of Randomized Controlled Trials. *Nutrients* **2018**, *10*, 373. [CrossRef] [PubMed]
41. Thomas, D.E.; Elliott, E.J. The use of low-glycaemic index diets in diabetes control. *Br. J. Nutr.* **2010**, *104*, 797–802. [CrossRef] [PubMed]
42. Food and Agricultural Organisation (FAO). *Carbohydrates in Human Nutrition. Report of a Joint FAO/WHO Expert Consultation*; FAO (Food and Nutrition Paper–66); FAO: Rome, Italy, 1998; Available online: http://www.fao.org/docrep/w8079e/w8079e00.htm (accessed on 16 August 2018).
43. Esfahani, A.; Wong, J.W.; Mirrahimi, A.; Villa, C.R.; Kendall, C.C. The application of the glycemic index and glycemic load in weight loss: A review of the clinical evidence. *IUBMB Life* **2011**, *63*, 7–13. [CrossRef]
44. Salmerón, J.; Hu, F.B.; Manson, J.E.; Stampfer, M.J.; Colditz, G.A.; Rimm, E.B.; Willett, W.C. Dietary fat intake and risk of type 2 diabetes in women. *Am. J. Clin. Nutr.* **2001**, *73*, 1019–1026. [CrossRef] [PubMed]
45. Kien, C.L. Dietary interventions for metabolic syndrome: Role of modifying dietary fats. *Curr. Diabetes Rep.* **2009**, *9*, 43–50. [CrossRef] [PubMed]
46. Haag, M.; Dippenaar, N.G. Dietary fats, fatty acids and insulin resistance: Short review of a multifaceted connection. *Med. Sci. Monit.: Int. Med. J. Exp. Clin. Res.* **2005**, *11*, RA359–RA367.
47. Kahleova, H.; Hlozkova, A.; Fleeman, R.; Fletcher, K.; Holubkov, R.; Barnard, N.D. Fat Quantity and Quality, as Part of a Low-Fat, Vegan Diet, Are Associated with Changes in Body Composition, Insulin Resistance, and Insulin Secretion. A 16-Week Randomized Controlled Trial. *Nutrients* **2019**, *11*, 615. [CrossRef]
48. Zhao, L.; Zhang, F.; Ding, X.; Wu, G.; Lam, Y.Y.; Wang, X.; Fu, H.; Xue, X.; Lu, C.; Ma, J.; et al. Gut bacteria selectively promoted by dietary fibers alleviate type 2 diabetes. *Science* **2018**, *359*, 1151–1156. [CrossRef] [PubMed]

49. Ebrahimzadeh Leylabadlo, H.; Sanaie, S.; Sadeghpour Heravi, F.; Ahmadian, Z.; Ghotaslou, R. From role of gut microbiota to microbial-based therapies in type 2-diabetes. *Infect. Genet. Evol.* **2020**, *81*, 104268. [CrossRef]
50. Makki, K.; Deehan, E.C.; Walter, J.; Bäckhed, F. The Impact of Dietary Fiber on Gut Microbiota in Host Health and Disease. *Cell Host Microbe* **2018**, *23*, 705–715. [CrossRef] [PubMed]
51. Davison, K.M.; Temple, N.J. Cereal fiber, fruit fiber, and type 2 diabetes: Explaining the paradox. *J. Diabetes Its Complicat.* **2018**, *32*, 240–245. [CrossRef] [PubMed]

nutrients

Review

The Effect of Prebiotics and Oral Anti-Diabetic Agents on Gut Microbiome in Patients with Type 2 Diabetes: A Systematic Review and Network Meta-Analysis of Randomised Controlled Trials

Omorogieva Ojo [1,*], Xiaohua Wang [2], Osarhumwese Osaretin Ojo [3], Joanne Brooke [4], Yiqing Jiang [2], Qingqing Dong [2] and Trevor Thompson [5]

1 School of Health Sciences, Avery Hill Campus, University of Greenwich, London SE9 2UG, UK
2 The School of Nursing, Soochow University, Suzhou 215006, China
3 Smoking Cessation Department, University Hospital, London SE13 6LH, UK
4 Faculty of Health, Education and Life Sciences, Birmingham City University, Birmingham B15 3TN, UK
5 School of Human Sciences, Avery Hill Campus, University of Greenwich, London SE9 2UG, UK
* Correspondence: o.ojo@greenwich.ac.uk

Abstract: Background: Nutritional interventions such as the use of prebiotics can promote eubiosis of gut microbiome and maintain glucose homeostasis in patients with type 2 diabetes (T2D). However, it would appear that results of the effects of prebiotics on the community of microbes in the gut are not consistent. Aim: To examine the effect of prebiotics and oral antidiabetic agents on gut microbiome in patients with T2D. Methods: The PRISMA Extension Statement for Systematic Reviews and Network Meta-analyses was used to conduct this review. Searches were carried out in EMBASE, EBSCO-host databases, Google Scholar and the reference lists of articles for studies that are relevant to the research question, from database inception to 15 August 2022. The search strategy was based on PICOS framework. Network Meta-analysis which allows the estimation of relative treatment effects by combing both direct trial evidence (e.g., treatment A vs. treatment B) and indirect evidence was conducted. Furthermore, pairwise meta-analysis was also carried out to estimate effect sizes based on head-to-head comparisons of treatments and/or control conditions. Results: Findings of the Network meta-analysis revealed that prebiotics significantly reduced HbA1c compared with control and the SMD was −0.43 [95% CI, −0.77, −0.08; $p = 0.02$], whereas there was no significant difference ($p > 0.05$) between the other treatments and control. In addition, anti-diabetic agents including glipizide and metformin also reduced HbA1C, although these were not significantly different ($p > 0.05$) from control. While prebiotics promoted *Bifidobacterium* and *Akkermansia*, the improvements were not significantly different ($p > 0.05$) from control. On the other hand, metformin decreased the relative abundance of *Bifidobacterium*, but increased *Lactobacillus* and *Akkermansia*, although the differences were not significant ($p > 0.05$) compared with control. With respect to fasting blood glucose and BMI, the effects of prebiotics and oral antidiabetic agents did not differ significantly ($p > 0.05$) from controls. Conclusions: The findings of the systematic review and Network meta-analysis demonstrated prebiotics were significantly ($p < 0.05$) more effective in reducing HbA1c than control in patients with T2D. However, the effects of prebiotics and oral antidiabetic agents did not differ significantly ($p > 0.05$) from the controls in relation to fasting blood glucose, post-prandial blood glucose, body mass index and the genera of gut bacteria examined. More studies are required to fully investigate the effects of prebiotics and oral antidiabetic agents in patients with T2D

Keywords: prebiotics; oral anti-diabetic agents; gut microbiome; glycated haemoglobin; type 2 diabetes; Network meta-analysis; meta-analysis

Citation: Ojo, O.; Wang, X.; Ojo, O.O.; Brooke, J.; Jiang, Y.; Dong, Q.; Thompson, T. The Effect of Prebiotics and Oral Anti-Diabetic Agents on Gut Microbiome in Patients with Type 2 Diabetes: A Systematic Review and Network Meta-Analysis of Randomised Controlled Trials. *Nutrients* **2022**, *13*, 5139. https://doi.org/10.3390/nu14235139

Academic Editor: Connie Weaver

Received: 3 November 2022
Accepted: 29 November 2022
Published: 2 December 2022

1. Introduction

The prevalence of type 2 diabetes (T2D) is increasing globally. It is estimated that by 2040, approximately 642 million people will have the condition worldwide [1]. Genetic predisposition and lifestyle factors such as lack of physical activities and poor nutritional intake which can lead to overweight and obesity are reported to be involved in the etiology of T2D [2]. Furthermore, diets with low fibre and high saturated fats and sugar, such as Western diets, may also influence gut microbial diversity and cause reduction in specific bacteria taxa and imbalance in gut microbiome [3,4].

There is evidence from meta-analysis of randomised controlled trials (RCTs) that T2D is associated with disequilibrium of gut microbial community and gut microbiota dysbiosis is implicated in the pathogenesis of type 2 diabetes [5]. Therefore, nutritional interventions including prebiotics have been used to promote eubiosis of gut microbiome and maintain glucose homeostasis in patients with T2D [6,7]. In addition, the role of oral anti-diabetic agents in modulating dysbiosis of gut microbiome may be a possible pathway by which these drugs regulate glucose balance [8,9].

1.1. Description of the Intervention

The concept of prebiotics relates to the food component that is not digestible. Dietary prebiotics must be resistant to gastric acid and should not be hydrolised by the effect of mammalian enzyme, and should be resistant to intestinal absorption. It should also be beneficial to the hosts through selective promotion of the growth of bacteria in the colon not causing negative effects to the hosts including not stimulating the growth of pathogenic microorganisms [10]. Prebiotics have also been recently defined as substrates (non-viable) which are used selectively by the host microorganisms which leads to benefits [10].

The definition of prebiotics has been revised to include ingredients that are selectively fermented and allows changes that are specific to the community and actions of microbes inhabiting the gastrointestinal tract which confers effect on the host which are beneficial physiologically [11]. Prebiotics are different from most dietary fibres including pectins, cellulose and xylans which promote the development of a broad variety of gut microbes [12].

Although prebiotics are not the only substrates that can affect the gut microbial community, a primary criterion that distinguishes prebiotics from other substrates is their selective utilisation by host microorganisms [12]. While a selective effect does not mean utilisation by just one microbial group, it may include several microbial groups, but not all the microbial groups [12].

Metformin is one of the oral anti-diabetic agents and it is a biguanide [13]. It is a first line medication for treating T2D and is effective in lowering body weight and cardiovascular risks [13]. Other oral anti-diabetic agents that are associated with modulation of gut microbiota include sulfonylurea and acarbose [9,14].

1.2. How This Intervention Might Work

Dysbiosis of intestinal microflora has been shown to have significant effect in the pathogenesis of metabolic disorders such as T2D [15]. Therefore, sustaining an ecosystem that is healthy and having good lifestyle and feeding habits are useful approaches in managing T2D [15]. The consumption of prebiotics may regulate gut microbiota dysbiosis and enable the growth of beneficial microbes [11]. There is evidence to suggest that prebiotic dietary fibre is a selective substrate that is utilised by bacteria which are beneficial to the host including *Bifidobacterium* and *Lactobacillus* that promote the health of the host [10]. For example, prebiotics may promote the growth of bacteria that are beneficial including *Lactobacillus*, *Bifidobacterium*, *Akkermansia*, *Eubacterium* and *Roseburia* [10].

Prebiotics are usually metabolised by the gut microbes through a process of fermentation to produce metabolites which are useful to the host [10]. The end product of metabolism of prebiotics is short chain fatty acids (SCFAs), which are primarily propionic, butyric and acetic acid [10]. SCFAs influence the integrity of the gut epithelium, immunity, glucose homeostasis, lipid profile and body weight [11]. In addition, SCFAs have effects on insulin resistance, suppress appetite and lipolysis, increase expenditure of energy and promote insulin sensitivity and production [9,15].

While butyrate is a good source of energy for colonocytes and enterocytes, propionate is a substrate for intestinal and hepatic gluconeogenesis, and the most abundant SCFA found in circulation is acetate [11]. The phenomenon of cross-feeding by other bacteria has also been discussed as possible mechanism employed in the production of SCFAs which are crucial for the intestinal health and other health benefits in areas distant to the gut [12,16]. Cross feeding is a process where a substrate stimulates the growth of members of the gut microbiota which produces metabolites which are utilised by other microbes to produce butyrate and other SCFAs [12].

Antidiabetic agents have also been shown to restore the richness and diversity of the gut microbial community to some level and have demonstrated ability to promote the growth of some useful bacteria [15]. In particular, anti-diabetic agents not only influence gut microbiota, in turn, microbiota affects how the individual responds to those drugs which explains the bidirectional relationship between microbes in the gut and anti-diabetic medications [17]. Metformin has been shown to reduce blood glucose in patients with T2D by interacting with microbes in the gut including altering the composition and diversity of gut microbiome [8,18].

1.3. *Why It Is Important to Do This Review*

The definition of prebiotics has been evolving over the years, therefore, a good knowledge of their effect on gut microbiome in patients with T2D will help in enriching our understanding of this concept, broaden their application and health related outcomes [10–12,16]. Furthermore, an understanding of gut microbial ecology in patients with T2D is useful in developing effective approaches to regulate gut microbiota dysbiosis for purposes that are preventive and therapeutic [15]. It has been suggested that the effectiveness of prebiotics in patients with T2D is based on the modulation of gut microbiome although the results are not consistent [19]. In addition, it seems the systematic reviews and/or meta-analysis conducted previously [20–22] have not focused on the effect of prebiotics on gut microbiome in patients with T2D. In other systematic reviews, studies involving probiotics [23] and prebiotics or symbiotics supplementations [24,25] were included. In addition, the review by Merkevicius et al. [24] included one animal study, but did not involve meta-analysis. The Bock et al. [25] review included patients with type 1 diabetes. In our previous systematic review [26], we examined the effect of dietary fibre in regulating the imbalance in the gut microbial community, but did not compare this with oral antidiabetic agents. In contrast, the current review is a systematic review and Network Meta-analysis (NMA) of RCTs which seeks to evaluate the impact of prebiotics and oral anti-diabetic agents on gut microbiota and metabolic parameters in patients with T2D.

1.4. *Research Questions*

Are prebiotics more effective than a control in managing patients with T2D?

What is the comparative effectiveness of prebiotic treatment or treatment with oral antidiabetic agents in patients with T2D?

Aim.

To examine the effect of prebiotics and oral anti-diabetic agents on gut microbiome in patients with T2D.

2. Methods

The Preferred Reporting Items for Systematic Reviews and Meta-Analyses Extension Statement for Reporting of Systematic Reviews Incorporating Network Meta-analyses of Health Care Interventions was used to conduct this systematic review and Network meta-analysis (PRISMA-ES for NMA) [27,28].

Registration: This systematic review and Network meta-analysis protocol was registered with Prospero and the Registration Number was CRD42022352060.

2.1. Studies Included

Only RCTs were selected for the review.

2.2. Participants of Interest

Patients with T2D were participants included in the review.

2.3. Types of Interventions

Pre-biotics and oral anti-diabetic agents were the interventions of choice.

2.4. Outcome Measures

The following were the outcomes of interest:

Gut Microbiome: *Lactobacillus, Bifidobacterium, Ruminococcus, Bacteroides, Roseburia, Clostridium* and *Akkermansia* (Relative abundance and genera only).

Blood Glucose Parameters: glycated haemoglobin (HbA1c), fasting blood glucose (FBG) and postprandial blood glucose.

Body Mass Index.

2.5. Search Strategy

EBSCOHost was searched for relevant articles using the Health Sciences Research Databases (which includes MEDLINE, APA PsycArticles, Academic Search Premier, CINAHL Plus with Full Text, Psychology and Behavioral Sciences Collection and APA PsycInfo databases). Furthermore, EMBASE and Google Scholar were additional databases searched. The reference lists of articles were searched for studies that were relevant to the research question. The searches were carried out from database inception to 15 August 2022. The Population, Intervention, Comparator, Outcomes, Studies (PICOS) tool was used to define the research question and establish the search strategy [29]. The search terms included synonyms and medical subject headings and these were combined with Boolean operators (OR/AND) (Table 1). OO and OOO conducted the searches separately and these were cross checked by X.W. and JB. Search results were transferred to EndNote (Analytics, Philadelphia, PA, USA) and duplicates of articles were deleted.

Table 1. Search Terms Based on PICOS Tool.

Patient/Population	Intervention	Outcome (Primary)	Study Designs	Combining Search Terms
Patients with diabetes	Prebiotics OR Oral anti-diabetic agents	Gut microbiome	Randomised controlled trial	
Diabetes mellitus, type 2 OR Diabetes complications OR Patients with diabetes OR diabetes mellitus OR type 2 diabetes OR Diabetes	Prebiotics OR Dietary fibre OR Fibre OR Polysaccharide OR Dietary carbohydrate OR Resistant Starch OR carbohydrate OR Oral anti-diabetic agents OR metformin or gliclazide OR acarbose	Microbiome OR Gastrointestinal microbiota OR Gut microbiota OR Microbiota	#1 Randomized OR Randomised controlled trial OR placebo OR controlled clinical trial OR therapy OR randomly OR drug OR trial OR groups #2 "Animals" NOT "Humans" #3 #1 NOT #2	1st Column + 2nd Column + 3rd Column + 4th Column

Abbreviation/Symbol: # (Number).

3. Collection of Data and Analysis

3.1. Study Selection

Criteria for Inclusion: Patients with T2D and those who were 18 years of age or older were selected for the review. Other inclusion criteria were studies involving prebiotics and/or oral antidiabetic agents as interventions and studies that meet the required outcomes, including; gut microbiome, glycaemic parameters and body mass index.

Criteria for Exclusion: Participants younger than 18 years of age, those with gestational diabetes, pre-diabetes and type 1 diabetes, and studies with probiotics and animal models were excluded from the review.

The PRISMA flow chart (Figure 1) provides details of studies included using the criteria for inclusion and exclusion previously outlined.

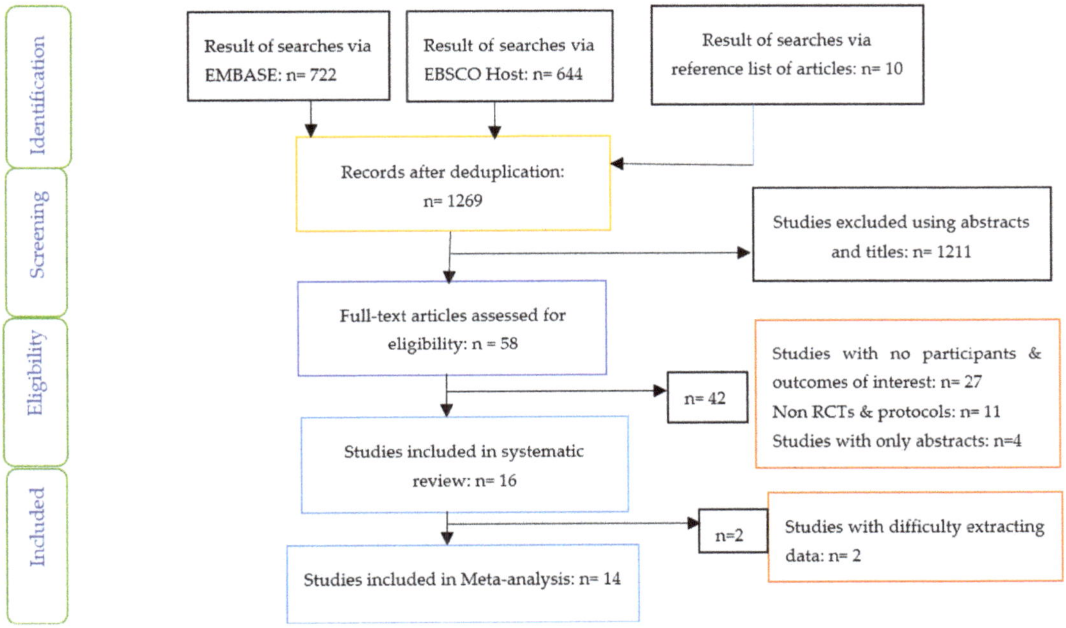

Figure 1. PRISMA flow chart on selection and inclusion of studies.

3.2. *Data Extraction and Management*

O.O., Y.J., Q.D. and X.W. extracted the data from included articles and these were cross-checked by all authors. Changes from baseline and final values of all parameters of interestwere used for the Network meta-analysis and pairwise meta-analyis [30]. The intervention group data were compared to the control group. Data from studies such as Medina-vera et al. [31] and Pedersen et al. [32] were extracted using the Engauge Digitizer [33]. Furthermore, the units of measurements were converted in some parameters such as fasting blood glucose (mmol/L), glycated haemoglobin (%) and Bifidobacterium (%). Means and standard deviations were calculated from median and 1st–3rd quartiles, respectively, in some parameters.

Risk of Bias Assessment of Studies.

The studies included were evaluated based on the established assessment tool [30]. The domains assessed were attrition bias, selection bias, detection bias, performance bias, reporting bias, and other bias [30]. The Review Manager 5.3 software [34] was used to assess the risk of bias.

4. Data Analysis

4.1. *Network Meta-Analysis (NMA)*

NMA was performed within a frequentist framework using the netmeta package [35] in R to compute standardised mean differences (SMDs). NMA allows the estimation of relative treatments effects by combing both direct trial evidence (e.g., treatment A vs. treatment B) and indirect evidence (e.g., in trials where A and B have not been directly compared but have all used a common comparator, e.g., placebo, allowing A and B to be compared indirectly).

Network plots for each outcome were first constructed and examined to identify any intervention comparisons which were disconnected from the main treatment network and which therefore could not be examined using NMA. We then performed NMA and constructed forest plots comparing each treatment to a reference condition (either placebo or inactive control according to what was most commonly employed for that outcome).

A key assumption of NMA is transitivity, which broadly speaking is that trials of different treatment comparisons are broadly similar on important methodological and sample characteristics (such as age, gender etc). If differences do exist that might cause the effect of a treatment to be amplified or diminished in a set of trials regardless of the particular treatment given (e.g., due to use of an older less treatment-responsive sample) then this assumption is violated. We assessed this by inspecting a summary table of key potential effect modifiers of sex, age, etc across the different sets of treatment comparisons. (Refer to table on mean age and sex distribution of treatments in the supplementary file). An alternative method of evaluating inconsistency is by comparing the differences between direct and indirect evidence for each comparison. However, we did not attempt to do this here as the data we examined allowed the computation of SMD exclusively from either direct or indirect evidence but not both.

Pairwise meta-analysis was also conducted to estimate effect sizes based on head-to-head comparisons of treatments and/or control conditions. Pairwise meta-analysis was conducted using Revman.

4.2. Meta-Analysis

The Review Manager (RevMan) 5.3 software [34] was used to conduct the meta-analysis. The measure of heterogeneity was the I^2 statistic [30], and statistical significance of heterogeneity was set at p < 0.10. The fixed effects model was used when heterogeneity was not important (I^2: 0–40%) and the random effects model was applied when heterogeneity was substantial or considerable (I^2: 40–100%) [30]. The SMD was used for the meta-analysis.

A subgroup analysis was carried out to examine the effect of prebiotics and oral antidiabetic agents in patients with T2D.

4.3. Effect Size

The result of the meta-analysis are depicted as forest plots and in terms of statistical significance, $p < 0.05$ was used to assess the overall effect of the intervention.

5. Results

Sixteen studies were included in the systematic review, while fourteen studies were included in the Network meta-analysis (Figure 1). The characteristics of the included studies including countries where studies were conducted, type of study, participants, sample size, mean age, mean diabetes duration, interventions and results/findings are outlined in Table 2. Four studies were carried out in China, three in Italy and two studies in Mexico. One study each was carried out in Japan, Korea, Norway, Canada, Netherlands, UK and Spain. All these studies were randomised controlled studies.

The network plots of the Network meta analysis can be found in the supplementary file (Figure S1 and Table S1).

Table 2. Description and characteristics of studies included.

Citation/Country of Study	Type of Study	Aim	Participants	Sample Size	Mean Age (Years)	Mean Diabetes Duration (Years)	Interventions	Results/Findings
Arias-Córdova et al. [36] Mexico	RCT	To assess the effects of NBS and HMS on GC and GV in patients with T2D when treatments were matched for digestible starch content.	All participants with T2D treated with metformin or a combination of glibenclamide and metformin	$n = 10$	48.5 ± 9.12	Not Applicable	NBS, HMS and DMS. including three treatment phases, each with a duration of 4 days, and washout period between treatments of 9-day.	The intake of NBS showed a reduction in fasting glycemia compared to DMS.
Birkeland et al. [37] Norway	RCT	To examine the effect of inulin-type fructans on faecal microbiota and SCFAs in patients with T2D.	T2D, with 2/3 of participants receiving glucose-lowering drugs	$n = 25$		4.7 (0.2–20.0)	16 g Inulin-type fructans versus 16 g maltodextrin. There was 4-week washout which separated the 6 weeks treatment period.	There was moderate, but significant increase in faecal levels of bifidobacteria in the group supplemented daily with inulin-type fructans.
Candela et al. [38] Italy	RCT	To explore the effect of microbiotic Ma-Pi 2 diet in modulating gut microbiota dysbiosis in patients with T2D.	Patients with T2D	Ma-Pi 2 diet: $n = 21$ Control diet: $n = 19$	66	Not Applicable	Fibre rich microbiotic Ma-Pi 2 diet is enriched with complex carbohydrates, legumes, fermented products, sea salt and green tea.	FBG and PBG were reduced significantly in both Ma-Pi 2 and control diets, although this was significantly higher in the Ma-Pi 2 diet compared to control. Both diets were also effective in supporting the recovery of health promoting SCFA producing bacteria including *Faecalibacterium*, *Roseburia*, *Bacteroides* and *Akkermansia*. Increases in *Collinsella* and *Streptococcus* were only counteracted by Ma-Pi 2 diet.
Gonai et al. [39] Japan	RCT	To explore the effects of GOS on on glycaemia, gut microbiota and metabolitic parameters in patients with T2D.	Patients with T2D	GOS group: 28 Placebo group: 27	GOS group: 55 ± 11 Placebo group: 54 ± 12	GOS group: 10 ± 8 Placebo group: 6 ± 5	10 g/d GOS syrup versus 10 g/d maltodextrin syrup. 4 weeks of treatment.	After consumption of GOS, Bifidobacteriaceae was significantly restored in patients with T2D, whereas lipopolysaccharide binding protein and glucose tolerance did not show improvement.
Gu et al. [40] China	RCT	To compare the effect of Acarbose versus sulfonylurea Glipizide on metabolic parameters, (e.g., glycaemic, control plasma BAs), and the intestinal microbiota, and discriminate such changes from disease-dependent alterations.	Patients with T2D	Acarbose group: 51 Glipizide group: 43	Acarbose group: 53 ± 7 Glipizide group: 54 ± 7	Not Applicable	Acarbose treatment versus Glipizide treatment. A 3-month treatment period.	Both the acarbose and glipizide groups improved glycemic control, with no significant differences. Acarbose increased the relative abundances of Lactobacillus and Bifidobacterium and depleted Bacteroides. However, Glipizide treatment did not affect the relative abundances at species-level. After 3 months of treatment, reductions in BW and BMI were more significant in the Acarbose group compared to the Glipizide group.

Table 2. *Cont.*

Citation/Country of Study	Type of Study	Aim	Participants	Sample Size	Mean Age (Years)	Mean Diabetes Duration (Years)	Interventions	Results/Findings
Medina-Vera et al. [31] Mexico	RCT	To assess functional food-based dietary intervention on biochemical parameters and faecal microbiota in patients with T2D.	Patients with T2D	DP group: n = 28 Placebo group: n = 25	DP group: 50.4 ± 8.7 Placebo group: 49.8 ± 10.6	DP group: 4.1 ± 3.5 Placebo group: 4.4 ± 3.9	A dietary portfolio, DP (14 g of dehydrated nopal, 4 g of chia seeds, 30 g of soy protein and 4 g of inulin) versus placebo (28 g of calcium caseinate and 15 g of maltodextrin). The treatment period was for 3 months.	Consumption of DP promoted the abundance of Bifidobacterium longum which has been reported to improve insulin sensitivity. There was significant reduction in the levels of HbA1c in patients with T2D in the DP group.
Pedersen et al. [32]	RCT	To investigate the effects of prebiotic supplementation on intestinal bacteria in patients with type 2 diabetes	Patients with type 2 diabetes	Prebiotic group: n = 14 Placebo group: n= 15	Prebiotic group (56.7 ± 6.0) Placebo group (58.1 ± 6.6)	Prebiotic group (4.6 ± 2.2) Placebo group (4.0 ± 3.1)	Prebiotic (galacto-oligosaccharide mixture) or placebo (maltodextrin supplements each given 5.5g/day for 12 weeks.	Prebiotic fibre supplementation did not improve glucose control or abundance of bacteria compared with control.
Reimer et al. [41] Canada	RCT	To assess the effect of the soluble viscous fibre PGX on glycemic control in adults withT2D.	T2D patients	PGX group: n = 147; Placebo group: n = 143	PGX group: 56.2 ± 8.6 Placebo group: 53.4 ± 9.9	Not Applicable	PGX (15–20 g/day) versus placebo (rice flour, 6.4–8.6 g/day) 52 weeks of treatment	PGX group increased Roseburia and led to a sustained reduction in HbA1c and FBG compared to placebo.
Shin et al. [42] Korea	RCT	To investigate whether the combination of SB and metformin influenced T2D symptoms.	T2D on 500 mg/day metformin	n = 12	SB + Metformin: 63.1 Placebo: 63.1	Not Applicable	SB (3.52 g SB extract) + metformin versus placebo+ metformin. A 4-week washout separated the 8 weeks of treatment	Lactobacillus and Akkermansia, showed significant increases after SB + metformin treatment. The glucose, HbA1c and BMI were not changed after 8 weeks of SB and placebo treatment.
Soare et al. [43] Italy	RCT	To compare the effects of the Ma-Pi 2 diet and the dietary guidelines for T2D recommended by professional societies in Italy on T2D patients.	Overweight or obese (BMI:27–45 kg/m^2), aged 40–75 years affected by T2D	Ma-Pi 2 diet:25 Control diet:26	Ma-Pi 2 diet: 67 ± 8.16 Control diet: 65 ± 7.28	Ma-Pi 2 diet:7 ± 7.793 Control diet:4.5 ± 8.845	Fibre-rich Ma-Pi 2 macrobiotic diet versus recommended diet of type 2 diabetes by professional societies. 3 weeks of treatment.	The patients that received Ma-Pi 2 diet showed significant reduction in FBG, PBG, HbA1c, and BMI compared to those receiving the recommended diet for T2D.
Soare et al. [44] Italy	RCT	To investigate whether the benefits of Ma-Pi 2 extended beyond the 21-day intensive dietary intervention.	Overweight or obese (BMI 27–45 kg/m^2), aged 40–75 years affected by T2D.		Ma-Pi 4 diet: 65 ± 8.89 Control diet:64 ± 8.15	Ma-Pi 4 diet: 7 ± 7.41 Control diet: 4 ± 6.67	Fibre-rich Ma-Pi 4 macrobiotic diet versus recommended diet of T2D diabetes by professional societies. 6 months of treatment.	The Ma-Pi 4 diet had great improvement in glycemic control, compared with the control group. Body weight loss was also observed in Ma-Pi 4 group, but was not significantly different compared to the control group.

Table 2. *Cont.*

Citation/Country of Study	Type of Study	Aim	Participants	Sample Size	Mean Age (Years)	Mean Diabetes Duration (Years)	Interventions	Results/Findings
Su et al. [45] China	RCT	To evaluate the effects of acarbose add-on therapy on gut microbiota and inflammatory cytokines among Chinese patients with T2D.	Patients with T2D that did not receive acarbose for at least 1 month.	Acarbose group: 59 Control group: 36	Acarbose group: 55.7 ± 11.0 Control group: 56.5 ± 10.2	Not Applicable	50 mg acarbose (t.i.d) a day with meals together with oral antidiabetic drugs and/or insulin or insulin analogs versus similar antidiabetic treatment to interventional group but without acarbose. Four weeks of treatment.	Treatment with acarbose can increase the abundance of Bifidobacterium longum in patients with T2D and improve glycemic control.
Tong et al. [46] China	RCT	To evaluate the role of gut microbiota during improvements in hyperglycemia and hyperlipidemia by two drugs: metformin and AMC for diabetic patients with hyperlipidemia.	Patients with T2D and Hyperlipidemia.	Metformin group: 100 AMC group: 100	Metformin group: 58.55 ± 9.17 AMC group: 59.00 ± 9.46	Not Applicable	AMC twice daily versus metformin tablets 0.25 g/time and 3 times/day. 12 weeks of treatment.	The effect of AMC in regulating the microbes in the gut and in improving HOMA-IR and triglyceride levels was more profound compared with metformin.
van Bommel et al. [47] Netherlands	RCT	To examine the effects of 12-week treatment with the SGLT2 inhibitor dapagliflozin and sulphonylurea gliclazide on gut microbiome composition in patients with T2D treated with metformin.	All participants with T2D treated with metformin as monotherapy	Dapagliflozin group: n = 24; Gliclazide group: n = 17.	Dapagliflozin group: 63 ± 7 Gliclazide group: 63 ± 7	Dapagliflozingroup: 9.8 ± 4.1 Gliclazide group: 10.7 ± 7.3	10 mg dapagliflozin and 30 mg gliclazide. Treatment for 12 weeks.	Both dapagliflozin and gliclazide reduced HbA1c and FBG, while BMI was reduced by dapagliflozin, but increased by gliclazide.
Wu et al. [18] Spain	RCT	To investigate the effect of metformin on the composition and function of the microbiota.	Individuals with type 2 diabetes	Metformin group: n = 22 Placebo group: n = 18	Metformin group: 52.6 ± 9.4 Placebo group: 54.9 ± 8.1	Not applicable	A start dose of 425 mg/day of metformin and increased progressively to reach 1700 mg/day or placebo (calorie restricted diet). Treatment was for four months.	Metformin and not calorie restricted diet had significant effect on composition and function of the gut microbiota and reduction in HbA1c and fasting blood glucose levels.

Table 2. *Cont.*

Citation/Country of Study	Type of Study	Aim	Participants	Sample Size	Mean Age (Years)	Mean Diabetes Duration (Years)	Interventions	Results/Findings
Zhao et al. [48] China	RCT	To examine the effect of dietary fibre on SCFA-producing strains in patients with type 2 diabetes.	Individuals with type 2 diabetes	High fibre diet group: $n = 27$ Control group: $n = 16$	High fibre diet group: 58.4 ± 32.2 Control group: 59.7 ± 24.0	High fibre diet group: 8.0 ± 30.1 Control group: 7.9 ± 20	High fibre diet composed of whole grains, traditional Chinese medicinal foods, and prebiotics.	The presence of SCFA producers in greater diversity and abundance by fibre. Participants had better improvement in HbA1c levels.

Abbreviations: BMI: Body Mass Index; T2D—type 2 diabetes; Ma-Pi —macrobiotic diet; FBG—fasting blood glucose; PBG—post-prandial blood glucose; HbA1c—glycated haemoglobin; GOS—galacto-oligosaccharides; AMC—an herbal formula consisting of eight herbs; HOMA-IR—homeostatic model assessment of insulin resistance; DP—dietary portfolio; SB—scutellaria baicalensis; SCFA—short chain fatty acids; PGX—PolyGlycopleX, a highly viscous polysaccharide complex; NBS—native banana starch; HMS—high-amylose maize starch; DMS—digestible maize starch; GC—glycemic control; GV—glycemic variability; BA—bile acid; BW—body weight; SGLT2—sodium-glucose-linked transporter-2.

6. Risk of Bias of Studies Included

Figure 2a,b show the risk of bias graph and risk of bias summary, respectively, of the studies in this review. There was low risk of bias in relation to incomplete outcome data (attrition bias), selective reporting (reporting bias), blinding of participants and personnel, and other bias in all the studies. Nine of the 16 studies demonstrated unclear risk of bias with respect to random sequence generation, while 11 studies demonstrated unclear risk of bias in relation to allocation concealment. In terms of blinding of outcome assessments, there were 3 studies with unclear risk of bias.

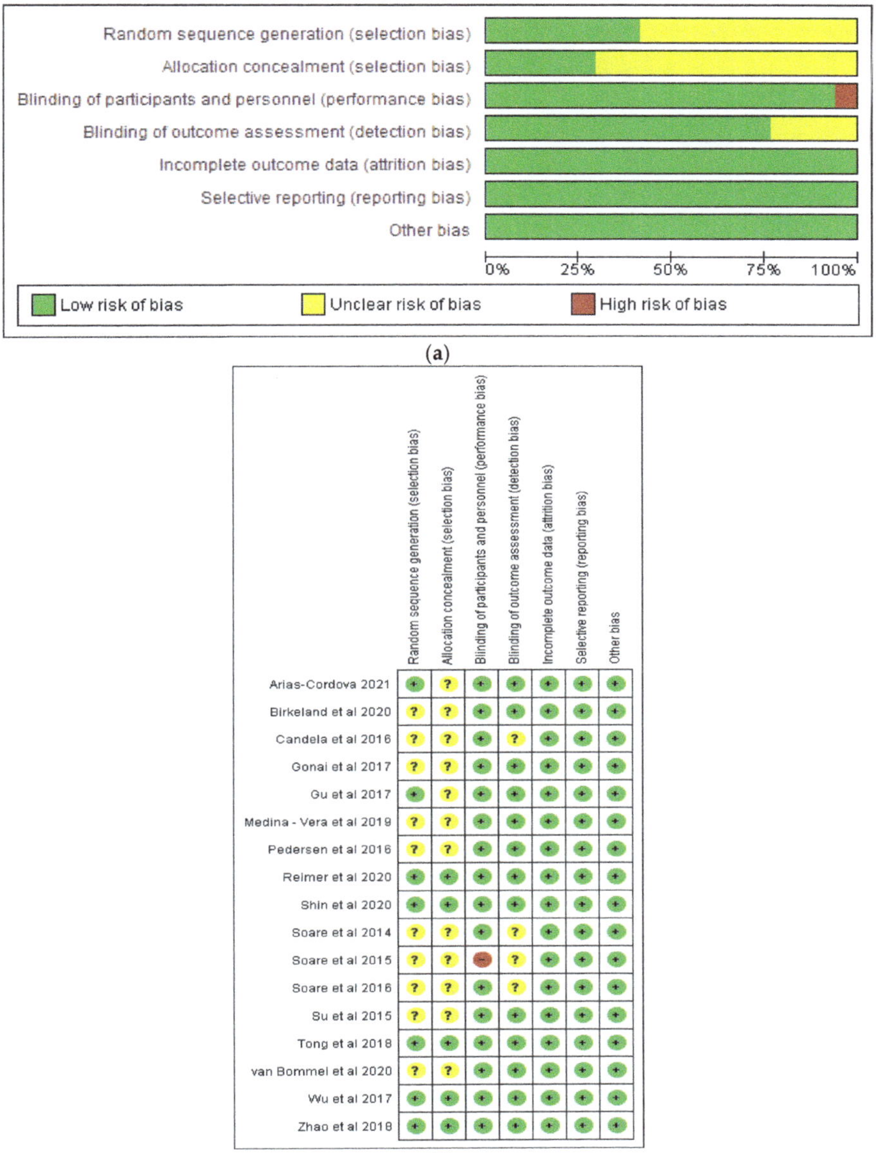

(b)

Figure 2. Graphs showing (**a**) risk of bias (**b**) risk of bias summary [18,31,32,36–48].

7. Effects of Interventions

Three distinct areas were identified based on the results of the systematic review and NWM, namely: Gut microbiome; Glycaemic control; and Body Mass Index (BMI).

Gut Microbiome.

The effects of prebiotics and oral antidiabetic agents on gut microbiome were varied (Table 3). For example, Birkeland et al. [37] found significant increase in faecal levels of *Bifidobacteria* following daily supplement of inulin-type fructans, while Gonai et al. [39] observed significant restoration of *Bifidobacteriaceae* in patients with T2D after the consumption of galacto-oligosaccharide. In addition, Zhao et al. [48] found high fibre diet promoted the growth of short chain fatty acid producing microbes in patients with diabetes. However, the effect of prebiotic treatment on *Bifidobacterium*, *Lactobacillus* and *Roseburia* was not significant in Pedersen et al. [32] study.

Table 3. Effects of prebiotics and Oral antidiabetic agents on gut microbiome.

Studies	Bifidobacterium	Lactobacillus	Roseburia	Bacteroides	Ruminococcus	Clostridium	Akkermansia
Birkeland et al. [37] Norway	There was moderate, but significant increase in faecal levels of bifidobacteria in the group supplemented daily with inulin-type fructans.	N/A	N/A	*Bacteroides ovatus* was enriched by the prebiotic fibre	N/A	N/A	N/A
Candela et al. [38] Italy	N/A	N/A	Ma-Pi 2 diet and control were effective in supporting the recovery of *Roseburia*	Ma-Pi 2 diet and control were effective in supporting the recovery of *Bacteroides*	Ma-Pi 2 diet and control supported the reduction of *Ruminococcus*	N/A	Ma-Pi 2 diet and control resulted in the increase of *Akkermansia*
Gonai et al. [39] Japan	*Bifidobacterium* was significantly restored after consumption of GOS	N/A	N/A	N/A	Ruminococcus was significantly lower after consumption of GOS	N/A	N/A
Gu et al. [40] China	The relative abundances of *Bifidobacterium* species increased in Acarbose group.	Acarbose group increased the relative abundances of *Lactobacillus* species.	N/A	The intervention of Acarbose depleted the relative abundances of Bacteroides species.	N/A	Acarbose group depleted the relative abundances of *Clostridium* species.	N/A
Medina-Vera et al. [31] Mexico	Consumption of dietary portfolio stimulated the abundance of *Bifidobacterium longum.*	N/A	N/A	N/A	N/A	N/A	Dietary portfolio increased *Akkermansia muciniphila*
Pedersen et al. [32]	The effect of prebiotic treatment on *Bifidobacterium* was not significant.	The effect of prebiotic treatment on *Lactobacillus* was not significant.	The effect of prebiotic treatment on *Roseburia* was not significant.	N/A	N/A	The effect of prebiotic treatment on *Clostridium* was not significant.	N/A
Reimer et al. [41] Canada	*Bifidobacterium* Spp. changed significantly over time after PGX.	*Lactobacillus* was greater in the placebo compared with the PolyGlycopleX	The relative abundance of *Roseburia* was significantly increased by the soluble viscous fibre PolyGlycopleX	N/A	N/A	*Clostridium coccoides* changed significantly over time after PGX	*Akkermansia muciniphila* changed significantly over time after PGX

Table 3. *Cont.*

Studies	Bifidobacterium	Lactobacillus	Roseburia	Bacteroides	Ruminococcus	Clostridium	Akkermansia
Shin et al. [42] Korea	The relative abundance of *Bifidobacterium* was significantly lower in the scutellaria baicalensis and metformin group compared to placebo.	*Scutellaria baicalensis* and metformin increased *Lactobacillus* significantly compared to placebo.	N/A	N/A	N/A	N/A	Scutellaria baicalensis and metformin increased *Akkermansia* significantly compared to placebo.
Su et al. [45] China	Acarbose treatment can increase the content of gut *Bifidobacterium longum* in type 2 diabetes mellitus patients.	N/A	N/A	N/A	N/A	N/A	N/A
Tong et al. [46] China	N/A	N/A	*Roseburia* was enhanced by herbal formula	N/A	N/A	N/A	There was decrease in *Akkermansia* in the metformin treated group
van Bommel et al. [47] Netherlands	N/A	N/A	N/A	N/A	N/A	N/A	*Akkermansia muciniphila* was not significantly affected by Dapagliflozin or Gliclazide treatment.
Wu et al. [18] Spain	There was increase in *Bifidobacterium adolescentis* after metformin treatment	N/A	N/A	N/A	N/A	N/A	N/A

Abbreviations: N/A (Not Applicable).

While Gu et al. [40] and Su et al. [45] found acarbose can increase the relative abundances of *Bifidobacterium* species, Wu et al. [18] noted *Bifidobacterium adolescentis* increased after metformin treatment.

8. Bifidobacterium

The Network meta-analysis for *Bifidobacterium* included 5 studies, 239 participants and 3 treatments. The result showed prebiotic treatment increased the relative abundance of *Bifidobacterium* although this was not significantly different compared with placebo with a SMD of 0.43 [95% CI, −0.69, 1.55; $p = 0.45$] (Figure 3a). In contrast, metformin treatment reduced the relative abundance of Bifidobacterium with a SMD of −1.81 [95% CI, −4.16, 0.54; $p = 0.13$] compared to placebo, but again this was not significant. Pairwise meta-analysis conducted to estimate effect sizes based on head-to-head comparisons of treatments and/or control conditions (Figure 3b) found no significant difference ($p > 0.05$) between prebiotic treatment and control on the relative abundance of *Bifidobacterium*. The effect of metformin was significant ($p < 0.05$).

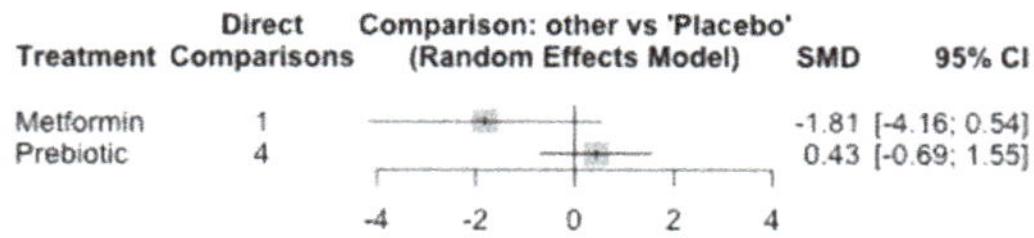

(**a**)

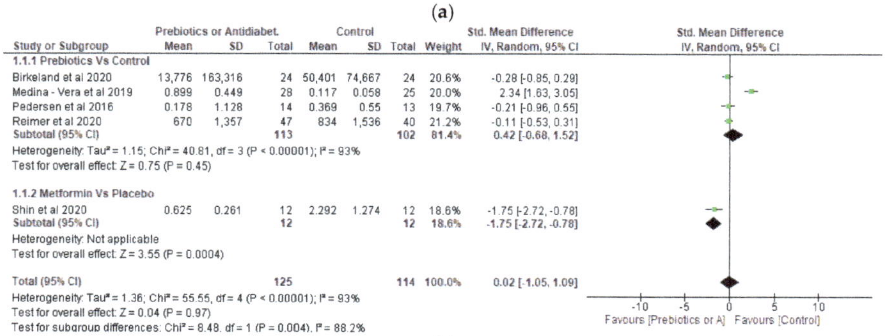

(**b**)

Figure 3. Network Meta-analysis (**a**) and Meta-analysis (**b**) of the effect of treatments versus control on *Bifidobacterium* [31,32,37,41,42].

9. Lactobaccilus

The Network meta-analysis for *Lactobaccilus* included 3 studies, involving 159 participants. The effect of metformin treament on the relative abundance of *Lactobaccilus* showed a significant increase with SMD of 1.43 [95% CI, 0.23, 2.64; p = 0.02] compared to placebo (Figure 4a). However, the effect of prebiotic compared to placebo was not significantly different with a SMD of −0.14 [95% CI, −0.81, 0.53; p = 0.68]. The meta-analysis (Figure 4b) also showed metformin significantly ($p < 0.05$) increased *Lactobaccilus* compared with control while differences between prebiotics and control did not differ significantly ($p > 0.05$).

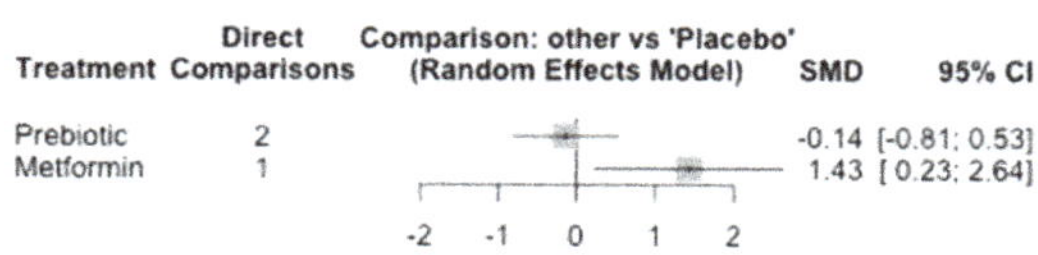

(**a**)

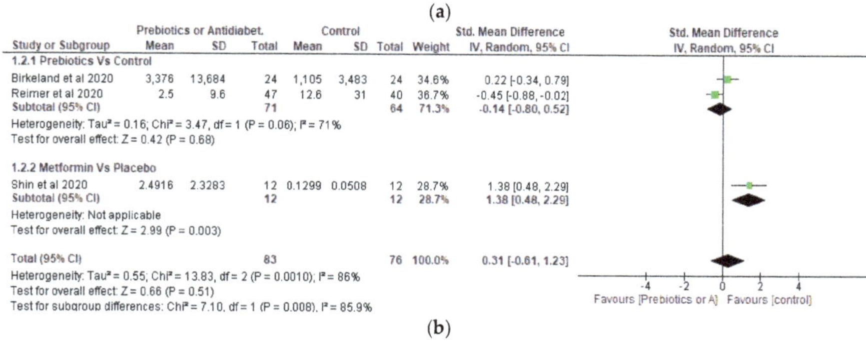

(**b**)

Figure 4. Network Meta-analysis (**a**) and Meta-analysis (**b**) of the effect of treatments versus control on *Lactobaccilus* [37,41,42].

10. Akkermansia

The Network meta-analysis for Akkermansia included 2 studies, 111 participants and 3 treatments. Both metformin and prebiotic treatments increased the relative abundance of *Akkermansia*, although the effects did not differ significantly ($p > 0.05$) compared to placebo (Figure 5a). The SMD was 0.10 [95% CI, −0.32, 0.52; $p = 0.64$] for prebiotic and 0.49 [95% CI, −0.33, 1.30; $p = 0.24$] for metformin treatments, respectively, compared with placebo. The results of the meta-analayis did not show any significant difference ($p > 0.05$) between the prebiotic and control, and metformin and control (Figure 5b).

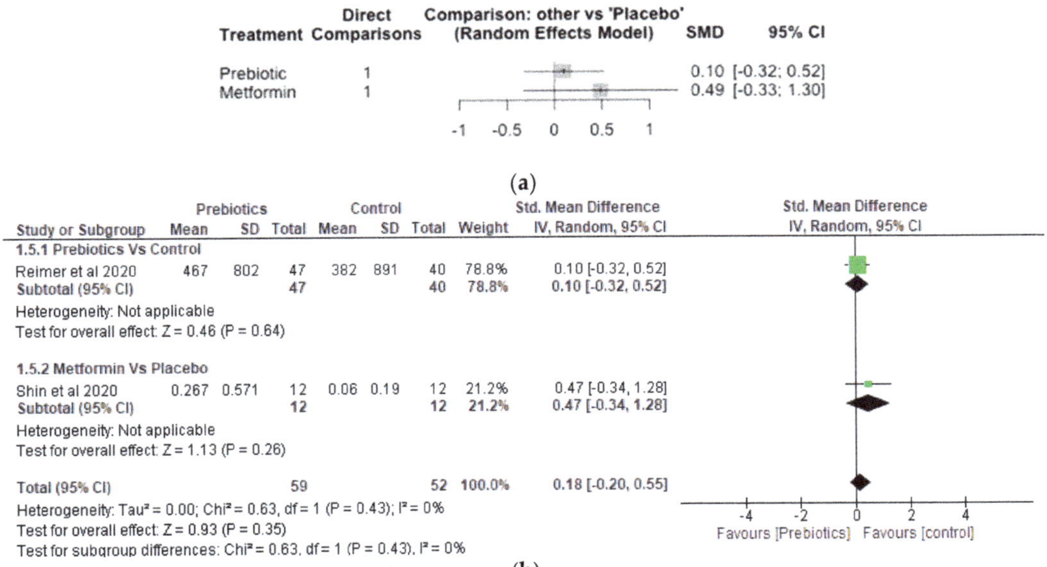

Figure 5. Network Meta-analysis (**a**) and Meta-analysis (**b**) of the effect of treatments versus control on *Akkermansia* [41,42].

11. Glycaemic Control

In the study by Arias-Córdova et al. [36], it was found that the native banana starch (NBS) with a content of 70.5% resistant starch and 10% digestible starch caused a reduction in fasting blood glucose from baseline compared with digestible maize starch with 100% digestible starch content. There was improvement in insulin sensitivity and significant improvement in glycaemic control including significant reduction in parameters such as HbA1c, postprandial blood glucose and fasting blood glucose levels in patients with type 2 diabetes who consumed prebiotic diets compared with control in some studies [31,38,41,43,44,48].

However, following the consumption of prebiotic diets, there was no improvement in glucose control in other studies [32,39].

With respect to the oral antidiabetic agents, Wu et al. [18] found metformin significantly reduced HbA1c and fasting blood glucose levels compared with calorie restricted diet. Furthermore, Tong et al. [46] reported metformin improved Homeostatic Model Assessment for Insulin Resistance (HOMA-IR) compared with control, while Su et al. [45] observed acarbose treatment improved glycemic control in patients with type 2 diabetes. Both dapagliflozin and gliclazide reduced HbA1c and fasting blood glucose levels in the study by van Bommel et al. [47]. Similarly, the acarbose and glipizide groups improved glycemic control, with no significant differences between the two groups [40].

However, Shin et al. [42] reported that *Scutellaria baicalensis* with metformin treatment or placebo did not change the glucose and HbA1c levels.

Glycated Haemoglobin (HbA1c).

The Network meta-analysis for HbA1c included 12 studies, 7 treatments and 1012 participants (number of observations). Compared with control, glipizide, herbal formula and metformin treatments reduced HbA1c although the difference was not significant ($p > 0.05$). In contrast, prebiotic treatment significantly reduced HbA1c compared to control with a SMD of −0.43 [95% CI, −0.77, −0.08; $p = 0.02$] (Figure 6a). The results of the meta-analysis demonstrated prebiotics significantly ($p < 0.05$) reduced HbA1c compared to control, whereas the differences between the other treatments and control were not significant ($p > 0.05$) (Figure 6b).

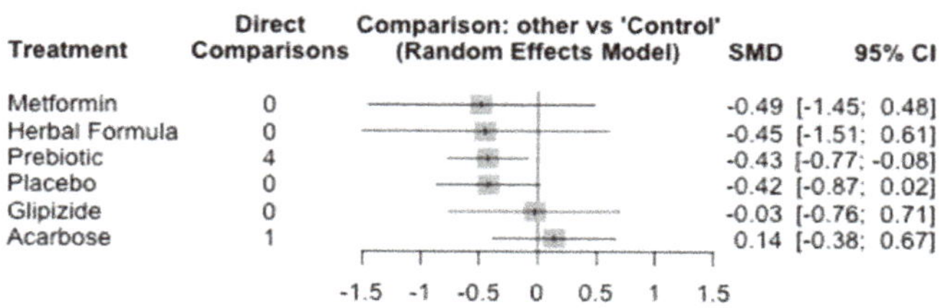

(**a**)

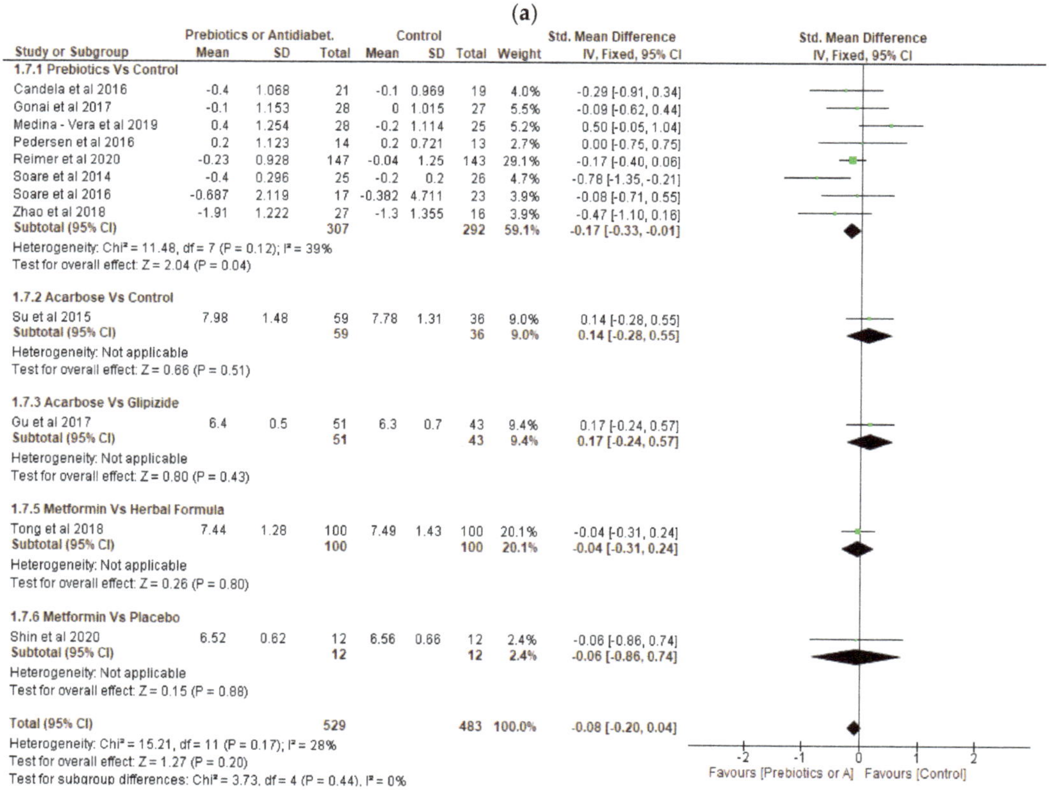

(**b**)

Figure 6. Network Meta-analysis (**a**) and Meta-analysis (**b**) of the effect of treatments versus control on glycated haemoglobin (HbA1c) [31,32,38–46,48].

Fasting Blood Glucose.

There were 9 studies, 731 participants or number of observations and 5 treatments involved in the Network meta-analysis of fasting blood glucose (Figure 7a). While prebiotic treatment reduced fasting blood glucose level with SMD of −0.10 [95% CI, −0.41, 0.21; $p = 0.52$], acarbose and glipizide increased fasting blood glucose with SMD of 0.15 [95% CI, −0.26, 0.57; $p = 0.48$] and 0.25 [95% CI, −0.33, 0.83; $p = 0.41$], respectively. However, differences between the various treatments (acarbose, glipizide and prebiotic) and the control were not significant ($p > 0.05$). The meta-analysis revealed that the various treatments did not differ significantly ($p > 0.05$) from control (Figure 7b).

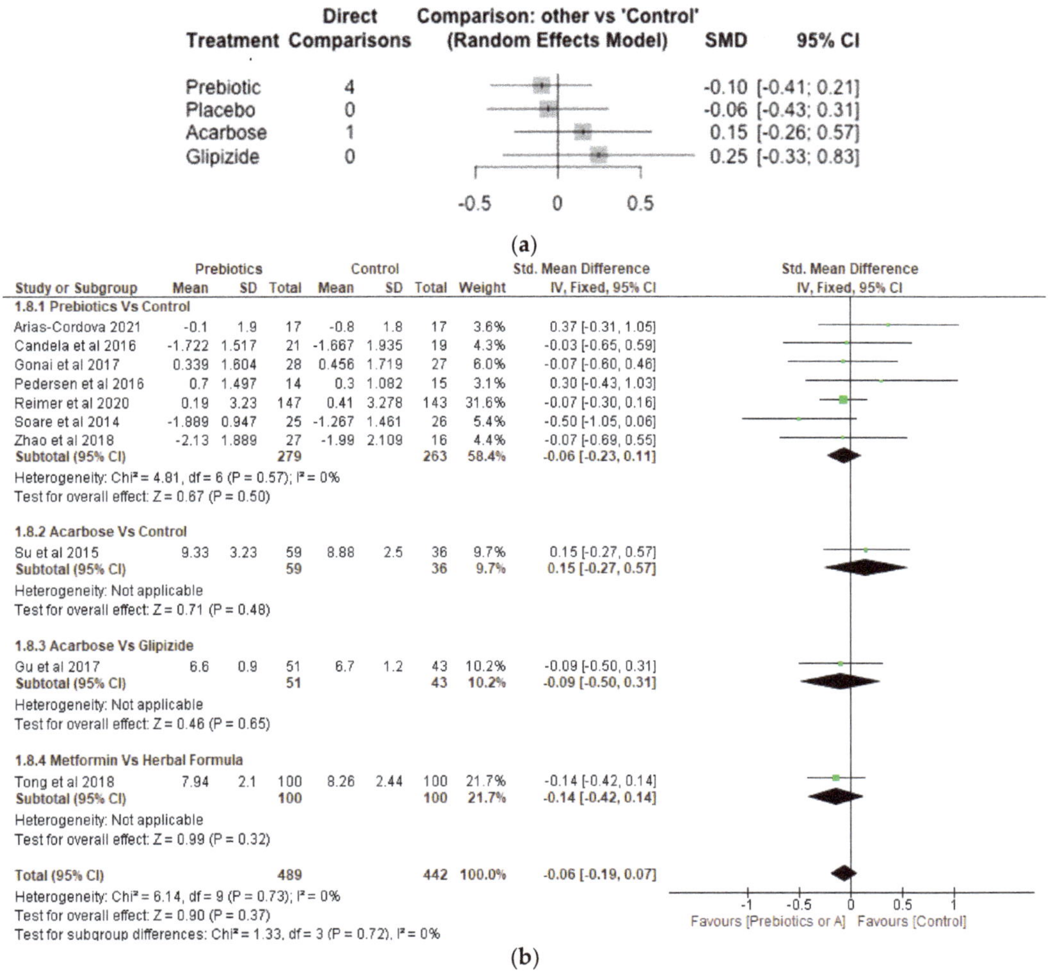

Figure 7. Network Meta-analysis (**a**) and Meta-analysis (**b**) of the effect of treatments versus control on Fasting Blood Glucose [32,36,38–41,43,45,46,48].

Postprandial Blood Glucose.

Two studies, 189 number of observations and 3 treatments were included in the Network meta-analysis of postprandial blood glucose (Figure 8a). While the difference between acarbose and control were not significant ($p > 0.05$), glipizide increased postprandial blood glucose significantly with SMD of 1.03 [95% CI, 0.44, 1.62; $p = 0.001$]. The result of the meta-analysis revealed that acarbose significantly ($p < 0.05$) reduced postprandial blood glucose compared to gliplizide, while the effect of acarbose compared to control, and metformin compared to herbal formula were not significantly different ($p > 0.05$) (Figure 8b).

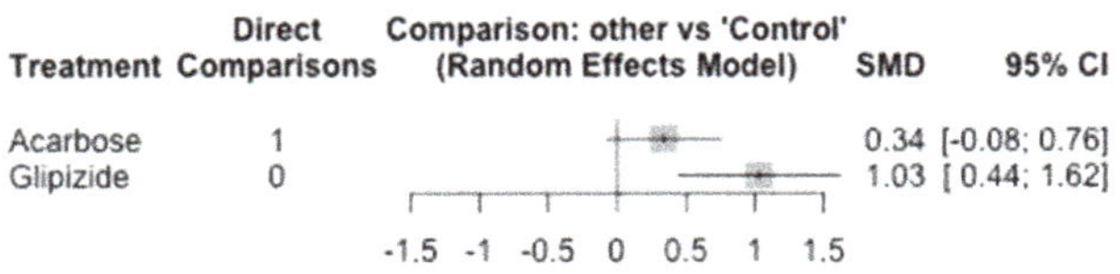

(**a**)

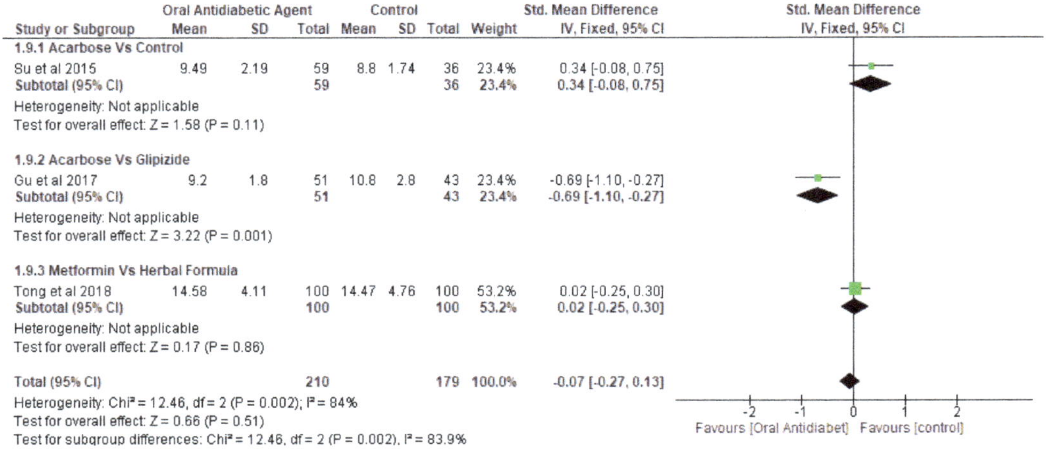

(**b**)

Figure 8. Network Meta-analysis (**a**) and Meta-analysis (**b**) of the effect of treatments versus control on Postprandial Blood Glucose [40,45,46].

12. Body Mass Index (BMI)

Soare et al. [43] reported prebiotic significantly reduced BMI compared with control in patients with type 2 diabetes. Similarly, after 3 months of treatment, reductions in body weight and body mass index were more pronounced in the acarbose group than in the Glipizide group [40]. Although BMI was reduced by dapagliflozin, it was increased by gliclazide [47]. Furthermore, *Scutellaria baicalensis* with metformin or placebo did not change the BMI after 8 weeks of treatment [42].

With respect to the Network meta-analysis, 7 studies, 496 participants and 5 treatments were included (Figure 9a). Although there were increases in BMI in the different treatments (herbal formula, metformin and prebiotic) compared with control, the differences were not significant ($p > 0.05$). The SMD was 0.04 [95% CI, −0.41, 0.49; $p = 0.86$] for prebiotic, 0.26 [95% CI, −0.75, 1.28; $p = 0.61$] for metformin and 0.47 [−0.61; 1.56; $p = 0.39$] for herbal formula, respectively, compared with control. The result of the meta-analysis showed the effects of prebiotics, acarbose and metformin were not significantly different ($p > 0.05$) from control with respect to BMI (Figure 9b).

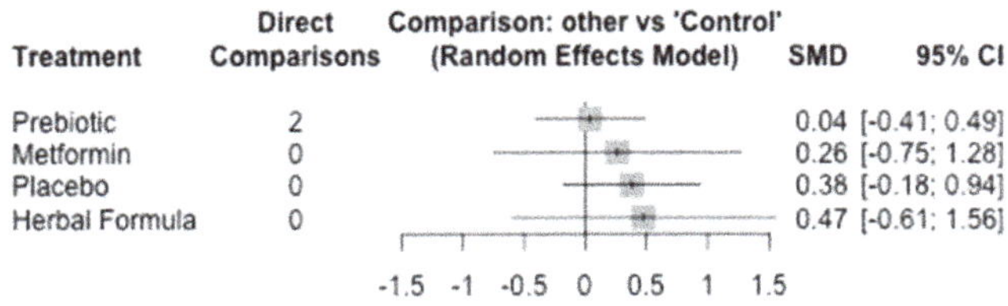

(**a**)

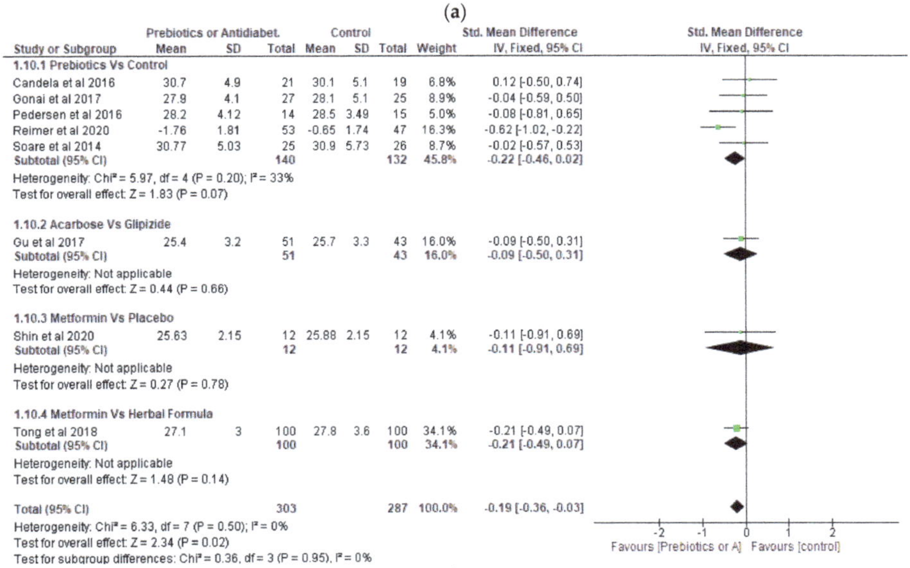

(**b**)

Figure 9. Network Meta-analysis (**a**) and Meta-analysis (**b**) of the effect of treatments versus control on Body Mass Index [32,38–43,46].

13. Discussion

The results of the Network meta-analysis demonstrated that prebiotics significantly reduced ($p < 0.05$) HbA1c in patients with T2D compared to control. In addition, antidiabetic agents including glipizide and metformin also reduced HbA1c, although these did not differ significantly ($p > 0.05$) compared to control.

While prebiotics increased the relative abundance of *Bifidobacterium* and *Akkermansia*, it did not differ significantly ($p > 0.05$) compared to control. On the other hand, metformin decreased the relative abundance of *Bifidobacterium*, but increased *Lactobacillus* and *Akkermansia*, although these did not differ significantly ($p > 0.05$) compared with control.

With respect to fasting blood glucose and BMI, the effects of prebiotics and oral antidiabetic agents did not differ significantly ($p > 0.05$) from controls.

The findings of this Network meta-analysis confirm the earlier results of the systematic review and meta-analysis carried out by Zhang et al. [21], Mahboobi et al. [21] and Wang et al. [22] which demonstrated prebiotics were effective in reducing glycated haemoglobin in patients with T2D. However, these earlier reviews did not include gut microbiota as one of the outcomes measured. Fallucca et al. [7] found microbiotic Ma-Pi 2 diet which is rich in carbohydrates, whole grains and vegetables significantly improved glycated haemoglobin in patients with T2D. It was reported that the diet could modulate the composition of gut microbiome [7].

According to Mahboobi et al. [21] the underlying mechanisms of action of prebiotics are based on the fact soluble fibres can delay gastric emptying, slow down glucose entry into the blood stream and reduce the rise of postprandial blood glucose. Furthermore, soluble fibres may alter the production of glucagon like peptide-1 (GLP-1) which is a gut hormone involved in the metabolism of glucose [21]. Soluble fibres may also lead to the production of SCFAs which may influence serum glucose and insulin levels [21]. With respect to glucose lowering agents, the mechanism of action on gut microbiome may relate to their role in lowering inflammatory cytokines and promoting production of SCFAs [14].

Patients with T2D have been shown to exhibit intestinal dysbiosis [31]. Decreases in *Bifidobacterium, Roseburia, Faecalibacterium* and *Akkermansia* have been associated with T2D [19,49,50]. Ghorbani et al. [19] reported that *Bifidobacterium* is inversely associated with T2D and that the role of *Lactobacillus* appears to be species dependent. For example, *Lactobacillus acidophilus* and *Lactobacillus salivarius* species positively correlated with T2D, while *Lactobacillus amylovorus* species are negatively associated with T2D [19]. *Akkermansia muciniphila* is reported to have a role in the homeostasis of glucose and in protecting against insulin resistance and T2D [19].

Diets high in fat such as Western diets may cause gut microbiota dysbiosis which can lead to increased levels of lipopolysaccharide, oxidative stress, pro-inflammatory cytokines, gut inflammation, gut permeability and insulin resistance [2,49].

Therefore, dietary intervention with prebiotics can substantially modulate gut and faecal microbiota through increases in alpha diversity and regulating the relative abundance of specific bacteria species, independent of antidiabetic drugs [31,37,38].

According to Ghorbani et al. [19], prebiotics are non-digestible fibres which can be fermented by the gut microbiome and can promote the growth of some bacteria. Prebiotic carbohydrates are composed mainly of inulin, fructo-oligosaccharide and galacto-oligosaccharides which are resistant to digestion in the small intestine [51]. However, they are fermented in the large intestine and have been reported to promote the abundance of *Bifidobacterium* and/or *Lactobacillus* [51]. Prebiotics promote eubiosis and attenuates pathological changes of dysbiosis, leading to promotion in the abundance of *Lactobacillus, Bifidobacterium, Faecalibacterium* and *Bacteroidetes* [49]. Other changes due to the effects of prebiotics include decreases in lipopolysaccharides, oxidative stress, proinflammatory cytokines and gut permeability, and improvements in gut motility and insulin sensitivity [49]. Prebiotics also promote GLP–1 and peptide YY [2]. Supplementation with prebiotics has been shown to improve appetite control of human subjects [2].

The SCFAs including propionate, butyrate, and acetate which are produced from the fermentation of complex carbohydrates including prebiotics are responsible for initiating the various metabolic pathways which regulate glycaemic control and inflammation [19,49]. Acetate has been reported to regulate appetite both directly and indirectly and can stimulate the production of GLP-1 and peptide YY which are appetite suppressing hormones from the L-cells of the intestine [19]. GLP-1 is an insulinotropic hormone which can regulate glucose homeostasis [19]. Propionate can also stimulate the production of GLP-1 and peptide YY, while propionate and butyrate can inhibit pro-inflammatory cytokines [19]. Butyrate, is useful in modulating intestinal barrier permeability and in ensuring pro-inflammatory products do not gain access from the lumen of the gut to the internal milieu [51]. This is important as it has been reported that the translocation of lipopolysaccharide promotes pro-inflammatory cytokines, low grade systemic inflammation, impairs glucose metabolism and increases insulin resistance and T2D [19].

Therefore, in order to promote an increase in the abundance of beneficial bacteria and ensure effective glycaemic control, it is essential that the type and amount of prebiotics consumed and the duration are considered [32]. For example, long term adherence to high fibre plant based diet and daily supplement with inulin type fructans have been reported to be effective in modulating gut microbiota and regulating glycaemic control [31,37]. Furthermore, combining different functional foods may modify human microbial community and improve glycaemic control [31].

Metformin has been reported to promote the growth of SCFA producing microbial species including *Bifidobacterium bifidum* and *Bifidobacterium adolescentis* and increased abundance of *Akkermansia muciniphila* and down regulating *Clostridia* [52]. The primary hypoglycemic effect of metformin is its role in inhibiting hepatic gluconeogenesis [52]. Gu et al. [40] reported acarbose impedes the breakdown and absorption of carbohydrates in the small intestine, and these provide the substrate for microbial fermentation in the large intestine and therefore promotes the abundance of saccharolytic bacteria such as *Lactobacillus* and *Bifidobacterium* species.

14. Limitations

The few studies available and the small sample sizes of some of the studies limit the power of this Network meta-analysis to detect statistical differences. While the current findings provide a foundation for assessing the relative effects of the different treatments, our results should be considered exploratory and that further studies are needed to fully examine the effects of prebiotics and oral anti-diabetic agents on gut microbiome and glycaemic control in patients with T2D.

15. Conclusions

The results of this systematic review and Network meta-analysis showed prebiotics were significantly ($p < 0.05$) more effective in reducing HbA1c than control in patients with T2D. However, the effects of prebiotics and oral antidiabetic agents did not differ significantly ($p > 0.05$) from the controls with respect to fasting blood glucose, post-prandial blood glucose, body mass index and the genera of gut bacteria examined.

More studies are required to fully investigate the effects of prebiotics and oral antidiabetic agents in patients with type 2 diabetes.

Supplementary Materials: The following supporting information can be downloaded at: https://www.mdpi.com/article/10.3390/nu14235139/s1, Figure S1: Network Plots; Table S1: Age & Sex Distribution of Treatments.

Author Contributions: Conceptualization, O.O.; methodology, O.O., O.O.O., X.W., Y.J., Q.D. and T.T.; validation, O.O., O.O.O., J.B. and X.W.; formal analysis, O.O. and T.T.; writing—original draft preparation, O.O.; writing—review and editing, O.O., O.O.O., X.W., Y.J., Q.D., J.B. and T.T. All authors have read and agreed to the published version of the manuscript.

Funding: No external funding was received for this research.

Institutional Review Board Statement: Not Applicable.

Informed Consent Statement: Not Applicable.

Data Availability Statement: Secondary data analysis of publicly available data was carried out.

Conflicts of Interest: There is no conflict of interest.

Abbreviations

AMC—herbal formula consisting of eight herbs; BA—bile acid; BMI—body Mass Index; BW—body weight; CI—confidence interval; DMS—digestible maize starch; DP—dietary portfolio; FBG—fasting blood glucose; GC—glycemic control; GLP-1—glucagon like peptide -1; GOS—galacto-oligosaccharides; GV—glycemic variability; HbA1c—glycated haemoglobin; HMS—high-amylose maize starch; HOMA-IR—homeostatic model assessment of insulin resistance; Ma-Pi —macrobiotic diet; N/A: Not Applicable; NBS—native banana starch; NMA—network meta-analysis; PBG—post-prandial blood glucose; PGX—PolyGlycopleX, a highly viscous polysaccharide complex; PICOS—Population, Intervention, Comparator, Outcomes, Studies; PRISMA—Preferred Reporting Items for Systematic Reviews and Meta-Analyses; RCT—randomised controlled trial; SB—scutellaria baicalensis; SCFA—short chain fatty acids; SGLT2—sodium-glucose-linked transporter-2; SMD—standardised mean difference; T2D—type 2 diabetes.

References

1. Roshanravan, N.; Alamdari, N.M.; Jafarabadi, M.A.; Mohammadi, A.; Shabestari, B.R.; Nasirzadeh, N.; Asghari, S.; Mansoori, B.; Akbarzadeh, M.; Ghavami, A.; et al. Effects of oral butyrate and inulin supplementation on inflammation-induced pyroptosis pathway in type 2 diabetes: A randomized, double-blind, placebo-controlled trial. *Cytokine* **2020**, *131*, 155101. [CrossRef] [PubMed]
2. Fallucca, F.; Porrata, C.; Fallucca, S.; Pianesi, M. Influence of diet on gut microbiota, inflammation and type 2 diabetes mellitus. First experience with macrobiotic Ma-Pi 2 diet. *Diabetes/Metab. Res. Rev.* **2014**, *30*, 48–54. [CrossRef]
3. Aliasgharzadeh, A.; Aliloo, A.; Ghotaslou, R.; Arbabi, S. Comparison of *bifidobacterium* spp. and *lactobacillus* spp. count in faeces of patients with type 2 diabetes mellitus and healthy people. *Middle East J. Fam. Med.* **2018**, *16*, 102–106.
4. Makki, K.; Deehan, E.C.; Walter, J.; Bäckhed, F. The impact of dietary fiber on gut microbiota in host health and disease. *Cell Host Microbe* **2018**, *23*, 705–715. [CrossRef] [PubMed]
5. Ojo, O.; Ojo, O.O.; Zand, N.; Wang, X. The Effect of Dietary Fibre on Gut Microbiota, Lipid Profile, and Inflammatory Markers in Patients with Type 2 Diabetes: A Systematic Review and Meta-Analysis of Randomised Controlled Trials. *Nutrients* **2021**, *13*, 1805. [CrossRef]
6. Fu, J.; Xu, K.; Ni, X.; Li, X.; Zhu, X.; Xu, W. Habitual Dietary Fiber Intake, Fecal Microbiota, and Hemoglobin A1c Level in Chinese Patients with Type 2 Diabetes. *Nutrients* **2022**, *14*, 1003. [CrossRef]
7. Fallucca, F.; Fontana, L.; Fallucca, S.; Pianesi, M. Gut microbiota and Ma-Pi 2 macrobiotic diet in the treatment of type 2 diabetes. *World J. Diabetes* **2015**, *6*, 403–411. [CrossRef] [PubMed]
8. He, D.; Han, H.; Fu, X.; Liu, A.; Zhan, Y.; Qiu, H.; Ma, L.; Zhang, X.; Wang, X. Metformin Reduces Blood Glucose in Treatment-Naive Type 2 Diabetes by Altering the Gut Microbiome. *Can. J. Diabetes* **2022**, *46*, 150–156. [CrossRef] [PubMed]
9. Lee, S.-E.; Choi, Y.; Jun, J.E.; Lee, Y.-B.; Jin, S.-M.; Hur, K.Y.; Ko, G.P.; Lee, M.-K. Additional effect of dietary fiber in patients with type 2 diabetes mellitus using metformin and sulfonylurea: An open-label, pilot trial. *Diabetes Metab. J.* **2019**, *43*, 422–431. [CrossRef] [PubMed]
10. Rezende, E.S.V.; Lima, G.C.; Naves, M.M.V. Dietary fibers as beneficial microbiota modulators: A proposed classification by prebiotic categories. *Nutrition* **2021**, *89*, 111217. [CrossRef] [PubMed]
11. Holscher, H.D. Dietary fiber and prebiotics and the gastrointestinal microbiota. *Gut Microbes* **2017**, *8*, 172–184. [CrossRef]
12. Gibson, G.R.; Hutkins, R.; Sanders, M.E.; Prescott, S.L.; Reimer, R.A.; Salminen, S.J.; Scott, K.; Stanton, C.; Swanson, K.S.; Cani, P.D.; et al. Expert consensus document: The International Scientific Association for Probiotics and Prebiotics (ISAPP) consensus statement on the definition and scope of prebiotics. *Nat. Rev. Gastroenterol. Hepatol.* **2017**, *14*, 491–502. [CrossRef] [PubMed]
13. Elbere, I.; Kalnina, I.; Silamiķelis, I.; Konrade, I.; Zaharenko, L.; Sekace, K.; Radovica-Spalvina, I.; Fridmanis, D.; Gudra, D.; Pirags, V.; et al. Association of metformin administration with gut microbiome dysbiosis in healthy volunteers. *PLoS ONE* **2018**, *13*, e0204317. [CrossRef]
14. Lv, Y.; Zhao, X.; Guo, W.; Gao, Y.; Yang, S.; Li, Z.; Wang, G. The Relationship between Frequently Used Glucose-Lowering Agents and Gut Microbiota in Type 2 Diabetes Mellitus. *J. Diabetes Res.* **2018**, *2018*, 1890978. [CrossRef] [PubMed]
15. Almugadam, B.S.; Liu, Y.; Chen, S.M.; Wang, C.H.; Shao, C.Y.; Ren, B.W.; Tang, L. Alterations of Gut Microbiota in Type 2 Diabetes Individuals and the Confounding Effect of Antidiabetic Agents. *J. Diabetes Res.* **2020**, *2020*, 7253978. [CrossRef] [PubMed]
16. Lordan, C.; Thapa, D.; Ross, R.P.; Cotter, P.D. Potential for enriching next-generation health-promoting gut bacteria through prebiotics and other dietary components. *Gut Microbes* **2020**, *11*, 1–20. [CrossRef]
17. Liu, W.; Luo, Z.; Zhou, J.; Sun, B. Gut Microbiota and Antidiabetic Drugs: Perspectives of Personalized Treatment in Type 2 Diabetes Mellitus. *Front. Cell. Infect. Microbiol.* **2022**, *12*, 853771. [CrossRef]
18. Wu, H.; Esteve, E.; Tremaroli, V.; Khan, M.T.; Caesar, R.; Mannerås-Holm, L.; Ståhlman, M.; Olsson, L.M.; Serino, M.; Planas-Fèlix, M.; et al. Metformin alters the gut microbiome of individuals with treatment-naive type 2 diabetes, contributing to the therapeutic effects of the drug. *Nat. Med.* **2017**, *23*, 850–858. [CrossRef]
19. Ghorbani, Y.; Schwenger, K.J.P.; Allard, J.P. Manipulation of intestinal microbiome as potential treatment for insulin resistance and type 2 diabetes. *Eur. J. Nutr.* **2021**, *60*, 2361–2379. [CrossRef]
20. Zhang, W.; Tang, Y.; Huang, J.; Yang, Y.; Yang, Q.; Hu, H. Efficacy of inulin supplementation in improving insulin control, HbA1c and HOMA-IR in patients with type 2 diabetes: A systematic review and meta-analysis of randomized controlled trials. *J. Clin. Biochem. Nutr.* **2020**, *66*, 176–183. [CrossRef] [PubMed]
21. Mahboobi, S.; Rahimi, F.; Jafarnejad, S. Effects of Prebiotic and Synbiotic Supplementation on Glycaemia and Lipid Profile in Type 2 Diabetes: A Meta-Analysis of Randomized Controlled Trials. *Adv. Pharm. Bull.* **2018**, *8*, 565–574. [CrossRef]
22. Wang, L.; Yang, H.; Huang, H.; Zhang, C.; Zuo, H.-X.; Xu, P.; Niu, Y.-M.; Wu, S.-S. Inulin-type fructans supplementation improves glycemic control for the prediabetes and type 2 diabetes populations: Results from a GRADE-assessed systematic review and dose–response meta-analysis of 33 randomized controlled trials. *J. Transl. Med.* **2019**, *17*, 410. [CrossRef]
23. Akbari, V.; Hendijani, F. Effects of probiotic supplementation in patients with type 2 diabetes: Systematic review and meta-analysis. *Nutr. Rev.* **2016**, *74*, 774–784. [CrossRef]
24. Merkevicius, K.; Kundelis, R.; Maleckas, A.; Velickiene, D. Microbiome Changes after Type 2 Diabetes Treatment: A Systematic Review. *Medicina* **2021**, *57*, 1084. [CrossRef] [PubMed]

25. Bock, P.M.; Telo, G.H.; Ramalho, R.; Sbaraini, M.; Leivas, G.; Martins, A.F.; Schaan, B.D. The effect of probiotics, prebiotics or synbiotics on metabolic outcomes in individuals with diabetes: A systematic review and meta-analysis. *Diabetologia* **2021**, *64*, 26–41. [CrossRef]
26. Ojo, O.; Feng, Q.-Q.; Ojo, O.O.; Wang, X.-H. The Role of Dietary Fibre in Modulating Gut Microbiota Dysbiosis in Patients with Type 2 Diabetes: A Systematic Review and Meta-Analysis of Randomised Controlled Trials. *Nutrients* **2020**, *12*, 3239. [CrossRef] [PubMed]
27. Hutton, B.; Salanti, G.; Caldwell, D.M.; Chaimani, A.; Schmid, C.H.; Cameron, C.; Ioannidis, J.P.A.; Straus, S.; Thorlund, K.; Jansen, J.P.; et al. The PRISMA Extension Statement for Reporting of Systematic Reviews Incorporating Network Meta-analyses of Health Care Interventions: Checklist and Explanations. *Ann. Intern. Med.* **2015**, *162*, 777–784. [CrossRef]
28. Page, M.J.; McKenzie, J.E.; Bossuyt, P.M.; Boutron, I.; Hoffmann, T.C.; Mulrow, C.D.; Shamseer, L.; Tetzlaff, J.M.; Akl, E.A.; Brennan, S.E.; et al. The PRISMA 2020 statement: An updated guideline for reporting systematic reviews. *BMJ* **2021**, *372*, 71. [CrossRef] [PubMed]
29. Methley, A.M.; Campbell, S.; Chew-Graham, C.; McNally, R.; Cheraghi-Sohi, S. PICO, PICOS and SPIDER: A comparison study of specificity and sensitivity in three search tools for qualitative systematic reviews. *BMC Health Serv. Res.* **2014**, *14*, 579. [CrossRef] [PubMed]
30. Higgins, J.P.T.; Green, S. *Cochrane Handbook for Systematic Reviews of Interventions*; Wiley-Blackwell: Hoboken, NJ, USA, 2009.
31. Medina-Vera, I.; Sanchez-Tapia, M.; Noriega-López, L.; Granados-Portillo, O.; Guevara-Cruz, M.; Flores-López, A.; Avila-Nava, A.; Fernández, M.L.; Tovar, A.R.; Torres, N. A dietary intervention with functional foods reduces metabolic endotoxaemia and attenuates biochemical abnormalities by modifying faecal microbiota in people with type 2 diabetes. *Diabetes Metab.* **2019**, *45*, 122–131. [CrossRef]
32. Pedersen, C.; Wu, H.; Jaiyeola, E.; Diribe, O.; La Ragione, R.; Robertson, M.D.; Wright, J.; Gallagher, E.; Horton, F.; Hinton, P.; et al. Host–microbiome interactions in human type 2 diabetes following prebiotic fibre (galacto-oligosaccharide) intake. *Br. J. Nutr.* **2016**, *116*, 1869–1877. [CrossRef]
33. Mitchell, M.; Muftakhidinov, B.; Winchen, T. Engauge Digitizer Software. 2020. Available online: http://markummitchell.github.io/engauge-digitizer (accessed on 21 August 2022).
34. The Nordic Cochrane Centre. *Review Manager (RevMan) [Computer Program]*; Version 5.3; The Nordic Cochrane Centre, The Cochrane Collaboration: Copenhagen, Denmark, 2014.
35. Rücker, G.; Krahn, U.; König, J.; Efthimiou, O.; Davies, A.; Papakonstantinou, T.; Schwarzer, G. Netmeta: Network Meta-Analysis Using Frequentist Methods. R Package Version 2.1-0. 2022. Available online: https://cran.r-project.org/web/packages/netmeta/netmeta.pdf (accessed on 2 November 2022).
36. Arias-Córdova, Y.; Ble-Castillo, J.L.; García-Vázquez, C.; Olvera-Hernández, V.; Ramos-García, M.; Navarrete-Cortes, A.; Jiménez-Domínguez, G.; Juárez-Rojop, I.E.; Tovilla-Zárate, C.A.; Martínez-López, M.C.; et al. Resistant Starch Consumption Effects on Glycemic Control and Glycemic Variability in Patients with Type 2 Diabetes: A Randomized Crossover Study. *Nutrients* **2021**, *13*, 4052. [CrossRef]
37. Birkeland, E.; Gharagozlian, S.; Birkeland, K.I.; Valeur, J.; Mage, I.; Rud, I.; Aas, A.M. Prebiotic effect of inulin-type fructans on faecal microbiota and short-chain fatty acids in type 2 diabetes: A randomised controlled trial. *Eur. J. Nutr.* **2020**, *59*, 3325–3338. [CrossRef] [PubMed]
38. Candela, M.; Biagi, E.; Soverini, M.; Consolandi, C.; Quercia, S.; Severgnini, M.; Peano, C.; Turroni, S.; Rampelli, S.; Pozzilli, P.; et al. Modulation of gut microbiota dysbioses in type 2 diabetic patients by macrobiotic Ma-Pi 2 diet. *Br. J. Nutr.* **2016**, *116*, 80–93. [CrossRef]
39. Gonai, M.; Shigehisa, A.; Kigawa, I.; Kurasaki, K.; Chonan, O.; Matsuki, T.; Yoshida, Y.; Aida, M.; Hamano, K.; Terauchi, Y. Galacto-oligosaccharides ameliorate dysbiotic Bifidobacteriaceae decline in Japanese patients with type 2 diabetes. *Benef. Microbes* **2017**, *8*, 705–716. [CrossRef] [PubMed]
40. Gu, Y.; Wang, X.; Li, J.; Zhang, Y.; Zhong, H.; Liu, R.; Zhang, D.; Feng, Q.; Xie, X.; Hong, J.; et al. Analyses of gut microbiota and plasma bile acids enable stratification of patients for antidiabetic treatment. *Nat. Commun.* **2017**, *8*, 1785. [CrossRef] [PubMed]
41. Reimer, R.A.; Wharton, S.; Green, T.J.; Manjoo, P.; Ramay, H.R.; Lyon, M.R.; Gahler, R.J.; Wood, S. Effect of a functional fibre supplement on glycemic control when added to a year-long medically supervised weight management program in adults with type 2 diabetes. *Eur. J. Nutr.* **2021**, *60*, 1237–1251. [CrossRef]
42. Shin, N.R.; Gu, N.; Choi, H.S.; Kim, H. Combined effects of Scutellaria baicalensis with metformin on glucose tolerance of patients with type 2 diabetes via gut microbiota modulation. *Am. J. Physiol. Metab. Endocrinol. Metab.* **2020**, *318*, E52–E61. [CrossRef]
43. Soare, A.; Khazrai, Y.M.; Del Toro, R.; Roncella, E.; Fontana, L.; Fallucca, S.; Angeletti, S.; Formisano, V.; Capata, F.; Ruiz, V.; et al. The effect of the macrobiotic Ma-Pi 2 diet vs. the recommended diet in the management of type 2 diabetes: The randomized controlled MADIAB trial. *Nutr. Metab.* **2014**, *11*, 39. [CrossRef] [PubMed]
44. Soare, A.; Del Toro, R.; Khazrai, Y.M.; Di Mauro, A.; Fallucca, S.; Angeletti, S.; Skrami, E.; Gesuita, R.; Tuccinardi, D.; Manfrini, S.; et al. A 6-month follow-up study of the randomized controlled Ma-Pi macrobiotic dietary intervention (MADIAB trial) in type 2 diabetes. *Nutr. Diabetes* **2016**, *6*, e222. [CrossRef]
45. Su, B.; Liu, H.; Li, J.; Sunli, Y.; Liu, B.; Liu, D.; Zhang, P.; Meng, X. Acarbose treatment affects the serum levels of inflammatory cytokines and the gut content of bifidobacteria in Chinese patients with type 2 diabetes mellitus. *J. Diabetes* **2015**, *7*, 729–739. [CrossRef] [PubMed]

46. Tong, X.; Xu, J.; Lian, F.; Yu, X.; Zhao, Y.; Xu, L.; Zhang, M.; Zhao, X.; Shen, J.; Wu, S.; et al. Structural alteration of gut microbiota during the amelioration of human type 2 diabetes with hyperlipidemia by metformin and a traditional chinese herbal formula: A multicenter, randomized, open label clinical trial. *mBio* **2018**, *9*, e02392-17. [CrossRef] [PubMed]
47. van Bommel, E.J.M.; Herrema, H.; Davids, M.; Kramer, M.H.H.; Nieuwdorp, M.; van Raalte, D.H. Effects of 12-week treatment with dapagliflozin and gliclazide on faecal microbiome: Results of a double-blind randomized trial in patients with type 2 diabetes. *Diabetes Metab.* **2020**, *46*, 164–168. [CrossRef] [PubMed]
48. Zhao, L.; Zhang, F.; Ding, X.; Wu, G.; Lam, Y.Y.; Wang, X.; Fu, H.; Xue, X.; Lu, C.; Ma, J.; et al. Gut bacteria selectively promoted by dietary fibers alleviate type 2 diabetes. *Science* **2018**, *359*, 1151–1156. [CrossRef] [PubMed]
49. Sharma, B.R.; Jaiswal, S.; Ravindra, P.V. Modulation of gut microbiota by bioactive compounds for prevention and management of type 2 diabetes. *Biomed. Pharmacother.* **2022**, *152*, 113148. [CrossRef] [PubMed]
50. Lê, K.-A.; Li, Y.; Xu, X.; Yang, W.; Liu, T.; Zhao, X.; Tang, Y.G.; Cai, D.H.; Go, V.L.W.; Pandol, S.; et al. Alterations in fecal *Lactobacillus* and *Bifidobacterium* species in type 2 diabetic patients in Southern China population. *Front. Physiol.* **2013**, *3*, 496. [CrossRef] [PubMed]
51. Robertson, M.D. Prebiotics and type 2 diabetes: Targeting the gut microbiota for improved glycaemic control? *Pract. Diabetes* **2020**, *37*, 133–137. [CrossRef]
52. Prattichizzo, F.; Giuliani, A.; Mensà, E.; Sabbatinelli, J.; De Nigris, V.; Rippo, M.R.; La Sala, L.; Procopio, A.D.; Olivieri, F.; Ceriello, A. Pleiotropic effects of metformin: Shaping the microbiome to manage type 2 diabetes and postpone ageing. *Ageing Res. Rev.* **2018**, *48*, 87–98. [CrossRef]

nutrients

MDPI

Review

Metabolic Syndrome: Is It Time to Add the Central Nervous System?

Milagros Rojas [1], Mervin Chávez-Castillo [2], Daniela Pirela [1], Heliana Parra [1], Manuel Nava [1], Maricarmen Chacín [3], Lissé Angarita [4], Roberto Añez [5], Juan Salazar [1], Rina Ortiz [6], Samuel Durán Agüero [7], Marbel Gravini-Donado [8], Valmore Bermúdez [9] and Edgar Díaz-Camargo [9,*]

1. Endocrine and Metabolic Diseases Research Center, School of Medicine, University of Zulia, Maracaibo 4004, Venezuela; migarocafi@gmail.com (M.R.); pirelacdaniela@gmail.com (D.P.); helianapp20@hotmail.com (H.P.); manuelnava_14@hotmail.com (M.N.); juanjsv18@hotmail.com (J.S.)
2. Psychiatric Hospital of Maracaibo, Maracaibo 4004, Venezuela; mervinch@hotmail.com
3. Facultad de Ciencias de la Salud, Universidad Simón Bolívar, Barranquilla 08002, Colombia; m.chacin@unisimonbolivar.edu.co
4. Escuela de Nutrición y Dietética, Facultad de Medicina, Universidad Andrés Bello, Sede Concepción 4260000, Chile; lisse.angarita@unab.cl
5. Departamento de Endocrinología y Nutrición, Hospital General Universitario Gregorio Marañón, 28007 Madrid, Spain; robertojose.anez@salud.madrid.org
6. Posgrado, Carrera de Medicina, Universidad Católica de Cuenca, Cantón de Cuenca 010101, Ecuador; rortiz@ucacue.edu.ec
7. Facultad de Ciencias Para el Cuidado de la Salud, Universidad San Sebastián, Los Leones 8420524, Chile; samuel.duran@uss.cl
8. Facultad de Ciencias Jurídicas y Sociales, Universidad Simón Bolívar, Barranquilla 080002, Colombia; mgravini1@unisimonbolivar.edu.co
9. Facultad de Ciencias Jurídicas y Sociales, Universidad Simón Bolívar, Cúcuta 540006, Colombia; v.bermudez@unisimonbolivar.edu.co

* Correspondence: e.diaz@unisimonbolivar.edu.co

Citation: Rojas, M.; Chávez-Castillo, M.; Pirela, D.; Parra, H.; Nava, M.; Chacín, M.; Angarita, L.; Añez, R.; Salazar, J.; Ortiz, R.; et al. Metabolic Syndrome: Is It Time to Add the Central Nervous System? *Nutrients* **2021**, *13*, 2254. https://doi.org/10.3390/nu13072254

Academic Editors: Omorogieva Ojo and Amanda Adegboye

Received: 21 May 2021
Accepted: 9 June 2021
Published: 30 June 2021

Publisher's Note: MDPI stays neutral with regard to jurisdictional claims in published maps and institutional affiliations.

Abstract: Metabolic syndrome (MS) is a set of cardio-metabolic risk factors that includes central obesity, hyperglycemia, hypertension, and dyslipidemias. The syndrome affects 25% of adults worldwide. The definition of MS has evolved over the last 80 years, with various classification systems and criteria, whose limitations and benefits are currently the subject of some controversy. Likewise, hypotheses regarding the etiology of MS add more confusion from clinical and epidemiological points of view. The leading suggestion for the pathophysiology of MS is insulin resistance (IR). IR can affect multiple tissues and organs, from the classic "triumvirate" (myocyte, adipocyte, and hepatocyte) to possible effects on organs considered more recently, such as the central nervous system (CNS). Mild cognitive impairment (MCI) and Alzheimer's disease (AD) may be clinical expressions of CNS involvement. However, the association between MCI and MS is not understood. The bidirectional relationship that seems to exist between these factors raises the questions of which phenomenon occurs first and whether MCI can be a precursor of MS. This review explores shared pathophysiological mechanisms between MCI and MS and establishes a hypothesis of a possible MCI role in the development of IR and the appearance of MS.

Keywords: metabolic syndrome; insulin resistance; diabetes mellitus type 2; mild cognitive impairment; Alzheimer's disease

1. Introduction

Metabolic syndrome (MS) is a serious public health problem. It affects about 25% of the general population and, more alarmingly, around 40% of adults over 40 years old worldwide [1,2]. The definition of this syndrome has recently evolved to include a group of at least three of five cardio-metabolic abnormalities. These conditions include high blood pressure, central obesity, insulin resistance (IR), elevated blood triglycerides, and atherogenic dyslipidemia [3], which together lead to an increased risk of cardio-metabolic pathologies [4–6], as well as other diseases, such as arthritis [7] and some types of cancer [8].

Likewise, the presence of MS-related disorders also plays a role in the pathophysiology of neurological disorders [9], reflecting the association between deficiencies in secretion and action of insulin and mild cognitive impairment (MCI) [10]. MCI is defined as cognitive dysfunction that exceeds what is typically expected for age and educational level but does not meet the criteria for a major neurocognitive disorder. General functionality is preserved in MCI [11] and it can be best described as an intermediate state between cognitive impairment characteristic of aging and major neurocognitive conditions, such as Alzheimer's disease (AD) [12,13].

Pathophysiological mechanisms of MCI have not been fully clarified. Thus, different hypotheses are proposed to explain the MCI/MS association [14]. A cyclical relationship seems to exist between IR and cognitive impairment, and the question of which phenomenon occurs first arises [15]. Additionally, if cognitive impairment precedes IR, it becomes a risk factor for developing MS. Therefore, this review briefly describes the history of MS and discusses clinical and preclinical findings that support the role of MS and IR as elements of pathophysiological mechanisms of cognitive impairment.

While most reviews on the topic focus on the MS-to-MCI relationship, this review goes beyond this by looking at the inverse relationship, examining available evidence regarding a new hypothesis that suggests that cognitive impairment could have a role in the development of IR and the appearance of MS. Among the mechanisms to be highlighted in this regard, the hyperphosphorylation of tau proteins and the formation of amyloid β (Aβ) plaques are proposed as alterations that go beyond the pathophysiology of Alzheimer's disease (AD), and their role in the pathophysiology of insulin alterations is examined.

2. Metabolic Syndrome: Historical Aspects

The first studies of MS started almost 100 years ago when Eskil Kylin, in 1921, and Gregorio Marañón, in 1922, independently published in the same journal (Zentralblatt für Innere Medizin) papers with the same title, "diabetes mellitus and hypertension" [16,17]. Yet, not until 1981 did Hanefeld and Leonhardt use the term "metabolic syndrome" for the first time [18].

In 1988, Gerald Reaven hypothesized that IR was a common etiological factor for a group of disorders he termed "Syndrome X" [19,20]. He used this name to emphasize its unknown origin. At this time, the fundamental pathophysiological role of IR was known. This mechanism had been studied by researchers, such as Randle [21]. In subsequent years, DeFronzo, Ferrannini, and others used the term "Insulin Resistance Syndrome", proposing that available evidence suggested its presence was the cause of MS [22].

The cause of MS and its components have been debated worldwide since the end of the 20th century. Many organizations, such as the World Health Organization (WHO) [23], the European Group for the Study of Insulin Resistance (EGIR) [24], the Adult Treatment Panel III (ATP-III) [25], the American Association of Clinical Endocrinologists (ACE/AACE) [26], and the International Diabetes Federation (IDF) [27], have proposed evolving diagnostic criteria. Some criteria have been progressively discarded and replaced with criteria that can be easily applied in daily clinical practice.

Finally, the IDF, the National Heart, Lung, and Blood Institute, the American Heart Association, the World Heart Federation, the International Atherosclerosis Society, and the International Association for the Study of Obesity made a joint statement in 2009 that concluded that a diagnosis of MS requires the presence of three or more of the following criteria: high abdominal circumference as defined for each geographical region, triacylglycerides (TAG) greater than or equal to 150 mg/dL, HDL levels less than 50 mg% in women or less than 40 mg% in men, systolic blood pressure (SBP) greater than or equal to 130 mmHg or diastolic blood pressure (DBP) greater than or equal to 85 mmHg, and glycemic levels greater than 100 mg/dL [3]. We now consider that the evolution of diagnostic criteria has reached a maturity level that makes it difficult to incorporate new criteria that are both easily recognized and provide useful clinical information [28]. IR continues to be the most widely accepted hypothesis to describe MS pathophysiology that involves various organs and associations with numerous diseases [29].

3. Mild Cognitive Impairment and Metabolic Syndrome: Molecular Basis

Epidemiological, clinical, and experimental evidence provides a solid basis for the hypothesis that IR is the pathophysiological origin for dyslipidemias, high blood pressure, and disorders in glucose homeostasis [30]. As studies of MS continued, a relationship emerged between obesity, the syndrome's most prevalent individual criterion, and neurological alterations [31]. This association necessitates additional consideration of the pathophysiological mechanisms involved and how they are interconnected.

3.1. From Metabolic Syndrome to Cognitive Impairment

Epidemiological [32], neuroimaging [33], and animal modeling studies are available to characterize MS pathophysiology, accompanying diseases, and individual components in the development of neurodegenerative diseases and associated cognitive impairment [34]. The connection between diabetes mellitus (DM) and AD and the connection between obesity and cognitive impairment are two areas that have been investigated in depth, including reports of statistically significant relationships [10,35,36].

The hippocampus plays an important role in learning and memory. The effect of IR on hippocampal function has been widely studied [37]. Lindqvist et al. administered high- and low-fat diets to different groups of rats, using bromodeoxyuridine (BrdU) to observe synaptogenesis after 4 weeks. Male rats fed the high-fat diet showed a significant decrease in neurogenesis in the hippocampus. The animals did not develop obesity [38]. This result could reflect the role of lipid alteration, a component of MS, in cognitive decline.

Karimi et al. studied long-term neuronal potentiation (LTP) in the dentate gyrus (DG) of the hippocampus in mice receiving different combinations of a high-fat diet and antioxidants. Mice fed a high-fat diet showed lower LTP levels compared with control animals. In contrast, mice that received antioxidants displayed elevated LTP [39]. These effects might be mediated by an increase in free radical production caused by the high-fat diet, leading to oxidative stress (OS). This hypothesis is supported by rescue of LTP levels through the administration of antioxidants [39].

3.1.1. Role of IR in the Formation of Amyloid-Beta Plaques

Production of reactive oxygen species (ROS) results in increased levels of amyloid precursor peptide-β (AβPP) and increased expression and accumulation of amyloid-β 42 (Aβ) [40]. This accumulation leads to the formation of amyloid-beta plaques, identified as a key element of AD [41]. One pathway for increased ROS production is increased insulin levels that lead to changes in normal NADPH oxidase (NOX4) pathway function. Elevated insulin levels induce phosphoinositol-3 kinase (PI3K) to phosphorylate Rac instead of phosphatidylinositol bisphosphate (PIP2). This alteration increases NOX4 activity and ROS levels. This aberrant metabolism is perpetuated because elevated ROS leads to activation of casein kinase 2 (CK2) and consequent activation of the retromer. This action signals the degradation of glucose receptor, GLUT4, leading to a continued increase in glucose levels in the blood and, therefore, increased production of insulin [42] (Figure 1).

The metabolic syndrome (MS) and its components cause brain alterations such as neuroinflammation, hyperphosphorylation of Tau, formation of beta amyloid plaques, and vascular changes (not represented in the figure). This is achieved through changes in the signaling of hormones such as adiponectin, leptin, and insulin. These changes are clinically expressed as mild cognitive impairment (MCI), Alzheimer's disease (AD), and major vascular neurocognitive disorder.

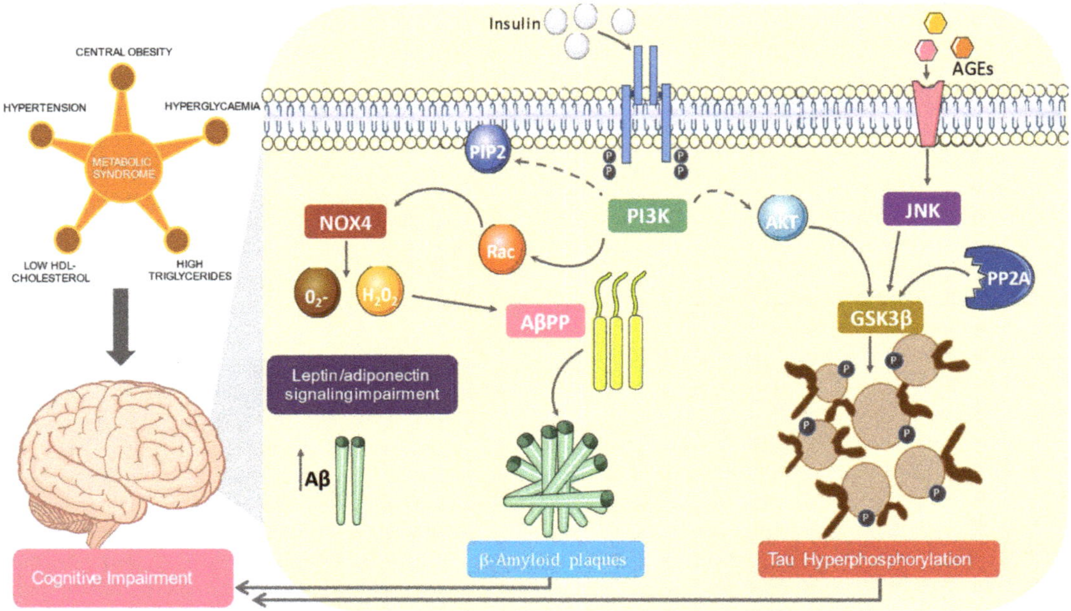

Figure 1. Impact of metabolic syndrome on cognitive impairment. PIP2: phosphatidylinositol bisphosphate; AKT: protein kinase B; PI3K: phosphoinositide kinase-3; JNK: c-Jun N-terminal kinase; AGEs: advanced glycation end products; PP2A: protein phosphatase 2A; GSK3β: glycogen synthase kinase 3 beta; NOX4: NADPH oxidase 4; 02-: superoxide; H_20_2: hydrogen peroxide; Aβ: amyloid beta. Solid lines mean activation; dashed lines mean inactivation.

This positive feedback produces an environment of neuronal inflammation, reported as a risk factor for AD. An associated hypothesis is that, along with neurofibrillary tangles and amyloid-beta plaques, inflammation has a critical role in the pathophysiology of the disease [43]. Insulin seems to play a significant role in the development of plaques, and hyperinsulinemia observed in IR leads to their formation. In vitro studies show that high levels of insulin affect the degradation and elimination of Aβ. Both insulin and Aβ are degraded by insulin-degrading enzyme (IDE). During hyperinsulinemia, IDE degrades insulin preferably to Aβ, promoting its oligomerization into insoluble aggregates [44]. In vivo experiments in rats corroborate these findings, showing that elimination of Aβ is reduced in the presence of high levels of insulin [45].

Insulin receptors provide an alternative explanation for IR effects on the hippocampus and other brain structures [46]. These receptors are abundant in metabolic active brain areas and exert their effects at the neuronal level via PI3K and mitogen-activated protein kinases (MAPK) pathways [47]. These pathways, when activated by insulin, promote angiogenesis in the brain. In the presence of IR, these pathways are not activated. This disruption might underlie the concomitant synaptic anomalies, memory disorders, decreases in neurogenesis at the hippocampus level, alterations in cognition, and decreases in levels of brain-derived neurotrophic factor (BDNF) [46].

3.1.2. Metabolic Syndrome, Insulin Resistance, and Tau Proteins

Conversely, tau protein (TP) helps stabilize microtubules and its alteration results in the formation of neurofibrillary tangles [48]. Further, IR induces hyperphosphorylation of TP and induces cognitive impairment in human and animal models [49,50]. Thus, IR is associated with poorer performance on cognitive tests and higher levels of phosphorylated PT in cerebrospinal fluid (CSF) in cognitively normal individuals and carriers of the APOE allele ε4 [10,51].

One mechanism underlying this phenomenon involves glycogen synthase kinase-3β (GSK3β), a tau kinase regulated by insulin via the protein kinase B (AKT) pathway. Decreased brain insulin signaling caused by IR induces chronic exposure of neurons to high levels of insulin or an eventual decrease in insulin levels, resulting in PI3K dysfunction and reduced AKT-dependent phosphorylation. Downstream, GSK3β is activated, and ultimately TP is hyperphosphorylated [52,53]. The production of advanced glycation end products (AGEs) from OS damage via GSK3β receptors (RAGE) also increases the activity of GSK3β by an alternate pathway involving c-Jun N-terminal kinase (JNK) [54].

A recent study showed protein kinase A (PKA) is a potent tau kinase and its activation increases TP hyperphosphorylation in an insulin-deficient animal model [55]. Moreover, insulin deficiency influences hyperphosphorylated TP level by decreasing the activity of protein phosphatase 2 (PP2A) [56]. PP2A is the primary tau phosphatase involved in AD and its deregulation is associated with TP hyperphosphorylation [57]. Similarly, hypothermia, common in chronic DM, also leads to inhibition of PP2A activity [58].

Another pathological mechanism in AD is truncation of TP by proteolytic enzymes, such as caspases, peptidases, and thrombins that promote tau aggregation and formation of the central component of neurofibrillary tangles (NFT) [59]. DM stimulates apoptosis through the activation of caspases in affected tissues. Through hyperglycemia, DM might increase tau aggregation by activating caspases, thus contributing to AD risk [60]. Kim et al. demonstrated such increased tau aggregation in the brain of db/db rats using in vivo and in vitro type 2 diabetes mellitus (T2DM) animal models [61].

3.1.3. Metabolic Syndrome, Leptin, Adiponectin, and Cognitive Disorders

Alterations in hormones involved in MS, such as leptin and adiponectin, are also linked to cognitive impairment [62]. Both hormones affect the metabolism of fatty acids and glucose as well as energy metabolism and food intake [63,64]. Their function in neuroplasticity, learning, and cognition [65,66] is now known via reports of leptin and adiponectin receptor expression in brain regions such as the hippocampus and neocortex [67].

Recent studies in animal models show that leptin deficiency or resistance is associated with cognitive disorders, such as reductions in LTP, long-term neuronal depression (LTD), and alterations in spatial memory [68]. Further, leptin modulates the production and elimination of Aβ in AD by inhibiting the formation of AβPP and increasing APOEε4-induced amyloid filament elimination. This activity may be mediated through the activation of AMP-activated protein kinase (AMPK) [69,70]. Leptin resistance in AD is associated with diminished activity in these pathways and increased cognitive impairment [71].

Additionally, adult rats deprived of adiponectin display several common characteristics of AD, including deposition of Aβ, TP phosphorylation, and neuroinflammation [72]. This observation is corroborated by Kim et al., who demonstrated that adiponectin receptor suppression also produced an AD-like phenotype [73] Thus, hormone deficiencies might be involved in AD pathogenesis. However, studies in humans are controversial since available information for the association of adiponectin and leptin levels in the blood and CSF with cognitive impairment is inconclusive [62].

3.1.4. Metabolic Syndrome, Microvasculature, and Cognitive Impairment

Micro- and macrovascular changes observed in MS, such as hypertension and DM, are also associated with brain alterations, such as vascular neurocognitive disorder. However, several recent studies note the contribution of vascular risk factors in AD. Mechanisms for this accelerating cognitive decline are not fully elucidated [74].

Hypertension leads to alterations observed in magnetic resonance imaging (MRI), such as white matter lesions (WML), lacunar infarcts, microhemorrhages, and microinfarcts. All these abnormalities are part of a spectrum called small vessel cerebral disease (SVD), which is common in AD [75]. SVD is characterized by loss of smooth muscle cells in the mid-tunic, deposition of fibro-hyaline material, reduced light, and thickening of vascular walls [76]. Moss et al. studied Rhesus monkeys using an aortic coarctation model.

Multiple microinfarcts and lesions in gray and white matter in hypertensive monkeys were associated with cognitive impairment [77]. Other mechanisms might involve large arteries via endothelial dysfunction that progresses to the formation of atherosclerotic plaques in the carotid or intracranial arteries. Such damage can cause ischemic events in brain regions related to cognition [78].

3.2. Exploring the Inverse Relationship—From Cognitive Impairment to Metabolic Syndrome

The consequences of MS on the development of cognitive impairment have been studied in depth [79], but the inverse relationship in which pathophysiological mechanisms of AD, such as hyperphosphorylation of TP and the formation of amyloid complexes-β, lead to the appearance of MS is largely unstudied.

3.2.1. Tau Proteins and Deficits in Insulin Signaling

Interestingly, TPs, in addition to microtubule stabilization, also interact with insulin signaling pathway components in the brain. The N-terminal portion of TP can bind to homology 3 (SH3) domains of the Src family of tyrosine kinases, including domains of the p85 alpha subunit of PI3K, a key protein in the insulin signaling pathway. Under pathological conditions, hyperphosphorylation of TPs can lead to loss of functionality, triggering alterations in insulin signaling that eventually generate altered fasting glycemia and DM. The ability of TPs to interact with SH3 domains is inversely correlated with the degree of phosphorylation, suggesting that scaffolding properties of TPs are regulated by their phosphorylation status [80].

Further, co-immunoprecipitation studies of mouse brain tissue and N1E115 cells indicate that TPs bind to phosphatase and tensin homologous protein (PTEN), a negative insulin signal translocation regulator that catalyzes dephosphorylation of phosphatidylinositol triphosphate (PIP3) to PIP2. Thus, TP, by interacting with and inhibiting PTEN, promotes insulin signaling [81]. These studies raise the possibility that insulin helps maintain adequate brain activity due to TP and, conversely, pathological forms of TP could be harmful due to a loss of protein function. This suggestion is supported by a study that showed that TP removal was accompanied by loss of inhibitory effects of insulin on PTEN in the hippocampus, resulting in brain IR.

Concurrently, the absence of TP reduced the anorexigenic effect of insulin in the hypothalamus after intracerebroventricular injection of TP [81]. Previously, such injection induced increased food intake, weight gain, adiposity, hyperinsulinemia, and glucose intolerance in rodents with insulin receptor deletion in the hypothalamus [82,83]. These effects produce alterations in energy metabolism that may increase the risk of suffering from obesity, DM, and MS.

TP is also highly expressed in pancreatic islet β-cells. However, its function in peripheral tissues is not fully understood [84,85]. Wijesekara et al. investigated TP actions on β-cell function and glucose homeostasis using a tau KO rat model. Rats showed weight gain, defects in glucose signaling, and IR, leading to DM and ultimately MS. Thus, TP might be crucial for normal energy metabolism in peripheral tissues [86].

3.2.2. Amyloid β and Insulinemic Alterations

In vitro and in vivo studies suggest that Aβ may also contribute to IR through various mechanisms. Aβ competitively inhibits the binding of insulin to its receptor [87] and activates the JAK2/STAT3/SOCS-1 signaling pathway to produce IR in the liver [88]. Further, the oligomer Aβ (AβO), a highly toxic species of Aβ, causes deregulation of N-methyl-D-aspartate (NMDA) receptors and leads to the production of excessive ROS. This effect is probably due to mitochondrial dysfunction [89].

This deregulation might lead to alterations in insulin signaling, since increased ROS activates several serine kinases, such as an inhibitor of the nuclear factor kappa-B kinase beta subunit (IKK-β), protein kinase C (PKC), and JNK. These kinases increase phosphorylation in Ser IRS-1 residues. ROS can cause OS and damage at mitochondrial and cellular levels. This stress generates mitophagy and, at high levels of stress, apoptosis. The elimination of mitochondria by mitophagy results in a decrease in oxidation and consequent accumulation of lipids, leading to IR and T2DM [90].

Conversely, AβO causes a rapid and substantial loss of insulin receptors in dendrites and inhibition of insulin receptor autophosphorylation associated with NMDA activity [91]. Additionally, an increase in levels of IR markers p(Ser)-IRS-1 and p-JNK were observed in neurons after intracerebroventricular injection of AβO in vivo in monkeys [92].

3.2.3. Amyloid β, Tau Protein, and Leptin

Aβ and TP have also been linked to alterations in leptin signaling. Bonda et al. showed that TP hyperphosphorylation leads to the formation of NFT and dysfunction in intracellular trafficking networks in the hippocampus. Thus, the leptin receptor in its long form (Ob-Rb) becomes unable to reach cell membranes, hindering its access to circulating free leptin and interrupting signaling. This activity might lead to increased food intake and weight gain with subsequent development of obesity and long-term MS [93]; leptin in the hippocampus is associated with regulating food intake and processing food-related memories [94].

Elevated levels of Aβ1-42 produced by beta-site amyloid cleaving enzyme 1 (BACE1) increase leptin resistance in the hypothalamus, which is associated with decreased sensitivity to exogenous leptin throughout the body and exacerbation of body weight gain in rats fed high-fat diets. Thus, countering BACE1 activity may be protective against metabolic disorders [95].

The above findings affirm cognitive impairment as a key trigger of alterations in insulin signaling in the hypothalamus. The latter region is the primary regulator of body weight via controlling food intake and peripheral metabolism [96]. Thus, cognitive impairment might lead to metabolic changes that precede the development of MS and its complications.

4. Mild Cognitive Impairment and Metabolic Syndrome: Epidemiological Basis

Evidence concerning the relationship of MS and its components with MCI has accumulated in the last few years to the point where grouping these disorders into a single clinical entity, the cognitive–metabolic syndrome, may be appropriate [97]. Below, we summarize clinical and epidemiological information on the MCI and MS relationship and its components.

4.1. From Metabolic Syndrome to Cognitive Impairment

Cardio-metabolic risk factors and MS affect cognition and increase the risk of major neurocognitive disorders [98–100]. Speed of processing, attention, and executive functions are the most frequently affected domains [101,102]. Thus, an association is often reported between risk factors, such as hyperlipidemia, T2DM, obesity, hypertension, and physical inactivity, and models of risks of cardiovascular disease (CVD) (e.g., Framingham Risk Score) with the risk of MCI and major neurocognitive disorders (Table 1) [103–106].

Strong evidence of a link between high blood pressure in middle age and poorer cognitive function in old age is available [107,108]. Different prospective studies in older people show that increased blood pressure is associated with worse cognitive function [109]. The risk of cognitive impairment can increase up to 2.8 times [110]. In older women, risks may increase by up to 20% [111]. Similar results were reported in individuals from Hispanic [112], Swedish [113], Asian [114], and North American communities [115,116].

Obesity, defined by a high abdominal circumference or a body mass index (BMI) ≥ 30, is also associated with poor cognitive function [117,118]. Individuals with high BMI during middle age show low scores among various cognitive tests [119]. Further, long-term obesity is linked to lower cognitive performance and an increased risk of neurocognitive impairment in older people [120–122].

Several epidemiological studies and meta-analyses provide evidence for an effect of hyperlipidemia, hypertriacylglycerolemia, and HDL-C levels on cognitive performance in individuals with and without major neurocognitive disorders. Elevated LDL-C levels are correlated with the degree of cognitive impairment [123] and decreased episodic memory (ECM) [124]. Further, hypertriacylglycerolemia is associated with low scores in verbal tests [124,125]. Low concentrations of this lipoprotein are associated with poor and decreased memory in middle-aged adults [126], while in older people, low levels are associated with major neurocognitive disorders [127]. In contrast, improvement in cognitive test performance is reported for subjects over 75 years old with high HDL-C [128,129], which is also associated with a significant decrease in the appearance of major neurocognitive disorders [130].

Hyperinsulinemia, glucose intolerance, and T2DM are other cardio-metabolic risk factors that recently have been associated with cognitive impairment and different major neurocognitive disorders [131]. Hyperinsulinemia and impaired glucose tolerance, both indicators of a prediabetic state and an increased risk of developing DM, are associated with cognitive dysfunction and an increased risk of developing MCI [132–135]. These premorbid states are associated with reduced long-term memory scores [136] and impaired verbal fluency [137]. Lower performance on psychomotor and memory tests is observed in diabetic individuals [138]. These lower scores correlate with an increased risk of developing cognitive impairment and MCI [139–143].

Several studies associate different elements of MS with cognitive functions. However, few studies of MS as a clinical entity and its relationship with MCI or its progression to major neurocognitive disorders are available. Roberts et al. reported a cross-sectional study in 1969 of 70 89-year-old individuals. Participants with MS showed non-amnestic MCI (naMCI) when accompanied by elevated C-reactive protein (CRP). The combination of inflammation and MS might be linked to specific subtypes of MCI [144]. Yaffe et al. conducted a longitudinal, multicenter study with 4895 women with an average age of 66.2 years. MS was associated with an increased risk of developing cognitive impairment in older women. Risk increased by an age-adjusted 23% for each increment in the number of MS components [145]. Similar findings were reported by Pal et al. [146] and Atti et al. [147], who concluded that MS is associated with an increased incidence of major neurocognitive disorders and an increased risk of progression from MCI to such disorders, respectively.

Table 1. Effects of components of MS on cognitive function, the risk of MCI, and major neurocognitive disorders.

MS Component	Authors (REF)	Methodology	Results
High blood pressure	McDonald et al. [109]	Longitudinal cohort study of the association between cognitive function and BP variability in adults $\geq$65 years.	After 5 years of monitoring, diurnal systolic BP variability was independently associated with a greater decrease in total CAMCOG (CV: 3.205; $p = 0.043$) and MMSE (CV: 3.985; $p = 0.020$) scores.
	Haring et al. [111]	Prospective study in 6426 cognitively intact older women of the relationship between hypertension and cognitive impairment.	Hypertension was associated with an increased risk for cognitive decline (HR: 1.20; 95% CI: 1.04–1.39; $p = 0.02$).
Obesity	Sabia et al. [121]	Longitudinal cohort study of the association between BMI and mid-age cognition throughout adult life in 5131 individuals.	Late midlife obesity was associated with lower scores on the MMSE and on memory and executive function scores compared with normal-weight individuals (mean difference (95% CI): −0.99 (−1.78–0.21), −0.82 (−1.57–0.08), and −0.80 (−1.49–0.12), respectively ($p < 0.05$)).
	Beydoun et al. [122]	Meta-analysis of the association between obesity and major neurocognitive disorder in older adults.	A significant U-shaped association was found between BMI and major neurocognitive disorder ($p = 0.034$), with an increased risk of disorder (OR (95% CI): 1.42 (0.93–2.18)) and increased incidence of AD (OR (95% CI): 1.80 (1.00–3.29)) in obese individuals.

Table 1. *Cont.*

MS Component	Authors (REF)	Methodology	Results
Dyslipidemias	de Frias et al. [124]	Longitudinal cohort study of the association between total cholesterol, triglycerides, and cognitive performance in older adults.	Hypertriglyceridemia was associated with a low score on the verbal memory tests ($\gamma = -2.31$; $p < 0.05$), while hypercholesterolemia was associated with a detrimental effect on facial recognition test score ($\gamma = -2.13$; $p < 0.05$).
	Singh-Manoux et al. [126]	Longitudinal cohort study of the relationship between HDL-c and verbal short-term memory in middle-aged adults.	After 5 years of monitoring, decreased HDL-C was associated with decreased verbal memory (OR = 1.61; 95% CI = 1.19–2.16).
DM	Elias et al. [117]	Longitudinal study of the effects of T2DM on cognitive performance of adult individuals.	The amount of time suffering from diabetes was associated with poorer cognitive performance ($\beta = -0.02$; $p < 0.02$).
	Kanaya et al. [137]	Longitudinal cohort study of changes in cognitive performance according to glucose tolerance status in older adults.	After 4 years, women with DM had a 4-fold increased risk of cognitive impairment (OR (95% CI): 4.38 (1.71–11.27); $p = 0.02$).
MS	Atti et al. [147]	Meta-analysis of the relationship between MS and progression to major neurocognitive disorder in individuals with MCI.	Having MS increased the risk of progression from MCI to major neurocognitive disorder (HR (95% CI): 2.69 (1.16–6.27); $p < 0.05$).
	Pal et al. [146]	Meta-analysis that quantified the relative risk of progression from MCI to major neurocognitive disorder in individuals with MS.	An increased risk of progression was found in individuals with MCI and SM (OR (95% CI): 2.95 (1.23–7.05) $p < 0.05$).

Abbreviations: MS: metabolic syndrome; MCI: mild cognitive impairment; BP: blood pressure; CAMCOG: Cambridge Cognitive Examination; MMSE: Mini-Mental State Examination; CV: coefficient of variation; BMI: body mass index; OR: odds ratio; HR: hazard ratio; CI: confidence interval; AD: Alzheimer's disease; HDL-c: high-density lipoprotein; DM: diabetes mellitus.

The impact of MS on cognitive function is not limited to adults. There is also evidence that suggests that MS components may be detrimental in younger populations. The presence of T2DM, obesity, and hypertension in children and adolescents is associated with poorer performance in overall functioning, and declines in executive function, memory, attention, and intelligence quotient (IQ) [148–152].

Cardiovascular and metabolic risk factors are modifiable and their timely identification and consequent management could prevent MCI or its progression to major neurocognitive disorders [101]. Thus, lifestyle changes, including increased physical activity and implementation of healthy diets, and antihypertensive, hypolipidemic, and insulin-sensitizing drugs are important considerations for the management of premorbid state characteristics of MS (Table 2) [153].

Changes in lifestyle and physical activity positively impact cognitive function. [154,155]. Physical activity is associated with better scores on tests of executive function, processing speed, and improvement in global cognitive function. These benefits were found both in healthy older subjects and in older subjects with MCI or major neurocognitive disorders [156–159]. More studies to elucidate types of exercise, times, and intensity needed to cause a positive impact on cognition are necessary; still, 150 min of physical activity per week is proposed to improve the brain health of individuals with MCI [160].

Better results are obtained if physical activity is combined with a healthy diet. Supplementation with B-vitamins, folic acid, docosahexaenoic acid (DHA), eicosapentaenoic acid (EPA), and flavonoids is associated with improved cognitive performance, particularly memory, in subjects with MCI [161]. Similarly, both cognitively normal individuals and those with MCI are reported to be at less risk of developing MCI or AD if they maintain high adherence to a Mediterranean diet [162]. Similar results are associated with Mediterranean-DASH diets [163], low-carbohydrate diets (keto-diet) [164], and fish PUFA diets [165].

Additionally, a causal relationship between antihypertensive drugs and improved cognitive function is supported by available evidence. Antihypertensive drugs, especially calcium channel blockers and renin–angiotensin system blockers, have a protective effect on cognitive decline and decrease the risk of AD and neurocognitive vascular disorders in older people [166]. Similarly, treatment with antihypertensive drugs reduces the risk of major neurocognitive disorders by 9% and shows improvement in all cognitive domains, except language [167]. Longitudinal studies that included older individuals without major neurocognitive disorders who were undergoing antihypertensive therapy produced supporting results [110,168].

Controlling glycemic concentrations and increasing peripheral insulin sensitivity are strategies that might positively affect cognitive function [146]. A recent meta-analysis showed that treatment with metformin or sulfonylureas is associated with a significant decrease in cognitive impairment in patients with T2DM. In contrast, the use of insulin aggravated the dysfunction [169]. Studies of metformin as monotherapy [170], or combined with vildagliptin [171], on the participants' cognitive function produced similar results. However, other studies show no association between the use of antidiabetic drugs and improvement in cognitive function [172,173]. One study linked the use of such drugs to the diagnosis of MCI [174].

Finally, unlike antihypertensive and antidiabetic drugs, hypolipidemics, such as statins, do not affect the risk of progression to MCI or major neurocognitive disorders of any kind [175,176]. Indeed, several clinical and epidemiological studies report no significant association between statin use and reduced cognitive impairment [177–180].

Table 2. Association between treatment of MS elements and MCI improvement.

Therapeutic Approach	Authors (REF)	Methodology	Results
Lifestyle changes	Karssemeijer et al. [157]	Meta-analysis of the effect of cognitive and physical exercise intervention in older adults with MCI or major neurocognitive disorder.	A positive effect of the combination of physical–cognitive interventions on global cognitive function was observed (MDS (95% CI) = 0.32 (0.17–0.47); $p < 0.05$). It was equally beneficial for individuals with MCI (MDS = 0.39 (0.15–0.63); $p < 0.05$), and for patients with major neurocognitive disorder (MDS = 0.36 (0.12–0.60); $p < 0.001$).
	Zhang et al. [165]	Meta-analysis of the association between risk of cognitive impairment and intake of fish and PUFAs.	Increased fish consumption was associated with decreased risk of major neurocognitive disorder (RR: 0.95; 95% CI: 0.90–0.99; $p = 0.042$). A significant curvilinear relationship was observed between PUFA consumption and MCI risk (p nonlinearity < 0.001).
	Krikorian et al. [164]	Prospective study of the effect of a ketogenic diet in older individuals with MCI.	An improvement in verbal memory performance was observed in patients on a low-carbohydrate diet ($p = 0.001$). Memory performance was positively correlated with ketone levels ($p = 0.04$).
Antihypertensives	Tzourio et al. [110]	A longitudinal study of the effect of antihypertensive drugs on the risk of cognitive decline in older individuals.	The risk of cognitive impairment was higher in untreated subjects (OR = 6.0 (95% CI: 2.4–15.0)), compared with subjects treated with antihypertensives (OR = 1.3 (95% CI: 0.3–4.9)).
	Guo et al. [168]	Prospective study that evaluated whether the use of antihypertensives affected the appearance and progression of major neurocognitive disorders in older adults.	The risk of major neurocognitive disorder was reduced in subjects receiving antihypertensive treatment and without major neurocognitive disorder at the beginning of the study (RR = 0.7 (95% CI: 0.6–1.0); $p = 0.03$).

Table 2. *Cont.*

Therapeutic Approach	Authors (REF)	Methodology	Results
Antidiabetics	Ng et al. [170]	Longitudinal study of the protective effect of metformin on the cognitive performance of older adults.	Use of metformin showed an inverse association with cognitive impairment (OR = 0.49 (CI 95%: 0.25–0.95); $p < 0.05$), and was associated with a low risk of cognitive impairment after 6 years of use (OR = 0.27 (CI 95%: 0.12–0.60); $p < 0.05$).
	Borzì et al. [171]	Retrospective study of the effect of vildagliptin on cognitive function in older diabetic adults with MCI.	The use of metformin as monotherapy or in combination with vildagliptin was associated with a significant reduction in MMSE score ($p < 0.001$).
Hypolipidemics	Bosch et al. [177]	Clinical trial on the effect of rosuvastatin in reducing cognitive impairment in older adults.	The mean difference in DSST score between rosuvastatin vs. placebo was −0.54 (95% CI: −1.88–0.80); $p < 0.05$.
	Bettermann et al. [178]	Clinical trial on the impact of statin use on delaying cognitive decline in patients with and without MCI.	Statins were associated with a decreased risk for increased neurocognitive impairment from all causes in patients who did not have MCI at the beginning of the study (HR = 0.79 (95% CI: 0.65–0.96) $p = 0.021$). In subjects with MCI, these protective effects were not observed.

Abbreviations: MS: metabolic syndrome; MCI: mild cognitive impairment; SMD: standardized mean difference scores; PUFAs: polyunsaturated fatty acids; MMSE: Mini-Mental State Examination; CV: coefficient of variation; OR: odds ratio; RR: relative risk; HR: hazard ratio; CI: confidence interval; DSST: Digit Symbol Substitution Test.

4.2. Exploring the Reverse Relationship—From Cognitive Disorder to Metabolic Syndrome

Epidemiological studies suggest the inverse relationship. MCI has been assessed as a contributor to the development of MS (Table 3). In a cross-sectional study of 3312 male and female participants aged 70 years and older in Japan, a higher prevalence of MS was observed in subjects with naMCI than in those with normal cognition. Moreover, women with naMCI had high blood pressure and high glucose levels more often, while men with naMCI showed only a higher frequency of high glucose levels compared with the control group. However, a causal relationship between the two could not be determined from this cross-sectional study [181]. Clinical evidence is still scarce and has focused more on specific components of MS, such as insulin/glucose alterations and T2DM, than on MS as an entity.

Alterations in insulin signaling have been reported in postmortem studies in brains from individuals with AD [182,183], as well as in patients with AD in clinical studies of plasma hyperinsulinemia and reductions in insulin levels in the CSF. These changes worsen as the disease progresses [184]. Animal models produce similar results [88,185]. Accordingly, Janson et al. used the Mayo Clinic Alzheimer's Disease Patient Registry to show a higher incidence of both T2DM and IR in 80% of AD patients. A greater increase in fasting plasma glucose (FPG) with age compared with the control group was also observed. AD patients might thus be at greater risk of developing a diabetic phenotype and suffering from T2DM [186].

Similarly, a longitudinal study using data from the Lothian Birth Cohort of 1936 (LBC1936) examined parameters, such as cognitive changes and glucose levels. This cohort consists of 1091 initially healthy individuals born in 1936. Individuals were assessed using glycosylated hemoglobin (HbA1c) data for four ages—70, 73, 76, and 79 years. Lower cognitive function at 70 years was associated with increased HbA1c in the following decade. Cognitive dysfunction is thus negatively correlated with increases in HbA1c. Maintaining high cognitive function could be a protective factor for the development of hyperglycemia and T2DM [187].

Likewise, Peng et al. initially conducted a cross-sectional study in 2126 participants, including 1063 patients recently diagnosed with T2DM and 1063 patients with standard glucose tolerance. Individuals with higher plasma concentrations of both Aβ40 and Aβ42 were more likely to have T2DM compared to subjects with the lowest concentrations [188]. In a follow-up study, the authors examined Tongii-Ezhou Cohort (TJEZ) data prospectively. One hundred and twenty-one individuals with T2DM and 242 healthy individuals were included. The same association was found, where the probability of T2DM was higher with higher plasma concentrations of Aβ, 3.79 (95% CI 1.81–7.94) for Aβ40 and 2.88 (95% CI 1.44–5.75) for Aβ42. The authors conclude that a positive association exists between Aβ and the risk of acquiring T2DM [188].

Table 3. Effect of cognitive dysfunction and suffering from MS or MS components.

Author (REF)	Methodology	Results
Janson et al. [186]	Longitudinal study where prevalence of T2DM in patients with AD was evaluated, along with the association between FPG and aging in these patients.	The prevalence of T2DM (34.6 vs. 18.1%; $p < 0.05$) and IFG (46.2 vs. 23.8%; $p < 0.01$) was higher in the AD group vs. the control group. A greater increase was seen in FPG per year in the AD group (0.83 vs. 0.57 mg/dL^{-1}; $p < 0.01$).
Bae et al. [181]	Cross-sectional study of the prevalence of MS by type of MCI in 3312 older adults and differences related to sex.	The prevalence of MS was higher in participants with naMCI (men: $p = 0.030$; women: $p = 0.040$) and the risk of MS was higher in men (OR = 2.45; 95% CI: 1.13–5.32) than in women (OR = 1.94; 95% CI: 1.12–3.39) compared with participants with normal cognition.
Altschul et al. [187]	Longitudinal cohort study of the association between cognitive function, HbA1c, and other variables in early and late life in 1091 adults.	High cognitive function at age 11 predicted low HbA1c levels at age 70 ($p < 0.001$). Additionally, high cognitive function at age 70 was associated with a smaller increase in HbA1c levels between age 70 and 79 ($p < 0.001$).
Peng et al. [188].	Study comparing 1063 newly T2DM diagnosed individuals with 1063 control individuals for an association between plasma concentrations of Aβ40 and Aβ42 with risk of T2DM.	The risk of T2DM was higher in individuals with the highest concentrations of Aβ40 and Aβ42 (OR = 2.96 (95% CI: 2.06–4.25)) compared with subjects with the lowest concentrations of Aβ.
Peng et al. [188].	Prospective study of the association between plasma concentrations of Aβ40 and Aβ42 with risk of T2DM.	A higher risk of T2DM was found in individuals with concentrations greater than that of Aβ (OR = 3.79 (95% CI: 1.81–7.94)) for Aβ40 and (OR = 2.88 (95% CI: 1.44–5.75)) for Aβ42.

Abbreviations: T2DM: diabetes mellitus type 2; HbA1c: glycated hemoglobin; FPG: fasting plasmatic glucose; IFG impaired fasting glucose; MCI: mild cognitive impairment; naMCI: non-amnestic mild cognitive impairment; Aβ: amyloid-beta; OR: odds ratio; CI: confidence interval.

These findings imply that therapeutic intervention aimed at MCI, especially AD, could be beneficial for treating MS and its components (Table 4). A drug approved for the treatment of moderate-to-severe AD is memantine, an NMDA receptor antagonist that reduces the accumulation of Aβ in AD patients [189]. Ettcheto et al. analyzed the effects of memantine in rats with model AD that were fed a high-fat diet. After 12 weeks of treatment with 30 mg/kg memantine, improvement of peripheral metabolic parameters, such as IR, was observed [190].

Similarly, Ahmed et al. investigated piracetam and memantine in the treatment of T2DM in 120 individuals. Piracetam is used to improve memory and brain function. Diabetic patients with AD treated with either drug showed a significant reduction in diabetic markers (GPA, HbA1c%, and insulin levels) compared to a symptomatic control group. Thus, agents used to treat MCI demonstrate a therapeutic potential for the treatment of metabolic disorder [191].

Another therapeutic strategy is based on reducing the activity of enzymes that promote the formation of Aβ, such as BACE1. The metabolic role of BACE1 is not fully understood, though loss of BACE1 in transgenic rats leads to increased sensitivity to insulin and decreased body weight [192]. Its mechanisms of action may involve leptin signaling and thermogenesis [95]. These results were extrapolated in a randomized clinical trial (RCT). Patients with AD treated with lanabecestat, a BACE1 inhibitor, showed greater weight loss than a placebo group after 104 weeks of treatment [193].

Additionally, immunotherapy against Aβ is used to improve insulin sensitivity and plasma glucose levels. Zhang et al. used an APP/PS1 EA rat model with increased plasma levels of Aβ40/42. Animals exhibited altered glucose/insulin tolerance and liver insulin signaling. After nine months of intraperitoneal injections of antibodies against Aβ, an improvement was observed in insulin sensitivity. Hepatic signaling of JAK2/STAT3/SOCS-1 compared to the control group was concurrently attenuated. Thus, neutralization of Aβ attenuates hyperglycemia and IR in vivo [194].

Table 4. Summary of preclinical and clinical studies exploring treatment of SM with anti-Alzheimer's drugs.

Author (REF)	Treatment	Methodology	Results
Ettcheto et al. [190]	Memantine	Preclinical study of the effects of MEM on learning and memory impairment in rats with familial AD and HFD-induced insulin resistance.	MEM prevented body weight increase in HFD-fed mice with APP/PS1 ($p < 0.001$). Hepatic IR protein levels showed a significant increase in APP/PS1 MEM mice compared to nontreated controls ($p < 0.05$), improving insulin function in the liver.
Zhang et al. [194]	Anti- Aβ Immunotherapy	Preclinical study of the effects of intraperitoneal injections of anti-Aβ antibodies in APP/PS1 rats on glucose metabolism.	After 9 months of treatment, neutralization of Aβ reduced fasting blood glucose level ($p < 0.001$), improved insulin sensitivity ($p < 0.05$), and inhibited hepatic JAK2/STAT3/SOCS1 signaling ($p < 0.05$) in APP/PS1 AD model rats.
Wessels et al. [193]	Lanabecestat	RCT that assessed whether lanabecestat slows the progression of AD compared with placebo in patients with early AD (mild cognitive impairment) and mild AD dementia.	Even though treatment with lanabecestat did not slow cognitive decline, patients who completed week 104 had a mean (SD) weight loss of 0 (4.7) kg for placebo, −0.8 (4.6) kg for patients treated with 20 mg lanabecestat, and −1.9 (5.2) kg for those treated with 50 mg.
Ahmed et al. [191]	MemantinePiracetam	Clinical study of the effect of piracetam and memantine on diabetes mellitus.	A significant decrease in all diabetic markers (FPG, HbA1c%, and insulin levels) in the diabetic and Alzheimer's patients was observed after treatment with memantine or piracetam compared to diabetic and Alzheimer's patients with symptomatic treatment ($p < 0.05$).

Abbreviations: MEM: memantin; AD: Alzheimer's disease; HFD: high-fat diet; Aβ: amyloid-beta; RCT: randomized clinical trial; FPG: fasting plasmatic glucose; HbA1c: glycated hemoglobin.

Numerous epidemiological and clinical studies and meta-analyses provide evidence that MS and its components have a substantial impact on the development of MCI. However, the inverse relationship, where MCI contributes to MS risk, is feasible, though some studies report the lack of association between these clinical entities [195–198]. A causal relationship between MS and MCI has yet to be conclusively identified.

5. Conclusions

The current epidemic of metabolic disorders, framed in what we term MS, increases in a society with unhealthy lifestyles, and many aspects of this condition are still unknown. Research on MS generates a constant stream of new information and a flood of debate on whether this pathology exists, its components, and the pathophysiological mechanisms that produce it.

Recently, a two-way relationship between MS and brain disorders, such as MCI and AD, has been observed, but without clarity regarding which phenomenon occurs first and which pathophysiological pathways are involved. Contrasting findings could be attributed to factors inherent in the complex nature of MS and MCI. Both disorders are multifactorial and display disparity in clinical manifestations. Further, research methodology is heterogeneous, reflecting the variability in criteria used to define MS, methods used for evaluating cognitive function, study design, and the presence of confounding factors. The latter factors might be varying characteristics of the populations studied, such as age, sex, race, educational status, socioeconomic status, and health–disease status. Large-scale studies with adequate power and longer follow-up periods will be necessary to establish a direct and accurate causal relationship between MS and MCI pathologies.

Author Contributions: Conceptualization: M.R., M.C.-C., E.D.-C. and V.B.; investigation: S.D.A.; M.R., M.C.-C., D.P., H.P. and M.N.; writing—original draft: M.G.-D.; S.D.A.; M.R., D.P., H.P. and M.N.; writing—review and editing: M.R., M.C.-C., M.C., L.A., R.A., J.S., V.B., R.O. and E.D.-C.; funding acquisition: V.B., E.D.-C., M.G.-D. and M.C. All authors have read and agreed to the published version of the manuscript.

Funding: This work was supported by research grant no. 2021810819 from La Universidad Simón Bolívar, Cúcuta, Colombia, and CC-0437-10-21-09-10 from Consejo de Desarrollo Científico, Humanístico y Tecnológico (CONDES), University of Zulia, and the research grant no. FZ-0058-2007 from Fundacite-Zulia.

Acknowledgments: This paper is derived from the doctoral dissertation work of Edgar Díaz-Camargo, student of the doctorate in Psychology at the Universidad Simón Bolívar.

Conflicts of Interest: The authors have no conflict of interest to disclose.

References

1. Hennekens, C.H.; Andreotti, F. Leading Avoidable Cause of Premature Deaths Worldwide: Case for Obesity. *Am. J. Med.* **2013**, *126*, 97–98. [CrossRef] [PubMed]
2. Bermúdez, V.; Añez, R.; Salazar, J.J.; Sanchez, H.; Castellanos, B.; Bello, L.; Villalobos, M. Comportamiento Epidemiológico del síndrome metabólico en el municipio Maracaibo-Venezuela. *Síndrome Cardiometabólico* **2013**, *3*, 31–42.
3. Alberti, K.G.M.M.; Eckel, R.H.; Grundy, S.M.; Zimmet, P.Z.; Cleeman, J.I.; Donato, K.A.; Smith, S.C., Jr. Harmonizing the metabolic syndrome: A joint interim statement of the International Diabetes Federation Task Force on Epidemiology and Prevention; National Heart, Lung, and Blood Institute; American Heart Association; World Heart Federation; International Atherosclerosis Society; and International Association for the Study of Obesity. *Circulation* **2009**, *120*, 1640–1645.
4. Morales Aguilar, R.; Lastre-Amell, G.; Pardo Vásquez, A. Estilos de vida relacionados con factores de riesgo cardiovascular. *Arch. Venez. Farmacol. Ter.* **2018**, *38*, 9.
5. Mente, A.; Yusuf, S.; Islam, S.; McQueen, M.J.; Tanomsup, S.; Onen, C.L.; Rangarajan, S.; Gerstein, H.C.; Anand, S.S. Metabolic syndrome and risk of acute myocardial infarction a case-control study of 26,903 subjects from 52 countries. *J. Am. Coll. Cardiol.* **2010**, *55*, 2390–2398. [CrossRef]
6. Espinoza Diaz, C.I.E.; Morocho Zambrano, A.A.; Pesantez Placencia, L.F.; Toala Guerrero, J.E.; Bravo Rey, P.J.; Garavito Martinez, A.M.; Carbo Tapia, A.D.; García Vargas, J.J. Prevalencia de síndrome metabólico y factores asociados en adultos mayores de la parroquia de Baños, Cuenca. *Arch. Venez. Farmacol. Ter.* **2018**, *39*, 6.
7. Kerekes, G.; Nurmohamed, M.T.; González-Gay, M.A.; Seres, I.; Paragh, G.; Kardos, Z.; Baráth, Z.; Tamási, L.; Soltész, P.; Szekanecz, Z. Rheumatoid arthritis and metabolic syndrome. *Nat. Rev. Rheumatol.* **2014**, *10*, 691–696. [CrossRef] [PubMed]
8. Uzunlulu, M.; Caklili, O.T.; Oguz, A. Association between Metabolic Syndrome and Cancer. *Ann. Nutr. Metab.* **2016**, *68*, 173–179. [CrossRef]
9. Bangen, K.J.; Armstrong, N.M.; Au, R.; Gross, A.L. Metabolic Syndrome and Cognitive Trajectories in the Framingham Offspring Study. *J. Alzheimer's Dis.* **2019**, *71*, 931–943. [CrossRef]

10. Laws, S.M.; Gaskin, S.; Woodfield, A.; Srikanth, V.; Bruce, D.; Fraser, P.E.; Porter, T.; Newsholme, P.; Wijesekara, N.; Burnham, S.; et al. Insulin resistance is associated with reductions in specific cognitive domains and increases in CSF tau in cognitively normal adults. *Sci. Rep.* **2017**, *7*, 1–11. [CrossRef]
11. Petersen, R.C.; Roberts, R.O.; Knopman, D.S.; Boeve, B.F.; Geda, Y.E.; Ivnik, R.J.; Smith, G.E.; Jack, C.R., Jr. Mild cognitive impairment: Ten years later. *Arch. Neurol.* **2009**, *66*, 1447–1455. [CrossRef] [PubMed]
12. Sanford, A.M. Mild Cognitive Impairment. *Clin. Geriatr. Med.* **2017**, *33*, 325–337. [CrossRef] [PubMed]
13. Vanegas, H. Buscando las bases moleculares de la enfermedad de Alzheimer. *Gac. Médica Caracas* **2017**, *125*, 4–11.
14. Biessels, G.J.; Despa, F. Cognitive decline and dementia in diabetes mellitus: Mechanisms and clinical implications. *Nat. Rev. Endocrinol.* **2018**, *14*, 591–604. [CrossRef] [PubMed]
15. Arnold, S.E.; Arvanitakis, Z.; Macauley-Rambach, S.L.; Koenig, A.M.; Wang, H.-Y.; Ahima, R.S.; Craft, S.; Gandy, S.; Buettner, C.; Stoeckel, L.E.; et al. Brain insulin resistance in type 2 diabetes and Alzheimer disease: Concepts and conundrums. *Nat. Rev. Neurol.* **2018**, *14*, 168–181. [CrossRef]
16. Kylin, E. Hypertonie and Zuckerkrankheit. *Zent. Inn. Med.* **1921**, *42*, 873–877.
17. Marañon, G. Über Hypertonie and Zuckerkrankheit. *Zent. Inn. Med.* **1922**, *43*, 169–176.
18. Hanefeld, M.; Leonhardt, W. Das Metabolische Syndrom. *Dt Gesundh Wesen.* **1981**, *36*, 545–551. [CrossRef]
19. Reaven, G.M. Banting lecture Role of insulin resistance in human disease. *Diabetes* **1988**, *37*, 1595–1607. [CrossRef]
20. Reaven, G.M. Why Syndrome X? From Harold Himsworth to the Insulin Resistance Syndrome. *Cell Metab.* **2005**, *1*, 9–14. [CrossRef]
21. Randle, P.; Garland, P.; Hales, C.; Newsholme, E. The glucose fatty-acid cycle its role in insulin sensitivity and the metabolic disturbances of diabetes mellitus. *Lancet* **1963**, *281*, 785–789. [CrossRef]
22. DeFronzo, R.A.; Ferrannini, E. Insulin resistance. A multifaceted syndrome responsible for NIDDM, obesity, hypertension, dyslipidemia, and atherosclerotic cardiovascular disease. *Diabetes Care* **1991**, *14*, 173–194. [CrossRef] [PubMed]
23. Alberti, K.G.; Zimmet, P.Z. Definition, diagnosis and classification of diabetes mellitus and its complications. Part 1: Diagnosis and classification of diabetes mellitus provisional report of a WHO consultation. *Diabet. Med. J. Br. Diabet. Assoc.* **1998**, *15*, 539–553. [CrossRef]
24. Balkau, B.; Charles, M.A. Comment on the provisional report from the WHO consultation. *Diabet. Med.* **1999**, *16*, 442–443. [CrossRef]
25. Reaven, G.M. The metabolic syndrome: Is this diagnosis necessary? *Am. J. Clin. Nutr.* **2006**, *83*, 1237–1247. [CrossRef]
26. Einhorn, D.; Reaven, G.M.; Cobin, R.H.; Ford, E.; Ganda, O.P.; Handelsman, Y.; Hellman, R.; Jellinger, P.S.; Kendall, D.; Krauss, R.M.; et al. American College of Endocrinology position statement on the insulin resistance syndrome. *Endocr. Pract.* **2003**, *9*, 237–252. [CrossRef] [PubMed]
27. Alberti, K.G.M.M.; Zimmet, P.; Shaw, J. Metabolic syndrome-a new world-wide definition. A Consensus Statement from the International Diabetes Federation. *Diabet. Med.* **2006**, *23*, 469–480. [CrossRef]
28. Alegría Ezquerra, E.; Castellano Vázquez, J.M.; Alegría Barrero, A. Obesity, metabolic syndrome and diabetes: Cardiovascular implications and therapy. *Rev. Esp. Cardiol.* **2008**, *61*, 752–764. [CrossRef]
29. Gunczler, P. Síndrome de resistencia a la insulina en niños y adolescentes. *Gac. Médica. Caracas* **2006**, *114*, 99–103.
30. Després, J.-P.; Lemieux, I.; Bergeron, J.; Pibarot, P.; Mathieu, P.; LaRose, E.; Rodés-Cabau, J.; Bertrand, O.F.; Poirier, P. Abdominal Obesity and the Metabolic Syndrome: Contribution to Global Cardiometabolic Risk. *Arter. Thromb. Vasc. Biol.* **2008**, *28*, 1039–1049. [CrossRef]
31. Jais, A.; Brüning, J.C. Hypothalamic inflammation in obesity and metabolic disease. *J. Clin. Investig.* **2017**, *127*, 24–32. [CrossRef]
32. Rönnemaa, E.; Zethelius, B.; Sundelöf, J.; Sundström, J.; Degerman-Gunnarsson, M.; Berne, C.; Lannfelt, L.; Kilander, L. Impaired insulin secretion increases the risk of Alzheimer disease. *Neurology* **2008**, *71*, 1065–1071. [CrossRef]
33. Cohen, A.D.; Klunk, W.E. Early detection of Alzheimer's disease using PiB and FDG PET. *Neurobiol. Dis.* **2014**, *72*, 117–122. [CrossRef]
34. Segura, B.; Jurado, M.Á.; Freixenet, N.; Albuin, C.; Muniesa, J.; Junque, C. Mental slowness and executive dysfunctions in patients with metabolic syndrome. *Neurosci. Lett.* **2009**, *462*, 49–53. [CrossRef]
35. Narváez López, E.J.; Bravo Peláez, J.A.; Almeida Lozano, K.A.; Alvarez Rivera, C.G.; Mendoza Argandoña, C.A.; Morales Sánchez, A.M.; Godos Rivera, D.T.; Del Salto Ocaña, T.E.; Catota Camacho, M.M. Implicación de polimorfismos de apolipoproteína en la fisiopatología de la ateroesclerosis y enfermedad de Alzheimer. *Rev. Latinoam. Hipertens.* **2018**, *13*, 6.
36. Anstey, K.J.; Cherbuin, N.; Budge, M.; Young, J. Body mass index in midlife and late-life as a risk factor for dementia: A me-ta-analysis of prospective studies. *Obes. Rev. Off. J. Int. Assoc. Study Obes.* **2011**, *12*, 426–437. [CrossRef]
37. Grillo, C.; Woodruff, J.L.; Macht, V.A.; Reagan, L.P. Insulin resistance and hippocampal dysfunction: Disentangling peripheral and brain causes from consequences. *Exp. Neurol.* **2019**, *318*, 71–77. [CrossRef]
38. Lindqvist, A.; Mohapel, P.; Bouter, B.; Frielingsdorf, H.; Pizzo, D.; Brundin, P.; Erlanson-Albertsson, C. High-fat diet impairs hippocampal neurogenesis in male rats. *Eur. J. Neurol.* **2006**, *13*, 1385–1388. [CrossRef]
39. Karimi, S.A.; Salehi, I.; Komaki, A.; Sarihi, A.; Zarei, M.; Shahidi, S. Effect of high-fat diet and antioxidants on hippocampal long-term potentiation in rats: An in vivo study. *Brain Res.* **2013**, *1539*, 1–6. [CrossRef] [PubMed]
40. Nguyen, T.T.; Ta, Q.T.H.; Nguyen, T.T.D.; Le, T.T.; Vo, V.G. Role of Insulin Resistance in the Alzheimer's Disease Progression. *Neurochem. Res.* **2020**, *45*, 1481–1491. [CrossRef] [PubMed]

41. Vander Zanden, C.M.; Chi, E.Y. Passive immunotherapies targeting amyloid beta and Tau oligomers in Alzheimer's disease. *J. Pharm. Sci.* **2020**, *109*, 68–73. [CrossRef]
42. Hurrle, S.; Hsu, W.H. The etiology of oxidative stress in insulin resistance. *Biomed. J.* **2017**, *40*, 257–262. [CrossRef] [PubMed]
43. Kinney, J.W.; Bemiller, S.M.; Murtishaw, A.S.; Leisgang, A.M.; Salazar, A.M.; Lamb, B.T. Inflammation as a central mechanism in Alz-heimer's disease. *Alzheimers Dement. Transl. Res. Clin. Interv.* **2018**, *4*, 575–590. [CrossRef]
44. Walker, J.M.; Harrison, F.E. Shared Neuropathological Characteristics of Obesity, Type 2 Diabetes and Alzheimer's Disease: Impacts on Cognitive Decline. *Nutrients* **2015**, *7*, 7332–7357. [CrossRef]
45. Shiiki, T.; Ohtsuki, S.; Kurihara, A.; Naganuma, H.; Nishimura, K.; Tachikawa, M.; Hosoya, K.; Terasaki, T. Brain insulin impairs amyloid-beta(1-40) clearance from the brain. *J. Neurosci. Off. J. Soc. Neurosci.* **2004**, *24*, 9632–9637. [CrossRef]
46. Liu, Z.; Patil, I.Y.; Jiang, T.; Sancheti, H.; Walsh, J.P.; Stiles, B.L.; Yin, F.; Cadenas, E. High-Fat Diet Induces Hepatic Insulin Resistance and Impairment of Synaptic Plasticity. *PLoS ONE* **2015**, *10*, e0128274. [CrossRef]
47. Zeng, Y.; Zhang, L.; Hu, Z. Cerebral insulin, insulin signaling pathway, and brain angiogenesis. *Neurol. Sci.* **2016**, *37*, 9–16. [CrossRef]
48. Liang, C.; Lam, P.; Martinez, S.; Mukherjee, J. Development of [18F]FAZIN3 for PET imaging of neurofibrillary tangles in Alz-heimer's Disease. *J. Nucl. Med.* **2020**, *61*, 1032.
49. Benedict, C.; Grillo, C. Insulin Resistance as a Therapeutic Target in the Treatment of Alzheimer's Disease: A State-of-the-Art Review. *Front. Neurosci.* **2018**, *12*, 215. [CrossRef]
50. Yarchoan, M.; Toledo, J.; Lee, E.B.; Arvanitakis, Z.; Kazi, H.; Han, L.-Y.; Louneva, N.; Lee, V.M.-Y.; Kim, S.F.; Trojanowski, J.Q.; et al. Abnormal serine phosphorylation of insulin receptor substrate 1 is associated with tau pathology in Alzheimer's disease and tauopathies. *Acta Neuropathol.* **2014**, *128*, 679–689. [CrossRef]
51. Starks, E.J.; Patrick O'Grady, J.; Hoscheidt, S.M.; Racine, A.M.; Carlsson, C.M.; Zetterberg, H.; Blennow, K.; Okonkwo, O.C.; Puglielli, L.; Asthana, S.; et al. Insulin resistance is associated with higher cerebrospinal fluid Tau levels in asymptomatic APOE ε4 Carriers. *J. Alzheimers Dis. JAD* **2015**, *46*, 525–533. [CrossRef]
52. Kim, B.; Sullivan, K.A.; Backus, C.; Feldman, E.L. Cortical Neurons Develop Insulin Resistance and Blunted Akt Signaling: A Potential Mechanism Contributing to Enhanced Ischemic Injury in Diabetes. *Antioxidants Redox Signal.* **2011**, *14*, 1829–1839. [CrossRef]
53. Zhang, Y.; Huang, N.-Q.; Yan, F.; Jin, H.; Zhou, S.-Y.; Shi, J.-S.; Jin, F. Diabetes mellitus and Alzheimer's disease: GSK-3β as a potential link. *Behav. Brain Res.* **2018**, *339*, 57–65. [CrossRef]
54. Esposito, G.; Scuderi, C.; Lu, J.; Savani, C.; De Filippis, D.; Iuvone, T.; Steardo, L., Jr.; Sheen, V.; Steardo, L. S100B induces tau protein hyperphosphorylation via Dickopff-1 up-regulation and disrupts the Wnt pathway in human neural stem cells. *J. Cell. Mol. Med.* **2008**, *12*, 914–927. [CrossRef]
55. van der Harg, J.M.; Eggels, L.; Bangel, F.N.; Ruigrok, S.R.; Zwart, R.; Hoozemans, J.J.M.; la Fleur, S.E.; Scheper, W. Insulin deficiency results in reversible protein kinase A activation and tau phosphorylation. *Neurobiol. Dis.* **2017**, *103*, 163–173. [CrossRef]
56. Gratuze, M.; Julien, J.; Petry, F.R.; Morin, F.; Planel, E. Insulin deprivation induces PP2A inhibition and tau hyperphosphorylation in hTau mice, a model of Alzheimer's disease-like tau pathology. *Sci. Rep.* **2017**, *7*, srep46359. [CrossRef]
57. Kins, S.; Crameri, A.; Evans, D.R.; Hemmings, B.A.; Nitsch, R.M.; Gotz, J. Reduced protein phosphatase 2A activity induces hyper-phosphorylation and altered compartmentalization of tau in transgenic mice. *J. Biol. Chem.* **2001**, *276*, 38193–38200. [CrossRef]
58. Planel, E.; Tatebayashi, Y.; Miyasaka, T.; Liu, L.; Wang, L.; Herman, M.; Yu, W.H.; Luchsinger, J.A.; Wadzinski, B.; Duff, K.E.; et al. Insulin dysfunction induces in vivo tau hyperphos-phorylation through distinct mechanisms. *J. Neurosci. Off. J. Soc. Neurosci.* **2007**, *27*, 13635–13648. [CrossRef]
59. Zilka, N.; Filipcik, P.; Koson, P.; Fialova, L.; Skrabana, R.; Zilkova, M.; Rolkova, G.P.; Kontsekova, E.; Novak, M. Truncated tau from sporadic Alzheimer's disease suffices to drive neurofibrillary degeneration in vivo. *FEBS Lett.* **2006**, *580*, 3582–3588. [CrossRef]
60. Kim, B.; Backus, C.; Oh, S.; Hayes, J.M.; Feldman, E.L. Increased Tau Phosphorylation and Cleavage in Mouse Models of Type 1 and Type 2 Diabetes. *Endocrinology* **2009**, *150*, 5294–5301. [CrossRef]
61. Kim, B.; Backus, C.; Oh, S.; Feldman, E.L. Hyperglycemia-Induced Tau Cleavage in vitro and in vivo: A Possible Link Between Diabetes and Alzheimer's Disease. *J. Alzheimer's Dis.* **2013**, *34*, 727–739. [CrossRef]
62. Forny-Germano, L.; De Felice, F.G.; Vieira, M.N.D.N. The Role of Leptin and Adiponectin in Obesity-Associated Cognitive Decline and Alzheimer's Disease. *Front. Neurosci.* **2019**, *12*, 1027. [CrossRef]
63. Suyama, S.; Maekawa, F.; Maejima, Y.; Kubota, N.; Kadowaki, T.; Yada, T. Glucose level determines excitatory or inhibitory effects of adiponectin on arcuate POMC neuron activity and feeding. *Sci. Rep.* **2016**, *6*, 30796. [CrossRef]
64. Friedman, J. The long road to leptin. *J. Clin. Investig.* **2016**, *126*, 4727–4734. [CrossRef]
65. Bouret, S.G. Neurodevelopmental actions of leptin. *Brain Res.* **2010**, *1350*, 2–9. [CrossRef]
66. Pousti, F.; Ahmadi, R.; Mirahmadi, F.; Hosseinmardi, N.; Rohampour, K. Adiponectin modulates synaptic plasticity in hippocampal dentate gyrus. *Neurosci. Lett.* **2018**, *662*, 227–232. [CrossRef]
67. Thundyil, J.; Pavlovski, D.; Sobey, C.G.; Arumugam, T.V. Adiponectin receptor signalling in the brain. *Br. J. Pharmacol.* **2011**, *165*, 313–327. [CrossRef]

68. Li, X.-L.; Aou, S.; Oomura, Y.; Hori, N.; Fukunaga, K.; Hori, T. Impairment of long-term potentiation and spatial memory in leptin receptor-deficient rodents. *Neuroscience* **2002**, *113*, 607–615. [CrossRef]
69. Pérez-González, R.; Alvira-Botero, M.X.; Robayo, O.; Antequera, D.; Garzón, M.; Martín-Moreno, A.M.; Brera, B.; De Ceballos, M.L.; Carro, E. Leptin gene therapy attenuates neuronal damages evoked by amyloid-β and rescues memory deficits in APP/PS1 mice. *Gene Ther.* **2014**, *21*, 298–308. [CrossRef]
70. Fewlass, D.C.; Noboa, K.; Pi-Sunyer, F.X.; Johnston, J.M.; Yan, S.D.; Tezapsidis, N. Obesity-related leptin regulates Alzheimer's Abeta. *FASEB J. Off. Publ. Fed. Am. Soc. Exp. Biol.* **2004**, *18*, 1870–1878.
71. Holden, K.F.; Lindquist, K.; Tylavsky, F.A.; Rosano, C.; Harris, T.B.; Yaffe, K. Serum leptin level and cognition in the elderly: Findings from the Health ABC Study. *Neurobiol. Aging* **2009**, *30*, 1483–1489. [CrossRef]
72. Ng, R.C.-L.; Chan, K.-H. Potential Neuroprotective Effects of Adiponectin in Alzheimer's Disease. *Int. J. Mol. Sci.* **2017**, *18*, 592. [CrossRef]
73. Kim, M.W.; Abid N bin Jo, M.H.; Jo, M.G.; Yoon, G.H.; Kim, M.O. Suppression of adiponectin receptor 1 promotes memory dysfunction and Alzheimer's disease-like pathologies. *Sci. Rep.* **2017**, *7*, 12435. [CrossRef]
74. Viswanathan, A.; Rocca, W.A.; Tzourio, C. Vascular risk factors and dementia: How to move forward? *Neurology* **2009**, *72*, 368–374. [CrossRef]
75. Borshchev, Y.Y.; Uspensky, Y.P.; Galagudza, M.M. Pathogenetic pathways of cognitive dysfunction and dementia in metabolic syndrome. *Life Sci.* **2019**, *237*, 116932. [CrossRef]
76. Pantoni, L. Cerebral small vessel disease: From pathogenesis and clinical characteristics to therapeutic challenges. *Lancet Neurol.* **2010**, *9*, 689–701. [CrossRef]
77. Moss, M.B.; Jonak, E. Cerebrovascular disease and dementia: A primate model of hypertension and cognition. *Alzheimer's Dement.* **2007**, *3*, S6–S15. [CrossRef]
78. Veglio, F.; Paglieri, C.; Rabbia, F.; Bisbocci, D.; Bergui, M.; Cerrato, P. Hypertension and cerebrovascular damage. *Atherosclerosis* **2009**, *205*, 331–341. [CrossRef]
79. Frisardi, V.; Solfrizzi, V.; Seripa, D.; Capurso, C.; Santamato, A.; Sancarlo, D.; Vendemiale, G.; Pilotto, A.; Panza, F. Metabolic-cognitive syndrome: A cross-talk between metabolic syndrome and Alzheimer's disease. *Ageing Res. Rev.* **2010**, *9*, 399–417. [CrossRef]
80. Reynolds, C.H.; Garwood, C.J.; Wray, S.; Price, C.; Kellie, S.; Perera, T.; Zvelebil, M.; Yang, A.; Sheppard, P.W.; Varndell, I.M.; et al. Phosphorylation Regulates Tau Interactions with Src Homology 3 Domains of Phosphatidylinositol 3-Kinase, Phospholipase Cγ1, Grb2, and Src Family Kinases. *J. Biol. Chem.* **2008**, *283*, 18177–18186. [CrossRef]
81. Marciniak, E.; Leboucher, A.; Caron, E.; Ahmed, T.; Tailleux, A.; Dumont, J.; Issad, T.; Gerhardt, E.; Pagesy, P.; Vileno, M.; et al. Tau deletion promotes brain insulin resistance. *J. Exp. Med.* **2017**, *214*, 2257–2269. [CrossRef] [PubMed]
82. Obici, S.; Feng, Z.; Karkanias, G.; Baskin, D.G.; Rossetti, L. Decreasing hypothalamic insulin receptors causes hyperphagia and insulin resistance in rats. *Nat. Neurosci.* **2002**, *5*, 566–572. [CrossRef] [PubMed]
83. Brüning, J.C.; Gautam, D.; Burks, D.J.; Gillette, J.; Schubert, M.; Orban, P.C.; Klein, R.; Krone, W.; Müller-Wieland, D.; Kahn, C.R. Role of brain insulin receptor in control of body weight and reproduction. *Science* **2000**, *289*, 2122–2125. [CrossRef]
84. Bharadwaj, P.; Wijesekara, N.; Liyanapathirana, M.; Newsholme, P.; Ittner, L.; Fraser, P.; Verdile, G. The Link between Type 2 Diabetes and Neurodegeneration: Roles for Amyloid-β, Amylin, and Tau Proteins. *J. Alzheimer's Dis.* **2017**, *59*, 421–432. [CrossRef]
85. Wijesekara, N.; Ahrens, R.; Sabale, M.; Wu, L.; Ha, K.; Verdile, G.; Fraser, P.E. Amyloid-β and islet amyloid pathologies link Alzheimer's disease and type 2 diabetes in a transgenic model. *FASEB J. Off. Publ. Fed. Am. Soc. Exp. Biol.* **2017**, *31*, 5409–5418. [CrossRef]
86. Wijesekara, N.; Gonçalves, R.A.; Ahrens, R.; De Felice, F.G.; Fraser, P.E. Tau ablation in mice leads to pancreatic β cell dysfunction and glucose intolerance. *FASEB J.* **2018**, *32*, 3166–3173. [CrossRef]
87. Xie, L.; Helmerhorst, E.; Taddei, K.; Plewright, B.; Van Bronswijk, W.; Martins, R. Alzheimer's beta-amyloid peptides compete for insulin binding to the insulin receptor. *J. Neurosci. Off. J. Soc. Neurosci.* **2002**, *22*, 221. [CrossRef]
88. Zhang, Y.; Zhou, B.; Zhang, F.; Wu, J.; Hu, Y.; Liu, Y.; Zhai, Q. Amyloid-β induces hepatic insulin resistance by activating JAK2/STAT3/SOCS-1 signaling pathway. *Diabetes* **2012**, *61*, 1434–1443. [CrossRef]
89. De Felice, F.G.; Velasco, P.T.; Lambert, M.P.; Viola, K.; Fernandez, S.J.; Ferreira, S.T.; Klein, W.L. Abeta oligomers induce neuronal oxidative stress through an N-methyl-D-aspartate receptor-dependent mechanism that is blocked by the Alzheimer drug memantine. *J. Biol. Chem.* **2007**, *282*, 11590–11601. [CrossRef]
90. Kim, J.-A.; Wei, Y.; Sowers, J.R. Role of Mitochondrial Dysfunction in Insulin Resistance. *Circ. Res.* **2008**, *102*, 401–414. [CrossRef] [PubMed]
91. Zhao, W.; De Felice, F.G.; Fernandez, S.; Chen, H.; Lambert, M.P.; Quon, M.J.; Krafft, G.A.; Klein, W.L. Amyloid beta oligomers induce impairment of neuronal insulin receptors. *FASEB J.* **2007**, *22*, 246–260. [CrossRef]
92. Bomfim, T.R.; Forny-Germano, L.; Sathler, L.B.; Brito-Moreira, J.; Houzel, J.C.; Decker, H.; Silverman, M.A.; Kazi, H.; Melo, H.M.; McClean, P.L.; et al. An anti-diabetes agent protects the mouse brain from defective insulin signaling caused by Alzheimer's disease–associated Aβ oligomers. *J. Clin. Investig.* **2012**, *122*, 1339–1353. [CrossRef]
93. Bonda, D.J.; Stone, J.G.; Torres, S.L.; Siedlak, S.L.; Perry, G.; Kryscio, R.; Jicha, G.; Casadesus, G.; Smith, M.A.; Zhu, X.; et al. Dysregulation of leptin signaling in Alzheimer disease: Evidence for neuronal leptin resistance. *J. Neurochem.* **2014**, *128*, 162–172. [CrossRef]

94. Kanoski, S.E.; Hayes, M.R.; Greenwald, H.S.; Fortin, S.M.; Gianessi, C.A.; Gilbert, J.R.; Grill, H.J. Hippocampal Leptin Signaling Reduces Food Intake and Modulates Food-Related Memory Processing. *Neuropsychopharmacology* **2011**, *36*, 1859–1870. [CrossRef]
95. Meakin, P.J.; Jalicy, S.M.; Montagut, G.; Allsop, D.J.P.; Cavellini, D.L.; Irvine, S.W.; McGinley, C.; Liddell, M.K.; McNeilly, A.D.; Parmionova, K.; et al. Bace1-dependent amyloid processing regulates hypothalamic leptin sensitivity in obese mice. *Sci. Rep.* **2018**, *8*, 55. [CrossRef] [PubMed]
96. Brief, D.J.; Davis, J.D. Reduction of food intake and body weight by chronic intraventricular insulin infusion. *Brain Res. Bull.* **1984**, *12*, 571–575. [CrossRef]
97. Panza, F.; Solfrizzi, V.; Logroscino, G.; Maggi, S.; Santamato, A.; Seripa, D.; Pilotto, A. Current epidemiological approaches to the met-abolic-cognitive syndrome. *J. Alzheimers Dis.* **2012**, *30*, S31–S75. [CrossRef] [PubMed]
98. Irimata, K.E.; Dugger, B.N.; Wilson, J.R. Impact of the Presence of Select Cardiovascular Risk Factors on Cognitive Changes among Dementia Subtypes. *Curr. Alzheimer Res.* **2018**, *15*, 1032–1044. [CrossRef] [PubMed]
99. Case, C.C.; Jones, P.H.; Nelson, K.; Smith, E.O.; Ballantyne, C.M. Impact of weight loss on the metabolic syndrome. *Diabetes Obes. Metab.* **2002**, *4*, 407–414. [CrossRef] [PubMed]
100. Yaffe, K.; Kanaya, A.; Lindquist, K.; Simonsick, E.M.; Harris, T.; Shorr, R.I.; Tylavsky, F.A.; Newman, A.B. The Metabolic Syndrome, Inflammation, and Risk of Cognitive Decline. *JAMA* **2004**, *292*, 2237–2242. [CrossRef] [PubMed]
101. Panza, F.; D'Introno, A.; Colacicco, A.M.; Capurso, C.; Del Parigi, A.; Capurso, S.A.; Caselli, R.J.; Pilotto, A.; Scafato, E.; Capurso, A.; et al. Cognitive frailty: Predementia syndrome and vascular risk factors. *Neurobiol. Aging* **2006**, *27*, 933–940. [CrossRef]
102. Dregan, A.; Stewart, R.; Gulliford, M.C. Cardiovascular risk factors and cognitive decline in adults aged 50 and over: A popula-tion-based cohort study. *Age Ageing* **2013**, *42*, 338–345. [CrossRef]
103. Harrison, S.L.; Ding, J.; Tang, E.Y.H.; Siervo, M.; Robinson, L.; Jagger, C.; Stephan, B.C.M. Cardiovascular Disease Risk Models and Longitudinal Changes in Cognition: A Systematic Review. *PLoS ONE* **2014**, *9*, e114431. [CrossRef] [PubMed]
104. Harrison, S.L.; De Craen, A.J.M.; Kerse, N.; Teh, R.; Granic, A.; Davies, K.; Wesnes, K.A.; Elzen, W.D.; Gussekloo, J.; Kirkwood, T.B.L.; et al. Predicting Risk of Cognitive Decline in Very Old Adults Using Three Models: The Framingham Stroke Risk Profile; the Cardiovascular Risk Factors, Aging, and Dementia Model; and Oxi-Inflammatory Biomarkers. *J. Am. Geriatr. Soc.* **2017**, *65*, 381–389. [CrossRef] [PubMed]
105. Purnell, C.; Gao, S.; Callahan, C.M.; Hendrie, H.C. Cardiovascular risk factors and incident Alzheimer disease: A systematic review of the literature. *Alzheimer Dis. Assoc. Disord.* **2009**, *23*, 1–10. [CrossRef]
106. Prasad, K.; Wiryasaputra, L.; Ng, A.; Kandiah, N. White Matter Disease Independently Predicts Progression from Mild Cognitive Impairment to Alzheimer's Disease in a Clinic Cohort. *Dement. Geriatr. Cogn. Disord.* **2011**, *31*, 431–434. [CrossRef]
107. Iadecola, C.; Yaffe, K.; Biller, J.; Bratzke, L.C.; Faraci, F.M.; Gorelick, P.B.; Gulati, M.; Kamel, H.; Knopman, D.S.; Launer, L.J.; et al. Impact of Hypertension on Cognitive Function: A Scientific Statement from the American Heart Association. *Hypertension* **2016**, *68*, e67–e94. [CrossRef] [PubMed]
108. Avila Vinueza, J.P.; Avila Vinueza, T.L.; Pesantez Calle, M.F.; Guaraca Pino, A.C.; Durazno Montesdeoca, G.C.; Cobos Alvarracin, M.Y. Frecuencia, factores de riesgo y hallazgos neuroimagenológicos de deterioro cognitivo leve en pacientes con hipertensión arterial. *Arch. Venez. Farmacol. Ter.* **2019**, *38*, 12.
109. McDonald, C.; Pearce, M.S.; Kerr, S.R.J.; Newton, J.L. Blood pressure variability and cognitive decline in older people: A 5-year longitudinal study. *J. Hypertens.* **2017**, *35*, 140–147. [CrossRef]
110. Tzourio, C.; Dufouil, C.; Ducimetiere, P.; Alperovitch, A. Cognitive decline in individuals with high blood pressure: A longitudinal study in the elderly. *Neurology* **1999**, *53*, 1948. [CrossRef]
111. Haring, B.; Wu, C.; Coker, L.H.; Seth, A.; Snetselaar, L.; Manson, J.E.; Rossouw, J.E.; Wassertheil-Smoller, S. Hypertension, Dietary Sodium, and Cognitive Decline: Results from the Women's Health Initiative Memory Study. *Am. J. Hypertens.* **2015**, *29*, 202–216. [CrossRef]
112. Tarraf, W.; Rodríguez, C.J.; Daviglus, M.L.; Lamar, M.; Schneiderman, N.; Gallo, L.; Talavera, G.A.; Kaplan, R.C.; Fornage, M.; Conceicao, A.; et al. Blood Pressure and Hispanic/Latino Cognitive Function: Hispanic Community Health Study/Study of Latinos Results. *J. Alzheimer's Dis.* **2017**, *59*, 31–42. [CrossRef]
113. Kilander, L.; Nyman, H.; Boberg, M.; Hansson, L.; Lithell, H. Hypertension is related to cognitive impairment: A 20-year follow-up of 999 men. *Hypertension* **1998**, *31*, 780–786. [CrossRef] [PubMed]
114. Launer, L.J. The association between midlife blood pressure levels and late-life cognitive function. The Honolulu-Asia Aging Study. *JAMA* **1995**, *274*, 1846–1851. [CrossRef] [PubMed]
115. Walker, K.A.; Sharrett, A.R.; Wu, A.; Schneider, A.L.C.; Albert, M.; Lutsey, P.L.; Bandeen-Roche, K.; Coresh, J.; Gross, A.L.; Windham, B.G.; et al. Association of Midlife to Late-Life Blood Pressure Patterns with Incident Dementia. *JAMA* **2019**, *322*, 535–545. [CrossRef] [PubMed]
116. Reitz, C.; Tang, M.-X.; Manly, J.; Mayeux, R.; Luchsinger, J.A. Hypertension and the Risk of Mild Cognitive Impairment. *Arch. Neurol.* **2007**, *64*, 1734–1740. [CrossRef] [PubMed]
117. Elias, M.F.; Elias, P.K.; Sullivan, L.M.; Wolf, P.A.; D'Agostino, R.B. Lower cognitive function in the presence of obesity and hyper-tension: The Framingham heart study. *Int. J. Obes. Relat. Metab. Disord. J. Int. Assoc. Study Obes.* **2003**, *27*, 260–268. [CrossRef] [PubMed]

118. Chacón, O.; Riaño-Garzón, M.E.; Bermúdez, V.; Quintero Sanguino, M.; Hernández Lalinde, J.D.; Mendoza Bernal, M.I. ¿Es la obesidad un factor de riesgo para el trastorno de déficit de atención con hiperactividad (TDAH)? *Rev. Latinoam Hipertens.* **2018**, *13*, 89–97.
119. Cournot, M.; Marquie, J.C.; Ansiau, D.; Martinaud, C.; Fonds, H.; Ferrieres, J.; Ruidavets, J.B. Relation between body mass index and cognitive function in healthy middle-aged men and women. *Neurology* **2006**, *67*, 1208–1214. [CrossRef]
120. Kloppenborg, R.P.; Berg, E.V.D.; Kappelle, L.J.; Biessels, G.J. Diabetes and other vascular risk factors for dementia: Which factor matters most? A systematic review. *Eur. J. Pharmacol.* **2008**, *585*, 97–108. [CrossRef]
121. Sabia, S.; Kivimaki, M.; Shipley, M.J.; Marmot, M.; Singh-Manoux, A. Body mass index over the adult life course and cognition in late midlife: The Whitehall II Cohort Study. *Am. J. Clin. Nutr.* **2008**, *89*, 601–607. [CrossRef]
122. Beydoun, M.A.; Wang, Y. Obesity and central obesity as risk factors for incident dementia and its subtypes: A systematic review and meta-analysis. *Obes. Rev.* **2008**, *9*, 204–218. [CrossRef] [PubMed]
123. Blom, K.; Emmelot-Vonk, M.H.; Koek, H.L. The influence of vascular risk factors on cognitive decline in patients with dementia: A systematic review. *Maturitas* **2013**, *76*, 113–117. [CrossRef]
124. de Frias, C.M.; Bunce, D.; Wahlin, A.; Adolfsson, R.; Sleegers, K.; Cruts, M.; van Broeckhoven, C.; Nilsson, L. Cholesterol and triglycerides moderate the effect of apolipoprotein E on memory functioning in older adults. *J. Gerontol. B Psychol. Sci. Soc. Sci.* **2007**, *62*, 112–118. [CrossRef] [PubMed]
125. Sims, R.; Madhere, S.; Callender, C.; Campbell, A. Patterns of Relationships between Cardiovascular Disease Risk Factors and Neurocognitive Function in African Americans. *Ethn. Dis.* **2008**, *18*, 471–476. [PubMed]
126. Singh-Manoux, A.; Gimeno, D.; Kivimaki, M.; Brunner, E.; Marmot, M.G. Low HDL cholesterol is a risk factor for deficit and decline in memory in midlife: The Whitehall II study. *Arterioscler Thromb. Vasc. Biol.* **2008**, *28*, 1556–1562. [CrossRef]
127. Zuliani, G.; Cavalieri, M.; Galvani, M.; Volpato, S.; Cherubini, A.; Bandinelli, S.; Corsi, A.M.; Lauretani, F.; Guralnik, J.M.; Fellin, R.; et al. Relationship Between Low Levels of High-Density Lipoprotein Cholesterol and Dementia in the Elderly. The InChianti Study. *J. Gerontol. Ser. A Biol. Sci. Med. Sci.* **2010**, *65*, 559–564. [CrossRef] [PubMed]
128. van Vliet, P.; van de Water, W.; de Craen, A.J.M.; Westendorp, R.G.J. The influence of age on the association between cholesterol and cognitive function. *Exp. Gerontol.* **2009**, *44*, 112–122. [CrossRef]
129. Kinno, R.; Mori, Y.; Kubota, S.; Nomoto, S.; Futamura, A.; Shiromaru, A.; Kuroda, T.; Yano, S.; Ishigaki, S.; Murakami, H.; et al. High serum high-density lipoprotein-cholesterol is associated with memory function and gyrification of insular and frontal opercular cortex in an elderly memory-clinic pop-ulation. *NeuroImage Clin.* **2019**, *22*, 101746. [CrossRef]
130. Bonarek, M.; Barberger-Gateau, P.; Letenneur, L.; Deschamps, V.; Iron, A.; Dubroca, B.; Dartigues, J.F. Relationships between cholesterol, apolipoprotein E polymorphism and dementia: A cross-sectional analysis from the PAQUID study. *Neuroepidemiology* **2000**, *19*, 141–148. [CrossRef]
131. Sanz, C.; Andrieu, S.; Sinclair, A.; Hanaire, H.; Vellas, B. For the REAL.FR Study Group Diabetes is associated with a slower rate of cognitive decline in Alzheimer disease. *Neurology* **2009**, *73*, 1359–1366. [CrossRef]
132. Solimany, F.; Mohammadi, E.; Omidfar, F. Comparison of cognitive abilities, depression and anxiety of type II diabetic patients with healthy individuals in Isfahan province in 2015. *Rev. Latinoam. Hipertens.* **2018**, *13*, 8.
133. Marseglia, A.; Fratiglioni, L.; Kalpouzos, G.; Wang, R.; Bäckman, L.; Xu, W. Prediabetes and diabetes accelerate cognitive decline and predict microvascular lesions: A population-based cohort study. *Alzheimer's Dement.* **2019**, *15*, 25–33. [CrossRef]
134. Rouch, I.; Roche, F.; Dauphinot, V.; Laurent, B.; Antérion, C.T.; Celle, S.; Krolak-Salmon, P.; Barthélémy, J.-C. Diabetes, impaired fasting glucose, and cognitive decline in a population of elderly community residents. *Aging Clin. Exp. Res.* **2012**, *24*, 377–383. [CrossRef] [PubMed]
135. Yaffe, K.; Blackwell, T.; Kanaya, A.M.; Davidowitz, N.; Barrett-Connor, E.; Krueger, K. Diabetes, impaired fasting glucose, and development of cognitive impairment in older women. *Neurology* **2004**, *63*, 658–663. [CrossRef] [PubMed]
136. Vanhanen, M.; Koivisto, K.; Kuusisto, J.; Mykkänen, L.; Helkala, E.-L.; Hänninen, T.; Riekkinen, P.; Soininen, H.; Laakso, M. Cognitive function in an elderly population with persistent impaired glucose tolerance. *Diabetes Care* **1998**, *21*, 398–402. [CrossRef] [PubMed]
137. Kanaya, A.M.; Barrett-Connor, E.; Gildengorin, G.; Yaffe, K. Change in cognitive function by glucose tolerance status in older adults: A 4-year prospective study of the Rancho Bernardo study cohort. *Arch. Intern. Med.* **2004**, *164*, 1327–1333. [CrossRef]
138. Miles, W.R.; Root, H.F. Psychologic Tests Applied to Diabetic Patients. *Arch. Intern. Med.* **1922**, *30*, 767–777. [CrossRef]
139. Grodstein, F.; Chen, J.; Wilson, R.S.; Manson, J.E. Nurses' Health Study. Type 2 diabetes and cognitive function in communi-ty-dwelling elderly women. *Diabetes Care.* **2001**, *24*, 1060–1065. [CrossRef] [PubMed]
140. Cukierman, T.; Gerstein, H.C.; Williamson, J.D. Cognitive decline and dementia in diabetes—systematic overview of prospective observational studies. *Diabetologia* **2005**, *48*, 2460–2469. [CrossRef]
141. Saczynski, J.S.; Jónsdóttir, M.K.; Garcia, M.E.; Jonsson, P.V.; Peila, R.; Eiriksdottir, G.; Olafsdottir, E.; Harris, T.B.; Gudnason, V.; Launer, L.J. Cognitive Impairment: An Increasingly Important Complication of Type 2 Diabetes: The Age, Gene/Environment Susceptibility-Reykjavik Study. *Am. J. Epidemiol.* **2008**, *168*, 1132–1139. [CrossRef]
142. Cukierman-Yaffe, T.; Gerstein, H.C.; Williamson, J.D.; Lazar, R.M.; Lovato, L.; Miller, M.E.; Coker, L.H.; Murray, A.; Sullivan, M.D.; Marcovina, S.M.; et al. Relationship between baseline gly-cemic control and cognitive function in individuals with type 2 diabetes and other cardiovascular risk factors: The action to control cardiovascular risk in diabetes-memory in diabetes (ACCORD-MIND) trial. *Diabetes Care* **2009**, *32*, 221–226. [CrossRef]

143. Elias, M.F.; Elias, P.K.; Sullivan, L.M.; Wolf, P.A.; D'Agostino, R.B. Obesity, diabetes and cognitive deficit: The Framingham Heart Study. *Neurobiol. Aging* **2005**, *26*, 11–16. [CrossRef] [PubMed]
144. Roberts, R.O.; Geda, Y.E.; Knopman, D.S.; Cha, R.H.; Boeve, B.F.; Ivnik, R.J.; Pankratz, V.S.; Tangalos, E.G.; Petersen, R.C. Metabolic syndrome, inflammation, and nonamnestic mild cognitive impairment in older persons: A population-based study. *Alzheimer Dis. Assoc. Disord.* **2010**, *24*, 11–18. [CrossRef] [PubMed]
145. Yaffe, K.; Weston, A.L.; Blackwell, T.; Krueger, K.A. The Metabolic Syndrome and Development of Cognitive Impairment Among Older Women. *Arch. Neurol.* **2009**, *66*, 324–328. [CrossRef]
146. Pal, K.; Mukadam, N.; Petersen, I.; Cooper, C. Mild cognitive impairment and progression to dementia in people with diabetes prediabetes and metabolic syndrome: A systematic review and meta-analysis. *Soc. Psychiatry Psychiatr. Epidemiol.* **2018**, *53*, 1149–1160. [CrossRef] [PubMed]
147. Atti, A.R.; Valente, S.; Iodice, A.; Caramella, I.; Ferrari, B.; Albert, U.; Mandelli, L.; De Ronchi, D. Metabolic Syndrome, Mild Cognitive Impairment, and Dementia: A Meta-Analysis of Longitudinal Studies. *Am. J. Geriatr. Psychiatry* **2019**, *27*, 625–637. [CrossRef] [PubMed]
148. Yau, P.L.; Javier, D.C.; Ryan, C.; Tsui, W.H.; Ardekani, B.A.; Ten, S.; Convit, A. Preliminary evidence for brain complications in obese adolescents with type 2 diabetes mellitus. *Diabetologia* **2010**, *53*, 2298–2306. [CrossRef]
149. Verdejo-Garcia, A.; Pérez-Expósito, M.; Schmidt-Río-Valle, J.; Fernández-Serrano, M.J.; Cruz, F.; Pérez-García, M.; López-Belmonte, G.; Martín-Matillas, M.; Martín-Lagos, J.A.; Marcos, A.; et al. Selective Alterations Within Executive Functions in Adolescents With Excess Weight. *Obesity* **2010**, *18*, 1572–1578. [CrossRef]
150. Lande, M.B.; Kaczorowski, J.M.; Auinger, P.; Schwartz, G.J.; Weitzman, M. Elevated blood pressure and decreased cognitive function among school-age children and adolescents in the United States. *J. Pediatr.* **2003**, *143*, 720–724. [CrossRef]
151. Li, Y.; Dai, Q.; Jackson, J.C.; Zhang, J. Overweight Is Associated with Decreased Cognitive Functioning Among School-age Children and Adolescents. *Obesity* **2008**, *16*, 1809–1815. [CrossRef]
152. Lozada, M.; Machado, S.; Manrique, M.; Martínez, D.; Suárez, O.; Guevara, H. Factores de riesgo asociados al síndrome metabólico en adolescentes. *Gac. Médica Caracas* **2008**, *116*, 323–329.
153. Bourdel-Marchasson, I.; Lapre, E.; Laksir, H.; Puget, E. Insulin resistance, diabetes and cognitive function: Consequences for pre-ventative strategies. *Diabetes Metab.* **2010**, *36*, 173–181. [CrossRef]
154. Pinillos Patiño, Y.; Herazo Beltrán, Y.; Vidarte Claros, J.A.; Quiroz, E.; Suarez Palacio, D. Niveles de Actividad Física y sus Deter-minantes en Mujeres Adultas de Barranquilla. *Cienc. Innov. Salud* **2014**, *2*, 10–17.
155. De La Cruz Vargas, J.A.; Dyzinger, W.; Herzog, S.; dos Santos, F.; Villegas, H.; Ezinga, M. Medicina del Estilo de Vida: Trabajando juntos para revertir la epidemia de las enfermedades crónicas en Latinoamérica. *Cienc. Innov. Salud* **2017**, *4*, 1–7. [CrossRef]
156. Frederiksen, K.S.; Verdelho, A.; Madureira, S.; Bäzner, H.; O'Brien, J.T.; Fazekas, F.; Scheltens, P.; Schmidt, R.; Wallin, A.; Wahlund, L.; et al. Physical activity in the elderly is associated with improved executive function and processing speed: The LADIS Study: Physical activity and cognitive function. *Int. J. Geriatr. Psychiatry.* **2015**, *30*, 744–750. [CrossRef] [PubMed]
157. Karssemeijer, E.G.A.; Aaronson, J.A.; Bossers, W.J.; Smits, T.; Olde Rikkert, M.G.M.; Kessels, R.P.C. Positive effects of combined cognitive and physical exercise training on cognitive function in older adults with mild cognitive impairment or dementia: A me-ta-analysis. *Ageing Res. Rev.* **2017**, *40*, 75–83. [CrossRef] [PubMed]
158. Groot, C.; Hooghiemstra, A.; Raijmakers, P.; Van Berckel, B.; Scheltens, P.; Scherder, E.; van der Flier, W.; Ossenkoppele, R. The effect of physical activity on cognitive function in patients with dementia: A meta-analysis of randomized control trials. *Ageing Res. Rev.* **2016**, *25*, 13–23. [CrossRef]
159. Díaz Cárdenas, S. Fomento de la Salud Física en Pacientes de la Facultad de Odontología de la Universidad de Cartagena: Sistematización de Experiencias. *Cienc. Innov. Salud* **2013**, *1*, 52–56. [CrossRef]
160. Lautenschlager, N.T.; Cox, K.L.; Ellis, K.A. Physical activity for cognitive health: What advice can we give to older adults with subjective cognitive decline and mild cognitive impairment? *Dialogues Clin. Neurosci.* **2019**, *21*, 61–68. [PubMed]
161. McGrattan, A.M.; McEvoy, C.; McGuinness, B.; McKinley, M.C.; Woodside, J.V. Effect of dietary interventions in mild cognitive impairment: A systematic review. *Br. J. Nutr.* **2018**, *120*, 1388–1405. [CrossRef] [PubMed]
162. Singh, B.; Parsaik, A.K.; Mielke, M.; Erwin, P.J.; Knopman, D.S.; Petersen, R.C.; Roberts, R.O. Association of Mediterranean Diet with Mild Cognitive Impairment and Alzheimer's Disease: A Systematic Review and Meta-Analysis. *J. Alzheimer's Dis.* **2014**, *39*, 271–282. [CrossRef] [PubMed]
163. Hosking, D.E.; Eramudugolla, R.; Cherbuin, N.; Anstey, K.J. MIND not Mediterranean diet related to 12-year incidence of cognitive impairment in an Australian longitudinal cohort study. *Alzheimer's Dement.* **2019**, *15*, 581–589. [CrossRef]
164. Krikorian, R.; Shidler, M.D.; Dangelo, K.; Couch, S.C.; Benoit, S.C.; Clegg, D.J. Dietary ketosis enhances memory in mild cognitive impairment. *Neurobiol. Aging* **2012**, *33*, 425.e19–425.e27. [CrossRef] [PubMed]
165. Zhang, Y.; Chen, J.; Qiu, J.; Li, Y.; Wang, J.; Jiao, J. Intakes of fish and polyunsaturated fatty acids and mild-to-severe cognitive impairment risks: A dose-response meta-analysis of 21 cohort studies1–3. *Am. J. Clin. Nutr.* **2015**, *103*, 330–340. [CrossRef] [PubMed]
166. Rouch, L.; Cestac, P.; Hanon, O.; Cool, C.; Helmer, C.; Bouhanick, B.; Chamontin, B.; Dartigues, J.-F.; Vellas, B.; Andrieu, S. Antihypertensive Drugs, Prevention of Cognitive Decline and Dementia: A Systematic Review of Observational Studies, Randomized Controlled Trials and Meta-Analyses, with Discussion of Potential Mechanisms. *CNS Drugs* **2015**, *29*, 113–130. [CrossRef] [PubMed]

167. Levi Marpillat, N.; Macquin-Mavier, I.; Tropeano, A.I.; Bachoud-Levi, A.-C.; Maison, P. Antihypertensive classes, cognitive decline and incidence of dementia: A network meta-analysis. *J. Hypertens.* **2013**, *31*, 1073–1082. [CrossRef]
168. Guo, Z.; Fratiglioni, L.; Zhu, L.; Fastbom, J.; Winblad, B.; Viitanen, M. Occurrence and progression of dementia in a community population aged 75 years and older: Relationship of antihypertensive medication use. *Arch. Neurol.* **1999**, *56*, 991–996. [CrossRef]
169. Zhang, Q.Q.; Li, W.S.; Liu, Z.; Zhang, H.L.; Ba, Y.G.; Zhang, R.X. Metformin therapy and cognitive dysfunction in patients with type 2 diabetes: A meta-analysis and systematic review. *Medicine* **2020**, *99*, 19378. [CrossRef]
170. Ng, T.P.; Feng, L.; Yap, K.B.; Lee, T.S.; Tan, C.H.; Winblad, B. Long-Term Metformin Usage and Cognitive Function among Older Adults with Diabetes. *J. Alzheimer's Dis.* **2014**, *41*, 61–68. [CrossRef]
171. Borzì, A.M.; Condorelli, G.; Biondi, A.; Basile, F.; Vicari, E.S.D.; Buscemi, C.; Luca, S.; Vacante, M. Effects of vildagliptin, a DPP-4 inhibitor, in elderly diabetic patients with mild cognitive impairment. *Arch. Gerontol. Geriatr.* **2019**, *84*, 103896. [CrossRef]
172. Areosa Sastre, A.; Vernooij, R.W.; González-Colaço Harmand, M.; Martínez, G. Effect of the treatment of Type 2 diabetes mellitus on the development of cognitive impairment and dementia. *Cochrane Database Syst. Rev.* **2017**, *6*, 003804. [CrossRef]
173. Koo, B.K.; Kim, L.; Lee, J.; Moon, M.K. Taking metformin and cognitive function change in older patients with diabetes. *Geriatr. Gerontol. Int.* **2019**, *19*, 755–761. [CrossRef] [PubMed]
174. Wennberg, A.M.V.; Hagen, C.E.; Edwards, K.; Roberts, R.O.; Machulda, M.M.; Knopman, D.S.; Petersen, R.C.; Mielke, M.M. Association of antidiabetic medi-cation use, cognitive decline, and risk of cognitive impairment in older people with type 2 diabetes: Results from the pop-ulation-based Mayo Clinic Study of Aging. *Int. J. Geriatr. Psychiatry.* **2018**, *33*, 1114–1120. [CrossRef] [PubMed]
175. Fink, H.A.; Jutkowitz, E.; McCarten, J.R.; Hemmy, L.S.; Butler, M.; Davila, H.; Ratner, E.; Calvert, C.; Barclay, T.R.; Brasure, M. Pharmacologic interventions to prevent cognitive decline, mild cognitive impairment, and clinical Alzheimer-type dementia: A systematic review. *Ann. Intern. Med.* **2018**, *168*, 39–51. [CrossRef]
176. Ligthart, S.A.; Moll van Charante, E.P.; Van Gool, W.A.; Richard, E. Treatment of cardiovascular risk factors to prevent cognitive decline and dementia: A systematic review. *Vasc. Health Risk Manag.* **2010**, *6*, 775–785. [CrossRef] [PubMed]
177. Bosch, J.; O'Donnell, M.; Swaminathan, B.; Lonn, E.M.; Sharma, M.; Dagenais, G.; Diaz, R.; Khunit, K.; Lewis, B.S.; Avezum, A.; et al. Effects of blood pressure and lipid lowering on cognition: Results from the HOPE-3 study. *Neurology* **2019**, *92*, 1435–1446. [CrossRef]
178. Bettermann, K.; Arnold, A.M.; Williamson, J.; Rapp, S.; Sink, K.; Toole, J.F.; Carlson, M.C.; Yasar, S.; DeKosky, S.; Burke, G.L. Statins, Risk of Dementia, and Cognitive Function: Secondary Analysis of the Ginkgo Evaluation of Memory Study. *J. Stroke Cerebrovasc. Dis.* **2012**, *21*, 436–444. [CrossRef]
179. Zandi, P.P.; Sparks, D.L.; Khachaturian, A.S.; Tschanz, J.; Norton, M.; Steinberg, M.; Welsh-Bohmer, K.A.; Breitner, J.C.S. Do Statins Reduce Risk of Incident Dementia and Alzheimer Disease? The Cache County Study. *Arch. Gen. Psychiatry* **2005**, *62*, 217–224. [CrossRef]
180. Rea, T.D.; Breitner, J.C.; Psaty, B.M.; Fitzpatrick, A.L.; Lopez, O.L.; Newman, A.B.; Hazzard, W.R.; Zandi, P.P.; Burke, G.L.; Lyketsos, C.G.; et al. Statin use and the risk of incident dementia: The cardiovascular health study. *Arch. Neurol.* **2005**, *62*, 1047–1051. [CrossRef]
181. Bae, S.; Shimada, H.; Lee, S.; Makizako, H.; Lee, S.; Harada, K.; Doi, T.; Tsutsumimoto, K.; Hotta, R.; Nakakubo, S.; et al. The Relationships Between Components of Metabolic Syndrome and Mild Cognitive Impairment Subtypes: A Cross-Sectional Study of Japanese Older Adults. *J. Alzheimer's Dis.* **2017**, *60*, 913–921. [CrossRef]
182. Steen, E.; Terry, B.M.; Rivera, E.J.; Cannon, J.L.; Neely, T.R.; Tavares, R.; Xu, X.J.; Wands, J.R.; De La Monte, S.M. Impaired insulin and insulin-like growth factor expression and signaling mechanisms in Alzheimer's disease—Is this type 3 diabetes? *J. Alzheimer's Dis.* **2005**, *7*, 63–80. [CrossRef] [PubMed]
183. Frölich, L.; Blum-Degen, D.; Bernstein, H.-G.; Engelsberger, S.; Humrich, J.; Laufer, S.; Muschner, D.; Thalheimer, A.; Türk, A.; Hoyer, S.; et al. Brain insulin and insulin receptors in aging and sporadic Alzheimer's disease. *J. Neural Transm.* **1998**, *105*, 423–438. [CrossRef] [PubMed]
184. Craft, S.; Peskind, E.; Schwartz, M.W.; Schellenberg, G.D.; Raskind, M.; Porte, D. Cerebrospinal fluid and plasma insulin levels in Alzheimer's disease: Relationship to severity of dementia and apolipoprotein E genotype. *Neurology* **1998**, *50*, 164–168. [CrossRef]
185. Mody, N.; Agouni, A.; Mcilroy, G.D.; Platt, B.; Delibegovic, M. Susceptibility to diet-induced obesity and glucose intolerance in the APP SWE/PSEN1 A246E mouse model of Alzheimer's disease is associated with increased brain levels of protein tyrosine phosphatase 1B (PTP1B) and retinol-binding protein 4 (RBP4), and basal phosphorylation of S6 ribosomal protein. *Diabetologia* **2011**, *54*, 2143–2151. [CrossRef] [PubMed]
186. Janson, J.; Laedtke, T.; Parisi, J.E.; O'Brien, P.; Petersen, R.C.; Butler, P.C. Increased Risk of Type 2 Diabetes in Alzheimer Disease. *Diabetes* **2004**, *53*, 474–481. [CrossRef] [PubMed]
187. Altschul, D.M.; Starr, J.M.; Deary, I.J. Cognitive function in early and later life is associated with blood glucose in older individuals: Analysis of the Lothian Birth Cohort of 1936. *Diabetologia* **2018**, *61*, 1946–1955. [CrossRef]
188. Peng, X.; Xu, Z.; Mo, X.; Guo, Q.; Yin, J.; Xu, M.; Peng, Z.; Sun, T.; Zhou, L.; Peng, X.; et al. Association of plasma β-amyloid 40 and 42 concentration with type 2 diabetes among Chinese adults. *Diabetologia* **2020**, *63*, 954–963. [CrossRef]
189. Lipton, S.A. Paradigm shift in NMDA receptor antagonist drug development: Molecular mechanism of uncompetitive inhi-bition by memantine in the treatment of Alzheimer's disease and other neurologic disorders. *J. Alzheimers Dis.* **2004**, *6*, 61–74. [CrossRef]

190. Ettcheto, M.; Sanchez-Lopez, E.; Gómez-Mínguez, Y.; Cabrera, H.; Busquets, O.; Beas-Zárate, C.; García, M.L.; Carro, E.; Casadesus, G.; Auladell, C.; et al. Peripheral and Central Effects of Memantine in a Mixed Preclinical Mice Model of Obesity and Familial Alzheimer's Disease. *Mol. Neurobiol.* **2018**, *55*, 7327–7339. [CrossRef]
191. Ahmed, A.S.; Elgharabawy, R.; Al-Najjar, A.H. Ameliorating effect of anti-Alzheimer's drugs on the bidirectional association between type 2 diabetes mellitus and Alzheimer's disease. *Exp. Biol. Med.* **2017**, *242*, 1335–1344. [CrossRef] [PubMed]
192. Meakin, P.J.; Harper, A.J.; Hamilton, D.L.; Gallagher, J.; McNeilly, A.D.; Burgess, L.A.; Vaanholt, L.M.; Bannon, K.A.; Latcham, J.; Hussain, I.; et al. Reduction in BACE1 decreases body weight, protects against diet-induced obesity and enhances insulin sensitivity in mice. *Biochem. J.* **2011**, *441*, 285–296. [CrossRef] [PubMed]
193. Wessels, A.M.; Tariot, P.N.; Zimmer, J.A.; Selzler, K.J.; Bragg, S.M.; Andersen, S.W.; Landry, J.; Krull, J.H.; Downing, A.M.; Willis, B.A.; et al. Efficacy and safety of Lanabecestat for treatment of early and mild Alzheimer disease: The AMARANTH and DAYBREAK-ALZ randomized clinical trials. *JAMA Neurol.* **2020**, *77*, 199–209. [CrossRef] [PubMed]
194. Zhang, Y.; Zhou, B.; Deng, B.; Zhang, F.; Wu, J.; Wang, Y.; Le, Y.; Zhai, Q. Amyloid-β Induces Hepatic Insulin Resistance In Vivo via JAK2. *Diabetes* **2012**, *62*, 1159–1166. [CrossRef] [PubMed]
195. Gunstad, J.; Spitznagel, M.B.; Paul, R.H.; Cohen, R.A.; Kohn, M.; Luyster, F.S.; Clark, R.; Williams, L.M.; Gordon, E. Body mass index and neuropsychological function in healthy children and adolescents. *Appetite* **2008**, *50*, 246–251. [CrossRef]
196. Muller, M.; Tang, M.-X.; Schupf, N.; Manly, J.J.; Mayeux, R.; Luchsinger, J.A. Metabolic Syndrome and Dementia Risk in a Multiethnic Elderly Cohort. *Dement. Geriatr. Cogn. Disord.* **2007**, *24*, 185–192. [CrossRef]
197. Forti, P.; Pisacane, N.; Rietti, E.; Lucicesare, A.; Olivelli, V.; Mariani, E.; Mecocci, P.; Ravaglia, G. Metabolic Syndrome and Risk of Dementia in Older Adults. *J. Am. Geriatr. Soc.* **2010**, *58*, 487–492. [CrossRef]
198. Feinkohl, I.; Janke, J.; Hadzidiakos, D.; Slooter, A.; Winterer, G.; Spies, C.; Pischon, T. Associations of the metabolic syndrome and its components with cognitive impairment in older adults. *BMC Geriatr.* **2019**, *19*, 1–11. [CrossRef]

nutrients

MDPI

Article

Ginkgo Biloba Leaf Extract Improves an Innate Immune Response of Peripheral Blood Leukocytes of Alzheimer's Disease Patients

Marta Sochocka [1,*], Michał Ochnik [1], Maciej Sobczyński [2], Katarzyna Gębura [3], Aleksandra Zambrowicz [4], Piotr Naporowski [5] and Jerzy Leszek [6]

1 Laboratory of Virology, Department of Immunology of Infectious Diseases, Hirszfeld Institute of Immunology and Experimental Therapy, Polish Academy of Sciences, 58-114 Wroclaw, Poland; michal.ochnik@hirszfeld.pl
2 Laboratory of Molecular Neurobiology, Nencki Institute of Experimental Biology of the Polish Academy of Sciences, 02-093 Warsaw, Poland; macsebsob@poczta.onet.pl
3 Laboratory of Clinical Immunogenetics and Pharmacogenetics, Hirszfeld Institute of Immunology and Experimental Therapy, Polish Academy of Sciences, 53-114 Wroclaw, Poland; katarzyna.gebura@hirszfeld.pl
4 Department of Animal Products Technology and Quality Management, Faculty of Biotechnology and Food Sciences, Wroclaw University of Environmental and Life Sciences, 51-630 Wroclaw, Poland; aleksandra.zambrowicz@upwr.edu.pl
5 Laboratory of Medical Microbiology, Department of Immunology of Infectious Diseases, Hirszfeld Institute of Immunology and Experimental Therapy, Polish Academy of Sciences, 53-114 Wroclaw, Poland; piotr.naporowski@hirszfeld.pl
6 Department of Psychiatry, Wroclaw Medical University, 50-367 Wroclaw, Poland; jerzy.leszek@umed.wroc.pl
* Correspondence: marta.sochocka@hirszfeld.pl; Tel.: +48-713709924

Citation: Sochocka, M.; Ochnik, M.; Sobczyński, M.; Gębura, K.; Zambrowicz, A.; Naporowski, P.; Leszek, J. Ginkgo Biloba Leaf Extract Improves an Innate Immune Response of Peripheral Blood Leukocytes of Alzheimer's Disease Patients. *Nutrients* **2022**, *14*, 2022. https://doi.org/10.3390/nu14102022

Academic Editors: Omorogieva Ojo and Amanda R Amorim Adegboye

Received: 1 April 2022
Accepted: 9 May 2022
Published: 11 May 2022

Abstract: Background: One of the main features of Alzheimer's disease (AD) pathology is failure in innate immune response and chronic inflammation. Lack of effective AD treatment means that more attention is paid to alternative therapy and drugs of natural origin, such as extract of *Ginkgo biloba* (EGb). The purpose of this study was to investigate the effect of EGb on the mechanisms of innate immune response of peripheral blood leukocytes (PBLs) in AD patients. Methods: In AD patients and healthy-age matched controls, the effect of EGb on two of innate immune reactions, i.e., PBLs resistance to viral infection ex vivo and production of cytokines, namely TNF-α, IFN-γ, IL-1β, IL-10, IL-15, and IFN-α, were investigated. The influence of EGb on inflammatory-associated genes expression that regulate innate immune response to viral infection and cytokine production, namely *IRF-3, IRF-7, tetherin, SOCS1, SOCS3, NFKB1, p65,* and *MxA* was also examined. Results: A beneficial effect of EGb especially in AD women was observed. EGb decreased production of TNF-α, IFN-γ, and IL-10 and increased IL-15 and IL-1β. The effect was more pronouncement in AD group. EGb also downregulated expression of investigated genes. Conclusions: EGb may have an advantageous properties for health management in elderly and AD sufferers but especially in women with AD. Improving peripheral innate immune cells' activity by adding EGb as accompanying treatment in AD may be, in the long term, a good course to modify the disease progression.

Keywords: extract of *Ginkgo biloba* (EGb); innate immunity; PBLs; Alzheimer's disease; cytokines

1. Introduction

Immunity and chronic inflammation play a key role in the survival of the older adults, and according to the latest knowledge, they also represent one of the main features of Alzheimer's disease (AD) pathology [1]. AD is one of the most important age-related health problems worldwide, and it is believed that the onset and progression of the disease may depend at least in part on optimal immune system functioning. Experimental studies highlight the pathological changes in the central and peripheral immune response in AD [2,3]. An increased levels of peripheral inflammatory markers, such as IL-6, TNF-α, or C-reactive protein (CRP), were found to be associated with future cognitive decline and dementia [2,4]. Currently, no effective drugs are available for the treatment of AD symptoms. An accessible pharmacotherapy aims to only slow disease progression and reduce cognitive symptoms. Some hope is associated with GV-971 (sodium oligomannate capsules), which improved cognitive functions in AD patients in China in a phase 3 trial and was approved for the treatment of AD in China [5]. Thus, more attention is paid to alternative therapy, such as using drugs of natural origin. Moreover, searching for natural compounds with immunoregulatory activity seems to be a good direction for future adjunct AD therapy.

Currently, phytomedicine is gaining its popularity, and many plant-derived phytotherapeutics with medicinal properties are used in the treatment of various diseases, including age-related diseases [6]. The phytomedicine of aging provide a wide range of bioactive compounds, such as flavonoids, terpenoids, or polyphenols with therapeutic effects. Health benefits consist mainly of acting as an immunity booster and exhibiting antioxidant, cardioprotective, and neuro-protective effects [7]. One of the most popular medicinal plants is *Ginkgo biloba*. Standardized extract of *G. biloba* (EGb) contains 24% ginkgo flavonoid glycosides, 6% terpene lactones, and up to 5 ppm ginkgolic acids [8]. The therapeutic potential of EGb is manifested in beneficial effect on the circulatory system (blood flow improvement, prevention of clot formation, reinforcing the walls of the capillaries) and nervous system with protection of nerve cells from injury [9]. Phytochemical constituents from *G. biloba,* such as flavonoids and terpenoids, showed beneficial effect in the treatment of concentration difficulties, memory impairment, and AD. Thus, EGb is considered as memory enhancer [10].

The use of phytotherapeutics/nutraceuticals as an adjunct therapy to classic drug therapies is highly recommended in many diseases. However, many more studies are still needed to evaluate the therapeutic potential and clarifying the mechanism of action of natural compounds, including EGb. It is believed that this could help to choose better phytotherapeutics as the accompanied treatment of neurodegenerative pathologies such as AD. EGb is already used in the treatment of AD and cognitive deficits acting as anti-aggregating and pro-cognitive preparation. It is implemented to improve memory impairment and cognitive decline [11]. However, less attention and research are concentrated on its effect on immune system functioning in AD patients. The purpose of this study was to investigate the effect of EGb on the mechanisms of innate immune response of peripheral blood leukocytes (PBLs) of AD sufferers.

2. Materials and Methods

2.1. Blood Samples

Peripheral venous blood was obtained from 39 Subjects: 22 AD patients (15 females, 7 males) and 17 healthy adult volunteers (10 females, 7 males) of an appropriate age (43–90 years) and collected in tubes containing anticoagulant EDTA or heparin. Patients were under the care of Department of Psychiatry of the Medical University in Wroclaw, Poland. Patients did not receive any anti-dementia and other drugs before blood venipuncture as well as any other immunomodulators. Among the patients, no infectious diseases occurred in the 3-month period before the inclusion to the study.

2.2. Ethics Approval and Consent to Participate

This study has been reviewed, approved, and conducted in accordance with the guidelines of the Ethics Committee of the Wroclaw Medical University (No. KB-349/2016). Signed consent was obtained from all participants of the study or their legal representative.

2.3. Clinical Examination

Patients were under psychiatric and neurological examinations as well as laboratory tests, electroencephalographic examinations (EEG), and computer tomography (CT) or magnetic resonance imaging (MRI) structural studies. Mini-Mental State Examination (MMSE) was used for the screening of dementia. All patients met DSM-V and NINCDA-ADRDA criteria for probable AD dementia. A diagnosis of AD was made when specific symptoms were present and by making sure other causes of dementia were absent, including anemia, brain tumor, chronic infection, intoxication from medication, severe depression, stroke, thyroid disease, and vitamin deficiencies. CT and MRI of the brain were performed as well to look for other causes of dementia, such as brain tumor or stroke. Semi-structured interview with the patient and informant, physical exam, evaluation of neurological status, and psychiatric exam were obtained. Vital signs and blood screening labs (hematology, chemistry panel, urinalysis, vitamin B12 (B12), thyrotropin (TSH)) were collected. *Exclusion criteria:* patients older than 90 years, any significant neurological disease such as Parkinson's disease, multi-infarct dementia, Huntington disease, normal pressure hydrocephalus, brain tumor, progressive supranuclear palsy, seizure disorder, subdural hematoma, multiple sclerosis, or history of significant head trauma followed by persistent neurologic defaults or known structural brain abnormalities. MRI scan with evidence of infection, infarction, or other focal lesions; subjects with multiple lacunes or lacunes in a critical memory structure; psychiatric disorder/psychotic features: major depression, bipolar disorder, agitation, or behavioral problems within the last 3 months; history of schizophrenia, alcohol abuse, history of alcohol or substance abuse or dependence within the past 2 years; any significant systemic illness or unstable medical condition; clinically significant abnormalities in B12, rapid plasma regain test (RPR), or TSH; and current use of specific psychoactive medications (e.g., certain antidepressants, neuroleptics, chronic anxiolytics, or sedative hypnotics, etc.). Patients were excluded if they did not agree to respond to the test questions and/or if they had life-threatening diseases other than AD.

2.4. Extract of G. Biloba (EGb)

Standardized dry extract from *G. biloba* leaves (GINKGONIS EXTRACTUM SICCUM RAFFINATUM ET QUANTIFICATUM PH. EUR. (European Pharmacopoeia)) provided by Martin Bauer Group, Finzelberg GmbH & Co. KG, Andernach, Germany, was investigated. EGb is a dry extract from *G. biloba* leaves. The extract is adjusted to 22.0–27.0% ginkgo flavonoids calculated as ginkgo flavone glycosides and 5.0–7.0% terpene lactones consisting of 2.8–3.4% ginkgolides A, B, and C and 2.6–3.2% bilobalide and contains less than 5 ppm ginkgolic acids.

EGb solution: Before each experiment, EGb was dissolved in dimethyl sulfoxide (DMSO) at a primary concentration of 20 mg/mL and mixed thoroughly until complete dissolution. Next, EGb solution in DMEM 2% FBS medium was prepared for experiments with PBLs. For antioxidant activity powder of EGb was dissolved in 96% ethanol (1 mg/mL).

2.5. Trypan Blue Staining for Cell Viability

The viability of PBLs was measured with 0.4% trypan blue staining. A total of 100 μL of cell suspension (1×10^6 cells/mL) was incubated with 100 μL of 0.4% trypan blue. After 15 min of incubation at room temperature, the viability of the cells was measured in a Bürker chamber with the use of light microscope (Olympus CX31). Dead cells were labeled with navy-blue, and live cells remained unstained.

2.6. Determination of Antioxidant Activity as the Ability to Scavenge DPPH Free Radicals

The antioxidant activity of the obtained hydrolysates was assessed on the basis of the radical scavenging effect of the stable 1,1-diphenyl-2-picrylhydrazyl (DPPH (Sigma, St. Louis, MO, USA, D21140-0)) free radical activity according to Yen and Chen with minor modifications [12]. The tested samples were dissolved in water to a final volume of 1 mL and mixed with 1 mL of ethanol (98%). The reaction was started by adding 0.5 mL of 0.3 M DPPH in ethanol. The mixtures were left for 30 min at room temperature, and the absorbance of the resulting solutions was measured at 517 nm. For calibration, aqueous solutions of known Trolox concentrations ranging from 2 to 20 μg (able to scavenge 500 μL of 0.3 mM DPPH radical solution) were used. Radical scavenging activity of the peptides was expressed as μM $Trolox_{eq}$/mg protein.

2.7. FRAP Method

The FRAP method (ferric-reducing antioxidant power) was used to determine the antioxidative capacity of hydrolysates according to Benzie and Strain [13]. A total of 3 mL of FRAP working solution (300 mM acetate buffer pH 3.6; 10 mM 2,4,6,tripyridyl-s-triazine (TPTZ) (Fluka, 93285) and 20 mM $FeCl_3 \times 6 H_2O$ (10:1:1 v/v)) was mixed with 1 mL of the sample. After 10 min of reaction, the absorbance was measured at λ = 593 nm. An aqueous solution of known Fe (II) concentration was used for calibration (in the range from 100 to 1000 μg). Results were expressed as μg Fe^{2+}/mg protein.

2.8. Determination of Fe (II) Ion Chelation

Chelation of iron ions by hydrolysates was estimated by the method of Xu et al. [14] with modifications. A 250 μL sample was mixed with 1250 μL H_2O and 110 μL 1 mM $FeCl_2$. After 2 min, 1 mL of 500 μM ferrozine (Sigma, 160601) aqueous solution was added and the mixture was allowed to react for 10 min. The absorbance of ferrous iron–ferrozine complex was measured spectrophotometrically at λ = 562 nm. A known concentration of $FeCl_2$ (0–20 μg) was used to generate a standard curve, and the ability to chelate iron ions was expressed as μg Fe^{2+}/mg protein.

2.9. Virus and Cell Line

A wild-type Indiana VSV (*Vesicular stomatitis virus, Rhabdoviridae*) serotype was used. VSV was obtained from Dr. C. Buckler (National Institutes of Health, Bethesda, MD, USA). Virus was grown and titrated in L_{929} cells. Viral titer was expressed with reference to the $TCID_{50}$ (tissue culture infectious dose) value, based on the cytopathic effect caused by this virus in approximately 50% of infected cells.

L_{929} (ATCC CCL1), a murine fibroblast-like cell line, was maintained in complete RPMI 1640 medium (HIIET, Wroclaw, Poland) with antibiotics (100 U/mL penicillin and 100 μg/mL streptomycin), 2 mM L-glutamine, and 2% fetal bovine serum (FBS) (all from Merck KGaA, Darmstadt, Germany).

2.10. Isolation of Peripheral Blood Leukocytes (PBLs)

PBLs were isolated according to a standard protocol from 10 mL of peripheral blood by gradient centrifugation in Gradisol G (Aqua-Med, Łódź, Poland) and maintained in RPMI 1640 medium (HIIET, Wroclaw, Poland) with antibiotics (100 U/mL penicillin and 100 μg/mL streptomycin), 2 mM L-glutamine, and 2% FBS (Merck KGaA, Darmstadt, Germany).

2.11. Determination of Resistance/Level of Innate Immunity of PBLs

Resistance/innate immunity was determined by infection of leukocytes (1×10^6 cells/mL) ex vivo with a VSV dose of 100 $TCID_{50}$. After 40 min of adsorption at room temperature (rt), the virus was washed out three times with RPMI medium with 2% FBS, and the cells were suspended in 1 mL of RPMI 2% FBS for investigations of the influence of EGb on PBLs resistance/innate immunity and cytokine production. A sample of the infected cells was kept at 4 °C and served as a control of the starting level of the virus. The rest of the cells were divided to two parts, i.e., VSV-infected and uninfected, and next were treated with 150 μg/mL EGb and incubated at 37 °C for 24 h. After that, time samples of the medium above the cells were collected and titrated in L_{929} cells. Viral titer was expressed in $TCID_{50}$. Resistance of PBLs to VSV infection was assessed as follows: a VSV titer $\geq$ 4 log $TCID_{50}$ was considered as a lack of resistance (deficiency in innate immunity), a titer of 2–3 log indicated partial resistance, and a titer of 0–1 log indicated complete resistance to VSV infection (high level of innate immunity).

2.12. Cytokine Measurement

The levels of IL-1β, IL-10, IL-15, IFN-α, IFN-γ, and TNF-α in supernatants from uninfected and VSV-infected leukocytes were detected using enzyme-linked immunosorbent assays (BD OptEIA TM human IL-1β, IL-10, IL-15, IFN-γ,TNF-α ELISA set, BD Biosciences; IFN-α Human ELISA Kit, Thermo Fisher, Carlsbad, CA, USA). The optical density was measured at 450 nm with λ correction 570 nm using a Multiskan RC spectrophotometric reader (Thermo Labsystems, Philadelphia, PA, USA). Cytokine concentrations were expressed in pg/mL.

2.13. RNA Isolation and Real-Time PCR

Real-time PCR was used to investigate mRNA expression. Total RNA (obtained from uninfected and VSV-infected cells treated with EGb 150 μg/mL) was extracted with the Relia Prep™ RNA Cell Miniprep System kit (Promega, Madison, WI, USA), and reverse transcription was performed using the High Capacity cDNA Reverse Transcription kit (Applied Biosystem, Thermo-fisher Scientific, Carlsbad, CA, USA) according to the manufacturer's instructions. Reaction was performed in T3000 Thermocycler (Biometra, Göttingen, Germany). Reaction volume was 20 μL, and final cDNA product was kept in −20 °C for a few weeks preceding quantitative PCR. Expression of interferon regulatory factor 3 and 7 (*IRF-3, IRF-7*), tetherin (*BST2*), suppressor of cytokine signaling 1 and 3 (*SOCS1, SOCS3*), nuclear factor NF-kappa-B p105 and p65 subunit (*NFKB1, p65* alias *RELA*), and interferon-induced GTP-binding protein Mx1 (*MxA*) was studied by quantitative PCR (qPCR) using Taq DNA Polymerase with Sybr Green I dye. Data were normalized to endogenous reference gene *18S*, which was confirmed to be stable across the groups. Reaction was performed using SG qPCR Master Mix (EURx, Gdańsk, Poland) in combination with appropriate primers shown in Table 1. Furthermore, primer specificity was confirmed empirically after qPCR by performing melting curve analysis of every sample. All experiments were performed using Light Cycler 480 II instrument (Roche Diagnostics GmbH, Basel, Switzerland). Every reaction was performed two times and differences in quantification cycle between both repeats were negligible (±0.2). Quantification cycles (Cq) for every reaction were calculated automatically by the Light Cycler 480 II instrument software (Roche Diagnostics GmbH, Basel, Switzerland). Relative changes in mRNA expression were calculated using the delta Cq method with the healthy control as the comparator.

Table 1. Primers used for real-time PCR.

Gene		Primer Sequence
IRF-3 [15]	F1	5′-ACC ACC CGT GGA CCA AGA G-3′
	R1	5′-TAC CAA GGC CCT GAG GCA C-3′
IRF-7 [15]	F2	5′-TGG TCC TGG TGA AGC TGG AA-3′
	R2	5′-GAT GTC ATA GAG GCT GTT GG-3′
BST2 [15]	F3	5′-AAG AAA GTG GAG CTT GAG G-3′
	R3	5′-CCT GGT TTT CTC TTC TCA GTC G-3′
SOCS1 [16]	F4	5′-GGA ACT GCT TTT TCG CCC TTA-3′
	R4	5′-AGC AGC TCG AAG AGG CAG TC-3′
SOCS3 [16]	F5	5′-CAA GGA CGG AGA CTT CGA TT-3′
	R5	5′-GGA GCC AGC CTG GAT CTG-3′
NFKB1 [17]	F6	5′-AGA AGT CTT ACC CTC AGG TCA-3′
	R6	5′-CAG TTA CAG TGC AGA TCC CA-3′
p65 [17]	F7	5′-GAA TGG CTC GTC TGT AGT GC-3′
	R7	5′-GCT GCT CAA TGA TCT CAA CAT-3′
MxA [15]	F8	5′-GCC GGC TGT GGA TAT GCT A-3′
	R8	5′-TTT ATC GAA ACA TCT GTG AAA GCA A-3′
18S [18]	F9	5′-GAA TGG CTC ATT AAA TCA GTT ATG G-3′
	R9	5′-TAT TAG CTC TAG AAT TAC CAC AGT TAT CC-3′

2.14. Statistical Analysis

To capture the central tendency (average) of the data, the Hodges–Lehman estimator—pseudo median—was used (in the following part called *Median*). Confidence intervals (CI95) at significance level $\alpha = 0.05$ were used to measure precision of estimation and testing some statistical hypothesis. CI95s were estimated with bootstrap method. Statistic $Sn = med\{med|x_i - x_j|; j = 1 \ldots n\}$ was used as robust measure of variability [19] as well as minimal and maximal observations. Sn is typical difference between two randomly sampled observations. Delta Δ is Hodges–Lehman estimator of the shift parameter between distributions of two independent populations. Delta Δ is median of differences between all pairs of observations, where one observation belongs to group A, and the second observation in the pair belongs to group B. Delta is shift parameter used also in Wilcoxon rank-sum test to comparison two independent populations (also known as Mann–Whitney test). Delta, CI95 for delta, and *p*-values of the delta were estimated numerically with bootstrap method because of a small sample sizes. In the case of investigation of an influence of EGb treatment on the level of innate immunity of AD patients and controls, an analysis of variance (ANOVA) was used. In the case of ANOVA, statistical effect size for every variable in the model was measured with partial eta-squared defined as $\eta^2 = \frac{SS_{variable}}{SS_{variable}+SS_{error}}$, where $SS_{variable}$ and SS_{error} is sum of squares in ANOVA for effect of the variable and sum of squares for the experimental error, respectively. Interpretation of the η^2 was typical, after Cohen.

For every investigated $i - th$ individual from AD patients and from controls, the effect of EGb treatment was calculated as $d_i = log_e \frac{level\ after\ treatment}{level\ before\ treatment}$ for all cytokine production and for all gene expression. An average effect in the group was estimated as pseudo median (*Median*) with Hodges–Lehman estimator (HL)f central tendency, so $EGb\ effect = HL(d)$.

3. Results

3.1. Blood Donors (Study Groups)

Table 2 presents basic characteristics of 39 investigated blood donors—AD patients and controls (age-matched, over 55 years old)—according to a few variables. There were 22—15 female and 7 male—among AD patients and 17—10 female and 7 male—among controls. MMSE score was evaluated. MMSE median for female was 18.75 and for male 18.5. Among the AD group, there were no differences between men and women in the level of dementia. Patients were collected randomly to estimate DGN and DSMV differences between sex. Chi-square test statistic $\chi^2_{df=2} = 0.598,\ p = 0.856$.

Table 2. Characteristics of blood donors—AD patients and controls.

Variable	Group	Median	Sn	Min–Max
Age	AD	68	11	32–80
	Control	61.5	9.5	42–90
Sex	**Group**	**Women**	**Men**	**% Men**
	AD	15	7	31.8
	Control	10	7	41.2
MMSE	**AD Group**	**Median**	**Sn**	**Min–Max**
	Women	18.75	3	12–24
	Men	18.5	3.5	12–23
DGN	**AD Group**	**Mild**	**Moderate**	**Serious**
	Women	4	6	5
	%	26.7	40.0	33.3
	Men	3	2	2
	%	42.9	28.6	28.6

Table 2. *Cont.*

	AD Group	Mild	Moderate	Serious
DSMV	Women	4	6	5
	%	26.7	40.0	33.3
	Men	3	2	2
	%	42.9	28.6	28.6

Sn, measure of variability; MMSE, mini-mental state examination; DGN, the diagnostic according to the guidelines of the German Society of Neurology; DSMV, fifth edition of diagnostic and statistical manual of mental disorders classification (criteria for major neurocognitive disorder).

3.2. EGb Characteristic (Cytotoxicity, Antioxidant Activity)

First, the starting solution of EGb was prepared. EGb at 20 mg/mL in DMSO was diluted to final concentrations of 25–500 μg/mL in RPMI 2% FBS. Freshly isolated PBLs were treated with several concentrations of EGb and incubated in 37 °C/5% CO_2 for 24 h. After that, time morphological changes of the cells were observed under the inverted microscope. Total number and viability of PBLs were determined by 0.4% trypan blue staining. The viable cells—with intact cell membranes—did not take up impermeable trypan blue (stayed non-colored), whereas dead cells (with damaged cell membranes, cell shadows, shrunken cells) were permeable and took up the dye (dyed with distinctive blue). Fresh EGb dilutions were prepared before each experiment. Experiments was performed three times in two independent repetitions each. It was noticed that EGb in concentration over 200 μg/mL resulted in cytotoxicity (cell viability below 90%). EGb in the range of 25–150 μg/mL was nontoxic for PBLs (cell viability over 90%). Control were PBLs incubated only with culture medium RPMI 2% FBS. The final concentrations of DMSO < 2% were nontoxic. Estimated cytotoxic concentration of EGb, which reduced viability of PBLs by 50% (CC_{50}), was calculated and equal to $CC_{50} \cong 743.5$ μg/mL. Based on cytotoxicity test for the future experiments of an influence of the extract on innate immune response of PBLs, the highest nontoxic concentration of EGb, – 150 μg/mL, was used. The cell viability for this concentration was about 95%.

An antioxidant activity (in vitro), i.e., ferric ion reducing antioxidant power, Fe^{2+} chelating ability, and DPPH radical scavenging activity of EGb were analyzed (Table 3). EGb preparation exerted strong DPPH scavenging activity; only 10 μg of preparation has the same effect as 0.03 μM Trolox. The antioxidant properties (inhibition concentration, IC_{50}) for DPPH scavenging activity reached the value 41.13 μg. EGb preparation also exerted strong concentration-dependent ferric ion reducing antioxidant power and Fe^{2+} chelating ability (Table 3).

Table 3. Concentration-dependent antioxidant activity (in vitro) of EGb.

EGb	DPPH		FRAP	Chelation
(μg)	(μMTrolox$_{eq}$)	(%) Inactivated DPPH	(μg Fe^{2+})	(μg Fe^{2+})
250	or	or	or	62.63
150	or	or	or	43.82
100	or	or	or	24.68
50	0.09	63.51	17.24	20.2
10	0.03	23.31	4.65	14.68
5	0.02	12.73	2.74	2.7

3.3. EGb Improves Innate Immune Response of PBLs

The level of innate immunity of AD patients and healthy age-matched controls was estimated based on the test with vesicular stomatitis virus (VSV) replication in freshly isolated PBLs ex vivo. To evaluate the effect of EGb (150 μg/mL) on innate immunity VSV titers were examined after EGb treatment in the collected supernatants. Results are presented in Table 4. The level of innate immunity/PBLs resistance to VSV infection was assessed with the scale: the lack of virus replication (0–1 log $TCID_{50}$/mL) indicated complete immunity; VSV replication over 1 log (about 2–3 log) indicated deficiencies and partial immunity; and VSV replication over 4 log evidenced high deficiency in innate immunity. The EGb effect in every patient was calculated as a difference between the level of innate immunity after EGb treatment and before EGb treatment (EGb effect = after−before). The average change of innate immunity in AD patients was −0.75; i.e., EGb increased the level of innate immunity of AD. The results were significant ($p = 0.0002$), with confidence interval CI95(-Inf; −0.41). In the control group an increase in innate immunity ($p = 0.0001$) was also observed. There was no difference, however, in the effect of EGb treatment between AD and controls ($p = 7539$). In summary, EGb expressed a beneficial effect on innate immune response of PBLs. EGb treatment significantly increased the level of innate immunity in AD patients as well as controls (Figure 1).

Table 4. Influence of EGb treatment on PBLs resistance/level of innate immunity in AD patients and healthy age-matched controls (n = 39).

PBLs Resistance/Level of Innate Immunity	Group	Median	Sn	Min; Max	Δ	*p*-Value
EGb effect	AD	−0.75	1	−2.5; 0.5	−0.75 CI95(−Inf; −0.41)	0.0002 [1]
	Control	−0.75	0.5	−2; 0	−0.75 CI95(−Inf; −0.55)	0.0001 [1]
	AD vs. Control				0.00 CI95(−0.49; 0.61)	0.7539
	ANOVA in AD group EGb effect depending on sex, DGN, MMSE, and age					
	Variable	**Effect size η^2 (interpret.)**			$F_{df=1;16}$	***p*-Value**
	Sex $\beta_{women} = -1.18$	0.3983 (large)			10.59	0.0049
	DGN (=DSMV)	0.0503 (small/medium)			0.42	0.6616
	Age	0.0026 (negligible)			0.042	0.8406
	MMSE	0.0000 (no effect)			0.0001	0.9923
	ANOVA in Control group Effect of EGb treatment depending on sex and age					
	Variable	**Effect size η^2 (interpret.)**			$F_{df=1;14}$	***p*-Value**
	Sex $\beta_{woman} = -0.21$	0.1202 (medium)			1.91	0.1883
	Age	0.0963 (medium)			1.49	0.2420

Δ, shift parameter of location (as a part of Wilcoxon rank test, Hodges-Lehman estimator of (pseudo) median); Sn, measure of variability (higher value means higher variability; robust estimator equivalent of standard deviation); [1], one side (left side) test and one side confidence interval; [2], Inf—infinity; β_{woman}, standardized difference between men and women adjusted with other variables present in ANOVA; MMSE, mini-mental state examination; DGN, the diagnostic according to the guidelines of the German Society of Neurology; DSMV, fifth edition of diagnostic and statistical manual of mental disorders classification (criteria for major neurocognitive disorder).

In addition, Table 4 presents an analysis of variance (ANOVA) where EGb effect was dependent variable in AD and control group. The analysis showed that among AD patients, the effect of EGb treatment on the level of innate immunity was not related to the severity of the disease. Interestingly, an important variable was sex. The statistic effect size here was partial eta-squared η^2 equals for sex effect $\eta\_Sex^2 = 0.3983$, which is commonly interpreted as a huge effect size according to Cohen's interpretation. Thus, EGb increased the level of innate immunity much stronger in women with AD than in men with AD ($p = 0.0049$).

Simultaneously, the sex differences in EGb effect was smaller and not statistically significant in the control group ($p = 0.1883$). The difference between AD women and AD men was more than 5.5 times higher than between women and men in the control group. Results are presented in Figure 2.

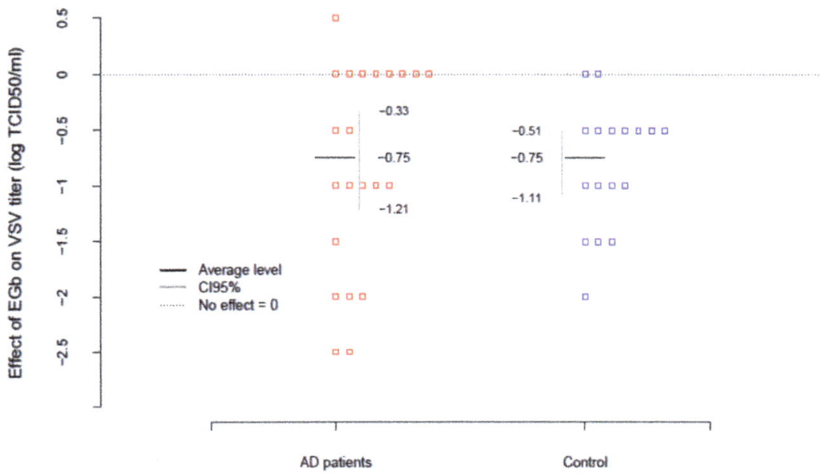

Figure 1. EGb effect on VSV replication (level of innate immunity) in PBLs from AD patients and healthy age-matched controls. EGb effect was measured as difference between VSV titer (log $TCID_{50}/mL$) after EGb treatment and VSV titer (log $TCID_{50}/mL$) before EGb treatment. Red and blue squares are individual observations.

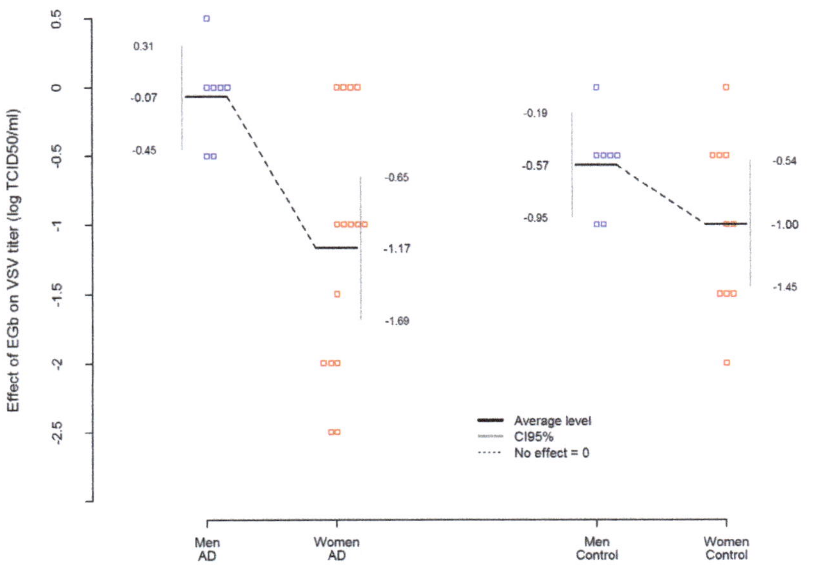

Figure 2. Sex-dependent differences in EGb effect on VSV replication/level of innate immunity in AD patients and healthy age-matched controls. EGb effect was measured as difference between VSV titer (log $TCID_{50}/mL$) after EGb treatment and VSV titer (log $TCID_{50}/mL$) before EGb treatment. Red and blue squares are individual observations.

3.4. Immunoregulatory Effect of EGb Treatment on Cytokine Production by PBLs

In the light of obtained results of beneficial effect of EGb on the PBLs resistance to viral infection/level of innate immunity, it became interesting to investigate the impact of the extract on cytokine production, which is engaged in innate immune response, as changes in viral replication are related to changes in cytokine balance produced by whole PBLs. Table 5 shows an average level of cytokine production, namely TNF-α, IFN-γ, IL-10, IL-1β, and IL-15, by uninfected (spontaneous release) and VSV-infected PBLs from AD patients and controls. The effect of EGb treatment is also presented. As shown in Table 5, VSV infection of PBLs resulted in high TNF-α but slight IL-1β and IL-15 production in both groups. PBLs infection with VSV also revealed increased IFN-γ release but only in the control group. In the case of IL-10, VSV infection resulted in decreased production of this cytokine in AD patients and controls. Interestingly, EGb treatment differentially influenced on all investigated cytokines.

Table 5. Uninfected (spontaneous) and VSV-infected cytokine production by PBLs. Effect of EGb treatment. Median is measure of average level. For every individual person and for every change in cytokine level after EGb, treatment equals $d = \log \frac{after}{before} = \log(after) - \log(before)$. EGb effect is average of all $d's$ in a group. EGb effect is than average of ratios $\frac{after}{before}$ expressed in logs. Delta Δ is measure of difference in EGb effects between AD and controls, and this statistic is the average difference between a randomly selected person from AD group and a randomly selected person from control group.

		Uninfected PBLs			VSV-Infected PBLs		
TNF−α		**PBLs**	**PBLs + EGb**	**EGb effect**	**PBLs + VSV**	**PBLs + VSV + EGb**	**EGb effect**
AD	Median	27.17	1.01	−3.16	60.47	1.1	−3.73
	CI95	9.55; 79.21	0.23; 3.54	−4.28; −2.03	19.58; 153.66	0.31; 3.74	−4.86; −2.59
Controls	Median	49.02	17.2	−0.95	212.83	22.56	−2.01
	CI95	24.88; 99.64	6.49; 41.73	−1.86; −0.36	100.85; 445.43	8.1; 63.35	−3.04; −1.28
AD vs. Controls	Δ	−8.14	−13.41	−1.96	−151.6	−20.93	−1.51
	CI95	−49.23; 31.88	−43.47; −2.09	−3.41; −0.56	−420.02; −7.61	−76.2; −2.35	−3.28; −0.06
		p-value (EGb effect AD vs. controls) = 0.0107			*p*-value (EGb effect AD vs. controls) = 0.0317		
IFN−γ		**PBLs**	**PBLs + EGb**	**EGb effect**	**PBLs + VSV**	**PBLs + VSV + EGb**	**EGb effect**
AD	Median	9.84	2.69	−1.31	9.86	2.83	−1.27
	CI95	6.84; 14.24	1.75; 4.28	−1.83; −0.8	6.95; 12.88	1.87; 4.21	−1.69; −0.89
Control	Median	3.29	3.31	−0.01	8.66	5.05	−0.56
	CI95	1.84; 5.08	1.69; 6.06	−0.36; 0.28	5.02; 13.68	3.09; 7.07	−0.96; −0.23
AD vs. Controls	Δ	6.03	−0.72	−1.28	0.45	−2.35	−0.69
	CI95	3.48; 9.41	−3.38; 1.17	−1.95; −0.55	−4.17; 4.83	−4.21; −0.21	−1.2; −0.15
		p-value (EGb effect AD vs. controls) = 0.0003			*p*-value (EGb effect AD vs. controls) = 0.0127		
IL−10		**PBLs**	**PBLs + EGb**	**EGb effect**	**PBLs + VSV**	**PBLs + VSV + EGb**	**EGB effect**
AD	Median	86.13	10.47	−1.9	43.49	2.37	−2.57
	CI95	34.77; 232.97	3.51; 34.75	−2.69; −1.18	15.73; 108.76	1; 6.58	−3.36; −1.75
Control	Median	162.87	28.76	−1.52	163.73	20.11	−1.99
	CI95	60.29; 356.65	10.13; 110.43	−2.27; −0.98	70.42; 369.71	8.93; 46.73	−2.63; −1.39

Table 5. *Cont.*

		Uninfected PBLs			VSV-Infected PBLs		
TNF−α		**PBLs**	**PBLs + EGb**	**EGb effect**	**PBLs + VSV**	**PBLs + VSV + EGb**	**EGb effect**
AD vs. Control	Δ	−70.75	−15.99	−0.26	−150.13	−17.55	−0.51
	CI95	−338.92; 57.65	−94.28; 14.73	−1.07; 0.52	−315.32; −13.16	−48.42; −2.18	−1.53; 0.41
		p-value (EGb effect AD vs. controls) = 0.5322			*p*-value (EGb effect AD vs. controls) = 0.3104		
IL−1β		**PBLs**	**PBLs + EGb**	**EGb effect**	**PBLs + VSV**	**PBLs + VSV + EGb**	**EGb effect**
AD	Median	367.64	328.83	−0.11	401.33	427.88	0.08
	CI95	238.78; 486.76	249.2; 424.48	−0.38; 0.17	302.81; 510.61	294.89; 552.79	−0.16; 0.29
Control	Median	288.6	496.12	0.57	550.54	604.51	0.12
	CI95	199.13; 399.27	329.64; 706.04	0.12; 0.96	403.5; 718.55	437.85; 830.44	−0.09; 0.36
AD vs. Control	Δ	73.16	−190.84	−0.71	−148.62	−180.34	−0.03
	CI95	−51.88; 211.25	−398.28; 0.32	−1.17; −0.2	−337.82; −6.34	−497.08; 34.26	−0.36; 0.26
		p-value (EGb effect AD vs. controls) = 0.005			*p*-value (EGb effect AD vs. controls) = 0.8316		
IL−15		**PBLs**	**PBLs + EGb**	**EGb effect**	**PBLs + VSV**	**PBLs + VSV + EGb**	**EGb effect**
AD	Median	22.7	39.59	0.55	23.7	39.11	0.48
	CI95	16.47; 31.27	30.9; 50.51	0.22; 0.91	18; 31.03	29.49; 54.57	0.21; 0.86
Control	Median	30.79	42.96	0.32	40.71	48.62	0.15
	CI95	23.05; 40.39	33.36; 56.07	0.04; 0.68	31.68; 54.52	35.66; 67.56	−0.08; 0.43
AD vs. Control	Δ	−8.58	−2.9	0.22	−15.44	−8.45	0.32
	CI95	−16.88; −0.29	−12.49; 5.47	−0.22; 0.69	−23.72; −9.17	−22.42; 3.89	−0.06; 0.69
		p-value (EGb effect AD vs. controls) = 0.3451			*p*-value (EGb effect AD vs. controls) = 0.0917		

EGb decreased **TNF-α** production by uninfected (spontaneous) and VSV-infected PBLs from AD patients and controls. Notably, this effect was considerably higher in AD patients. The average level of spontaneous TNF-α production by PBLs from AD patients was $Med_{TNF-\alpha;AD}^{PBLs;\,before} = 27.17$ pg/mL, and after EGb treatment, it decreased to $Med_{TNF-\alpha;AD}^{PBLs;\,after} = 1.01$ pg/mL. This change (effect) was statistically significant at $\alpha = 0.05$. In the control group, the average level of spontaneous TNF-α production by PBLs before EGb treatment was $Med_{TNF-\alpha;Control}^{PBLs;\,before} = 49.02$ pg/mL, and after EGb treatment, it decreased to $Med_{TNF-\alpha;Control}^{PBLs;\,after} = 17.2$ pg/mL. This change (effect) was also statistically significant. There was significant difference in this effect between AD patients and controls. The decrease of spontaneous TNF-α production by PBLs from AD patients after EGb treatment was higher than in the control group ($p = 0.0107$). Similarly, the average level of TNF-α produced by VSV-infected PBLs from AD was $Med_{TNF-\alpha;AD}^{PBLs+VSV;\,before} = 60.47$ pg/mL, and after EGb treatment, it decreased to $Med_{TNF-\alpha;AD}^{PBLs+VSV;\,after} = 1.1$ pg/mL. This change (effect) was statistically significant at $\alpha = 0.05$. For control group, the average level of TNF-α produced by VSV-infected PBLs was $Med_{TNF-\alpha;Control}^{PBLs+VSV;\,before} = 212.83$ pg/mL, and after EGb treatment, it decreased to $Med_{TNF-\alpha;Control}^{PBLs+VSV;\,after} = 22.56$ pg/mL. This change (effect) was also statistically significant at $\alpha = 0.05$. There was significant difference in this effect between AD patients and controls. The decrease of TNF-α produced by VSV-infected PBLs from AD patients after EGb treatment was higher than in the control group ($p = 0.0317$).

In the case of **IFN-γ**, we can thereby see that EGb decreased spontaneous production of this cytokine in AD group from $Med_{INF-\gamma-\alpha;AD}^{PBLs;\,before} = 9.84$ pg/mL to $Med_{IFN-\gamma-\alpha;AD}^{PBLs;\,after} = 2.69$ pg/mL. The change (effect) was statistically significant at $\alpha = 0.05$. This effect was not observed in the control group probably due to a small sample size. Thus, there was significant difference in

EGb effect between AD patients and controls.The decrease of spontaneous INF-γ production in AD patients after EGb treatment was higher than in the control group ($p = 0.0003$). Similarly, after EGb treatment, the decrease in IFN-γ production by VSV-infected PBLs from AD patients was observed—from $Med_{IFN-\gamma-\alpha;AD}^{PBLs+VSV;\,before} = 9.86$ pg/mL to $Med_{IFN-\gamma-\alpha;AD}^{PBLs+VSV;\,after} = 2.83$ pg/mL. The change (effect) was statistically significant at $\alpha = 0.05$. For control group, the average level of INF-γ produced by VSV-infected PBLs was $Med_{IFN-\gamma-\alpha;Control}^{PBLs+VSV;\,before} = 8.66$ pg/mL and after EGb treatment, it decreased to $Med_{IFN-\gamma-\alpha;Control}^{PBLs+VSV;\,after} = 5.05$ pg/mL. The change (effect) was also statistically significant at $\alpha = 0.05$. Significant difference in EGb effect between AD patients and controls was noticed. A decrease of INF-γ production by VSV-infected PBLs in AD patients after EGb treatment was higher than in the control group ($p = 0.0127$).

The reducing effect of EGb treatment was shown either for anti-inflammatory cytokine **IL-10** production. EGb decreased spontaneous IL-10 production in AD group from $Med_{IL-10-\alpha;AD}^{PBLs;\,before} = 86.13$ pg/mL to $Med_{IL-10-\alpha;AD}^{PBLs;\,after} = 10.47$ pg/mL. This change was statistically significant. In the control group, EGb decreased IL-10 production from $Med_{IL-10-\alpha;AD}^{PBLs;\,before} = 162.87$ pg/mL to $Med_{IL-10-\alpha;AD}^{PBLs;\,after} = 28.76$ pg/mL. This change was also statistically significant. In the case of IL-10 production after VSV-infection, EGb decreased the level of this cytokine in AD patients from $Med_{IL-10-\alpha;Control}^{PBLs+VSV;\,before} = 43.49$ pg/mL to $Med_{IL-10-\alpha;Control}^{PBLs+VSV;\,after} = 2.37$ pg/mL. This change was statistically significant. In the control group, EGb also decreased IL-10 production from $Med_{IL-10-\alpha;Control}^{PBLs+VSV;\,before} = 163.73$ pg/mL to $Med_{IL-10-\alpha;Control}^{PBLs+VSV;\,after} = 20.11$ pg/mL. This change was also statistically significant. A decrease of IL-10 by uninfected and VSV-infected PBLs after EGb treatment was the same in AD patients and controls.

In contrast, EGb increased IL-1β and IL-15 production by uninfected and VSV-infected PBLs from AD and controls. EGb increased **IL-1β** production by uninfected PBLs from control group. The average level of spontaneous IL-1β production was $Med_{IL1-\beta-\alpha;AD}^{PBLs;\,before} = 288.6$ pg/mL, and after EGb treatment, it was $Med_{IL-1\beta-\alpha;AD}^{PBLs;\,after} = 496.12$ pg/mL. The change (effect) was statistically significant at $\alpha = 0.05$. This effect was not observed in AD group probably due to a small sample size. Similarly, the average level of IL-1β produced by VSV-infected PBLs from AD was $Med_{IL-1\beta-\alpha;AD}^{PBLs+VSV;\,before} = 401.33$ pg/mL, and after EGb treatment, it increased to $Med_{IL-1\beta-\alpha;AD}^{PBLs+VSV;\,after} = 427.88$ pg/mL. For control group, the average level of IL-1β produced by VSV-infected PBLs was $Med_{IL-1\beta-\alpha;Control}^{PBLs+VSV;\,before} = 550.54$ pg/mL and after EGb treatment, it increased to $Med_{IL-1\beta-\alpha;Control}^{PBLs+VSV;\,after} = 604.51$ pg/mL. An increase of IL-1β by VSV-infected PBLs after EGb treatment was the same in AD patients and controls.

EGb treatment influenced positively and statistically significant on **IL-15** production by uninfected PBLs from AD and controls. An increase of this cytokine was observed, respectively, in AD from $Med_{IL-15-\alpha;AD}^{PBLs;\,before} = 22.7$ pg/mL to $Med_{IL-15-\alpha;AD}^{PBLs;\,after} = 39.59$ pg/mL and in controls from $Med_{IL-15-\alpha;AD}^{PBLs;\,before} = 30.79$ pg/mL to $Med_{IL-15-\alpha;AD}^{PBLs;\,after} = 42.96$ pg/mL. In the case of IL-15 production by VSV-infected PBLs, an increased effect of EGb treatment was noticed, respectively, in AD from $Med_{IL-15-\alpha;Control}^{PBLs+VSV;\,before} = 23.7$ pg/mL to $Med_{IL-15-\alpha;Control}^{PBLs+VSV;\,after} = 39.11$ pg/mL and in controls from $Med_{IL-15-\alpha;Control}^{PBLs+VSV;\,before} = 40.71$ pg/mL to $Med_{IL-15-\alpha;Control}^{PBLs+VSV;\,after} = 48.62$ pg/mL. An increase of IL-15 by uninfected and VSV-infected PBLs after EGb treatment was the same in AD patients and controls.

Additionally, the level of **IFN-α** was also investigated. Results showed that PBLs from AD and controls did not produce spontaneous IFN-α, which was expected. In every investigated person in both groups, spontaneous IFN-α production was undetectable. Infection with VSV resulted in high IFN-α production by PBLs in both groups. However, leukocytes from healthy age-matched subjects produced $Med_{IFN-\alpha;Control}^{PBLs+VSV;\,before} = 88.47$ pg/mL—over eight times more IFN-α than leukocytes from AD patients $Med_{IFN-\alpha;AD}^{PBLs+VSV;\,before} = 10.49$ pg/mL.

EGb treatment decreased IFN-α, and this effect was very strong, to an undetectable level, in the case of all AD patients and controls. Obtained results were very interesting; however, data were not included in the Table 5 due to experiments that were performed only for half of the participants and need to be confirmed in a larger sample size.

In summary, EGb showed immunoregulatory activity resulted in a different influence on selected pro- and anti-inflammatory cytokine production by PBLs. It decreased TNF-α, IFN-γ, and IL-10 production but increased IL-1β and IL-15 production by uninfected and VSV-infected leukocytes of AD patients and controls. The effects of EGb treatment, however, were much stronger in AD patients. Results of EGb treatment on cytokine production are presented in Figure 3.

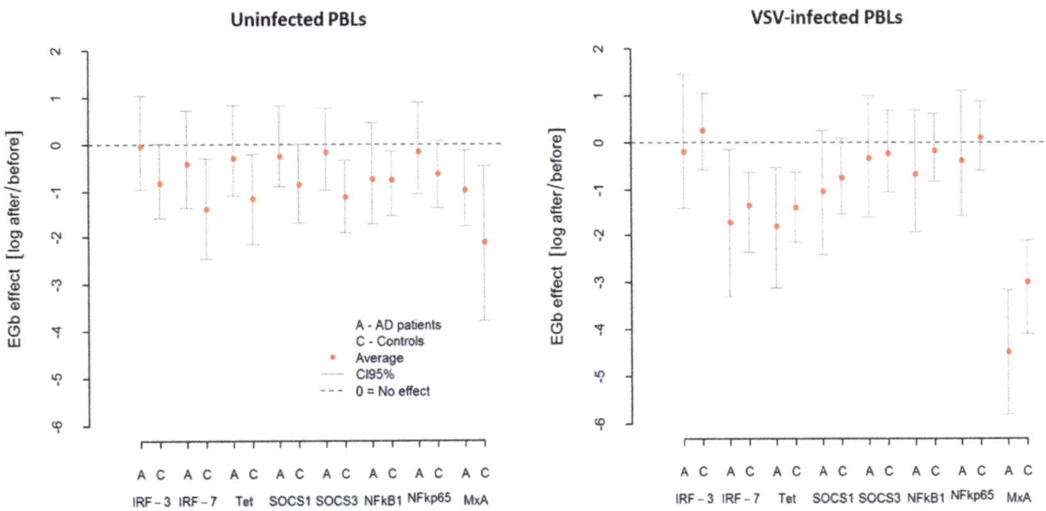

Figure 3. Influence of EGb on pro- and anti-inflammatory cytokine production by uninfected (spontaneous) and VSV-infected PBLs from AD patients and healthy age-matched controls. EGb effect was measured as natural logarithm of ratio after/before cytokine level. Zero means no effect. Points represent average change after treatment with confidence interval CI95% for the average.

3.5. EGb Treatment Down-Regulates Inflammatory-Associated Genes Expression

To detect and respond to viral infection host cell activates multiple cellular signaling networks. Viral proteins and nucleic acids ultimately drive an antiviral response and activating transcription factors. Therefore, to study the influence of EGb on innate immune mechanisms (PBLs resistance to VSV infection/level of innate immunity and cytokine production), we investigated the expression of *IRFs-3* and *-7* (interferon regulatory factors) mRNA as well as ISGs (IFN-stimulated genes)—*tetherin* (encoding bone marrow stromal cell antigen 2) and *MxA* (encoding myxovirus resistance protein 1). *NF-κB* transcription factors and *SOCS* (suppressor of the cytokine signaling proteins) were also examined.

As was suspected, VSV infection resulted in upregulation in all investigated genes in AD patients and controls. However, this effect was more pronounced in AD patients. From Table 6, we can see that *MxA* was the most highly expressed gene with over 40 relative fold change in AD, while in controls, it was over 16 relative fold change. Similarly, after VSV infection in AD, a marked induction (3–8.5 relative fold change) of *IRF-7*, *tetherin,* and *SOCS1* was demonstrated. As presented in Figure 4 and Table 6, in both groups (AD and controls), EGb treatment decreased expression of all investigated genes (average effect < 0) in uninfected and VSV-infected PBLs. In uninfected PBLs from AD patients, average reduction in all investigated genes expression after EGb treatment was about $EGb\ effect^{PBLs}_{all\ genes;AD} \approx -0.4$; i.e., the average level of expression was

$\frac{1}{e^{-0.4}} = 1.5$ times lower compared to the expression before EGb treatment. In the case of controls, it was $EGb\ effect^{PBLs}_{all\ genes;Controls} \approx -1.1$; i.e., average reduction in genes expression in controls after EGb treatment was $\frac{1}{e^{-1.1}} = 3$ times lower compared to the expression before EGb treatment. Interestingly, in VSV-infected PBLs the effect of EGb treatment was stronger in AD group for all genes than for controls. In PBLs from AD patients, average reduction of all investigated genes expression was about $EGb\ effect^{PBLs+VSV}_{all\ genes;AD} \approx -1$; i i.e., the average level of expression was $\frac{1}{e^{-1}} = 2.7$ times lower compared, to the genes expression before EGb treatment. In PBLs from control group it was $EGb\ effect^{PBLs+VSV}_{all\ genes;Controls} \approx -0.7$; i.e., the average expression in VSV-infected leukocytes was $\frac{1}{e^{-0.7}} = 2$ times lower compared to genes expression before EGb treatment. Table 6 presents level of investigated genes expressions before and after EGb treatment, EGb effects, and comparison of EGb effects in both groups. Confidence intervals (CI95) in Figure 4 and in Table 6 show greater precision of expression measurement than testing for statistical significance.

Table 6. Genes expression before and after EGb treatment and measure of EGb effect in AD patients and control group. Median is measure of average level. For every individual person and for every gene change of expression level after, treatment equals $d = \log \frac{after}{before} = \log(after) - \log(before)$. EGb effect is average of all *d*s in a group. EGb effect is the average of ratios $\frac{after}{before}$ expressed in logs. Delta Δ is measure of difference in EGb effects between AD and controls, and this statistic is the average difference between a randomly selected person from AD group and a randomly selected person from control group.

		Uninfected PBLs			**VSV-Infected PBLs**		
IRF−3		**PBLs**	**PBLs + EGb**	**EGb effect**	**PBLs + VSV**	**PBLs + VSV + EGb**	**EGB effect**
AD	Median	1.65	1.78	−0.04	2.88	2.21	−0.2
	CI95	0.76; 4.01	0.59; 4.78	−0.95; 1.04	0.61; 9.7	1.04; 6.61	−1.41; 1.45
Control	Median	2.62	1.17	−0.82	1.95	2.53	0.25
	CI95	1.08; 6.3	0.61; 2.25	−1.6; 0.02	1.05; 3.73	0.98; 6.14	−0.58; 1.05
AD vs. Control	delta	−1.1	0.92	0.81	2.07	−0.18	−0.56
	CI95	−6.43; 1.22	−0.79; 3.75	−0.57; 2.13	−1.28; 9.98	−4.61; 6.35	−1.99; 1.37
	p-value		*p* = 0.2371			*p* = 0.4612	
IRF−7		**PBLs**	**PBLs + EGb**	**EGb effect**	**PBLs + VSV**	**PBLs + VSV + EGb**	**EGB effect**
AD	Median	1.39	1.06	−0.42	11.78	1.83	−1.73
	CI95	0.58; 4.08	0.36; 3.01	−1.37; 0.72	3.31; 32.97	0.67; 7.28	−3.31; −0.14
Control	Median	2.1	0.52	−1.39	7.84	2.1	−1.34
	CI95	0.74; 5.9	0.29; 0.96	−2.47; −0.3	4.44; 14.53	0.71; 5.13	−2.37; −0.63
AD vs. Control	delta	−1.06	0.77	1.02	7.81	0.18	−0.35
	CI95	−5.65; 2.38	−0.37; 3.19	−0.62; 2.54	−5.26; 39.16	−3.96; 16.53	−2.12; 1.79
	p-value		*p* = 0.1921			*p* = 0.7206	
Tet		**PBLs**	**PBLs + EGb**	**EGb effect**	**PBLs + VSV**	**PBLs + VSV + EGb**	**EGB effect**
AD	Median	1.32	1.12	−0.3	8.27	1.16	−1.81
	CI95	0.63; 2.94	0.38; 2.58	−1.09; 0.83	2.12; 24.23	0.58; 3.58	−3.13; −0.53
Control	Median	2.12	0.63	−1.16	6.66	1.79	−1.39
	CI95	0.85; 5.1	0.3; 1.28	−2.16; −0.21	3.89; 11.87	0.69; 4.59	−2.16; −0.63

Table 6. *Cont.*

		Uninfected PBLs			VSV-Infected PBLs		
IRF−3		PBLs	PBLs + EGb	EGb effect	PBLs + VSV	PBLs + VSV + EGb	EGB effect
AD vs. Control	delta	−1.23	0.59	0.88	3.76	−0.89	−0.43
	CI95	−5.11; 0.87	−0.44; 1.81	−0.53; 2.39	−5.16; 19.04	−4.57; 3.69	−2.12; 1.26
	p-value		*p* = 0.2264			*p* = 0.5552	
SOCS1		PBLs	PBLs + EGb	EGb effect	PBLs + VSV	PBLs + VSV + EGb	EGB effect
AD	Median	3.51	3.09	−0.26	11.94	3.63	−1.04
	CI95	1.6; 8.11	1.2; 6.88	−0.9; 0.82	3.64; 32.57	1.54; 15.29	−2.42; 0.26
Control	Median	2.28	0.97	−0.86	4.57	2.37	−0.75
	CI95	0.94; 5.72	0.55; 1.68	−1.7; 0	2.89; 7.79	0.97; 5.17	−1.53; 0.09
AD vs. Control	delta	0.84	2.73	0.63	10.7	2.62	−0.3
	CI95	−2.86; 5.41	0.16; 6.75	−0.58; 1.89	−1.6; 34.36	−3.24; 36.4	−2; 1.48
	p-value		*p* = 0.2868			*p* = 0.6482	
SOCS3		PBLs	PBLs + EGb	EGb effect	PBLs + VSV	PBLs + VSV + EGb	EGB effect
AD	Median	2.52	2.34	−0.17	3.93	2.43	−0.34
	CI95	1.29; 5.7	0.86; 5.73	−0.97; 0.77	0.9; 14.48	1.09; 10.14	−1.61; 0.99
Control	Median	2.26	0.68	−1.12	1.29	1.22	−0.23
	CI95	0.87; 4.98	0.34; 1.38	−1.92; −0.34	0.81; 2.61	0.5; 2.78	−1.06; 0.68
AD vs. Control	delta	−0.06	1.9	0.92	4.61	2	−0.17
	CI95	−2.96; 3.8	0.16; 5.99	−0.22; 2.17	−0.65; 17.48	−1.32; 23.93	−1.73; 1.56
	p-value		*p* = 0.1284			*p* = 0.8019	
NFkB1		PBLs	PBLs + EGb	EGb effect	PBLs + VSV	PBLs + VSV + EGb	EGB effect
AD	Median	2.99	1.63	−0.74	3.87	1.77	−0.68
	CI95	1.41; 7.73	0.55; 4.45	−1.72; 0.46	0.78; 15.27	0.79; 6.55	−1.93; 0.69
Control	Median	2.09	0.87	−0.76	1.88	1.75	−0.18
	CI95	0.99; 4.11	0.45; 1.72	−1.54; −0.14	1.14; 3.26	0.78; 3.82	−0.83; 0.62
AD vs. Control	delta	0.88	0.94	−0.01	5.12	0.5	−0.55
	CI95	−1.87; 7.59	−0.62; 4.33	−1.25; 1.56	−1.06; 21.44	−3.2; 11.06	−2.02; 1.13
	p-value		*p* = 0.9873			*p* = 0.4645	
NFkp65		PBLs	PBLs + EGb	EGb effect	PBLs + VSV	PBLs + VSV + EGb	EGB effect
AD	Median	2.63	2.5	−0.16	4.71	2.9	−0.39
	CI95	1.16; 6.81	0.92; 6.18	−1.05; 0.89	0.98; 16.35	1.38; 10.06	−1.56; 1.1
Control	Median	1.93	0.99	−0.63	2.06	2.46	0.1
	CI95	0.88; 4.11	0.56; 1.75	−1.36; 0.08	1.22; 3.46	1.01; 5.68	−0.6; 0.88
AD vs. Control	delta	0.77	1.9	0.42	5.28	0.79	−0.55
	CI95	−2.05; 6.57	−0.12; 5.4	−0.79; 1.84	−0.94; 20.13	−4.27; 17	−1.99; 1.17
	p-value		*p* = 0.5612			*p* = 0.4365	
MxA		PBLs	PBLs + EGb	EGb effect	PBLs + VSV	PBLs + VSV + EGb	EGB effect
AD	Median	0.47	0.2	−0.98	21.82	0.22	−4.49
	CI95	0.17; 1.32	0.07; 0.37	−1.77; −0.12	6.87; 64.67	0.1; 0.62	−5.81; −3.17
Control	Median	1.32	0.15	−2.13	21.69	1.07	−3.01
	CI95	0.27; 6.13	0.07; 0.32	−3.78; −0.46	14.08; 33.2	0.34; 3.57	−4.1; −2.12
AD vs. Control	delta	−2.23	0.02	1.19	6.3	−1.54	−1.5
	CI95	−8.97; 0.34	−0.42; 0.44	−1.14; 3.17	−17.07; 73.42	−3.86; 0.03	−3.15; 0.07
	p-value		*p* = 0.3098			*p* = 0.069	

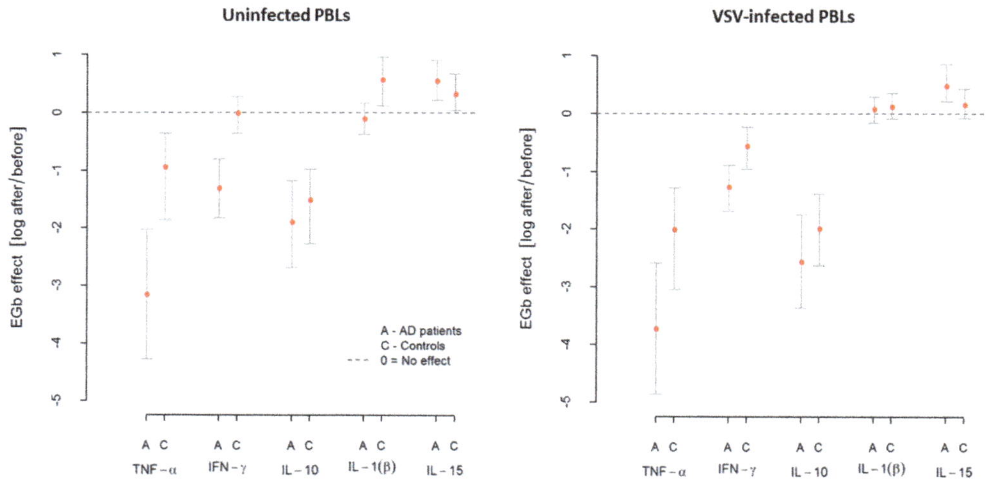

Figure 4. Changes in relative expression of inflammatory-associated genes after EGb treatment in uninfected and VSV-infected PBLs from AD patients and healthy age-matched controls. EGb effect was measured as natural logarithm of ratio after/before gene expression. Zero means no effect. Points represent average change after treatment with confidence interval CI95% for the average.

4. Discussion

With numbers of sufferers with cognitive impairments still increasing, dementia such as Alzheimer's disease (AD) is observed. AD is an age-related disorder; however, neurodegenerative changes may begin many years before clinical manifestation of the disease. Therefore, brain health is becoming very important for the adults and the elderly. The lack of effective AD treatment means that those affected by this disease are still looking for any alternative approaches. Nutritional science and dietary supplements, such as extracts of *Ginkgo biloba* (EGb), offer an approach (preventative and restorative) to adding phytonutrients to daily consumption of food and drink [20]. EGb is the most widely studied natural compound with proven beneficial effect on cognitive functions (improving memory and concentration) in both healthy adults and patients with mild cognitive impairment (MCI) or dementia. However, its effect on the immune system is less explored, with little literature data available. We have previously reported that EGb increased an innate immune response and regulated cytokine production in healthy young people [21,22]. We suggest the possibility of using EGb for the treatment of immune deficiencies. As AD is considered as a systemic disease with evidence for changes in immune system functioning, we examined the effect of EGb on innate immune mechanisms in AD patients. In the present study, we showed that EGb is a good immunoregulator increasing an innate immune response of AD patients. We also examined the effect of EGb treatment on several inflammatory-associated genes expression.

Currently, many EGb preparation are available, with different compositions resulting in various therapeutic features [23]. We used a standardized extract, prepared according to the European Pharmacopoeia (Ph. Eur. 8.0); it is easily accessible, which is a great benefit. In addition, an advantage of herbal drugs is that they have very low side-effects compared to chemical drugs [6]. Indeed, we observed immuno-enhancing activity of EGb in an absolutely nontoxic concentration for human peripheral blood leukocytes (PBLs). The leaves of maidenhair tree (*G. biloba*) are also excellent sources of antioxidants. Ethanol extracts of that plant are well-characterized and contain about 22–27% flavonol glycosides, including polyphenols such as tannins and terpene lactones (5–7%). The high phenolics content determines strong ferric ion activity, reducing antioxidant power, copper chelating ability, peroxyl radical scavenging activity, and radical scavenging activity for *G. biloba* extracts [24–26]. Our results correspond to the results of a study on antioxidant properties of various plant extracts carried out by Szerlauth and others (2019) [24]. It was found that the plant extracts exhibit remarkable antioxidant properties (IC_{50}) from 0.44 (*Alliumsativum*) to 44.18 (*Juglansregia*) AAEQ values [24].

The observed age-related progressive decline in immune system functioning contributes to a development of chronic states of inflammation resulting in systemic diseases, including AD. This phenomenon—dysregulation of the immune system—is not limited to the one mechanism, but it concerns aging of the innate and adaptive immune cells, alterations in circulating inflammatory mediators, and changes in lymphoid and non-lymphoid tissues [4]. In AD patients, dysregulation of innate immune mechanisms were also observed. PBLs resistance to viral infection is a good indicator of the innate immune system condition. We previously established that the level of innate immunity, measured with the test based on vesicular stomatitis virus (VSV) replication in human PBLs, was remarkably correlated with clinical advancement of AD. The higher VSV titer means a lower level of innate immunity. More severe patients were characterized with a lower level of innate immunity [27]. We also showed as a potential therapeutic the oral administration of proline-rich polypeptide complex (PRP) isolated from bovine colostrum, a nutraceutical intended to boost an immune system by increasing an innate immune response in AD patients [27]. Here, we present for a first time the strong immune-enhancement activity of natural, herbal preparation, EGb, in an ex vivo model of PBLs from AD patients. As we suspected, EGb was capable of significantly improving innate immune response by decreasing VSV replication in PBLs of AD patients but also controls. The most interesting, however, was that EGb notably increased an innate immune response in AD women. This sex discrepancy in EGb effect was not observed in the control group. Similarly, in the above-mentioned study of EGb effect on the immune system functioning in healthy young people, sex differences were not observed. Thus, EGb may be promising for immune improving in AD patients, especially in AD women. It is important due to the fact that women are more afflicted with the frequency, prevalence, and clinical manifestation of the disease [28].

According to recent data in animal models, EGb showed wide-ranging anti-inflammatory and antioxidant properties [29]. The immune effect of EGb was showed by Wan at al. [30] in the model of mouse microglia cells. Microglia, brain resident macrophages, play an important role in the development of many central nervous system (CNS) diseases. Reactive microglia that produce large amounts of inflammatory mediators influence the development of AD [31]. Significantly reduced production of pro-inflammatory cytokines, such as IL-6 and TNF-α, and increased anti-inflammatory IL-4, IL-13, or TGF-β were found in the brains of animals supplemented with EGb [30]. Similar results were presented by Tao et al. [32] in allergic mice. Administration of EGb showed significant decrease in release of pro-inflammatory IL-4, IL-5, IL-6, IL-8, and TNF-α. In our study with human leukocytes model ex vivo from AD patients and healthy age-matched controls, EGb presented immunoregulatory activity. EGb significantly decreased production of pro-inflammatory TNF-α and IFN-γ as well as anti-inflammatory IL-10 by uninfected and VSV-infected PBLs of AD patients and controls. At the same time, EGb significantly increased production of IL-15 and slightly increased IL-1β in both groups of participants. Interestingly, this effect was more pronounced among AD patients.

It was established that VSV infection of PBLs induced secretion of IFN-α [33]. In the present study, IFN-α was also investigated. IFN-α was not assessed for all study participants, and this investigation needs to be continued. However, interesting observations were obtained. Although, as suspected, there was no spontaneous secretion of IFN-α, after VSV infection, leukocytes of both groups responded with production of this cytokine. After incubation with EGb, the production of IFN-α decreased to no detectable level in all AD patients as well as in controls. IFN-α was the only cytokine strongly inhibited by the extract.

EGb also downregulated expression of several genes that regulate innate immune response to viral infection and cytokine production, such as interferon regulatory factors (IRFs) *IRF-3* and *-7*, which are primary transcriptional factors regulate the type I IFN response after RNA virus infections [34]; IFN-stimulated antiviral genes *MxA* and *tetherin*, critical for controlling VSV infection [33]; but also *NFkB* transcription factors that mediate induction of various pro-inflammatory genes in innate immune cells; and *SOCS*, the main regulators of antimicrobial innate immune response [35,36]. It was presented earlier that *G. biloba* extract has anti-oxidative as well as anti-inflammatory properties. A marked suppression of transcription factor NFκB and pro-inflammatory cytokines (TNF-α, IL-1α, 1L-6) was shown [37].

5. Conclusions

Therefore, EGb may have an advantageous properties for health management in elderly and AD sufferers but especially in women afflicted with AD. Female sex is a major risk factor for developing late-onset AD, which is suggested to be implicated in the menopause transition [38]. The observed beneficial effect of EGb on innate immune response/increase PBLs resistance to VSV infection may be at least partially explained by its antioxidant activity and differential influence on cytokine production. Even though the most important antiviral response is mediated with IFN I, the role of other cytokines should not be diminished. In our study, EGb decreased IFNs production but increased IL-15 and IL-1β. IL-15 is known as playing an important role in promoting the development and homeostasis of NK cells and CD8 T; however, IL-15 also mediates the anti-viral responses of these cell populations during an active immune response [39]. IL-1β, next to type I IFNs, is also a central mediator driving innate antiviral immunity and inflammation [40]. It was suggested that activation status of peripheral innate immune cells may be a good biomarker of AD pathology [3]. Thus, improving their activity by adding EGb as accompanying treatment may be a good long-term course to modify the disease progression.

Author Contributions: J.L., conception and study design, clinical examination of the patients, and drafted and revised the manuscript; M.S. (Marta Sochocka), conception and study design, laboratory experiments, acquired and analyzed and interpreted all data, drafted and revised the manuscript, and edited language; M.O., laboratory experiments and drafted the manuscript; P.N. and K.G., laboratory experiments; A.Z., laboratory experiments and drafted the manuscript; M.S. (Maciej Sobczyński), acquired and analyzed and interpreted all data. All authors have read and agreed to the published version of the manuscript.

Funding: This work was supported by internal research funds of the Wroclaw Medical University and Hirszfeld Institute of Immunology and Experimental Therapy, Polish Academy of Sciences.

Institutional Review Board Statement: The study was conducted in accordance with the Declaration of Helsinki, and approved by the Ethics Committee of the Wroclaw Medical University (No. KB 349/2016).

Informed Consent Statement: Informed consent was obtained from all subjects involved in the study.

Acknowledgments: The authors wish to acknowledge the nursing staff of the Wroclaw Medical University for the support in recruiting participants.

Conflicts of Interest: The authors report no conflict of interest, financial or otherwise.

References

1. Kinney, J.W.; BeMiller, S.M.; Murtishaw, A.S.; Leisgang, A.M.; Salazar, A.M.; Lamb, B.T. Inflammation as a central mechanism in Alzheimer's disease. *Alzheimer's Dementia: Transl. Res. Clin. Interv.* **2018**, *4*, 575–590. [CrossRef] [PubMed]
2. Bettcher, B.M.; Tansey, M.G.; Dorothée, G.; Heneka, M.T. Peripheral and central immune system crosstalk in Alzheimer disease—A research prospectus. *Nat. Rev. Neurol.* **2021**, *17*, 689–701. [CrossRef] [PubMed]
3. Le Page, A.; Dupuis, G.; Frost, E.H.; Larbi, A.; Pawelec, G.; Witkowski, J.M.; Fulop, T. Role of the peripheral innate immune system in the development of Alzheimer's disease. *Exp. Gerontol.* **2018**, *107*, 59–66. [CrossRef] [PubMed]
4. Lutshumba, J.; Nikolajczyk, B.S.; Bachstetter, A.D. Dysregulation of Systemic Immunity in Aging and Dementia. *Front. Cell. Neurosci.* **2021**, *15*, 652111. [CrossRef] [PubMed]
5. Syed, Y.Y. Sodium Oligomannate: First Approval. *Drugs* **2020**, *80*, 441–444. [CrossRef]
6. Bhattacharya, T.; Dey, P.S.; Akter, R.; Kabir, T.; Rahman, H.; Rauf, A. Effect of natural leaf extracts as phytomedicine in curing geriatrics. *Exp. Gerontol.* **2021**, *150*, 111352. [CrossRef]

7. Ullah, A.; Munir, S.; Badshah, S.L.; Khan, N.; Ghani, L.; Poulson, B.G.; Emwas, A.-H.; Jaremko, M. Important Flavonoids and Their Role as a Therapeutic Agent. *Molecules* **2020**, *25*, 5243. [CrossRef]
8. Sierpina, V.S.; Wollschlaeger, B.; Blumenthal, M. Ginkgo biloba. *Am. Fam. Phys.* **2003**, *68*, 923–926.
9. Fumia, A.; Cicero, N.; Gitto, M.; Nicosia, N.; Alesci, A. Role of nutraceuticals on neurodegenerative diseases: Neuroprotective and immunomodulant activity. *Nat. Prod. Res.* **2021**, *265*, 1–18. [CrossRef]
10. Singh, S.K.; Srivastav, S.; Castellani, R.J.; Plascencia-Villa, G.; Perry, G. Neuroprotective and Antioxidant Effect of Ginkgo biloba Extract Against AD and Other Neurological Disorders. *Neurotherapeutics* **2019**, *16*, 666–674. [CrossRef]
11. Nowak, A.; Kojder, K.; Zielonka-Brzezicka, J.; Wróbel, J.; Bosiacki, M.; Fabiańska, M.; Wróbel, M.; Sołek-Pastuszka, J.; Klimowicz, A. The Use of Ginkgo *Biloba* L. as a Neuroprotective Agent in the Alzheimer's Disease. *Front. Pharmacol.* **2021**, *12*, 775034. [CrossRef] [PubMed]
12. Yen, G.-C.; Chen, H.-Y. Antioxidant activity of various tea extracts in relation to their antimutagencity. *J. Agric. Food Chem.* **1995**, *43*, 27–32. [CrossRef]
13. Benzie, I.F.F.; Strain, J.J. The ferric reducing ability of plasma (FRAP) as a measure of "antioxidant power": The FRAP assay. *Anal. Biochem.* **1996**, *239*, 70–76. [CrossRef] [PubMed]
14. Xu, X.; Katayama, S.; Mine, Y. Antioxidant activity of tryptic digests of hen egg yolk phosvitin. *J. Sci. Food Agric.* **2007**, *87*, 2604–2608. [CrossRef]
15. Liu, M.-Q.; Zhao, M.; Kong, W.-H.; Tang, L.; Wang, F.; Zhu, Z.-R.; Wang, X.; Qiu, H.-Y.; Zhou, D.-J.; Wang, X.; et al. Combination antiretroviral therapy (cART) restores HIV-1 infection-mediated impairment of JAK-STAT signaling pathway. *Oncotarget* **2017**, *8*, 22524–22533. [CrossRef]
16. McCormick, S.M.; Gowda, N.; Fang, J.X.; Heller, N.M. Suppressor of Cytokine Signaling (SOCS)1 Regulates Interleukin-4 (IL-4)-activated Insulin Receptor Substrate (IRS)-2 Tyrosine Phosphorylation in Monocytes and Macrophages via the Proteasome. *J. Biol. Chem.* **2016**, *291*, 20574–20587. [CrossRef]
17. Grassi, M.A.; Rao, V.R.; Chen, S.; Cao, D.; Gao, X.; Cleary, P.A.; Huang, R.S.; Paterson, A.D.; Natarajan, R.; Rehman, J.; et al. Lymphoblastoid Cell Lines as a Tool to Study Inter-Individual Differences in the Response to Glucose. *PLoS ONE* **2016**, *11*, e0160504. [CrossRef]
18. O'Reilly, S.; Ciechomska, M.; Fullard, N.; Przyborski, S.; van Laar, J.M. IL-13 mediates collagen deposition via STAT6 and microRNA-135b: A role for epigenetics. *Sci Rep.* **2016**, *26*, 25066. [CrossRef]
19. Rousseeuw, P.J.; Croux, C. Alternatives to the Median Absolute Deviation. *J. Am. Stat. Assoc.* **1993**, *88*, 1273–1283. [CrossRef]
20. Lewis, J.E.; Poles, J.; Shaw, D.P.; Karhu, E.; Khan, S.A.; Lyons, A.E.; Sacco, S.B.; McDanie, H.R. The effects of twenty-one nutrients and phytonutrients on cognitive function: A narrative review. *J. Clin. Transl. Res.* **2021**, *7*, 575–620.
21. Sochocka, M.; Zaczyńska, E.; Tabol, A.; Czarny, A.; Leszek, J.; Sobczyński, M. The influence of donepezil and EGb 761 on the innate immunity of human leukocytes: Effect on the NF-κB system. *Int. Immunopharmacol.* **2010**, *10*, 1505–1513. [CrossRef] [PubMed]
22. Sochocka, M.; Taboł, A.; Sobczyński, M.; Zaczyńska, E.; Czarny, A.; Leszek, J. Innate antiviral immunity of human PBLs and immunoregulatory activity of EGb 761. *Open Life Sci.* **2014**, *9*, 359–366. [CrossRef]
23. Isah, T. Rethinking Ginkgo *Biloba* L.: Medicinal uses and conservation. *Pharmacogn. Rev.* **2015**, *9*, 140–148. [CrossRef] [PubMed]
24. Szerlauth, A.; Muráth, S.; Viski, S.; Szilagyi, I. Radical scavenging activity of plant extracts from improved processing. *Heliyon* **2019**, *5*, e02763. [CrossRef] [PubMed]
25. Pohl, F.; Lin, P.K.T. The Potential Use of Plant Natural Products and Plant Extracts with Antioxidant Properties for the Prevention/Treatment of Neurodegenerative Diseases: In Vitro, In Vivo and Clinical Trials. *Molecules* **2018**, *23*, 3283. [CrossRef]
26. Li, J.; Zhang, Y.-C.; Chen, G. Effect of Ginkgo biloba Extract EGb761 on Hippocampal Neuronal Injury and Carbonyl Stress of D-show="" $132#?=""Gal-Induced Aging Rats. *Evid.-Based Complement. Altern. Med.* **2019**, *2019*, 5165910.
27. Sochocka, M.; Ochnik, M.; Sobczyński, M.; Siemieniec, I.; Orzechowska, B.; Naporowski, P.; Leszek, J. New therapeutic targeting of Alzheimer's disease. Potential use of proline-rich polypeptide complex to modulate an innate immune response-preliminary study. *J. Neuroinflamm.* **2019**, *16*, 137. [CrossRef]
28. Ferretti, M.T.; Iulita, M.F.; Cavedo, E.; Chiesa, P.A.; Schumacher Dimech, A.; Santuccione Chadha, A.; Baracchi, F.; Girouard, H.; Misoch, S.; Giacobini, E.; et al. Sex differences in Alzheimer disease—The gateway to precision medicine. *Nat. Rev. Neurol.* **2018**, *14*, 457–469. [CrossRef]
29. Achete de Souza, G.; de Marqui, S.V.; Matias, J.N.; Guiguer, E.L.; Barbalho, S.M. Effects of Ginkgo biloba on Diseases Related to Oxidative Stress. *Planta Med.* **2020**, *86*, 376–386. [CrossRef]
30. Wan, W.; Zhang, C.; Danielsen, M.; Li, Q.; Chen, W.; Chan, Y.; Li, Y. EGb761 improves cognitive function and regulates inflammatory responses in the APP/PS1 mouse. *Exp. Gerontol.* **2016**, *81*, 92–100. [CrossRef]
31. Li, Y.; Tan, M.-S.; Jiang, T.; Tan, L. Microglia in Alzheimer's disease. *BioMed. Res. Int.* **2014**, *2014*, 437483. [CrossRef] [PubMed]
32. Tao, Z.; Jin, W.; Ao, M.; Zhai, S.; Xu, H.; Yu, L. Evaluation of the anti-inflammatory properties of the active constituents in Ginkgo biloba for the treatment of pulmonary diseases. *Food Funct.* **2019**, *10*, 2209–2220. [CrossRef] [PubMed]
33. Tomczyk, T.; Wróbel, G.; Chaber, R.; Siemieniec, I.; Piasecki, E.; Krzystek-Korpacka, M.; Orzechowska, B. Immune Consequences of in vitro Infection of Human Peripheral Blood Leukocytes with Vesicular Stomatitis Virus. *J. Innate Immun.* **2018**, *10*, 131–144. [CrossRef] [PubMed]
34. Jefferies, C.A. Regulating IRFs in IFN Driven Disease. *Front. Immunol.* **2019**, *10*, 325. [CrossRef] [PubMed]

35. Huang, S.; Liu, K.; Cheng, A.; Wang, M.; Cui, M.; Huang, J.; Zhu, D.; Chen, S.; Liu, M.; Zhao, X.; et al. SOCS Proteins Participate in the Regulation of Innate Immune Response Caused by Viruses. *Front. Immunol.* **2020**, *11*, 558341. [CrossRef]
36. Liu, T.; Zhang, L.; Joo, D. NF-κB signaling in inflammation. *Signal Transduct. Target. Ther.* **2017**, *2*, 17023. [CrossRef]
37. Kaur, S.; Sharma, N.; Nehru, B. Anti-inflammatory effects of Ginkgo biloba extract against trimethyltin-induced hippocampal neuronal injury. *Inflammopharmacology* **2018**, *26*, 87–104. [CrossRef]
38. Scheyer, O.; Rahman, A.; Hristov, H.; Berkowitz, C.; Isaacson, R.S.; Diaz Brinton, R.; Mosconi, L. Female Sex and Alzheimer's Risk: The Menopause Connection. *J. Prev. Alzheimers Dis.* **2018**, *5*, 225–230. [CrossRef]
39. Verbist, K.; Klonowski, K.D. Functions of IL-15 in anti-viral immunity: Multiplicity and variety. *Cytokine* **2012**, *59*, 467–478. [CrossRef]
40. Aarreberg, L.D.; Wilkins, C.; Ramos, H.J.; Green, R.; Davis, M.A.; Chow, K.T.; Gale, M. Interleukin-1β Signaling in Dendritic Cells Induces Antiviral Interferon Responses. *MBio* **2018**, *9*, e00342-18. [CrossRef]

nutrients

MDPI

Article

Association between Dietary Habits and *Helicobacter pylori* Infection among Bahraini Adults

Fatema Habbash [1,2,*], Tariq Abdulkarim Alalwan [3], Simone Perna [3,*], Naila Ahmed [4,5], Omar Sharif [4,5], Adel Al Sayyad [1,6], Clara Gasparri [7], Cinzia Ferraris [8] and Mariangela Rondanelli [9,10,*]

1 Department of Family and Community Medicine, Arabian Gulf University, Manama 329, Bahrain
2 Family and Community Medicine, Internal Medicine Department, King Abdullah Medical City, Manama 328, Bahrain
3 Department of Biology, College of Science, University of Bahrain, Sakhir P.O. Box 32038, Bahrain
4 Department of Gastroenterology, King Hamad University Hospital, Muharraq 228, Bahrain
5 Department of Internal Medicine, The Royal College of Surgeons in Ireland, Muharraq 228, Bahrain
6 Public Health, Ministry of Health, Manama 323, Bahrain
7 Endocrinology and Nutrition Unit, Azienda di Servizi alla Persona "Istituto Santa Margherita", University of Pavia, 27100 Pavia, Italy
8 Laboratory of Food Education and Sport Nutrition, Department of Public Health, Experimental and Forensic Medicine, University of Pavia, 27100 Pavia, Italy
9 IRCCS Mondino Foundation, 27100 Pavia, Italy
10 Unit of Human and Clinical Nutrition, Department of Public Health, Experimental and Forensic Medicine, University of Pavia, 27100 Pavia, Italy
* Correspondence: fatma.h@agu.edu.bh (F.H.); simoneperna@hotmail.it (S.P.); mariangela.rondanelli@unipv.it (M.R.); Tel.: +973-39872001 (F.H.)

Abstract: Helicobacter pylori (*H. Pylori*) infection is the main bacterial cause of several gastrointestinal disorders. This study aims to estimate the prevalence of *H. pylori* infection in a population of Bahraini adults seeking care in gastroenterology clinics in a tertiary care hospital in the Kingdom of Bahrain and examine the association between dietary habits and other factors with *H. pylori* infection. The study is a hospital-based retrospective, cross-sectional analytical study that included 200 participants. *H. pylori* infection prevalence among the studied group was 55.5%, and it was significantly higher among participants with a high school education or less (44.1%). Among dietary habits, the mean of frequency of green tea, coffee and honey intake was significantly lower among the *H. pylori* infected participants compared to their non-infected counterparts. *H. pylori* infection was significantly higher among participants with vitamin D deficiency (63.6%) compared to participants with normal vitamin D (30%) ($p = 0.001$) and each unit decrease in serum vitamin D was associated with an increased risk of infection by 1.1 times (OR = 1.1; 95% CI: 1.05, 1.18; $p < 0.001$). The study revealed that high educational levels, consumption of honey, green tea, and coffee, as well as normal serum vitamin D level, were independent protectors against *H. pylori* infection. Additional studies are needed to estimate the prevalence and predisposing factors of *H. pylori* infection in the general population.

Keywords: dietary habits; *Helicobacter pylori*; socio-demographic factors; biochemical measurement; vitamin D; Bahrain

Citation: Habbash, F.; Alalwan, T.A.; Perna, S.; Ahmed, N.; Sharif, O.; Al Sayyad, A.; Gasparri, C.; Ferraris, C.; Rondanelli, M. Association between Dietary Habits and *Helicobacter pylori* Infection among Bahraini Adults. *Nutrients* **2022**, *14*, 4215. https://doi.org/10.3390/nu14194215

Academic Editors: Omorogieva Ojo and Amanda R Amorim Adegboye

Received: 12 September 2022
Accepted: 30 September 2022
Published: 10 October 2022

1. Introduction

Helicobacter pylori (*H. pylori*) is a Gram-negative spiral-shaped bacterium, which colonizes and grows in human gastric epithelial tissue and mucosa [1]. More than 50% of the global population are infected by *H. pylori* especially in developing countries and among populations with low socioeconomic status [2]. *H. pylori* is usually transmitted through the feco-oral route due to ingestion of contaminated water or food, but it can be transmitted through direct contact with saliva and vomitus [3,4]. The microorganism was classified as a group 1 carcinogen [5] and it causes various upper gastrointestinal (GI) disorders including gastritis, gastroduodenal ulcer diseases, and gastric adenocarcinoma [6]. The latter was recognized to be the fourth-leading cause of cancer-related deaths worldwide in 2020 [7].

Acquisition and various disease outcomes of *H. pylori* infection are intermediated by complex interactions between bacterial virulence, host, and environmental factors [8]. There is a high level of disparity in *H. pylori* genetic recombination and these genetic differences might exist even in *H. pylori* colonizing the same individual [9]. Host genetic background might contribute to protection from infection with *H. pylori* infection [9]. Some factors associated with *H. pylori* infection include age, gender, ethnicity, educational level, and household income [10]. Furthermore, the crowding index, living standards which include sanitation and hygiene, and the source of drinking water have all been shown to be associated with H. *pylori* infection [11]. Findings related to the relationship between smoking and *H. pylori* infection in previous studies are conflicting [12–15].

Over the past years, epidemiological studies have found that diet plays a significant role in the development of *H. pylori* infection and investigated the association between the intake of certain foods and nutrients and the development of such infection [16–18]. Some studies have reported that salty, pickled, fermented, or smoked foods increased the risk of *H. pylori* infection [19,20]. On the other hand, other studies have shown that antioxidant-rich fruits and vegetables were protective against *H. pylori* infection [21–23]. Moreover, it was reported that lower intakes of raw vegetables were significantly associated with a higher risk of *H. pylori* infection [17]. Similarly, meat consumption and consumption of restaurant food were associated in some studies with an increased risk of *H. pylori* infection, while chili pepper intake was shown to have a protective effect [12,17]. In addition, some studies revealed a protective effect of honey and green tea consumption against *H. pylori* infection [24–26]. Coffee consumption has been linked to several health benefits and some studies found an inverse association between coffee consumption and the systemic levels of some inflammatory markers [27]. However, it was reported that frequent consumption of coffee was associated with an increased rate of *H. pylori* infection and exacerbation of *H. pylori*-related gastritis symptoms [18,28]. Some studies did not find any association between coffee consumption and *H. pylori* infection [10,29,30]. In addition, the relationship between *H. pylori* infection and several modifiable cardio-metabolic risk factors was reported in the literature [31–33].

Risk factors associated with *H. pylori* infection, especially lifestyle and dietary habits, have not been investigated thoroughly in the Kingdom of Bahrain. Given the high burden of *H. pylori* infection in developing countries and the high prevalence of modifiable cardio-metabolic risk factors in the Middle East and North Africa (MENA) region, including the Kingdom of Bahrain, a study investigating the relationship between predisposing factors to *H. pylori* infection including dietary habits is warranted. This study aims to provide preliminary data regarding *H. pylori* infection prevalence and predisposing factors among a group of Bahraini adults followed in the Gastroenterology (GE) unit in a tertiary care hospital. The findings of this study will help in the future planning of appropriate preventive, diagnostic, and treatment strategies for *H. pylori* infection.

2. Materials and Methods

2.1. Study Design, Setting and Duration

This hospital-based retrospective cross-sectional, analytical study was conducted in the Gastroenterology unit in King Hamad University Hospital (KHUH) in the Kingdom of Bahrain between the period of January and September 2021. It combines data from medical records for *H. pylori* status, other comorbidities, and biochemical parameters with sociodemographic, lifestyle, and dietary habits information using a tele-interview.

2.2. Study Participants

Participants were recruited if they were 18 years or above, Bahraini, following treatment in GE unit in KHUH, and had done *H. pylori* testing within the previous 18 months with either upper GI tract endoscopy biopsy testing or UBT, or both. Patients with updated medical records within the past 18 months were eligible. Patients were excluded if they had a history and/or documentation of *H. pylori* eradication therapy prior to *H. pylori* testing, had a previous diagnosis of cancer, inflammatory diseases such as coeliac disease, inflammatory bowel disease, or certain food allergies, or had a history of gastric or intestinal surgery or previous gastric perforation or hemorrhage. Patients who were tested for *H. pylori* status by methods other than upper GI tract endoscopy biopsy testing or UBT were not included. Women who were pregnant at the time of the study or previously pregnant within the past 18 months were excluded since their dietary habits might be changed during pregnancy.

2.3. Sample Size

Sample size was calculated assuming the following parameters: alpha error = 0.05, power = 80%, expected effect size: odds ratio (OR) = 1.4 (for the diet as a risk factor), prevalence of *H. pylori* (outcome) = 0.50. A total of 200 patients were included in the study.

2.4. Research Tools

A validated structured questionnaire was used as the instrument for data collection.

The questionnaire contains four sections to collect data on sociodemographic, lifestyle, dietary habits, and medical conditions and biomarkers. Assessment of dietary habits and the frequency of consumption of food and beverages items were assessed by combining a validated short version FFQ that was used in previous studies [10,34], the Bahraini FFQ which is in process of validation, and some food and beverages included based on findings from the literature (honey, green tea, and soft drinks). The frequency of consumption for the past 18 months was assessed by selecting one of five categories "less than once per month/none", "1–2 times/month", "1–2 times/week", "3–4 times/week" and "every day". The last section involved data collected from the participant's medical records which included documentation of comorbidities such as type 2 diabetes (T2D), hypertension and hyperlipidemia, systolic and diastolic blood pressure (BP) (mmHg), height (cm), and weight (kg), FBS (mmol/L), HbA1c as a percentage (DCCT unit), total cholesterol (mmol/L), low-density lipoprotein (LDL) (mmol/L), high-density lipoprotein (HDL) (mmol/L), triglycerides (TG) (mmol/L) and vitamin D level (ng/mL). *H. pylori* status (positive vs. negative) was determined upon the result of either UBT, upper GI tract biopsy testing, or both, which had been done within the previous 18 months. The method used for diagnosis was recorded.

2.5. Data Collection and Procedures

A structured electronic questionnaire was used by five trained interviewers during telephone interviews to document the consent of participants and collect data related to sociodemographic, lifestyle, and dietary habits. *H. pylori* status and medical data were retrieved from the medical records and documented in the electronic questionnaire form prior to submission. All questionnaires were collected centrally by the main investigator to ensure confidentiality.

2.6. Data Entry

Information obtained on anthropometric and biochemical measurements were recoded as binary (normal level versus not) based on widely known cutoff levels for each parameter. These BMI categorization and cut-point values for biomarkers were based on international and national recommendations [35–39].

2.7. Statistical Analysis

All data were entered and analyzed using the Statistical Package for the Social Sciences (SPSS) version 26 (Chicago, IL, USA) software. Categorical variables were computed as frequencies and percentages, and continuous (numerical) variables were computed as mean and standard deviation. Student's t-test was conducted to examine differences of means. The Chi-Square test was used to compare frequency distributions of categorical variables. The frequency of food and beverages consumption per week was converted into a numerical scale (less than once per month/None = 0, 1–2 times/month = 1, 1–2 times/week = 2, 3–4 times/week = 4, and everyday = 7). The mean for each item was calculated for *H. pylori* positive and negative groups. Univariate logistic regression was employed to evaluate the crude association between dietary factors and *H. pylori* status. Binary logistic regression was used to explore the risk factors that affect the presence of *H. pylori*. The odds ratio (OR) was calculated using a 95% confidence interval (CI). A *p*-value of $\leq$0.05 was statistically significant in all statistical tests used. The internal consistency reliability using Cronbach's alpha coefficient was used to verify the reliability of the food frequency questionnaire.

3. Results

3.1. The Prevalence of H. pylori Infection among the Study Participants

In the present study, 200 participants were recruited. The prevalence of *H. pylori* infection among the study population was 55.5% (95% CI: 48.7%, 62.3%). *H. pylori* status was determined in more than half of the participants (51%) by gastric biopsy testing, 35.5% by a urea breath test, and 13.5% by both methods (Table S1).

3.2. Sociodemographic Characteristics of the Study Participants

Table 1 presents the sociodemographic characteristics of the study participants and the association between the sociodemographic characteristics and *H. pylori* infection. The age range of the participants was between 18 to 79 years and the mean age was 51.4 years (95% CI: 49.5–53.3). The proportion of females was larger than males (57% vs. 43%). The majority of participants were married (80%) and with a high school education or less (70.5%).

Table 1. Sociodemographic characteristics and their association with *H. pylori* status among the study participants.

Sociodemographic Characteristics	*n* (%)	*H. pylori* Status		Chi-Square *p*-Value
		Positive *n* (%)	Negative *n* (%)	
Age				
18–30 years	17 (8.5)	10 (58.8)	7 (41.2)	0.980
31–45 years	42 (21.0)	23 (54.8)	19 (45.2)	
46–60 years	83 (41.5)	45 (54.2)	38 (45.8)	
Older than 60 years	58 (29.0)	33 (56.9)	25 (43.1)	
Gender				
Male	86 (43.0)	48 (55.8)	38 (44.2)	0.938
Female	114 (57.0)	63 (55.3)	51 (44.7)	
Marital status				
Married	160 (80.0)	88 (55)	72 (45)	0.776
Unmarried	40 (20.0)	23 (57.5)	17 (42.5)	
Educational level				
High school or below	141 (70.5)	85 (60.3)	56 (39.7)	**0.035**
College/university	59 (29.5)	26 (44.1)	33 (55.9)	
Employment status				
Employed	49 (24.5)	24 (49)	25 (51)	0.277
Unemployed	81(40.5)	43 (53.1)	38 (46.9)	
Retired	70 (35.0)	44 (62.9)	26 (37.1)	

Table 1. *Cont.*

Sociodemographic Characteristics	*H. pylori* Status			Chi-Square *p*-Value
	n (%)	Positive	Negative	
		n (%)	*n* (%)	
Household income				
BHD 300 or less	17(8.5)	16 (59.3)	11 (40.7)	0.201
BHD 301–600	61 (30.5)	35 (57.4)	26 (42.6)	
BHD 601–900	59 (29.5)	37 (62.7)	22 (37.3)	
BHD 900 or more	53 (26.5)	23 (43.4)	30 (56.6)	
Number of household members				
1–4	79 (39.5)	44 (55.7)	35 (44.3)	0.493
5–8	90 (45.0)	47 (52.2)	43 (47.8)	
8 or greater	31 (15.5)	20 (64.5)	11 (35.5)	
Number of household rooms				
1–3	69 (34.5)	37 (53.6)	32 (46.4)	0.858
4–6	96 (48.0)	54 (56.3)	42 (43.8)	
6 or more	32 (16.0)	19 (59.4)	13 (40.6)	

Data in bold are statistically significant.

3.3. Association of Sociodemographic Characteristics and Lifestyle Factors with H. pylori Infection

H. pylori infection was significantly higher among participants with high school education or less compared to those with college/university education (p = 0. 035) (Table 1). *H. pylori* infection was more prevalent among smokers (73.9%) compared to non-smokers (53.1%); however, the difference was not statistically significant (Table 2). There was no significant association between *H. pylori* infection and any of the lifestyle factors investigated in this study (Table 2).

Table 2. Lifestyle factors and their association with *H. pylori* status among the study participants.

Lifestyle Factors	*H. pylori* Status			Chi-Square *p*-Value
	n (%)	Positive	Negative	
		n (%)	*n* (%)	
Smoking status				
Non-smoker	177 (88.5)	94 (53.1)	83 (46.9)	0.059
Smoker	23 (11.5)	17 (73.9)	6 (26.1)	
Number of cigarettes smoked/day				
10 or less	17 (36.9)	8 (47.1)	9 (52.9)	0.560
11–20	17 (37.0)	10 (58.8)	7 (41.2)	
20 or more	12 (26.1)	8 (66.7)	4 (33.3)	
Level of perceived stress				
Less than 3 (Low)	30 (15)	18 (60)	12 (40)	0.861
3–6 (Moderate)	89 (44.5)	49 (55.1)	40 (44.9)	
7–10 (High)	81 (40.5)	44 (54.3)	37 (45.7)	
Number of hours of sleep per night				
Less than 5 h	39 (19.5)	25 (64.1)	14 (35.9)	0.186
5–7 h	123 (61.5)	62 (50.4)	61 (49.6)	
More than 7 h	38 (19)	24 (63.2)	14 (36.8)	

Table 2. *Cont.*

Lifestyle Factors	n (%)	*H. pylori* Status Positive n (%)	Negative n (%)	Chi-Square p-Value
Number of times/week of being physically active				
None	75 (37.5)	42 (56)	33 (44)	0.427
Less than 1 time per week	22 (11.0)	12 (54.5)	10 (45.5)	
1–3 times per week	21 (10.5)	15 (71.4)	6 (28.6)	
More than 3 times per week	82 (41.0)	42 (51.2)	40 (48.8)	
Duration of physical activity				
Less than 20 min	21 (16.8)	11 (52.4)	10 (47.6)	0.776
20 min or more	104 (83.2)	58 (55.8)	46 (44.2)	

3.4. Association between Dietary Habits and H. pylori Infection

Table 3 demonstrates the differences and associations between the frequency of consumption of food and beverages in *H. pylori* positive and negative participants. There was a significant negative relationship between the mean level of green tea ($p = 0.012$), honey ($p = 0.018$), and coffee consumption ($p = 0.007$) with *H. pylori* infection. The mean of frequency of green tea, coffee, and honey intake was significantly lower among the *H. pylori* infected participants compared to their non-infected counterparts. There was no significant association between the mean level of frequency of consumption of other food and beverage items and *H. pylori* infection.

Table 4 shows that the *H. pylori* positivity rate was significantly lower (38.6%) in green tea consumers $\geq$ 1 day/week compared with their counterparts (60.3%) ($p = 0.0011$). Logistic regression analysis showed a lower risk of *H. pylori* infection in participants who consume green tea $\geq$ 1 day/week (OR, 0.011; 95% CI, 0.23–0.92). *H. pylori* positivity rate was lower (50%) in coffee consumers $\geq$ 1 day/week compared with their counterparts (63.4%) and in honey consumers $\geq$ 1 day/week (48.8%) compared with the other participants (60.5%), but this difference was not statistically significant (Table 4).

As shown in Table 5, *H. pylori* infection was more prevalent among participants who consumed well water during childhood as the main source of drinking water (66%), chili peppers (58.8%), salty foods (60.6%), and restaurant meals more than three times a week (59%); however, the differences were not statistically significant.

3.5. Association of Some Medical Conditions and Biomarkers of the Study Participants with H. pylori Infection

Table 6 shows some of the medical conditions and biochemical markers of the study participants and their association with *H. pylori* infection. *H. pylori* positivity rate among participants who were overweight (58.9%) or obese (50.50.5%) was higher compared to participants with normal BMI (48.1%); however, this difference was not statistically significant. The proportion of *H. pylori* was higher among participants with hyperlipidemia (61.7%), and abnormally high levels of LDL (57.7%) as compared to their counterparts. *H. pylori* infection was significantly more prevalent among participants with vitamin D deficiency (63.6%) compared to participants with normal vitamin D levels (30%) ($p = 0.001$).

Table 3. Mean level of frequency of consumption of food and beverage items and their association with *H. pylori* status.

Food	*H. pylori* Status		Mann–Whitney *p*-Value
	Positive	Negative	
	Mean ± SD	Mean ± SD	
Grains	6.3 ± 1.6	6.0 ± 1.9	0.374
Green vegetables	4.7 ± 2.6	5.4 ± 2.3	0.057
Tuberous vegetables	4.4 ± 2.5	4.9 ± 2.3	0.153
Fish	2.2 ± 1.4	2.2 ± 1.2	0.746
Chicken	3.9 ± 2.1	4.3 ± 2.0	0.166
Red meat	1.6 ± 1.2	1.8 ± 1.1	0.139
Sausage	0.3 ± 0.8	0.2 ± 0.5	0.740
Hot dog	0.1 ± 0.5	0.2 ± 0.5	0.220
Salami or ham	0.1 ± 0.5	0.3 ± 0.9	0.134
Hamburger	0.8 ± 1.1	0.8 ± 1.2	0.922
Milk	3.3 ± 2.9	3.8 ± 3.0	0.263
Yogurt	3.4 ± 2.7	3.6 ± 2.6	0.374
Salty cheese	3.6 ± 2.7	4.1 ± 2.7	0.182
Fresh fruits	4.9 ± 2.5	5.1 ± 2.4	0.651
Legumes	1.8 ± 1.5	1.7 ± 1.8	0.215
Eggs	3.1 ± 2.3	3.2 ± 2.2	0.571
Nuts and dried fruits	2.5 ± 2.3	3.2 ± 2.7	0.154
Salted fish	0.3 ± 0.7	0.6 ± 1.1	0.130
Pickled vegetables	1.0 ± 1.9	0.7 ± 1.3	0.508
Onion	4.4 ± 2.8	4.4 ± 2.8	1.000
Garlic	4.0 ± 2.9	4.0 ± 3.0	0.926
Tomato	4.8 ± 2.6	4.7 ± 2.8	0.828
Butter and ghee	1.4 ± 2.0	1.6 ± 2.2	0.377
Vegetable oils	5.4 ± 2.4	5.6 ± 2.3	0.561
Deserts	2.7 ± 2.6	2.7 ± 2.5	0.835
Tea	4.6 ± 3.0	5.1 ± 2.7	0.214
Green tea	0.8 ± 1.8	1.3 ± 2.1	**0.012**
Coffee	2.7 ± 2.9	3.7 ± 3.0	**0.007**
Soft drinks	1.4 ± 2.2	1.1 ± 1.8	0.350
Honey	1.8 ± 2.3	2.8 ± 2.9	**0.018**

Data in bold are statistically significant.

Table 4. Association between *H. pylori* status with green tea, coffee, and honey consumption.

	H. pylori Status		Chi-Square *p*-Value	OR	*p*-Value	95% CI for OR
	Positive	Negative				
	n (%)	*n* (%)				
Green tea						
<1 day weekly	94 (60.3)	62 (39.7)	**0.011**	0.46	**0.029**	(0.23, 0.92)
≥1 day weekly	17 (38.6)	27 (61.4)				
Coffee						
<1 day weekly	52 (63.4)	30 (36.6)	0.060	0.62	0.107	(0.34, 1.11)
≥1 day weekly	59 (50)	59 (50)				
Honey						
<1 day weekly	69 (60.5)	45 (39.5)	0.100	0.65	0.149	(0.37, 1.17)
≥1 day weekly	42 (48.8)	44 (51.2)				

Data in bold are statistically significant.

Table 5. Dietary habits of the study participants and their association with *H. pylori* status.

Dietary Factors	N (%)	*H. pylori* Status		Chi-Square *p*-Value
		Positive	Negative	
		n (%)	*n* (%)	
Source of drinking water during childhood				
Tap water	115 (57.5)	62 (53.9)	53 (46.1)	0.117
Well water	53 (26.5)	35 (66)	18 (34)	
Filtered or mineral water	32 (16.0)	14 (43.8)	18 (56.3)	
Consumption of chili pepper				
Yes	102 (51.0)	60 (58.8)	42 (41.2)	0.335
No	98 (49.0)	51 (52)	47 (48)	
Salt status of consumed dishes				
Salty	99 (49.5)	60 (60.6)	39 (39.4)	0.150
Less salty/Salt free	101 (50.5)	51 (50.5)	50 (49.5)	
Speed of meals consumption				
Fast	61 (30.5)	33 (54.1)	28 (45.9)	0.883
Normal	78 (39.0)	45 (57.7)	33 (42.3)	
Slow	61 (30.5)	33 (54.1)	28 (45.9)	
Temperature status of meals consumed				
Cool/warm	83 (41.5)	45 (54.2)	38 (45.8)	0.758
Hot	117 (58.5)	66 (56.4)	51 (43.6)	
Frequency of consuming meals prepared outside home				
<1 time per month/Never	54 (27.0)	30 (55.6)	24 (44.4)	0.897
1–3 times/month	60 (30.0)	34 (56.7)	26 (43.3)	
1–2 times per week	47 (23.5)	24 (51.1)	23 (48.9)	
3–7 times per week	39 (19.5)	23 (59)	16 (41)	

Table 6. Medical condition of the study participants and its association with *H. pylori* status.

Medical Characteristics	N (%)	*H. pylori* Status		Chi-Square *p*-Value
		Positive	Negative	
		n (%)	*n* (%)	
Diagnosis of Hypertension				
Yes	67 (33.5)	38 (56.7)	29 (43.3)	0.806
No	133 (66.5)	73 (54.9)	60 (45.1)	
Diagnosis of Diabetes				
Yes	62 (31.0)	35 (56.5)	27 (43.5)	0.856
No	138 (69.0)	76 (55.1)	62 (44.9)	
Diagnosis of Hyperlipidemia				
Yes	60 (30.0)	37 (61.7)	23 (38.3)	0.251
No	140 (70.0)	74 (52.9)	66 (47.1)	
BMI				
Normal weight (18.5–24.9)	27 (15.3)	13 (48.1)	14 (51.9)	0.530
Overweight (25.0–29.9)	56 (31.6)	33 (58.9)	23 (41.1)	
Obese (≥30.0)	93 (52.5)	47 (50.5)	46 (49.5)	
Blood pressure				
Normal	46 (23.2)	25 (54.3)	66 (45.5)	0.851
Above normal	152 (76.8)	85 (55.9)	67 (44.1)	

Table 6. *Cont.*

Medical Characteristics	N (%)	*H. pylori* Status Positive *n* (%)	Negative *n* (%)	Chi-Square *p*-Value
Cholesterol				
Normal (<5.6 mmol/L)	67 (68.4)	35 (52.2)	32 (47.8)	0.402
Above normal (≥5.6 mmol/L)	31 (31.6)	19 (61.3)	12 (38.7)	
TG				
Normal (<1.7 mmol/L)	70 (72.2)	38 (54.3)	32 (45.7)	0.829
Above normal (≥1.7 mmol/L)	27 (27.8)	14 (51.9)	13 (48.1)	
LDL				
Normal (<2.6 mmol/L)	45 (46.4)	22 (48.9)	23 (51.1)	0.386
Above normal (≥2.6 mmol/L)	52 (53.6)	30 (57.7)	22 (42.3)	
HDL				
Normal (>1.5 mmol/L)	26 (26.8)	17 (65.4)	9 (34.6)	0.159
Below normal (≤1.5 mmol/L)	71 (73.2)	35 (49.3)	36 (50.7)	
FBS				
Normal (<5.6 mmol/L)	18 (45.0)	10 (55.6)	8 (44.4)	0.525
Above normal (≥5.6 mmol/L)	22 (55.0)	10 (45.5)	12 (54.5)	
HbA1c				
Normal (<5.7%)	23 (24.2)	12 (52.2)	11 (47.8)	0.688
Above normal (≥5.7%)	72 (75.8)	41 (56.9)	31 (43.1)	
Vitamin D				
Normal (≥20 nmol/L)	40 (34.2)	12 (30)	28 (70)	**0.001**
Below normal (<20 nmol/L)	77 (65.8)	49 (63.6)	28(36.4)	

BMI: Body Mass Index, TG: Triglycerides, HDL: High-Density Lipoprotein, LDL: Low-Density Lipoprotein, FBS: Fasting Blood Sugar, HbA1: Glycated hemoglobin. Analysis was done in accordance with the available data. BMI = 177, Blood pressure = 198, Cholesterol = 98, Triglycerides = 97, LDL = 97, HDL = 97, FBS = 40, HbA1c = 95, and vitamin D = 117. The data in bold is statistically significant.

3.6. Bivariate Logistic Regression Analysis of Some of the Variables

Table 7 gives an overview of the results of the bivariate logistic regression analysis. Participants with a high school degree or below were 1.38 times more likely to develop *H. pylori* infection compared to participants with a college/university degree, but the difference is not statistically significant ($p = 0.474$). In addition, smokers were 3.58 times more likely to develop *H. pylori* compared to non-smoker participants, but the difference is not statistically significant ($p = 0.129$). Finally, the risk of *H. pylori* infection increases by 1.11 per one unit decrease of vitamin D (OR: 1.11, 95% CI of 1.05, 1.18, $p < 0.001$).

Table 7. Binary logistic regression of educational level, smoking status, and Vitamin D level on *H. pylori* status.

	Odd Ratio	*p*-Value	95% CI for Odd Ratio
Educational level			
High school or below	1.38	0.474	(0.57, 3.35)
College/university	Reference		
Smoking status			
Smoker	3.58	0.129	(0.69, 18.51)
Non-smoker	Reference		
Vitamin D	1.11	**<0.001**	(1.05, 1.18)

Data in bold is statistically significant.

4. Discussion

The present study estimated the prevalence of *H. pylori* infection in a population of Bahraini adults seeking care at tertiary level and investigated the relationship between several factors including dietary habits and *H. pylori* infection. The overall prevalence of *H. pylori* infection found in this study was 55.5%. This is comparable to the prevalence of 59.4% reported by Alshaikh et al. (2021) in a recent retrospective study conducted in the Kingdom of Bahrain [40]. Interestingly, previous studies conducted more than 20 years back on samples of dyspeptic adult patients who underwent gastroscopy in a tertiary care hospital in Bahrain revealed prevalence ranges between 75% and 79.4% [41,42]. These findings could suggest a decreasing trend of *H. pylori* infection among symptomatic patients specifically, which raises a question if that the decreasing trend of *H. pylori* infection observed in symptomatic patients applies to the general population of Bahrain. The prevalence found in this study is near to the prevalence of 52.4% reported by Assaad et al. (2018) in a study conducted in Lebanese patients referred for upper GI endoscopy [10]. However, a lower prevalence was reported in studies conducted in dyspeptic patients in Oman (41%) and in Jazan Province in Saudi Arabia (46.5%) [30,43]. Many studies conducted in the Middle East/North Africa (MENA) region including Iran, Egypt, and Turkey reported higher infection rates which reached 86.8% in Iran [14,31,44]. The prevalence of *H. pylori* among subjects with dyspepsia in the United States, Brazil, and China was 28.9%, 57%, and 84% respectively [43]. The difference in the prevalence of *H. pylori* infection observed in different studies might be due to variation in the study design, sample size, study setting, the period in which the study was conducted, participants' characteristics, ethnicities of the sample, and testing methods used to determine *H. pylori* status. In addition, variation in bacterial virulence and stereotypes, antibiotic resistance, environment, living standards, socioeconomic and lifestyle factors, and dietary habits in different contexts could affect this prevalence. The prevalence reported in this study cannot be generalized to reflect *H. pylori* prevalence in the Kingdom of Bahrain, since it represents only the rate of the infection in a relatively small cohort of patients who were following treatment in one of the tertiary hospitals in the country.

H. pylori infection was significantly higher among participants with lower educational levels (high school degree or below) (60.3% positivity rate) compared to subjects with higher educational levels (college/university degrees) (44.1% positivity rate). Participants with university degrees might be more knowledgeable/aware of health-related issues and have a healthier lifestyle compared to those with lesser educational levels. In agreement with this study's finding, participants with higher educational levels were less likely to have *H. pylori* infection in studies conducted in Turkey, Korea, and China [14,45].

None of the lifestyle factors studied was associated with *H. pylori* infection. This finding could be due to the small sample size, participants' characteristics as patients with certain health-seeking behavior, and the study design which is prone to recall bias. Furthermore, the data was collected during the COVID-19 pandemic in which the lifestyle of the majority of the population has been changed due to the quarantine and social distancing precautions [45]. Consistent with this finding, Assaad et al. (2018) in Lebanon found no association between *H. pylori* infection and any of the lifestyle factors studied which include smoking, alcohol consumption, physical activity, number of sleep hours per night, and perceived level of stress [10].

Findings in this study revealed that the *H. pylori* infection rate was lesser among participants with higher consumption of green tea and honey. *H. pylori* infection rate was significantly lesser among participants who consume green tea one time or more per week. Green tea and honey have been shown to exhibit antibacterial activity to inhibit the growth of *H. pylori* and gastric mucosal inflammation [46,47]. Honey has a potent antibacterial activity due to certain characteristics as low pH, high osmolarity and hydrogen peroxide content [47,48]. Consistent with this finding, a study conducted in Bulgaria to assess the dietary habits of 150 patients with dyspepsia revealed that honey intake at least once a week (OR: 0.38) and green/black tea consumption for at least one day or more a week (OR: 0.45) were significantly associated with lower prevalence of *H. pylori* infection [24]. Similarly, Mard et al. (2014) and Yordanov et al. (2017) found a significant negative correlation between the intake of honey and *H. pylori* infection [25,26]. This study also showed that *H. pylori* infection was associated with lower frequencies of coffee consumption. Coffee consumption has been linked to several health benefits as lowering the risk of some diseases such as cardiovascular diseases, type 2 diabetes, obesity, and some types of cancers [49]. Coffee is rich in polyphenols which are known to affect immune function and chronic inflammation [27]. It also contains arabinogalactan proteins which are a type of polysaccharide that exhibits prebiotic and immunomodulatory properties [49]. Loftfield et al. (2015) found an inverse association between coffee consumption and the systemic levels of some inflammatory markers [27]. Findings from the literature on the relationship between coffee consumption and *H. pylori* infection are inconsistent. Alebie et al. (2016) in a study that included 145 Ethiopian students with gastritis found that consumption of coffee exacerbates *H. pylori* related gastritis symptoms [28]. Monno et al. (2019) in a retrospective study conducted in Italy revealed that the frequent consumption of coffee increases the *H. pylori* infection rate [18], while other studies did not find any association between coffee consumption and *H. pylori* infection [10,29,30]. The relationship between coffee consumption and *H. pylori* infection might be attributed to the differences in the type of coffee consumed and preparation methods. Another reason for the observed finding could be an intentional reduction of coffee consumption among *H. pylori* infected participants due to their personal beliefs or health care workers' instructions. It might be important in future studies to include data about the type and amount of coffee consumed, preparation methods and if the participant intentionally altered consumption for any reason.

It is essential to study details about dietary patterns, since evaluating food items in isolation might not provide a full view of nutrients' interaction. Moreover, dietary habits were assessed during the COVID-19 pandemic, during which some dietary habits might have been altered [50]. The COVID-19 pandemic affected eating behaviors and limited access to fresh food due to quarantine precautions and lockdown which led to increased consumption of processed and fast foods that are rich in salt, sugar, and saturated fat [51].

This study showed that *H. pylori* infection was more prevalent among participants with vitamin D deficiency. Evidence from the recent literature indicates that vitamin D possesses immunoregulatory functions that exhibit an effect on susceptibility to infections in general and to *H. pylori* specifically [52]. Vitamin D might decrease the risk of infection by various mechanisms; vitamin D improves innate immunity by modulating the production of antimicrobial peptides and cytokine response [52]. Furthermore, Vitamin D helps to enhance the activity of monocytes and macrophages and contributes to systemic antimicrobial effects [53]. Consistent with this study's findings, a multi-centric study reported that *H. pylori* infected participants had significantly lower serum vitamin D levels compared to the non-infected group [54]. Assaad et al. (2018) in Lebanon, reported that *H. pylori* infection risk was significantly higher among participants with vitamin D deficiency (OR = 29.14) compared to participants with normal vitamin D levels [10]. A recent study conducted in Turkey revealed that vitamin D deficiency was associated with increased odds of *H. pylori* infection by almost 3 times [55]. Yang et al. (2019) revealed that vitamin D had a protective effect against *H. pylori* infection and improved the success rate of *H. pylori* eradication [52]. The relationship between vitamin D and *H. pylori* infection is worth more investigation in the context of Bahrain, as many factors might be involved including diet and comorbidities. Considering vitamin D supplementation as part of prevention and treatment plans of *H. pylori* infection for certain groups in the population might be effective.

5. Strengths and Limitations

This study had several strengths and limitations. First, to the best of our knowledge, this is the first study in the Kingdom of Bahrain and one of few in the region to evaluate the association between sociodemographics, lifestyle, dietary habits, and some medical conditions with *H. pylori* infection. Second, a short version-13 item-FFQ previously validated by Yassibas [31] was used to assess dietary intake, FFQ is considered one of the best dietary tools to assess the relationship between diet and disease. Furthermore, the internal consistency and reliability of this tool were improved by adding items from the Bahraini FFQ that is in process of validation and other food items and beverages that were related to *H. pylori* infection in previous studies. Third, the data were collected through telephone interviews and not self-administered, so that the interviewer might clarify any misunderstandings if needed and minimize missing information. Fourth, *H. pylori* status was determined upon upper GI endoscopy biopsy testing and/or UBT, both of which have high diagnostic accuracy. Finally, medical data were retrieved from the patients' medical records, minimizing any self-reporting or categorization bias. Some limitations regarding this study should be considered when interpreting the results. The data collection was conducted during the COVID-19 pandemic; within this period some lifestyle and dietary habits might be affected. In addition, due to the regulations related to that period, some non-urgent investigations/procedures were rescheduled, which affected our reach to the targeted population. Due to that reason, we included any patient who had done the *H. pylori* testing within the previous 18 months by either UBT or upper GI endoscopy biopsy testing. The convenience sampling method used to select the participants and this subgroup characteristics might limit the ability to generalize the results to the general population. Moreover, the use of FFQ might represent some limitations. Food intake in the previous 18 months of the interview was self-reported by participants with no measure for verification, which might lead to possible recall and information bias. In addition, intake of food and beverage items was assessed without specifying quantities or portion size. However, the variation in portion size between participants is smaller than the variation in the frequency of consumption, which will have a limited effect on the findings. Some medical data were missing or not updated for a group of participants. This could contribute to the final findings. Finally, an inference of causality cannot be generated due to the cross-sectional study design.

6. Conclusions

H. pylori infection is a major public health problem that affects more than half of the world population leading to a range of GI and extra-gastric problems. This study is the first in Bahrain and one of few in the region to investigate the relationship between diet and *H. pylori* infection. *H. pylori* infection was significantly higher among participants with lower educational levels (high school degree or below) compared to those with higher educational levels (university degree). Intake of honey, green tea, and coffee was found to be protective against *H. pylori* infection. In addition, vitamin D deficiency was a risk factor for *H. pylori* infection. Including diet, in prevention measures and in support of treatment options of *H. pylori* infection will provide an acceptable convenient approach to control *H. pylori* with reasonable cost, high availability, and lesser side effects compared to medications.

Supplementary Materials: The following are available online at https://www.mdpi.com/article/10.3390/nu14194215/s1, Table S1: Frequencies and distribution of H. pylori diagnostic methods used; Table S2: Percent distribution of frequency of consumption of food and beverage items by the study participants.

Author Contributions: Conceptualization, F.H., T.A.A. and S.P.; methodology, F.H., T.A.A. and S.P.; software, F.H.; validation, F.H.; formal analysis, F.H. and S.P.; investigation, F.H., N.A. and O.S.; Methodology, F.H., T.A.A., S.P., N.A., C.G., C.F. and M.R.; data curation, F.H.; Project administration, F.H. and T.A.A.; writing—original draft preparation, F.H.; writing—review and editing, F.H., T.A.A., A.A.S., S.P., C.G., C.F. and M.R.; visualization, F.H.; All authors have read and agreed to the published version of the manuscript.

Funding: This research received no external funding.

Institutional Review Board Statement: The study was conducted in accordance with the Declaration of Helsinki, and the protocol was approved by the Research and Ethics Committee at King Hamad University Hospital in the Kingdom of Bahrain (protocol code 20-386, date of approval 15 December 2020).

Informed Consent Statement: Informed consent was obtained from all subjects involved in the study.

Data Availability Statement: Data supporting reported results is not publicly archived, if required, it can be provided by the principal investigator.

Acknowledgments: We would like to thank Afnan Freije, Salwa Al-Thawadi for their administrative support, and Dalal Al Rumaihi, Abdulrahman Muhoorfi, Jawaher Alsaqer, Heba Abdallaa and Sharifa Ahmed for their contribution in facilitating and supporting data collection.

Conflicts of Interest: The authors declare no conflict of interest.

References

1. Hooi, J.K.; Lai, W.Y.; Ng, W.K.; Suen, M.M.; Underwood, F.E.; Tanyingoh, D.; Malfertheiner, P.; Graham, D.Y.; Wong, V.W.; Wu, J.C.; et al. Global prevalence of Helicobacter pylori infection: Systematic review and meta-analysis. *Gastroenterology* **2017**, *153*, 420–429. [CrossRef] [PubMed]
2. Mitchell, H.; Katelaris, P. Epidemiology, clinical impacts and current clinical management of *Helicobacter pylori* infection. *Med. J. Aust.* **2016**, *204*, 376–380. [CrossRef] [PubMed]
3. Diaconu, S.; Predescu, A.; Moldoveanu, A.; Pop, C.S.; Fierbințeanu-Braticevici, C. *Helicobacter pylori* infection: Old and new. *J. Med. Life* **2017**, *10*, 112–117. [PubMed]
4. Amaral, O.; Fernandes, I.; Veiga, N.; Pereira, C.; Chaves, C.; Nelas, P.; Silva, D. Living conditions and *Helicobacter pylori* in adults. *BioMed Res. Int.* **2017**, *2017*, 9082716. [CrossRef] [PubMed]
5. Schistosomes. *Liver Flukes and* Helicobacter pylori; International Agency for Research on Cancer IARC: Lyon, France, 1994; Volume 61.
6. Moss, S.F. The clinical evidence linking *Helicobacter pylori* to gastric cancer. *Cell. Mol. Gastroenterol. Hepatol.* **2017**, *3*, 183–191. [CrossRef] [PubMed]
7. Sung, H.; Ferlay, J.; Siegel, R.L.; Laversanne, M.; Soerjomataram, I.; Jemal, A.; Bray, F. Global cancer statistics 2020: GLOBOCAN estimates of incidence and mortality worldwide for 36 cancers in 185 countries. *CA Cancer J. Clin.* **2021**, *71*, 209–249. [CrossRef]
8. Kao, C.; Sheu, B.; Wu, J. *Helicobacter pylori* infection: An overview of bacterial virulence factors and pathogenesis. *Biomed. J.* **2016**, *39*, 14–23. [CrossRef]
9. Capparelli, R.; Iannelli, D. Genetics of host protection against *Helicobacter pylori* infections. *Int. J. Mol. Sci.* **2021**, *22*, 3192. [CrossRef]
10. Assaad, S.; Chaaban, R.; Tannous, F.; Costanian, C. Dietary habits and *Helicobacter pylori* infection: A cross sectional study at a Lebanese hospital. *BMC Gastroenterol.* **2018**, *18*, 48. [CrossRef]
11. Krueger, W.S.; Hilborn, E.D.; Converse, R.R.; Wade, T.J. Environmental risk factors associated with *Helicobacter pylori* seroprevalence in the United States: A cross-sectional analysis of NHANES data. *Epidemiol. Infect.* **2015**, *143*, 2520–2531. [CrossRef]
12. Mhaskar, R.S.; Ricardo, I.; Azliyati, A.; Laxminarayan, R.; Amol, B.; Santosh, W.; Boo, K. Assessment of risk factors of *Helicobacter pylori* infection and peptic ulcer disease. *J. Glob. Infect. Dis.* **2013**, *5*, 60–67. [CrossRef] [PubMed]
13. Murray, L.J.; McCRUM, E.E.; Evans, A.E.; Bamford, K.B. Epidemiology of *Helicobacter pylori* infection among 4742 randomly selected subjects from Northern Ireland. *Int. J. Epidemiol.* **1997**, *26*, 880–887. [CrossRef] [PubMed]
14. Ozaydin, N.; Turkyilmaz, S.A.; Cali, S. Prevalence and risk factors of *Helicobacter pylori* in Turkey: A nationally-representative, cross-sectional, screening with the 13 C-Urea breath test. *BMC Public Health* **2013**, *13*, 1215. [CrossRef] [PubMed]
15. Ogihara, A.; Kikuchi, S.; Hasegawa, A.; Kurosawa, M.; Miki, K.; Kaneko, E.; Mizukoshi, H. Relationship between *Helicobacter pylori* infection and smoking and drinking habits. *J. Gastroenterol. Hepatol.* **2000**, *15*, 271–276. [CrossRef] [PubMed]
16. Shu, L.; Zheng, P.; Zhang, X.; Feng, Y. Dietary patterns and *Helicobacter pylori* infection in a group of Chinese adults ages between 45 and 59 years old: An observational study. *Medicine* **2019**, *98*, e14113. [CrossRef]
17. Eslami, O.; Shahraki, M.; Shahraki, T.; Ansari, H. Association of *Helicobacter pylori* infection with metabolic parameters and dietary habits among medical undergraduate students in southeastern of Iran. *J. Res. Med. Sci. Off. J. Isfahan Univ. Med. Sci.* **2017**, *22*, 12.
18. Monno, R.; De Laurentiis, V.; Trerotoli, P.; Roselli, A.M.; Ierardi, E.; Portincasa, P. *Helicobacter pylori* infection: Association with dietary habits and socioeconomic conditions. *Clin. Res. Hepatol. Gastroenterol.* **2019**, *43*, 603–607. [CrossRef]
19. Tsugane, S.; Tei, Y.; Takahashi, T.; Watanabe, S.; Sugano, K. Salty food intake and risk of *Helicobacter pylori* infection. *J. Cancer Res.* **1994**, *85*, 474–478.
20. Fox, J.G.; Dangler, C.A.; Taylor, N.S.; King, A.; Koh, T.J.; Wang, T.C. High-salt diet induces gastric epithelial hyperplasia and parietal cell loss, and enhances *Helicobacter pylori* colonization in C57BL/6 mice. *Cancer Res.* **1999**, *59*, 4823–4828.
21. Shinchi, K.; Ishii, H.; Imanishi, K.; Kono, S. Relationship of cigarette smoking, alcohol use, and dietary habits with *Helicobacter pylori* infection in Japanese men. *Scand. J. Gastroenterol.* **1997**, *32*, 651–655. [CrossRef]
22. Hwang, H.; Dwyer, J.; Russell, R.M. Diet, *Helicobacter pylori* infection, food preservation and gastric cancer risk: Are there new roles for preventative factors? *Nutr. Rev.* **1994**, *52*, 75–83. [CrossRef] [PubMed]

23. Jarosz, M.; Rychlik, E.; Siuba, M.; Respondek, W.; Ryżko-Skiba, M.; Sajór, I.; Gugała, S.; Błażejczyk, T.; Ciok, J. Dietary and socio-economic factors in relation to *Helicobacter pylori* re-infection. *World J. Gastroenterol. WJG* **2009**, *15*, 1119–1125. [CrossRef] [PubMed]
24. Boyanova, L.; Ilieva, J.; Gergova, G.; Vladimirov, B.; Nikolov, R.; Mitov, I. Honey and green/black tea consumption may reduce the risk of *Helicobacter pylori* infection. *Diagn. Microbiol. Infect. Dis.* **2015**, *82*, 85–86. [CrossRef] [PubMed]
25. Mard, S.A.; Khadem Haghighian, H.; Sebghatulahi, V.; Ahmadi, B. Dietary factors in relation to *Helicobacter pylori* infection. *Gastroenterol. Res. Pract.* **2014**, *2014*, 826910. [CrossRef] [PubMed]
26. Yordanov, D.; Boyanova, L.; Markovska, R.; Ilieva, J.; Andreev, N.; Gergova, G.; Mitov, I. Influence of dietary factors on *Helicobacter pylori* and CagA seroprevalence in Bulgaria. *Gastroenterol. Res. Pract.* **2017**, *2017*, 9212143. [CrossRef] [PubMed]
27. Loftfield, E.; Shiels, M.S.; Graubard, B.I.; Katki, H.A.; Chaturvedi, A.K.; Trabert, B.; Pinto, L.A.; Kemp, T.J.; Shebl, F.M.; Mayne, S.T.; et al. Associations of coffee drinking with systemic immune and inflammatory markers. *Cancer Epidemiol. Prev. Biomark.* **2015**, *24*, 1052–1060. [CrossRef]
28. Alebie, G.; Kaba, D. Prevalence of *Helicobacter pylori* infection and associated factors among gastritis students in Jigjiga University, Jigjiga, Somali regional state of Ethiopia. *J. Bacteriol. Mycol.* **2016**, *3*, 00060. [CrossRef]
29. Ali, A.; Riaz Ahmad, M.; Iqbal, Z.; Basit, A. Identification of the Risk Factors Associated with *Helicobacter pylori* Infection in Lahore. Pakistan. *J. Biom. Biostat.* **2017**, *8*, 1000348. [CrossRef]
30. Altheeb AlKalbani, S.R.; FT, N.A.; Al-Hinai, M.; AlMuniri, A. Diet and lifestyle factors and the risk of *H. pylori* infection in Omani patients attending SQUH daycare for OGD. *J. Fam. Med. Commun. Health* **2016**, *3*, 1077.
31. Devrajani, B.R.; Shah, S.Z.A.; Soomro, A.A.; Devrajani, T. Type 2 diabetes mellitus: A risk factor for *Helicobacter pylori* infection: A hospital based case-control study. *Int. J. Diabetes Dev. Ctries.* **2010**, *30*, 22–26. [CrossRef]
32. Bener, A.; Micallef, R.; Afifi, M.; Derbala, M.; Al-Mulla, H.M.; Usmani, M.A. Association between type 2 diabetes mellitus and *Helicobacter pylori* infection. *Turk. J. Gastroenterol.* **2007**, *18*, 225–229. [PubMed]
33. Gravina, A.G.; Zagari, R.M.; De Musis, C.; Romano, L.; Loguercio, C.; Romano, M. *Helicobacter pylori* and extragastric diseases: A review. *World J. Gastroenterol.* **2018**, *24*, 3204–3221. [CrossRef]
34. Yassibas, E.; Arslan, P.; Yalcin, S. Evaluation of dietary and life-style habits of patients with gastric cancer: A case-control study in Turkey. *Asian Pac. J. Cancer Prev.* **2012**, *13*, 2291–2297. [CrossRef] [PubMed]
35. American Diabetes Association. Standards of Medical Care in Diabetes-2021 Abridged for Primary Care Providers. *Clin. Diabetes A Publ. Am. Diabetes Assoc.* **2021**, *39*, 14–43. [CrossRef]
36. Arnett, D.K.; Blumenthal, R.S.; Albert, M.A.; Buroker, A.B.; Goldberger, Z.D.; Hahn, E.J.; Himmelfarb, C.D.; Khera, A.; Lloyd-Jones, D.; McEvoy, J.W.; et al. 2019 ACC/AHA guideline on the primary prevention of cardiovascular disease: Executive summary: A report of the American College of Cardiology/American Heart Association Task Force on Clinical Practice Guidelines. *J. Am. Coll. Cardiol.* **2019**, *74*, 1376–1414. [CrossRef] [PubMed]
37. Hossein-nezhad, A.; Holick, M.F. Vitamin D for health: A global perspective. *Mayo Clin. Proc.* **2013**, *88*, 720–755. [CrossRef] [PubMed]
38. Bakris, G.; Ali, W.; Parati, G. ACC/AHA versus ESC/ESH on hypertension guidelines: JACC guideline comparison. *J. Am. Coll. Cardiol.* **2019**, *73*, 3018–3026. [CrossRef]
39. Jensen, M.D.; Ryan, D.H.; Apovian, C.M.; Ard, J.D.; Comuzzie, A.G.; Donato, K.A.; Hu, F.B.; Hubbard, V.S.; Jakicic, J.M.; Kushner, R.F.; et al. 2013 AHA/ACC/TOS guideline for the management of overweight and obesity in adults: A report of the American College of Cardiology/American Heart Association Task Force on Practice Guidelines and The Obesity Society. *J. Am. Coll. Cardiol.* **2014**, *63*, 2985–3023. [CrossRef]
40. Alshaikh, S.; Ahmed Al Sharakhat, M.; Toorani, Z.; Farid, E.; Abdulla, M. Prevalence and Diagnosis of *Helicobacter pylori* Infection in a Tertiary Hospital in the Kingdom of Bahrain. *Bahrain Med. Bull.* **2021**, *43*, 547–550. [CrossRef]
41. Fakhro, A.R.E.; Fateha, B.E.D.; Farid, I.M.A.; Jamsheer, H.M. The association between *Helicobacter pylori* infection and lymphoid reaction in patients suffering from dyspepsia in Bahrain. *Saudi J. Gastroenterol.* **1999**, *5*, 129–133.
42. Kamath, R.; Al-Qamish, J.; Yousif, A.; Fakro, A.R.; John, S. Prevalence of *Helicobacter pylori* among dyspeptic patients in Bahrain. *Bahrain Med. Bull.* **1995**, *17*, 50–52.
43. Akeel, M.; Elmakki, E.; Shehata, A.; Elhafey, A.; Aboshouk, T.; Ageely, H.; Mahfouz, M.S. Prevalence and factors associated with *H. pylori* infection in Saudi patients with dyspepsia. *Electron. Physician* **2018**, *10*, 7279–7286. [CrossRef] [PubMed]
44. Khedmat, H.; Karbasi-Afshar, R.; Agah, S.; Taheri, S. *Helicobacter pylori* Infection in the general population: A Middle Eastern perspective. *Casp. J. Intern. Med.* **2013**, *4*, 745–753.
45. Lim, S.H.; Kwon, J.; Kim, N.; Kim, G.H.; Kang, J.M.; Park, M.J.; Yim, J.Y.; Kim, H.U.; Baik, G.H.; Seo, G.S. Prevalence and risk factors of *Helicobacter pylori* infection in Korea: Nationwide multicenter study over 13 years. *BMC Gastroenterol.* **2013**, *13*, 104. [CrossRef]
46. Stoicov, C.; Saffari, R.; Houghton, J. Green tea inhibits Helicobacter growth in vivo and in vitro. *Int. J. Antimicrob. Agents* **2009**, *33*, 473–478. [CrossRef] [PubMed]
47. Hołubiuk, Ł.; Imiela, J. Diet and *Helicobacter pylori* infection. *Prz. Gastroenterol.* **2016**, *11*, 150–154. [CrossRef] [PubMed]
48. Ndip, R.N.; Takang, A.E.; Echakachi, C.M.; Malongue, A.; Akoachere, J.; Ndip, L.M.; Luma, H.N. In-vitro antimicrobial activity of selected honeys on clinical isolates of *Helicobacter pylori*. *Afr. Health Sci.* **2007**, *7*, 228–231.
49. Gökcen, B.B.; Şanlier, N. Coffee consumption and disease correlations. *Crit. Rev. Food Sci. Nutr.* **2019**, *59*, 336–348. [CrossRef]
50. Lim, S.; Kong, A.P.; Tuomilehto, J. Influence of COVID-19 pandemic and related quarantine procedures on metabolic risk. *Prim. Care Diabetes* **2021**, *15*, 745–750. [CrossRef]

51. Papazisis, Z.; Nikolaidis, P.T.; Trakada, G. Sleep, physical activity, and diet of adults during the second lockdown of the COVID-19 pandemic in Greece. *Int. J. Environ. Res. Public Health* **2021**, *18*, 7292. [CrossRef]
52. Yang, L.; He, X.; Li, L.; Lu, C. Effect of vitamin D on *Helicobacter pylori* infection and eradication: A meta-analysis. *Helicobacter* **2019**, *24*, e12655. [CrossRef]
53. El Shahawy, M.S.; Hemida, M.H.; El Metwaly, I.; Shady, Z.M. The effect of vitamin D deficiency on eradication rates of *Helicobacter pylori* infection. *JGH Open* **2018**, *2*, 270–275. [CrossRef] [PubMed]
54. Han, C.; Ni, Z.; Yuan, T.; Zhang, J.; Wang, C.; Wang, X.; Ning, H.B.; Liu, J.; Sun, N.; Liu, C.F. Influence of serum vitamin D level on *Helicobacter pylori* eradication: A multi-center, observational, prospective and cohort study. *J. Dig. Dis.* **2019**, *20*, 421–426. [CrossRef] [PubMed]
55. Surmeli, D.M.; Surmeli, Z.G.; Bahsi, R.; Turgut, T.; Oztorun, H.S.; Atmis, V.; Varli, M.; Aras, S. Vitamin D deficiency and risk of *Helicobacter pylori* infection in older adults: A cross-sectional study. *Aging Clin. Exp. Res.* **2019**, *31*, 985–991. [CrossRef] [PubMed]

nutrients

MDPI

Systematic Review

Prevalence of Zinc Deficiency in Inflammatory Bowel Disease: A Systematic Review and Meta-Analysis

Roberta Zupo [1,*,†], Annamaria Sila [1], Fabio Castellana [1], Roberto Bringiotti [2], Margherita Curlo [3], Giovanni De Pergola [4,†], Sara De Nucci [1,4], Gianluigi Giannelli [5], Mauro Mastronardi [3] and Rodolfo Sardone [1]

1 Unit of Data Sciences and Technology Innovation for Population Health, National Institute of Gastroenterology "Saverio de Bellis", Research Hospital, 70013 Castellana Grotte, Italy
2 U.O.S.D. Digestive Endoscopy Ospedale Di Venere, 70131 Bari, Italy
3 Section of Gastroenterology II, National Institute of Research "Saverio De Bellis", 70013 Castellana Grotte, Italy
4 Unit of Geriatrics and Internal Medicine, National Institute of Gastroenterology "Saverio de Bellis", Research Hospital, 70013 Castellana Grotte, Italy
5 Scientific Direction, National Institute of Research "Saverio De Bellis", 70013 Castellana Grotte, Italy
* Correspondence: roberta.zupo@irccsdebellis.it
† On behalf of the Nutrition and Nutraceuticals Committee of the Italian Association of Medical Endocrinologists (AME).

Abstract: Malabsorptive disorders are closely associated with micronutrient deficiencies. In inflammatory bowel disease (IBD), trace element deficiencies pose a clinical burden from disease onset throughout its course, contributing to morbidity and poor quality of life. We aimed to conduct a systematic review and meta-analysis of the prevalence of zinc deficiency in IBD. Literature screening was performed on six electronic databases until 1 May 2022. Two independent investigators assessed the 152 retrieved articles for inclusion criteria, met by only nine, that included 17 prevalence entries for Crohn's disease (CD) (n = 9) and ulcerative colitis (UC) (n = 8). No exclusion criteria were applied to language, deficiency cut-offs, population age, general health status, country, or study setting (cohort or cross-sectional). The prevalence of zinc deficiency in blood was scored positive if due to a single disease, not cumulative factors. Zinc deficiency prevalence across selected studies showed higher values in CD than in UC. Pooled analyses by the IBD subgroup showed a total population of 1677 with CD, for an overall mean zinc deficiency prevalence of 54% and 95% confidence intervals (CI) ranging from 0.51 to 0.56, versus 41% (95%CI 0.38–0.45) in the UC population (n = 806). The overall prevalence at meta-analysis was estimated at 50% (95%CI 0.48–0.52), but with high heterogeneity, I^2 = 96%. The funnel plot analysis failed to show any evidence of publication bias. The risk of bias across selected studies was moderate to low. In IBD contexts, one of two patients suffers from zinc deficiency. Mismanagement of micronutrient deficiencies plays a role in inflammation trajectories and related cross-pathways. Clinicians in the field are advised to list zinc among trace elements to be monitored in serum.

Keywords: zinc deficiency; inflammatory bowel disease; meta-analysis

Citation: Zupo, R.; Sila, A.; Castellana, F.; Bringiotti, R.; Curlo, M.; De Pergola, G.; De Nucci, S.; Giannelli, G.; Mastronardi, M.; Sardone, R. Prevalence of Zinc Deficiency in Inflammatory Bowel Disease: A Systematic Review and Meta-Analysis. *Nutrients* **2022**, *14*, 4052. https://doi.org/10.3390/nu14194052

Academic Editors: Omorogieva Ojo and Amanda R Amorim Adegboye

Received: 13 August 2022
Accepted: 23 September 2022
Published: 29 September 2022

1. Introduction

Zinc is among the trace inorganics that are found in body fluids and tissues in small amounts but are essential for body growth and function. About 85% of zinc in the body is found in muscle and bone, 11% in skin and liver, and the rest in all other tissues. Interestingly, no single test reflects the zinc status in the whole body; however, tests for plasma or serum zinc are the most widely used [1]. Zinc in plasma is bound nearly 60% to albumin, 40% to macroglobulins, and 3% to amino acids and the renal ultrafiltration fraction [2]. Human metabolic pathways show that zinc is involved in the function of many enzymes, being an integral component of nearly 10 percent of the human proteome (e.g., of several key enzymes and transcription factors). According to recent dietary guidelines, an adult daily intake of 11 mg (males) and 8 mg (females) is recommended [2]. Dietary zinc sources include a wide range of edible sources [3]. Oysters have the greatest zinc concentration per serving, although red meat and poultry supply most of the zinc in the diet. Beans, nuts, different types of shellfish (such as crabs and lobsters), whole grains, fortified breakfast cereals, and dairy products are other good sources [4]. Phytates, found in whole-grain bread, cereals, legumes, and other foods, bind to zinc and prevent its absorption [5,6]. Consequently, zinc bioavailability from cereal and plant diets is lower than from animal foods, despite the significant zinc content in many cereal and plant foods. Citric acid may improve absorption, whereas iron, copper, calcium, fiber, and phytates may inhibit it. Zinc is an essential element for the integrity of bodily structures and activities. Zinc acts as a cofactor for various enzymes involved in growth, cell signaling pathways, cellular activities, immune function, and tissue repair.

Once ingested, zinc is absorbed in the small intestine, both the distal duodenum and proximal jejunum. However, research has yet to shed light on zinc homeostasis in enterocytes and the molecular processes intrinsic to intestinal absorption. In particular, the transfer of zinc through enterocytes upon absorption, its subsequent basolateral release into the bloodstream, and the involvement of zinc-binding or zinc-transport proteins in this process need to be elucidated, apart from the already known metallothionein. In addition, the involvement of zinc-transporters in the cytoplasmic organelles of enterocytes (such as ZnT-2, ZnT-4, ZnT-6, and ZnT-7) in cellular zinc trafficking and homeostasis needs to be investigated in intestinal cell models in vitro to understand the regulation of zinc transit at the enterocyte level. Zinc levels are often low in patients with chronic diarrhea and malabsorptive disorders [7].

This is why trace elements deficiency is common in patients with inflammatory bowel disease (IBD) during both active disease and remission [8,9]. Increased zinc losses occur mostly in conjunction with diarrhea, ostomies, and high-exit fistulas, often experienced in IBD. In conjunction with the chronic malabsorption state in cases of intestinal inflammation, micronutrient leaks are likely responsible for zinc deficiency from the disease onset. Reports indicate that subclinical zinc deficiency may lead to mucosal inflammation in these patients, as well as exacerbate colitis, and increase the production of pro-inflammatory cytokines [10].

Of note, biologically speaking, zinc homeostasis is strongly affected by a balance between the zinc-binding protein metallothionein and the expression of two zinc transporters. Because albumin is the zinc transporter, a low albumin level, mainly common to IBD patients experiencing malnutrition, malabsorption, an increased fractional catabolic rate of albumin, and increased albumin transfer out of the vascular system, may affect zinc levels.

Extensive reports so far substantiate the burden of micronutrient deficiencies in malabsorptive settings. However, studies investigating zinc deficiency in IBD patients are few, heterogeneous, and were performed in small patient subsets. The findings are often fragmented, whereas the deficiencies spectrum is broad. Here, we conducted a systematic review and meta-analysis of available data to estimate the prevalence of zinc deficiency in IBD, looking at the pattern of prevalence profiles across the two well-known forms of IBD, presumed to reflect the intrinsic difference in the inflammatory site.

2. Methods

2.1. Search Strategy, Selection Criteria, and Data Extraction

A computerized literature search of MEDLINE and the Cochrane database did not identify any previous systematic reviews on the prevalence of zinc deficiency in IBD. The present systematic review followed the Preferred Reporting Items for Systematic reviews and Meta-Analyses (PRISMA) guidelines, adhering to the PRISMA 27-item checklist [11]. An *a priori* protocol for the search strategy and inclusion criteria was established and recorded, with no particular changes to the information provided at registration on PROSPERO, a prospective international registry of systematic reviews (CRD42023330824). We performed separate searches in the US National Library of Medicine (PubMed), Medical Literature Analysis and Retrieval System Online (MEDLINE), EMBASE, Scopus, Ovid, and Google Scholar to retrieve original articles investigating serum zinc levels and the prevalence of zinc deficiency in IBD populations. The primary objective was to assess a pooled prevalence of a plasma zinc concentration deficit in IBD settings. We also considered the gray literature using the massive preprint archive https://arxiv.org/in (accessed on 1 August 2020) in the study selection phase, and the database http://www.opengrey.eu/to (accessed on 1 August 2020) to access notable conference abstracts and other non-peer-reviewed material. No exclusion criteria were applied to language, the defined deficiency status cut-off, nor population age, general health status, country, recruitment settings (hospital, community, or home care), and study setting (trials, cohort, or cross-sectional). We used only original articles investigating IBD populations and providing disease-specific prevalence data separately for Crohn's and ulcerative colitis, as an inclusion criterion.

The research strategy used in PubMed and MEDLINE and adapted to the other four electronic sources included the keywords "zinc", "inflammatory bowel disease", "Crohn's disease", and "ulcerative colitis" combined through the use of Boolean indicators such as "AND" and "OR". The search strategy used the Boolean indicator "NOT" to rule out letters, revisions, and meta-analyses. The literature search had no time restriction, and papers were retrieved until 1 May 2022. No language restrictions were made. Two researchers (RZ, AS) conducted the searches, reviewed titles and abstracts of articles retrieved separately and in duplicate, checked full texts, and selected the papers for inclusion in the study. Technical reports, letters to the editor, and systematic and narrative review articles were excluded. Inter-rater reliability (IRR) was used to estimate inter-coder agreement and the κ statistic to measure accuracy and precision. In accordance with PRISMA concepts and the quality assessment steps, a coefficient k of at least 0.9 was obtained in all data extraction steps [12].

2.2. Data Analysis

Two investigators (RZ, AS) extracted the following information separately and in duplicate in piloted form: author, publication year, survey year, country, and design (longitudinal, cross-sectional). Researchers tabulated data by IBD type of disease (Crohn's disease, CD, and ulcerative colitis, UC) to retrieve information on (1) sample size (n), (2) age (expressed as mean ± standard deviation, SD, or interquartile range, IQR), (3) male and female representativeness (expressed as n and %), serum zinc levels (according to IBD type, where possible, and expressed as mean ± SD or median and IQR), (4) threshold value used to assess zinc deficiency, (5) prevalence of zinc deficiency by IBD, and (6) summary of study findings. All references selected for retrieval from the databases were managed with the MS Excel data collection software platform by an expert biostatistician (FC). Finally, the data extracted from the selected studies and stored in the database were structured as evidence tables.

The quality of the studies included in the meta-analysis was evaluated using a tool developed by Hoy and colleagues [13]. Each study was assigned a score of one (yes) or zero (no) for each of the ten criteria. Based on the total score, studies were categorized as having a low (>8), moderate (6–8), or high ($\leq$5) risk of bias. Disagreements between the two investigators on the methodological quality of the included studies were addressed by discussion, involving a third investigator in the final agreement (RS). All data analyses were performed using R, version 2021.09.1; our biostatistician (FC) used the meta-package to conduct meta-analyses of the zinc deficiency prevalence (%), subdivided according to IBD illness type (CD, UC). A common-effects model was used to calculate the prevalence and 95% confidence intervals (CI) (Figures 1 and 2). The Higgins and colleagues [14] I^2 test was used to estimate percentage heterogeneity between studies that cannot be explained by chance. The closer this value is to zero, the less the variability between studies. Negative values are comparable to zero and indicate that there is no heterogeneity. Values below 25% suggest a low, between 25% and 50% moderate, and above 50% high heterogeneity among studies. The funnel plot shown in Figure 3 was used as a visual tool when investigating publication bias. The horizontal axis shows the scatter of treatment effects estimated from individual studies, while the vertical axis shows study size.

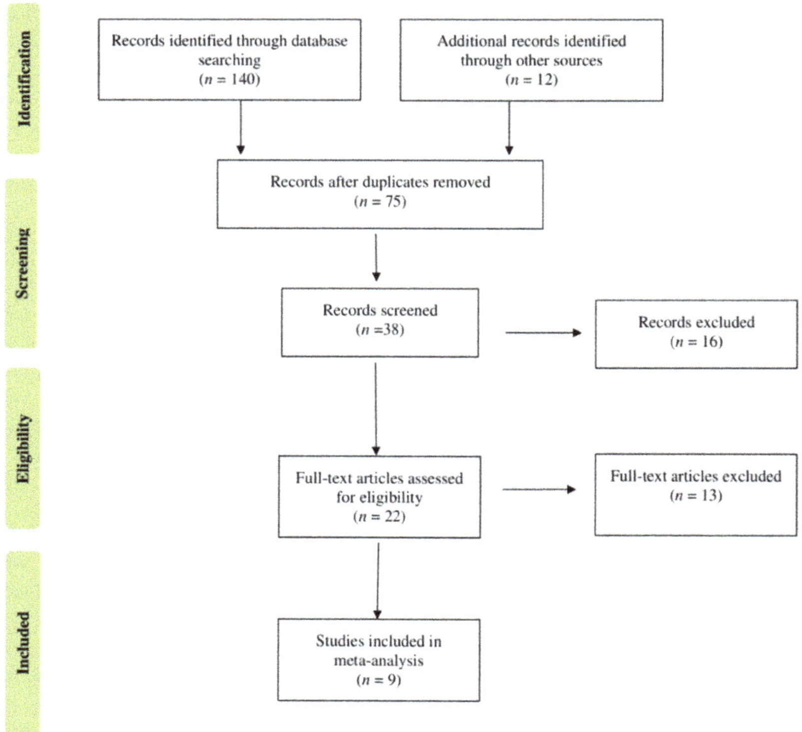

Figure 1. Flow diagram of literature screening process.

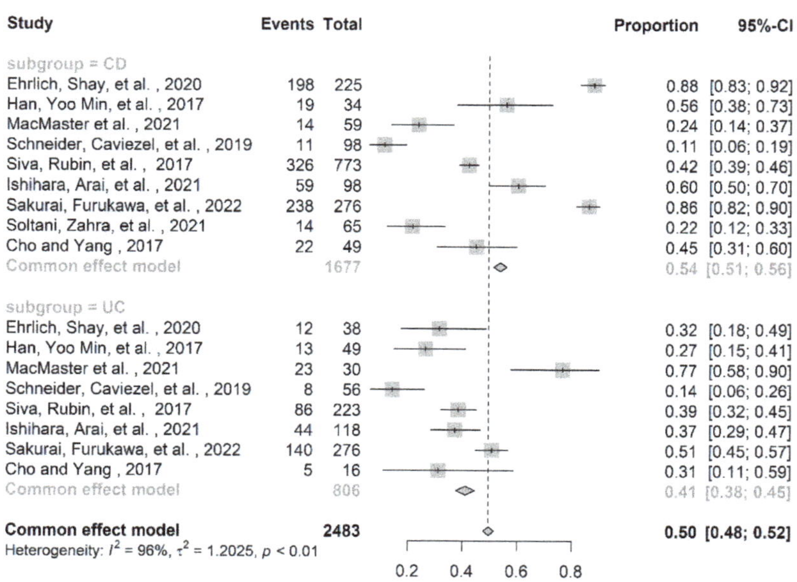

Figure 2. Pooled and grouped prevalence of zinc deficiency in IBD [15–23].

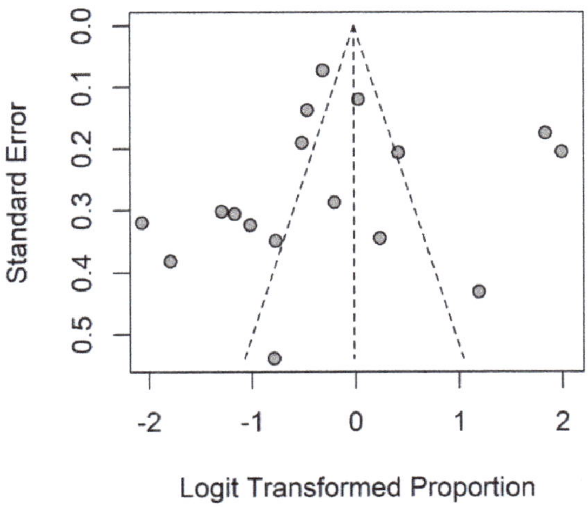

Figure 3. Funnel plot for assessment of publication bias across selected studies. (n = 9).

3. Results

The first systematic search of the literature yielded 152 entries. After excluding duplicates, 75 were classified as potentially relevant and selected for the title and abstract analysis. Then, 37 were excluded for failure to meet the characteristics of the approach or the review goal. After reviewing the full text of the remaining 38 records, only 9 met the inclusion criteria and were included in the meta-analysis [15–23]. The Preferred Reporting Items for Systematic Reviews and Meta-analyses (PRISMA) flow chart illustrating the number of studies at each stage of the review is shown in Figure 1. The final study base included nine articles reporting zinc deficiency prevalence by IBD condition (CD, UC).

Details of the design (cohort or cross-sectional), sample size (n) and gender ratio (%), survey year, study population, age range, serum zinc levels at recruitment, zinc deficiency cutoffs, country, and summary of findings are provided in Table 1. The cross-sectional design (67%, n = 6) predominated over the longitudinal (33%, n = 3). Recruitment settings were all community-based, and the geographic distribution of studies favored Asia (67%, n = 6), followed by Europe (22%, n = 2) and America (11%, n = 1). Following the inclusion criteria, all subjects had IBD. Only one study investigated subjects with Crohn's disease alone, whereas the other eight analyzed both IBD conditions. In total, this meta-analysis analyzed 736 subjects suffering from UC and 1677 with CD, resulting in a total IBD population of 2413 subjects. About 46% of the CD population and ~47% of the UC population were female. The prevalence of zinc deficiency across the selected studies showed higher mean values in the CD than in the UC population. Clustered analyses by IBD subgroup (CD, UC) showed a total population of 1677 for CD, with an overall zinc deficiency mean prevalence of 54% (95%CI 0.51 to 0.56). Notably, within this subgroup, the largest reports by Ehrlich and colleagues [15] and Sakurai and colleagues [21] reported a higher prevalence of 88% (95%CI 0.83 to 0.92) and 86% (95%CI 0.82 to 0.90), respectively (Table 2). In contrast, the total UC population of 806 subjects showed an overall 41% prevalence (95%CI 0.38 to 0.45) of zinc deficiency. The prevalence was more evenly distributed within this subgroup, apart from in the MacMaster and colleagues study [17], which reported significantly higher numbers (77%, 95%CI 0.58 to 0.90). The meta-analysis produced an overall estimate of 50% (95%CI 0.48 to 0.52) zinc deficiency prevalence, with high heterogeneity I^2 = 96% (Table 2). The funnel plot analysis showed no evidence of publication bias (Figure 2). According to the 10-item quality assessment checklist for prevalence studies by Hoy and colleagues [13], we found a moderate (n = 5) [16–18,21,22] to low (n = 4) [15,19,20,23] risk of bias across selected studies (Table 1).

Table 1. Descriptive of studies selected from the literature screening process. ($n = 9$).

Authors, Year [Ref.]	Survey Year	Country	Study Design	Age (Years)		Sample Size		Sex (Female)		Serum Zinc Levels		Deficiency Cut-Off	Summary of Findings	Overall Risk of Bias
				CD	UC	CD	UC	CD	UC	CD	UC			
Ehrlich, Shay, et al., 2020 [15]	2000–2016	Asia (Israel)	Longitudinal, 7-year	14.1 (12–16) *	13.5 (10.8–15.7) *	225	38	96/225 (43%)	18/38 (46%)	CD + UC: 70.5 ± 16.3 mcg/dL		≤70 mcg/dL	The prevalence of zinc deficiency in patients with CD at diagnosis was 88% (CD) and 31.6% (UC) in patients with IBD.	Low risk
Han, Yoo Min, et al., 2017 [16]	2013–2015	Asia (Korea)	Cross-sectional	CD + UC: 32 (16–70) ♯		34	49	CD + UC: 19/83 (22.9%)		CD+ UC: 76.6 ± 14.9 mcg/dL		≤70 mcg/dL	Many Korean patients with IBD have zinc deficiencies, suggesting the need to monitor levels of these micronutrients.	Moderate risk
Macmaster, Dami-anopoulou, et al., 2021 [17]	2017–2018	Europe (UK)	Cross-sectional	48.0 (19.5–78.4) ♯	47.2 (21.0–78.5) ♯	59	30	37 (63%)	16 (53%)	NA		Laboratory range (not specified)	Zinc deficiencies had been found in 23.7% (CD) and 76.6% (UC) of subjects with IBD	Moderate risk
Schneider, Caviezel, et al., 2020 [18]	2016–2017	Europe (Switzerland)	Cross-sectional	41.32 (14.5) ‡	41.6 (13.7) ‡	98	56	48 (49%)	31 (55%)	NA		<10.7 µmol/L	In this study, insufficient serum zinc concentrations were observed in 11.2% of patients with CD and in 14.3% of patients with UC	Moderate risk
Siva, Rubin, et al., 2017 [19]	2000–2015	America (USA)	Longitudinal, 3-year	NA	NA	773	223	421/773 (54%)	107/223 (48%)	NA		<0.66 mg/mL	Patients with IBD with serum zinc deficiency are more likely to have adverse disease-specific outcomes	Low risk
Ishihara, Arai, et al., 2021 [20]	2018	Asia (Japan)	Cross-sectional	13 (4–16) ♯	11 (1–16) ♯	98	118	30/98 (31%)	53/118 (45%)	64 (33–124) µg/dL *	69 (41–177) µg/dL *	<70 µg/dL	Prevalence of zinc deficiency in pediatric patients with IBD was 60.2% (CD) and 37.3% (UC)	Low risk
Sakurai, Furukawa, et al., 2022 [21]	2017–2020	Asia (Japan)	Longitudinal, 20 weeks	39.5 (23–63) ♯	56.0 (28–87) ♯	276	206	NA		57.5 (31–74) µg/dL	63 (46–74) µg/dL	<80 µg/dL	Zinc deficiencies had been found in 86.2% (CD) and 50.7% (UC) of IBD subjects	Moderate risk
Soltani, Zahra, et al., 2021 [22]	2018–2019	Asia (Iran)	Cross-sectional	CD: 39.2 ± 13.4	42 ± 16.2	CD: 65		CD: 49 (75.4%)		CD: 86.2 ± 17.0 ng/dL		Laboratory range (not specified)	Zinc deficiency was observed in 21.5% of a CD sample	Moderate risk
Cho And Yang, 2018 [23]	2012–2016	Asia (Korea)	Cross-sectional	14.4 (5.0–17.4) ♯	14.2 (9.9–17.4) ♯	49	16	16/49 (33%)	9/16 (56%)	71.5 (32.0–105.0) µg/dL ♯	77.0 (55.0–106.0) µg/dL ♯	<70 µg/dL	Zinc deficiencywas found in 44.9% (CD) and 31.2% (UC) of IBD sample	Low risk

Abbreviations: CD (Crohn's disease), UC (ulcerative colitis). * median (IQR), ♯ median (range), ‡ mean (standard deviation.

Table 2. Description of selected studies for meta-analysis.

n	Authors, Year	Disease	Deficiency Cases	Total Cases	Prevalence (%)	CI 95%
1	Ehrlich, Shay, et al., 2020 [15]	CD	198	225	88.00	0.83 to 0.92
		UC	12	38	31.58	0.18 to 0.49
2	Han, Yoo Min, et al., 2017 [16]	CD	19	34	55.88	0.56 to 0.73
		UC	13	49	26.53	0.15 to 0.41
3	MacMaster, Damianopoulou, et al., 2021 [17]	CD	14	59	23.73	0.14 to 0.37
		UC	23	30	76.67	0.58 to 0.90
4	Schneider, Caviezel, et al., 2020 [18]	CD	11	98	11.22	0.06 to 0.19
		UC	8	56	14.29	0.06 to 0.26
5	Siva, Rubin, et al., 2017 [19]	CD	326	773	42.17	0.39 to 0.46
		UC	86	223	38.57	0.32 to 0.45
6	Ishihara, Arai, et al., 2021 [20]	CD	59	98	60.20	0.50 to 0.70
		UC	44	118	37.29	0.29 to 0.47
7	Sakurai, Furukawa, et al., 2022 [21]	CD	238	276	86.23	0.82 to 0.90
		UC	140	276	50.72	0.45 to 0.57
8	Soltani, Zahra, et al., 2021 [22]	CD	14	65	21.54	0.12 to 0.33
9	Cho and Yang, 2018 [23]	CD	22	49	44.90	0.31 to 0.60
		UC	5	16	31.25	0.11 to 0.59

Abbreviations: CD (Crohn's Disease), UC (Ulcerative Colitis), CI (Confidence Interval).

4. Discussion

The present systematic review and meta-analysis aimed to provide a revised estimate of the prevalence of zinc deficiency in IBD populations, without restrictions as to the country, patients age, and study design. We clustered the 17 entries from the nine studies conducted in three different countries (Asia, Europe, and America) that had reported single prevalence data by type of IBD within the population examined. To the best of our knowledge, no meta-analytic report has yet been published in the literature on this topic. Given the growing concern about the management of individuals with chronic malabsorption diseases, zinc deficiency in CD and UC is among the major sensitive issues. As the main finding, this meta-analysis produced an overall zinc deficiency estimate of 50% (95%CI 0.48–0.52), with high heterogeneity I^2 = 96% and a moderate to low risk of bias across selected reports. Funnel plot analysis showed no evidence of publication bias. The clustered meta-analysis by the IBD group (CD and UC) showed a higher overall prevalence of zinc deficiency in the CD group than in the UC group (54%, 95%CI 0.51 to 0.56, versus 41%, 95%CI 0.38 to 0.45). In fact, while CD is known to affect any part of the gastrointestinal tract, including the mouth, esophagus, stomach, small and large intestines, rectum, and anus, UC compromises the colon and rectum. Moreover, the more malabsorptive CD affects all layers of the intestinal wall discontinuously, whereas in UC, the inflammation occurs in the innermost lining of the intestinal wall and is a continuous stretch within the colon.

Of the trace element deficiencies so far reported in the literature, zinc deficiency in IBD patients may result from poor oral intake and especially from the intrinsic malabsorptive nature of IBD. As proven by preclinical models and translational studies in humans, the relationship between trace zinc and chronic malabsorptive disease must be considered bidirectional since low serum zinc concentrations may also exacerbate inflammation through dysfunction and deficient epithelial barrier reconstruction, altered mucosal immunity, and increased pro-inflammatory cytokines [24–26]. In support of these mechanistic hypotheses, a recent report suggested that zinc supplementation might favor permeability modifications in CD patients in remission. Thus, an improved intestinal barrier function may help reduce the recurrence risk, especially in CD [9]. Indeed, the enhanced gastrointestinal epithelial barrier function driven by zinc may play an essential role In potential therapeutic actions, especially in CD, which is more malabsorptive than UC. This latter point is also corroborated by the known proximal intestinal sites of zinc absorption, i.e., the duodenum and jejunum [27].

Furthermore, previous research suggested that zinc may have some efficacy in modulating the immune system through an improved response to pathogens, reduced inflammatory response, and improved atopic/allergic reactions. Zinc is also involved in cell cycle regulation, particularly apoptosis, and hence has potential anticarcinogenic effects. All these effects have a "symbiotic" relationship with the gut microbiota [28]. From a prognostic perspective, zinc deficiency may predispose to growth retardation in young populations, and to loss of appetite, impaired immune function, and structural impairment of the intestinal endothelium. In severe cases, it may also drive hair loss, diarrhea, delayed sexual maturation, impotence, hypogonadism in males, and eye and skin lesions [29]. All these aspects highlight the importance of early nutritional preventive management in IBD settings, from better quality of life and healthcare burden perspectives. As to the loss of appetite reported in zinc deficiency [30], lower trace zinc values may also facilitate unintentional weight loss, malnutrition [31], sarcopenia, and cachectic states, mainly in restrictive dietary settings. With regard to the structural impairment of the intestinal endothelium, intervention studies on CD patients observed that zinc supplementation has some potential to reduce transmucosal leakage in these patients [9]. This latter finding is critical because intestinal epithelial barrier dysfunction may allow leukocytes to pass through, causing exposure to a "storm" of luminal antigens, a hallmark of IBD activity. Therefore, zinc supplementation may reduce the inflammatory response and maintain remission in CD. However, this empirical evidence is not yet supported by specific guidelines regulating supplementation [32].

In order to supplement the biological explanation of our prevalence data, we looked at information on inflammation in the selected reports, considering CRP values and disease activity scores in relation to zinc deficiency. On the one hand, we found considerable inconsistency in the data, as a minority of 33% indicated a significant correlation between zinc deficiency and elevated CRP [15,18,19,21], while a statistical relation or data collection was lacking for the remainder. On the other hand, a clearer trend was found for the activity score in relation to zinc deficiency, although the scores used were heterogeneous across reports. However, available data are still insufficient to comprehensively analyze the prevalence categorized by the activity score variable, although this covariate is certainly to be considered in the fluctuations of the prevalence of zinc deficiency in IBD.

The limited data and biographical heterogeneity of the study populations reduce the reliability of this meta-analysis in qualitative terms. In addition, designs differed among the selected studies: the cross-sectional was the most common. However, this is the first meta-analytic study to be conducted on the prevalence of zinc deficiency in IBD. Further investigation is needed to corroborate and better define these data and how they may fluctuate in relation to disease activity.

5. Conclusions

Prevention of micronutrient deficiencies has the potential to reduce the risk of disease-related disability. Still, more evidence is needed to corroborate these first prevalence metadata. The present research highlights the importance of considering zinc as a micronutrient to be monitored, because every second IBD patient shows a deficiency. According to our results, zinc deficiency is more prevalent in the CD population, probably due to the more severely malabsorptive nature of this condition and also in light of the proximal site of zinc absorption in the intestine. In light of this, the latest ESPEN micronutrient guidelines [33] point out that the dietary reference intakes (DRI) of zinc for adults should be 8–15 mg. However, in malabsorption conditions, such as short bariatric surgery, cystic fibrosis, chronic pancreatitis, and IBD, the need for higher amounts of zinc (30–40 mg daily) to maintain the zinc balance needs to be considered.

Author Contributions: R.Z., F.C., A.S. and R.S.: conceptualization, research, resource provision, data collection, writing original version, and visualization. R.S. and G.G.: review and correction. R.Z., S.D.N., G.D.P., M.M., M.C. and R.B.: research and data collection. R.S.: conceptualization, validation, review, and correction. R.S. and F.C.: conceptualization, validation, review and correction, and visualization. All authors have read and agreed to the published version of the manuscript.

Funding: The Italian Ministry of Health with "Ricerca Corrente 2020" funds.

Institutional Review Board Statement: Not applicable.

Informed Consent Statement: Not applicable.

Data Availability Statement: The original contributions presented in the study are included in the article, further inquiries can be directed to the corresponding authors.

Conflicts of Interest: The authors declare no conflict of interest.

References

1. Hess, S.Y.; Peerson, J.M.; King, J.C.; Brown, K.H. Use of Serum Zinc Concentration as an Indicator of Population Zinc Status. *Food Nutr. Bull.* **2007**, *28*, S403–S429. [CrossRef] [PubMed]
2. Zinc. Available online: https://www.hsph.harvard.edu/nutritionsource/zinc/ (accessed on 10 May 2022).
3. Institute of Medicine; Food and Nutrition Board; Standing Committee on the Scientific Evaluation of Dietary Reference Intakes; Subcommittee of Interpretation and Uses of Dietary Reference Intakes; Subcommittee on Upper Reference Levels of Nutrients. *Panel on Micronutrients Dietary Reference Intakes for Vitamin A, Vitamin K, Arsenic, Boron, Chromium, Copper, Iodine, Iron, Manganese, Molybdenum, Nickel, Silicon, Vanadium, and Zinc*; National Academies Press: Washington, DC, USA, 2002; ISBN 9780309072908.
4. US Department of Agriculture; A.R.S. FoodData Central. 2019. Available online: https://fdc.nal.usda.gov/ (accessed on 10 May 2022).
5. Sandström, B. Bioavailability of Zinc. *Eur. J. Clin. Nutr.* **1997**, *51* (Suppl. 1), S17–S19. [PubMed]
6. Wise, A. Phytate and Zinc Bioavailability. *Int. J. Food Sci. Nutr.* **1995**, *46*, 53–63. [CrossRef] [PubMed]

7. Michielan, A.; D'Incà, R. Intestinal Permeability in Inflammatory Bowel Disease: Pathogenesis, Clinical Evaluation, and Therapy of Leaky Gut. *Mediat. Inflamm.* **2015**, *2015*, 1–10. [CrossRef]
8. Iwaya, H.; Kashiwaya, M.; Shinoki, A.; Lee, J.-S.; Hayashi, K.; Hara, H.; Ishizuka, S. Marginal Zinc Deficiency Exacerbates Experimental Colitis Induced by Dextran Sulfate Sodium in Rats. *J. Nutr.* **2011**, *141*, 1077–1082. [CrossRef]
9. Sturniolo, G.C.; Di Leo, V.; Ferronato, A.; D'Odorico, A.; D'Incà, R. Zinc Supplementation Tightens "Leaky Gut" in Crohn's Disease. *Inflamm. Bowel Dis.* **2001**, *7*, 94–98. [CrossRef]
10. Wong, C.P.; Rinaldi, N.A.; Ho, E. Zinc Deficiency Enhanced Inflammatory Response by Increasing Immune Cell Activation and Inducing IL6 Promoter Demethylation. *Mol. Nutr. Food Res.* **2015**, *59*, 991–999. [CrossRef]
11. Page, M.J.; Moher, D.; Bossuyt, P.M.; Boutron, I.; Hoffmann, T.C.; Mulrow, C.D.; Shamseer, L.; Tetzlaff, J.M.; Akl, E.A.; Brennan, S.E.; et al. PRISMA 2020 Explanation and Elaboration: Updated Guidance and Exemplars for Reporting Systematic Reviews. *BMJ* **2021**, *372*, n160. [CrossRef]
12. Belur, J.; Tompson, L.; Thornton, A.; Simon, M. Interrater Reliability in Systematic Review Methodology: Exploring Variation in Coder Decision-Making. *Sociol. Methods Res.* **2021**, *50*, 837–865. [CrossRef]
13. Hoy, D.; Brooks, P.; Woolf, A.; Blyth, F.; March, L.; Bain, C.; Baker, P.; Smith, E.; Buchbinder, R. Assessing Risk of Bias in Prevalence Studies: Modification of an Existing Tool and Evidence of Interrater Agreement. *J. Clin. Epidemiol.* **2012**, *65*, 934–939. [CrossRef]
14. Higgins, J.P.T.; Thompson, S.G.; Deeks, J.J.; Altman, D.G. Measuring Inconsistency in Meta-Analyses. *BMJ* **2003**, *327*, 557–560. [CrossRef] [PubMed]
15. Ehrlich, S.; Mark, A.G.; Rinawi, F.; Shamir, R.; Assa, A. Micronutrient Deficiencies in Children with Inflammatory Bowel Diseases. *Nutr. Clin. Pract.* **2020**, *35*, 315–322. [CrossRef] [PubMed]
16. Han, Y.M.; Yoon, H.; Lim, S.; Sung, M.-K.; Shin, C.M.; Park, Y.S.; Kim, N.; Lee, D.H.; Kim, J.S. Risk Factors for Vitamin D, Zinc, and Selenium Deficiencies in Korean Patients with Inflammatory Bowel Disease. *Gut Liver* **2017**, *11*, 363–369. [CrossRef] [PubMed]
17. MacMaster, M.J.; Damianopoulou, S.; Thomson, C.; Talwar, D.; Stefanowicz, F.; Catchpole, A.; Gerasimidis, K.; Gaya, D.R. A Prospective Analysis of Micronutrient Status in Quiescent Inflammatory Bowel Disease. *Clin. Nutr.* **2021**, *40*, 327–331. [CrossRef]
18. Schneider, T.; Caviezel, D.; Ayata, C.K.; Kiss, C.; Niess, J.H.; Hruz, P. The Copper/Zinc Ratio Correlates with Markers of Disease Activity in Patients with Inflammatory Bowel Disease. *Crohns Colitis 360* **2020**, *2*, otaa001. [CrossRef] [PubMed]
19. Siva, S.; Rubin, D.T.; Gulotta, G.; Wroblewski, K.; Pekow, J. Zinc Deficiency Is Associated with Poor Clinical Outcomes in Patients with Inflammatory Bowel Disease. *Inflamm. Bowel Dis.* **2017**, *23*, 152–157. [CrossRef] [PubMed]
20. Ishihara, J.; Arai, K.; Kudo, T.; Nambu, R.; Tajiri, H.; Aomatsu, T.; Abe, N.; Kakiuchi, T.; Hashimoto, K.; Sogo, T.; et al. Serum Zinc and Selenium in Children with Inflammatory Bowel Disease: A Multicenter Study in Japan. *Dig. Dis. Sci.* **2021**, *67*, 2485–2491. [CrossRef] [PubMed]
21. Sakurai, K.; Furukawa, S.; Katsurada, T.; Otagiri, S.; Yamanashi, K.; Nagashima, K.; Onishi, R.; Yagisawa, K.; Nishimura, H.; Ito, T.; et al. Effectiveness of Administering Zinc Acetate Hydrate to Patients with Inflammatory Bowel Disease and Zinc Deficiency: A Retrospective Observational Two-Center Study. *Intestig. Res.* **2022**, *20*, 78–89. [CrossRef]
22. Soltani, Z.; Rafiei, F.; EBRAHIMi, A.; Rafiei, R. The Prevalence of Zinc Deficiency in Crohn's Disease Patients. *Maedica* **2021**, *16*, 29–33.
23. Cho, J.M.; Yang, H.R. Hair Mineral and Trace Element Contents as Reliable Markers of Nutritional Status Compared to Serum Levels of These Elements in Children Newly Diagnosed with Inflammatory Bowel Disease. *Biol. Trace Elem. Res.* **2018**, *185*, 20–29. [CrossRef]
24. Prasad, A.S. Effects of Zinc Deficiency on Th1 and Th2 Cytokine Shifts. *J. Infect. Dis.* **2000**, *182* (Suppl. 1), S62–S68. [CrossRef] [PubMed]
25. Ranaldi, G.; Ferruzza, S.; Canali, R.; Leoni, G.; Zalewski, P.D.; Sambuy, Y.; Perozzi, G.; Murgia, C. Intracellular Zinc Is Required for Intestinal Cell Survival Signals Triggered by the Inflammatory Cytokine TNFα. *J. Nutr. Biochem.* **2013**, *24*, 967–976. [CrossRef] [PubMed]
26. Maggini, S.; Wintergerst, E.S.; Beveridge, S.; Hornig, D.H. Selected Vitamins and Trace Elements Support Immune Function by Strengthening Epithelial Barriers and Cellular and Humoral Immune Responses. *Br. J. Nutr.* **2007**, *98* (Suppl. 1), S29–S35. [CrossRef] [PubMed]
27. Lee, H.H.; Prasad, A.S.; Brewer, G.J.; Owyang, C. Zinc Absorption in Human Small Intestine. *Am. J. Physiol.* **1989**, *256*, G87–G91. [CrossRef]
28. Scarpellini, E.; Balsiger, L.M.; Maurizi, V.; Rinninella, E.; Gasbarrini, A.; Giostra, N.; Santori, P.; Abenavoli, L.; Rasetti, C. Zinc and Gut Microbiota in Health and Gastrointestinal Disease under the COVID-19 Suggestion. *Biofactors* **2022**, *48*, 294–306. [CrossRef]
29. Sandstead, H.H. Zinc Deficiency. A Public Health Problem? *Am. J. Dis. Child.* **1991**, *145*, 853–859. [CrossRef]
30. Shay, N.F.; Mangian, H.F. Neurobiology of Zinc-Influenced Eating Behavior. *J. Nutr.* **2000**, *130*, 1493S–1499S. [CrossRef]
31. Zupo, R.; Castellana, F.; Bortone, I.; Griseta, C.; Sardone, R.; Lampignano, L.; Lozupone, M.; Solfrizzi, V.; Castellana, M.; Giannelli, G.; et al. Nutritional Domains in Frailty Tools: Working towards an Operational Definition of Nutritional Frailty. *Ageing Res. Rev.* **2020**, *64*, 101148. [CrossRef]
32. Poursadegh, F.; Ahadi, M.; Vosoughinia, H.; Salehi, M.; Beheshti Namdar, A.; Farzanehfar, M.R.; Memar, B.; Ziaolhagh, R. A STROBE Compliant Observational Study on Trace Elements in Patients with Ulcerative Colitis and Their Relationship with Disease Activity. *Medicine* **2018**, *97*, e13523. [CrossRef]
33. Berger, M.M.; Shenkin, A.; Schweinlin, A.; Amrein, K.; Augsburger, M.; Biesalski, H.-K.; Bischoff, S.C.; Casaer, M.P.; Gundogan, K.; Lepp, H.-L.; et al. ESPEN Micronutrient Guideline. *Clin. Nutr.* **2022**, *41*, 1357–1424. [CrossRef]

nutrients

Review

Chronic Kidney Disease: Role of Diet for a Reduction in the Severity of the Disease

Tania Naber [1] and Sharad Purohit [2,3,4,*]

[1] Department of Interdisciplinary Research, College of Allied Health Sciences, Augusta University, Augusta, GA 30912, USA; tnaber@augusta.edu
[2] Department of Undergraduate Health Professionals, College of Allied Health Sciences, Augusta University, Augusta, GA 30912, USA
[3] Department of Gynecology and Obstetrics, Medical College of Georgia, Augusta University, Augusta, GA 30912, USA
[4] Center for Biotechnology and Genomic Medicine, Augusta University, Augusta, GA 30912, USA
* Correspondence: spurohit@augusta.edu

Abstract: Chronic kidney disease affects ~37 million adults in the US, and it is often undiagnosed due to a lack of apparent symptoms in early stages. Chronic kidney disease (CKD) interferes with the body's physiological and biological mechanisms, such as fluid electrolyte and pH balance, blood pressure regulation, excretion of toxins and waste, vitamin D metabolism, and hormonal regulation. Many CKD patients are at risk of hyperkalemia, hyperphosphatemia, chronic metabolic acidosis, bone deterioration, blood pressure abnormalities, and edema. These risks may be minimized, and the disease's progression may be slowed through careful monitoring of protein, phosphorus, potassium, sodium, and calcium, relieving symptoms experienced by CKD patients. In this review, the current Kidney Disease Outcomes Quality Initiative (KDOQI) recommendations are highlighted, reflecting the 2020 update, including explanations for the pathophysiology behind the recommendations. The Dietary Approaches to Stop Hypertension, the Mediterranean diet, and the whole foods plant-based diet are currently being examined for their potential role in delaying CKD progression. Biological explanations for why the whole foods plant-based diet may benefit CKD patients compared to diets that include animal products are examined. Strong evidence continues to support the importance of diet meeting the daily requirement in the prevention and progression of kidney disease, and medical nutrition therapy with a registered dietitian is a critical aspect in medical intervention for CKD.

Keywords: diabetes; chronic kidney disease; proteinuria; diabetes; inflammation; diet; nutrition; plant-based foods; medical nutrition therapy

Citation: Naber, T.; Purohit, S. Chronic Kidney Disease: Role of Diet for a Reduction in the Severity of the Disease. *Nutrients* **2021**, *13*, 3277. https://doi.org/10.3390/nu13093277

Academic Editors: Omorogieva Ojo, Amanda Adegboye and Pietro Manuel Ferraro

Received: 10 August 2021
Accepted: 17 September 2021
Published: 19 September 2021

1. Introduction

The kidneys control many biological mechanisms such as fluid, electrolyte, pH balance, blood pressure, excretion of toxins and waste, vitamin D metabolism, and hormone synthesis. About thirty-seven million US adults are estimated to have chronic kidney disease (CKD), which is more than one in seven [1] Even more astonishing, nine in ten adults do not know they have the disease, and half of the adults with little kidney function who are not on dialysis are unaware they have CKD [1]. Chronic kidney disease often goes undiagnosed due to a lack of apparent symptoms in early stages. An estimated 94% with mild to moderate decline in renal function and about 48% of individuals with severe renal dysfunction go undiagnosed [2].

The kidneys are responsible for a series of life-sustaining mechanisms (Figure 1). The primary functions of the kidneys are to sustain and maintain fluid and electrolyte and metabolic acid–base balance, which is accomplished through solute and fluid regulation, conservation of nutrients, and excretion of metabolic bodily waste [3]. The kidneys have endocrine and exocrine functions regulating and maintaining critical biological mechanisms in the body [4]. The exocrine functions involve fluid and electrolyte balance [5], acid–base regulation [6], and excretion of body waste [7] (Figure 1). The endocrine functions include the activation of vitamin D for the incorporation of calcium into bones [8], and hormone synthesis for the regulation of blood pressure and synthesis of red blood cells [8,9].

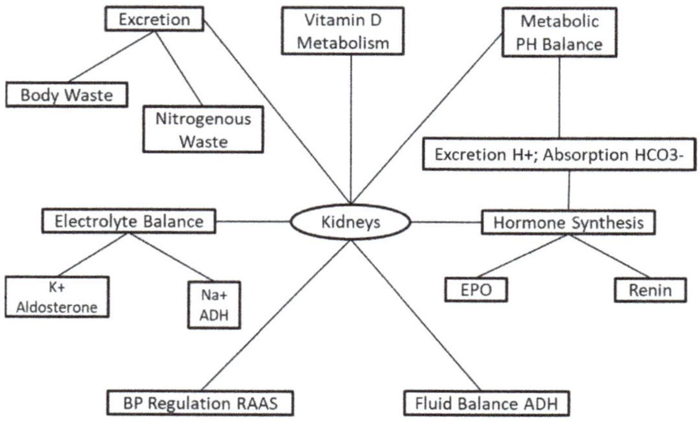

Figure 1. The physiological functions of kidneys.

The National Kidney Foundation (NKF) defines CKD as either a decline in glomerular filtration rate (GFR) to <15 mL/min/1.73 m^2 or the presence of kidney damage persisting for at least three months [10]. The prevalence of diabetes and hypertension is growing exponentially, predicting that CKD will continue to rise [11]. CKD patients are at increased risks for other health conditions, including acute kidney injury (AKI), T2DM, and mortality [12]. Chronic kidney disease is nationally incorporated into health promotion and disease-prevention programs to reduce its prevalence [13]. The US Department of Health and Human Services Healthy People 2020 had a target goal to minimize CKD prevalence from 14.8% in 2001 to 13.3% by 2020 [14].

Medical nutrition therapy is imperative for CKD patients because it may slow the progression of the disease through careful monitoring of protein, calcium, phosphorus, potassium, and sodium [15], relieving symptoms experienced in CKD patients while not restricting too many nutrients that would put the patient at high risk for malnutrition [16]. This review covers CKD pathophysiology, the most current diet recommendations, and the mechanisms that may delay the progression of the disease. In addition, the mechanisms of the newly explored whole food plant-based diet (WFPBD) are explained for its possible advantages in CKD prevention and progression. We performed a literature search on PubMed using "medical nutrition therapy", "chronic kidney disease", "clinical trials", and "outcomes of medical nutrition therapy in chronic kidney diseases" from January 2021 to May 2021. Published articles reporting clinical trials were selected for writing this review, and the information from these papers were incorporated as tables. To be included in this narrative review, the paper had to be a clinical trial on: (a) type of protein intake and its relevance to CKD, (b) maintaining calcium, phosphate, and vitamin D (VD) levels, and (c) electrolyte balance in CDK patients. All other articles were excluded. The main contribution in this review is to provide current clinical research to dieticians and physicians in a concise manner that introduces possibilities in acquiring an appropriate CKD diet that widens dietary variation by including foods with less nutrient bioavailability than animal products and additives. In addition, we provide points for future research needed, such as RCTs, which may produce data that may support the efficacy of a whole food plant-based diet on ameliorating CKD progression.

2. Medical Nutrition Therapy

The NKF published the first Kidney Disease Outcomes Quality Initiative (KDOQI), which is a set of nutritional guidelines for patients with end-stage renal disease in 1996 [17]. Since then, the KDOQI guidelines have gone through revisions and expanded to include nutrition recommendations for each stage of CKD, dialysis, and pre/post-kidney transplant [17,18]. Recommendations provided in this review are from the recent KDOQI Clinical Guideline for Nutrition in CKD: 2020 Update, which was developed with the Academy of Nutrition and Dietetics.

2.1. Protein and Renal Function

The federal Dietary Guidelines for Americans recommend an amount of 0.8 g/kg/body wt/d dietary protein intake for healthy adults [19]. Exceeding the recommended dietary allowance (RDA) may increase the risk of health complications even for healthy adults [19]. Protein intake recommendations for CKD patients are dependent on the stage of the disease, which is determined by declining GFR function [18].

The effects of high-protein diets (HPDs) on renal health have been investigated since the 1920s when rats given a HPD presented with increased kidney weight [20]. Data suggest that chronic protein intake (more than 1.2 g/kg/body weight/d) [21] leads to increased pressure and glomerular morphologic changes, resulting in renal dysfunction [22]. Glomerular hyperfiltration is defined as modifying renal hemodynamics through glomerular capillary hyperemia and increasing intraglomerular pressure [21]. HPDs induce glomerular hyperfiltration, hyperemia, and increased hydraulic pressure, resulting in vasodilation of the afferent arteriole [22]. HPDs contribute to progressive glomerular damage, which, combined with the renal deterioration from diseased kidneys may contribute to CKD progression. The Modification of Diet in Renal Disease (MDRD) was the largest RCT to examine the hypothesis that dietary protein restriction delays the progression of CKD [PMID 10541304]. The study found proteinuria to be one of the two strongest predictors in the rate of CKD progression in two studies [23]. Oba et al. collected 43 healthy (non-diseased) kidneys from live human donors to examine the effect of an HPD on the single-nephron GFR (SNGFR) [24]. This study concluded that an HPD might increase SNGFR and induce glomerular hyperfiltration; however, this study is unique by identifying that the analysis of human SNGFR is an exemplary parameter to alterations in renal hemodynamics at the single-nephron level [24]. The exact mechanism for renal hemodynamic responses to heightened protein intake is not yet understood [25].

Low-protein diets (LPDs) have been shown to improve hyperfiltration, reduce nitrogenous waste, and ease the renal workload by decreasing glomerular pressure [21]. Proteinuria declined by 20–50% in CKD patients who adhered to a LPD [26,27]. Although LPDs provide direct benefits to CKD patients, healthcare professionals are concerned about protein-energy malnutrition and protein-energy wasting (PEW) in CKD patients due to inadequate energy intake [26,27]. When determining the estimated energy requirements for CKD patients, 25–35 kcal/kg/body weight/d is recommended to maintain energy and nitrogen balance and avoid risk for malnutrition [16] (Table 1).

Table 1. Protein and energy requirements and recommendations for adult chronic kidney disease (CKD) patients.

Nutrients	Damage in CKD	Recommendation	Outcome	Ref
Protein CKD 3–5 patients. Not on Dialysis/Without Diabetes	Proteinuria/glomerular sclerosis/hyperfiltration/intraglomerular hypertension and hyperperfusion	0.55–0.6 g/kg body wt/day	Reduce uremia, uremic toxins, and hyperfiltration Improve organ function and renal hemodynamics	[18,21,22]
CKD 3–5 Pts Not on Dialysis and w/Diabetes	Proteinuria/glomerular sclerosis/hyperfiltration/intraglomerular hypertension and hyperperfusion	0.6–0.8 g/kg body wt/day	Reduce uremia, uremic toxins, and hyperfiltration Improve organ function, and renal hemodynamics	[18,25,26]
HD and PD Pts w/ and without Diabetes	Proteinuria/glomerular sclerosis/hyperfiltration/intraglomerular hypertension and hyperperfusion	1.0–1.2 g/kg body wt/day	Reduce uremia, uremic toxins, and hyperfiltration Improve organ function, and renal hemodynamics.	[18,25,26]
Energy Intake CKD 1–5D or post transplantation	Inadequate intake ↑ risk PEW, ↑ risk malnutrition. Excessive intake ↑ risk CVD, ↑ risk diabetes	25–35 kcals/kg body wt/day	Maintain neutral nitrogen balance and body composition.	[18,28]

Recommendations are for metabolically stable patients under strict clinical supervision. ↑ increased/high.

2.2. Very Low-Protein Diet

Low-protein diets and very low-protein diets (VLPD) (0.28–0.43 g/kg/body wt/d) may be achieved with nutrition supplementation with essential amino acids (EAAs) and ketoanalogues [27] to safeguard against PEW. The KDOQI guidelines recommend restricting protein to slow ESRD progression and improve quality of life (QoL) by reducing symptoms for metabolically stable patients [18]. The NKF defines metabolically stable as being absent from inflammatory or infectious diseases, poorly controlled diabetes, consumptive diseases, antibiotics or immunosuppressive medications, significant short-term loss of body weight, and no hospitalizations within two weeks [18,29]. For patients with CKD stage 3–5, a protein restriction providing 0.55–0.60 g/kg/body wt/d or a VLPD with supplementation with ketoacid analogues is recommended [18,29]. Diabetic adults with CKD 3–5 are recommended a protein diet providing 0.6–0.8 g/kg/body weight/d, and patients on maintenance hemodialysis (HD) or peritoneal dialysis (PD) with or without diabetes are recommended a protein intake providing 1.0–1.2 g/kg/body weight/d [20,30]. Diet modifications, such as reducing protein from heme sources and including plant-based proteins, protect against metabolic acidosis by lowering acid production; these effects are mostly seen with a VLPD (0.3–0.5 g/kg/body weight/d) with supplementation with ketoacid analogues [27,29]. Conservative reductions in protein intake as small as 0.1–0.2 g/kg/day have shown significant effects in preserving kidney function, hence slowing CKD progression [31]. A randomly controlled trial (RTC) reported a vegetarian (VLPD) (0.3 g/kg/body wt/d) supplemented with ketoanalogues compared with a standard LPD (0.6 g/kg/body wt/d) ameliorated kidney function decline over time and reduced the need for renal replacement therapy (RRT) [32].

An alternate protein source may be more beneficial to the patient's health than restricting the amount of protein alone; the protein source may be of greater importance than the quantity [18,29]. Plant proteins are typically ingested along with fiber, phytonutrients, and antioxidants, although animal proteins are ingested along with saturated fat and cholesterol [2]; this may be one reason plant proteins are associated with a more vast decline in blood pressure compared to animal protein, as shown from the INTERMAP Study on micronutrients and macronutrients on blood pressure [33]. Additionally, animal protein is associated with decreased insulin sensitivity, increased reactive oxygen species (ROS) [34], and induced hyperfiltration [35]; ingesting an equal amount of plant protein does not promote the same effects [36]. Most of the food within plant-based diets come from plant sources [37,38]. These types of diets are generally lower in protein and saturated fat, contain higher levels of potassium and phosphorus, are richer in fiber, and provide the body with additional nutrients in the form of vitamins, minerals, and phytochemicals. Adopting a plant-based diet has been shown to have numerous health benefits, such as a reduction in atherosclerotic plaque buildup, decreased risk of cardiovascular disease, decreased BMI, reduced body weight, and lower blood pressure [39,40], which are parameters that are clinically relevant for management of CKD patients [39].

Reductions in daily protein have produced some evidence in slowing CKD progression [41] by retarding the rate of kidney function decline [42]. However, determining an optimal amount of protein for CKD is complicated, especially when assessing the patient's individual circumstances [43]. When considering a protein-restricted diet, the patient's individual nutritional status should be evaluated with caution [41]. All previous protein-restricted diet studies are inconclusive [23,44].

2.3. Vitamin D

The primary role of vitamin D (VD) is to activate intestinal calcium reabsorption [45], but as kidney disease progresses, alterations in the biological mechanism occur. Low levels of active VD in ESRD patients are associated with increased bone reabsorption and reduced bone mineral density [46]. Studies report a progressive decline in VD of more than 80% from CKD 1–5, dialysis [47], and transplant patients [48]. Vitamin D metabolism is interrupted by the inability for the second hydroxylation step of 25-hydroxyvitamin D to occur, which converts it to the active form 1,25 dihydroxy vitamin D, which occurs in the kidneys [49]. Inhibition of 1,25 dihydroxy vitamin D induces hypocalcemia, which stimulates the parathyroid gland to release parathyroid hormone at persistent circulating levels [50,51]. Over time, this may result in renal osteodystrophy, including secondary parathyroidism, osteitis fibrosa, osteomalacia, and adynamic bone disease [45].

The current KDOQI guidelines for CKD nutrition state ergocalciferol or cholecalciferol effectively treats VD deficiency/inefficiency; however, specific dosing should be individualized and derived through a step-by-step approach [17]. This step-by-step approach includes monitoring 25(OH)D serum levels and serum calcium and serum phosphorus, which helps the healthcare team recommend specific dosing veered to the patient's individual requirements [18]. Supplementation with ergocalciferol or cholecalciferol is essential in treating and preventing BMD disease in CKD [50,51]. A meta-analysis performed by Kandula et al. [52] suggests that supplementation of 1,25 dihydroxy vitamin D in CKD leads to increase in the serum levels and improves biochemical end-points. The study failed to observe any clinically significant outcomes due to observed improvements in biochemical end-points [52].

2.4. Calcium

Calcium balance is regulated by intestinal calcium absorption, kidney reabsorption, and calciotropic hormones that activate calcium exchange from the bone when serum calcium levels are low [18]. Insufficient calcium absorption and chronic calcium deficiency result in increased risk for hyperthyroidism and osteitis [17]. However, excessive calcium poses an increased risk for calcification, resulting in comorbidities and higher mortality [17]. Alterations in calcium metabolism are multifactorial and include the use of active vitamin D analogues. Research shows that ingesting about 800–1000 mg/d of calcium may be sufficient to maintain calcium balance for patients with CKD 3–4 in the absence of vitamin D analogues [17] (Table 2). However, calcium recommendations for early stages of CKD typically follow the RDA (1000–1200 mg/d) for adults because the level of kidney function has not yet disrupted calcium balance, although this is in individualized circumstances. Maintaining calcium balance is more complicated for CKD patients on dialysis, and hypercalcemia is relatively standard. Vitamin D is an important factor in maintaining calcium balance. VD supplementation therapy is prescribed to CKD patients with inefficient active VD levels to increase calcium reabsorption and prevent high serum para-thyroid hormone (PTH) and bone turnover [53]. Massart et al. [54] and Jean et al. [53] reported increased 1,25(OH)2D levels after cholecalciferol supplementation. Strong evidence shows the importance of adequate active VD for calcium balance, and it is achieved with VD supplementation in CKD patients [18].

2.5. Phosphorus

Phosphorus plays a critical role in bone formation, acid–base balance, and energy production [48]. The body's ability to maintain phosphate balance is achieved by excreting excess phosphate in the urine. As CKD progresses, declining renal function prevents the kidneys from excreting enough phosphorus needed for phosphorus homeostasis [18]. The 2020 NKF guidelines recommended CKD 1–5 and HD patients receive an intake of phosphorus that keeps serum phosphorus levels within normal ranges (3.4–4.5 mg/dL) and to restrict dietary phosphate in the case of hyperphosphatemia [18,55] (Table 2). Hyperphosphatemia may lead to critical pathogenic consequences, including renal osteodystrophy, cardiovascular and soft tissue calcification, secondary hyperthyroidism, cardiac disease, and mortality in ESRD patients [56]. Phosphorus requirements depend on the stage of renal failure combined with the consideration to not restrict phosphorus intake to the point of malnutrition, which is mainly relevant to HD patients [57]. Despite the KDOQI revision for phosphorus intake in CKD, nephrologists recommend a phosphorus restriction of 800–1000 mg/d [10]; however, adequate studies are lacking that demonstrate the efficacy of 800–1000 mg dietary phosphorus restriction and the outcomes in CKD patients [18].

The three sources of dietary phosphorus are organic phosphorus from plant foods (bioavailability 20–40%), organic phosphorus from animal protein (bioavailability 40–60%), and inorganic phosphorus found in additives and processed foods (bioavailability ≈100%) [58]. Humans lack phytase, which is the enzyme that degrades phytates in plant foods, and this is why the bioavailability is the lowest of the three sources [58]. Inorganic phosphorus (additives) is almost entirely absorbed and may add up to 1000 mg/d of phosphorus from additives alone [26]. Choosing phosphorus-containing foods lower in bioavailability and without phosphate additives is recommended [17]. A study by Moe et al. that included CKD-4 patients reported lower phosphate levels in patients fed a 7-day vegetarian diet than patients fed a 7-day animal-based diet [25]. About 100 mg of phosphorus is found in 100 mL of milk and >500 mg per 100 g of cheese; the content of phosphorus is high in dairy products [59]. One study reported higher dietary phosphorus intake and a higher phosphorus to protein ratio in HD patient's diets was associated with increased mortality risk in the preceding years, even after adjusting for phosphate binders [60]. Sources containing only organic phosphorus are more nutrient-dense than foods with phosphate additives, which are usually processed and high in sodium [30].

Table 2. Daily requirements for electrolytes in chronic kidney disease (CKD) patients.

Electrolytes	Damage in CKD	Recommendation	Outcome	Ref
Total calcium CKD 3–4 w/no use of taking active vitamin D analogues	Ca2+ deficiency ↑ risk secondary hyperparathyroidism and bone disorders. Excessive Ca2+ ↑ risk extraosseous calcification and CVD	800–1000 mg/day	Maintain Ca2+ balance	[18,29,61]
CKD 5 w/use of active vitamin D analogues	Ca2+ deficiency ↑ risk secondary hyperparathyroidism and bone disorders. Excessive Ca2+ ↑ risk extraosseous calcification and CVD	Individualize Ca2+ restriction based on the use of vitamin D analogues	Maintain Ca2+ balance and prevent hypercalcemeia	[18,29,62]
Dietary Phosphorus * CKD 1–5	High dietary phosphorus intake associated w/ accelerated progression of disease and greater 5-year mortality risk	adjust dietary phosphorus intake to maintain normal serum phosphate levels between 3.4–4.5 mg/dL	Maintain Ca2+ and PTH balance. ↓ Secondary hyperparathyroidism mineral and bone disorders. Slow progression of CKD	[29,63]
Dietary Potassium CKD1–5 or post-transplantation	Hyper/hypokalemia associated w/muscular weakness, hypertension, ventricular arrhythmias, and death. Hypokalemia associated w/peripheral neuropathy.	adjust dietary K+ intake to maintain serum potassium within 3.5–5.5 mEq/L	Slow progression of CKD. Prevention of peripheral neuropathy and other nerve related dysfunction.	[29,64,65]
Sodium (Na+) CKD 1–5 or post transplantation	↑ BP excessive fluid retention/increased weight	<2300 mg/day	↓ BP and normalize fluid balance/weight reduction/may ↓ proteinuria	[29,66–68]

* Phosphate recommendations recently changed; previously 800 mg, ↑ increased/high, ↓: decreased/lowered.

2.6. Potassium

Potassium (K) is the most abundant intracellular ion with a concentration of about 98%; it has many biological functions such as cellular metabolism and acid–base homeostasis [69]. It is also vital for cardiac function, neural transmission, muscular contractions, and glucose metabolism [67,70]. If potassium balance is disrupted by increased serum potassium, the patient is at risk for developing hyperkalemia (Table 2). Hyperkalemia is a severe metabolic condition that is often experienced in patients with CKD. The kidneys' ability to excrete potassium is inversely related to a GFR function [69]. Hyperkalemia alters the nervous system's function, causing electrophysiological dysfunctions [64,71], presenting clinical manifestations such as muscle weakness, paresthesia, paralysis, nausea, hypotension, cardiac arrhythmias, and cardiac arrest [67,70]. As CKD progresses, potassium levels are monitored closely; patients are advised to limit dietary potassium intake to maintain serum potassium levels within normal range (3.5–5.5 mEq/L) [17]. Potassium is rich in many foods such as vegetables, dark leafy greens, potatoes, tomatoes, fruit, coffee and tea, and citrus. CKD nutrition therapy recommends vegetables and fruits that are low in potassium and high in fiber along with [17] other nutrients, and to boil vegetables to decrease potassium concentration [17].

The ideal potassium intake is difficult to determine because of factors that influence serum potassium levels, such as medications, hydration level, acid–base status, glycemic control, adrenal function, and gastrointestinal complications [17]. It is essential to consider these factors when assessing the appropriate intake of potassium for a CKD patient, as the recommendations for potassium are individualized based on other preexisting health conditions the patient might have or be at risk for. The DASH diet is widely used as nutrition therapy for hypertension because of its effectiveness in lowering blood pressure, preventing and managing hypertension, and reducing cardiovascular risk [72]. The DASH diet may be protective against the progression of CKD, but its effectiveness in delaying the progression of the disease in CKD patients has not been established [72]. The DASH diet is high in potassium and low in sodium; it suggests four to five servings of fruits and vegetables a day, which sums up to about 4700 mg/d of potassium [72]. Studies on the DASH diet with CKD patients are scarce, and the few existing studies include CKD patients with serum potassium levels in normal range at the start of the study [62]. This is a limitation of the study for determining the efficacy of the DASH diet for CKD patients [73]. Another diet currently being studied for its benefits in CKD is the Mediterranean Diet (MedDiet). Instead of its focus being on low sodium and high potassium, it focuses on healthy fats, lean meats, and plant-based foods, which naturally offer a diet low in sodium. MedDiet studies began in the 1960s, and since then, increasing evidence supports the MediDiet to be protective against CKD and DM [51]. The MedDiet is rich in plant-based foods and low in processed and red meat [74]. It is moderate in seafood, eggs, dairy, and red wine; and olive oil is the main source of added fat [75]. Adherence to the MedDiet helps prevent and manage CVD and DM [71,76], which would in turn help prevent CKD. However, the role of the MedDiet in delaying CKD progression remains uncertain due to insufficient data on patients with pre-existing CKD or dialysis [70].

Additionally, Kalemic control is further compounded by extensive use of the renin–angiotensin–aldosterone system inhibitor (RAASI) therapy in CKD patients [77]. Development of hyperkalemia in CKD patients requires lowering the dose or discontinuation of the RAASI therapy to protect patients from developing cardiovascular events and end stage kidney disease.

The true benefit of potassium restriction in CKD is not clear, considering that a diet with a high content of potassium-rich foods, such as plant-based low-protein diets, can be as beneficial on the prognosis. Potassium levels in serum can further be improved using the new K-binders, whose benefits and efficacy are shown in randomized control trials [78,79], allowing implementing plant-based low-protein diets with lower risk of hyperkalemia. Further research investigating the effect of a low-potassium diet and the progression of renal disease are required. It is unclear whether a potassium-restricted diet can slow CKD progression; however, research shows that it may reduce all-cause mortality in CKD [79].

2.7. Sodium

Sodium overload in advanced CKD patients induces extracellular volume, which may lead to hypertension and heart failure. Hypertension is a known risk factor for the progression and mortality of CVD; however, the effect of sodium on the advancement of CKD remains inconclusive [18]. A recent working hypothesis suggests that the accumulation of sodium in interstitial space induces inflammatory toxicity that is independent of volume, and it is mediated by immune cells [80]. Sodium accumulation in the body increases as the GFR declines over time [81].

A low sodium diet is central to the management of hydro-saline homeostasis, reducing systolic and diastolic blood pressure as well as proteinuria [82]. Nevertheless, a low-salt diet must be carefully monitored in older patients, considering they are at higher risk for acute kidney injury and damaged renal autoregulation [51]. The efficacy of low sodium intake and the reduction in BP in hypertensive patients dates to 1948 [83,84], currently reaching a worldwide understanding of the relationship between sodium and hypertension [85,86]. Patients with hypertension have a 75% increased risk of developing CKD than normotensive individuals [83,87] and a 25% increased risk of developing a decline in GFR among pre-hypertension patients [84,87]. The McMahon et al. study assessed the effects of high- vs. low-sodium diets on BP, 24 h protein and albumin excretion, and fluid status in 20 hypertensive stage 3–4 CKD adult patients [66]. The study concluded that the low-sodium diet resulted in statistical and clinically significant declines in BP, extracellular fluid volume, albuminuria, and proteinuria in study patients [66].

Nonetheless, sodium restriction is protective for the onset of hypertension. There is plentiful and strong evidence in the efficacy to prescribing a sodium-restricted diet for disease management in CKD [85]. For CKD stages 3–5, the most recent sodium intake recommendation is a maximum of 2.3 g/d and to make sodium restriction a lifestyle for controlling fluid volume and maintaining a desirable weight for CKD 3–5D [17] (Table 2). Effective habits to reduce sodium may be achieved by identifying high-sodium foods such as processed foods, canned vegetables, pickled and fermented foods, soups, chips, salted nuts and seeds, processed foods, and restaurant items. Simple modifications such as choosing unprocessed foods, choosing frozen over canned vegetables, avoiding soups and pickled and fermented foods, choosing unsalted nuts and seeds, and requesting no additional salt when ordering out are beneficial for achieving a restricted sodium diet.

2.8. Whole Food Plant-Based Diet

Studies report that a whole food plant-based (WFPB) diet reduces the risk for T2DM and CVD in CKD patients [2]. A WFPB diet is more restrictive than a vegan diet by the exclusion of processed and refined foods such as isolated vegetable oils, bleached flours, and cane and beet root sugar; the diet focuses on increased fiber and nutrient-dense foods that are low in protein and energy [2]. Whole grains, nuts, seeds, legumes, monosaturated oils, fruits and vegetables, and tubers make up the foods in a WFPB diet [88,89].

WFPB diets provide about 75% of carbohydrates (CHO), emphasizing dietary fiber [90,91]. Fiber intake of about 27 g/d reduces serum urea and creatine in CKD; high serum urea and creatine indicate abnormal GFR [90,91]. High fiber intake shifts the gut microbiota by increasing the amount of gut microflora that break down and process fiber [92]. Soluble fiber intake such as apples and oats reduces serum cholesterol, postprandial glucose, insulin response [92,93], and induce satiety from delayed gastric emptying [93,94]. Insoluble fiber such as whole grains and legumes increase motility and transit time by softening stool and promoting regular bowel movement, which is especially critical for CVD and CKD patients as they commonly experience slowed colonic transit time [92,93]. WFPB diets are significantly higher in fiber than other diets resulting in several health benefits for just fiber alone [65,92].

WFPB diets do not restrict fat intake; however, the foods promoted are made up of monounsaturated and polyunsaturated fats and limit processed oils and saturated fat [2]. Previous studies show a daily caloric intake of total fat to be less than 15% in WFPB diets, which is protective against CVD [91]. It is well established that omega-3 fatty acids reduce inflammation [94,95], blood pressure, and increase HDL cholesterol [95,96]. Plant-based omega-3s are in foods such as flaxseeds, chia seeds, walnuts, olives, and some dark green vegetables [2]. The consumption of 1.5–3 g/d of omega-3s is associated with CVD prevention in CKD patients [94].

It is challenging for patients to comply with a restricted phosphorus diet because it is found in most foods [60]. Many fruits and vegetables contain a slight phosphorus trace, while its content is higher in seeds, nuts, and legumes; and it is even higher in animal products [68]. However, plant foods contain phytates that limit phosphorus's gastrointestinal absorption, decreasing the bioavailability of phytate-based phosphorus [68]. Additionally, a WFPB diet restricts processed foods and sugar, including restructured meat and soft drinks, which contain inorganic phosphorus-based additives for preservation. These additives generally go unnoticed due to their complex and unrecognizable names, with inorganic phosphorus having the highest absorption rate, at more than 90% [97,98].

A WFPB diet is naturally low in sodium due to the restriction of processed foods, assisting the patient with maintaining appropriate sodium levels. Additionally, WFPB diets are generally lower in energy and may be beneficial for weight management. However, caution and careful planning are critical to avoid inadequate energy intake and PEW, which could worsen the patient's health, increasing their risk for morbidity and mortality [57]. A wide variety of plant-based foods need to make up the diet and increased consumption of starchy vegetables, fruits, and legumes to meet the RDAs for protein and energy [2]. A drawback in the WFPB diet is the need to supplement with vitamin B12, because sufficient vitamin B12 intake is only met through the consumption of animal-based foods [2]. Although evidence is growing that supports the positive health benefits of WFPB diets, there is a need for more research to determine any nutritional deficiencies or other adverse health effects from a WFPB diet in a clinical population with CKD patients [2].

3. The Role of a Registered Dietitian

Dietary education and patient counseling provided by a registered dietitian (RD) is essential for preventing and managing CKD. Careful and detailed dietary planning, frequent assessment of nutritional status, and dietary monitoring compliance are critical for successful dietary management.

The progressive decline in GFR is a risk factor for the development of metabolic acidosis. The main goal of therapy is to prevent or correct this metabolic acidosis, which has been shown to slow down the progression of CKD to end-stage renal disease [99]. The biggest contributor to this acid pool is the consumption of a diet higher in animal proteins [100]. The simplest treatment for this metabolic acidosis includes dietary management by reducing the protein in the diet or switching the diet to an increase in plant-based proteins [101]. It has been shown that dietary intervention of lowering protein intake or switching to plant-based protein reduces metabolic acidosis in stage 3–4 CKD patients [63].

Primary and secondary studies out of the MDRD study suggest that dietary interventions such as a low-protein diet reduce the rate of kidney function decline and lower the risk of ESKD in CKD patients [13,102]. Dietary interventions, such as low-protein diet, have been shown to retard the progression of CKD [102]. The dietary restriction of protein and phosphorous are shown to reduce the decline of kidney function and has been observed in type 1 diabetes patients [103]. The consensus among clinicians is that dietary interventions slow the rate of kidney function and potentially reduce the risk of end-stage kidney disease in patients with diabetes and CKD.

CKD patients often have or are at risk for comorbidities that entail specific diet management recommendations; this can be challenging and overwhelming. Additionally, CKD diet recommendations alter depending on the disease stage; this can create confusion for the patient. The dietitian has a more significant role than just providing dietary advice and recommendations for the patient. Counseling should be individualized and altered to the patient's overall health, pre-existing conditions, and personal preferences. Adopting and adhering to a new diet requires the ability to motivate and inspire patients to make changes that will improve their health and prevent morbidities, although the changes may be uncomfortable for the patient. Adequate education about the rationale of the recommendations and how the patient will benefit are essential to convey. Equally as important is to assess the retention and understanding of the patient from the nutrition education. Through a thorough patient assessment and evaluation, the dietician may help prevent kidney disease by carefully monitoring their diabetic, hypertensive, and CVD patients by ordering the appropriate screening labs. It is imperative to regularly screen the patient for CVD, T2DM, malnutrition, and anemia, as they are at high risk of developing them. Providing alternative food options tailored to the patient's likes and dislikes to replace restricted foods is more productive than focusing on the restrictions. Providing substitution education to the patient is essential to attain and maintain patient compliance and achieve successful dietary management.

The major limitation of this review is that although it is a literature review, we did not performed a systematic review of the literature. The literature survey was performed for a narrative review of the currently available studies to attempt to compile available studies under this review. To keep the review within limits, the search strategy was not comprehensive, and the studies were not assessed critically. Furthermore, this review was limited to protein, calcium, phosphate, and VD and electrolytic balance; it does not provide more comprehensive information on the clinical management of CKD.

4. Future Research and Clinical Practice

Secondary analysis of the MDRD study showed that patients with low protein intake during follow-up began experiencing uremic symptoms at lower GFR than patients with higher protein intake [103]. The reduced risk of end-stage renal failure reported may be from a delay in starting dialysis due to improved uremic symptoms rather than delayed kidney decline [103]. In addition, the study included 200 (24%) polycystic kidney disease (PKD) patients who may have contributed to data showing a delay in renal dysfunction due to the differences in the course of disease progression between CKD and PKD [103]. The INTERMAP Study lacked the use of "gold standard" diet assessments, food variation among different countries, and variation in dietary intake, which weakens the associations between nutrient intake and blood pressure [33].

Despite the large number of clinical trials being performed in the clinical and nutritional management of CKD, very few of these have translocated into clinical practices due to the lack of strong associations, not so clear research design, or low number of study subjects. There is a demand for future research to provide conclusive information that will assist clinicians and dietitians to make the most appropriate recommendations for their patients. Evaluating the impact of MNT on CKD progression by analysis of associated risk factors in patients with comorbidities is needed [18]. The clarity regarding which stage of CKD is most appropriate to alter protein intake is necessary. Future VD studies are required to determine the correct dosing and type of VD supplements for CKD patients. Future studies examining, comparing, and contrasting WFPBD, Mediterranean diet, and DASH diet in CKD patients to determine their effects on clinical outcomes are needed. Another challenge in CKD patients is not following the dietary recommendations. Research should be focus on boosting patient diet compliance by developing methods that will improve compliance and long-term adherence to nutrition prescriptions.

5. Conclusions

Chronic kidney disease is a growing health crisis in the U.S. Diabetes and hypertension are the leading causes of CKD development; as the US is experiencing an increasing prevalence of both, CKD is expected to remain a critical national health issue. At ESRD, the kidneys have lost their ability function, and as a result, a series of malfunctions occur that lead to adverse health problems and health outcomes. Once diagnosed with ESRD, the patient either will be on dialysis for the rest of their life or receive a kidney transplant. Medical nutrition therapy with a RD is a critical aspect in the intervention for CKD because it is almost solely through nutrition that aids in the delay of the disease's progression and the prevention of comorbidities and mortality.

Author Contributions: Conceptualization, T.N.; resources, S.P.; writing—original draft preparation, T.N.; writing—review and editing, T.N. and S.P.; visualization, T.N. and S.P.; supervision, S.P. All authors have read and agreed to the published version of the manuscript.

Funding: S.P. was supported by Postdoctoral Fellowships (2-2011-153, 10-2006-792) and Career Development Award (3-2004-195) from Juvenile Diabetes Research Foundation NY, USA. T.N. is recipient of the Mildred B. Davis Fellowship (2021) awarded from American Association of Family and Consumer Sciences.

Institutional Review Board Statement: Not applicable.

Informed Consent Statement: Not applicable.

Data Availability Statement: Not applicable.

Conflicts of Interest: The authors declare no conflict of interest.

References

1. Centers for Disease Control, Chronic Kidney Disease in the United States. 2021. Available online: https://www.cdc.gov/kidneydisease/pdf/Chronic-Kidney-Disease-in-the-US-2021-h.pdf (accessed on 23 February 2021).
2. Adair, K.E.; Bowden, R.G. Ameliorating Chronic Kidney Disease Using a Whole Food Plant-Based Diet. *Nutrients* **2020**, *12*, 1007. [CrossRef] [PubMed]
3. Wallace, M.A. Anatomy and Physiology of the Kidney. *Aorn J.* **1998**, *68*, 799–820. [CrossRef]
4. Regan, M.C.; Young, L.S.; Geraghty, J.; Fitzpatrick, J.M. Regional Renal Blood Flow in Normal and Disease States. *Urol. Res.* **1995**, *23*, 1–10. [CrossRef]
5. McLafferty, E.; Johnstone, C.; Hendry, C.; Farley, A. Fluid and Electrolyte Balance. *Nurs. Stand.* **2014**, *28*, 42. [CrossRef]
6. Shioji, N.; Hayashi, M.; Morimatsu, H. Kidney, Fluid, and Acid-Base Balance. *Masui* **2016**, *65*, 503–510.
7. Drábková, N.; Hojná, S.; Zicha, J.; Vaněčková, I. Contribution of Selected Vasoactive Systems to Blood Pressure Regulation in Two Models of Chronic Kidney Disease. *Physiol. Res.* **2020**, 405–414. [CrossRef]
8. Kulda, V. Vitamin D metabolism. *Vnitr. Lek.* **2012**, *58*, 400–404.
9. Palmer, S.C.; Saglimbene, V.; Mavridis, D.; Salanti, G.; Craig, J.C.; Tonelli, M.; Wiebe, N.; Strippoli, G.F. Erythropoiesis-stimulating Agents for Anaemia in Adults with Chronic Kidney Disease: A Network Meta-analysis. *Cochrane Database Syst. Rev.* **2014**, *2014*. [CrossRef] [PubMed]
10. National Kidney Foundation K/DOQI Clinical Practice Guidelines for Chronic Kidney Disease: Evaluation, Classification, and Stratification. *Am. J. Kidney Dis.* **2002**, *39*, S1–S266.
11. Coresh, J.; Selvin, E.; Stevens, L.A.; Manzi, J.; Kusek, J.W.; Eggers, P.; Van Lente, F.; Levey, A.S. Prevalence of Chronic Kidney Disease in the United States. *JAMA* **2007**, *298*, 2038–2047. [CrossRef] [PubMed]
12. Hsu, C.; Ordoñez, J.; Chertow, G.; Fan, D.; McCulloch, C.; Go, A. The Risk of Acute Renal Failure in Patients with Chronic Kidney Disease. *Kidney Int.* **2008**, *74*, 101–107. [CrossRef]
13. Levey, A.S.; Andreoli, S.P.; DuBose, T.; Provenzano, R.; Collins, A.J. Chronic Kidney Disease: Common, Harmful, and Treatable—World Kidney Day 2007. *J. Am. Soc. Nephrol.* **2007**, *18*, 374–378. [CrossRef]
14. McCullough, K.P.; Morgenstern, H.; Saran, R.; Herman, W.H.; Robinson, B.M. Projecting ESRD Incidence and Prevalence in the United States through 2030. *J. Am. Soc. Nephrol.* **2019**, *30*, 127–135. [CrossRef] [PubMed]
15. Evans, P.D.; Taal, M.W. Epidemiology and Causes of Chronic Kidney Disease. *Medicine* **2015**, *43*, 450–453. [CrossRef]
16. Kovesdy, C.P.; Kopple, J.D.; Kalantar-Zadeh, K. Management of Protein-Energy Wasting in Non-Dialysis-Dependent Chronic Kidney Disease: Reconciling Low Protein Intake with Nutritional Therapy. *Am. J. Clin. Nutr.* **2013**, *97*, 1163–1177. [CrossRef] [PubMed]
17. Ketteler, M.; Block, G.A.; Evenepoel, P.; Fukagawa, M.; Herzog, C.A.; McCann, L.; Moe, S.M.; Shroff, R.; Tonelli, M.A.; Toussaint, N.D.; et al. Executive Summary of the 2017 KDIGO Chronic Kidney Disease-Mineral and Bone Disorder (CKD-MBD) Guideline Update: What's Changed and Why It Matters. *Kidney Int.* **2017**, *92*, 26–36. [CrossRef]
18. Ikizler, T.A.; Burrowes, J.D.; Byham-Gray, L.D.; Campbell, K.L.; Carrero, J.-J.; Chan, W.; Fouque, D.; Friedman, A.N.; Ghaddar, S.; Goldstein-Fuchs, D.J.; et al. KDOQI Clinical Practice Guideline for Nutrition in CKD: 2020 Update. *Am. J. Kidney Dis.* **2020**, *76*, S1–S107. [CrossRef] [PubMed]
19. Delimaris, I. Adverse Effects Associated with Protein Intake above the Recommended Dietary Allowance for Adults. *ISRN Nutr.* **2013**, *2013*, 126929. [CrossRef] [PubMed]
20. MacKay, E.M.; MacKay, L.L.; Addis, T. Factors Which Determine Renal Weight. *Am. J. Physiol.-Leg. Content* **1928**, *86*, 466–470. [CrossRef]
21. Kalantar-Zadeh, K.; Fouque, D. Nutritional Management of Chronic Kidney Disease. *N. Engl. J. Med.* **2017**, *377*, 1765–1776. [CrossRef] [PubMed]
22. Meyer, T.W.; Anderson, S.; Brenner, B.M. Dietary Protein Intake and Progressive Glomerular Sclerosis: The Role of Capillary Hypertension and Hyperperfusion in the Progression of Renal Disease. *Ann. Intern. Med.* **1983**, *98*, 832–838. [CrossRef]

23. Modification of Diet in Renal Disease Study Group; Hunsicker, L.G.; Adler, S.; Caggiula, A.; England, B.K.; Greene, T.; Kusek, J.W.; Rogers, N.L.; Teschan, P.E.; Beck, G. Predictors of the Progression of Renal Disease in the Modification of Diet in Renal Disease Study. *Kidney Int.* **1997**, *51*, 1908–1919. [CrossRef]
24. Oba, R.; Kanzaki, G.; Sasaki, T.; Okabayashi, Y.; Haruhara, K.; Koike, K.; Kobayashi, A.; Yamamoto, I.; Tsuboi, N.; Yokoo, T. Dietary Protein Intake and Single-Nephron Glomerular Filtration Rate. *Nutrients* **2020**, *12*, 2549. [CrossRef]
25. Moe, S.M.; Zidehsarai, M.P.; Chambers, M.A.; Jackman, L.A.; Radcliffe, J.S.; Trevino, L.L.; Donahue, S.E.; Asplin, J.R. Vegetarian Compared with Meat Dietary Protein Source and Phosphorus Homeostasis in Chronic Kidney Disease. *Clin. J. Am. Soc. Nephrol.* **2011**, *6*, 257–264. [CrossRef]
26. Bell, R.R.; Draper, H.H.; Tzeng, D.Y.M.; Shin, H.K.; Schmidt, G.R. Physiological Responses of Human Adults to Foods Containing Phosphate Additives. *J. Nutr.* **1977**, *107*, 42–50. [CrossRef]
27. Fouque, D.; Chen, J.; Chen, W.; Garneata, L.; Hwang, S.; Kalantar-Zadeh, K.; Kopple, J.D.; Mitch, W.E.; Piccoli, G.; Teplan, V.; et al. Adherence to Ketoacids/Essential Amino Acids-Supplemented Low Protein Diets and New Indications for Patients with Chronic Kidney Disease. *BMC Nephrol.* **2016**, *17*. [CrossRef]
28. Kopple, J.D.; Shinaberger, J.H.; Coburn, J.W.; Sorensen, M.K.; Rubini, M.E. Optimal Dietary Protein Treatment during Chronic Hemodialysis. *Trans. Am. Soc. Artif. Intern. Organs* **1969**, *15*, 302–308. [PubMed]
29. Lynch, K.E.; Lynch, R.; Curhan, G.C.; Brunelli, S.M. Prescribed Dietary Phosphate Restriction and Survival among Hemodialysis Patients. *Clin. J. Am. Soc. Nephrol.* **2011**, *6*, 620–629. [CrossRef] [PubMed]
30. Gutiérrez, O.M. Sodium- and Phosphorus-Based Food Additives: Persistent but Surmountable Hurdles in the Management of Nutrition in Chronic Kidney Disease. *Adv. Chronic Kidney Dis.* **2013**, *20*, 150–156. [CrossRef]
31. Fouque, D.; Aparicio, M. Eleven Reasons to Control the Protein Intake of Patients with Chronic Kidney Disease. *Nat. Rev. Nephrol.* **2007**, *3*, 383–392. [CrossRef] [PubMed]
32. Garneata, L.; Stancu, A.; Dragomir, D.; Stefan, G.; Mircescu, G. Ketoanalogue-Supplemented Vegetarian Very Low–Protein Diet and CKD Progression. *JASN* **2016**, *27*, 2164–2176. [CrossRef] [PubMed]
33. Elliott, P.; Stamler, J.; Dyer, A.R.; Appel, L.; Dennis, B.; Kesteloot, H.; Ueshima, H.; Okayama, A.; Chan, Q.; Garside, D.B.; et al. Association between Protein Intake and Blood Pressure: The INTERMAP Study. *Arch. Intern. Med.* **2006**, *166*, 79–87. [CrossRef]
34. Evenepoel, P.; Meijers, B.K. Dietary Fiber and Protein: Nutritional Therapy in Chronic Kidney Disease and Beyond. *Kidney Int.* **2012**, *81*, 227–229. [CrossRef] [PubMed]
35. Simon, A.H.; Lima, P.R.; Almerinda, M.; Alves, V.F.; Bottini, P.V.; de Faria, J.B. Renal Haemodynamic Responses to a Chicken or Beef Meal in Normal Individuals. *Nephrol. Dial. Transpl.* **1998**, *13*, 2261–2264. [CrossRef] [PubMed]
36. Nakamura, H.; Takasawa, M.; Kashara, S.; Tsuda, A.; Momotsu, T.; Ito, S.; Shibata, A. Effects of Acute Protein Loads of Different Sources on Renal Function of Patients with Diabetic Nephropathy. *Tohoku J. Exp. Med.* **1989**, *159*, 153–162. [CrossRef]
37. Attini, R.; Leone, F.; Parisi, S.; Fassio, F.; Capizzi, I.; Loi, V.; Colla, L.; Rossetti, M.; Gerbino, M.; Maxia, S.; et al. Vegan-Vegetarian Low-Protein Supplemented Diets in Pregnant CKD Patients: Fifteen Years of Experience. *BMC Nephrol.* **2016**, *17*, 132. [CrossRef]
38. Cases, A.; Cigarrán-Guldrís, S.; Mas, S.; Gonzalez-Parra, E. Vegetable-Based Diets for Chronic Kidney Disease? It Is Time to Reconsider. *Nutrients* **2019**, *11*, 1263. [CrossRef]
39. Moorthi, R.N.; Vorland, C.J.; Gallant, K.M.H. Diet and Diabetic Kidney Disease: Plant versus Animal Protein. *Curr. Diab. Rep.* **2017**, *17*, 15. [CrossRef]
40. Kahleova, H.; Levin, S.; Barnard, N. Cardio-Metabolic Benefits of Plant-Based Diets. *Nutrients* **2017**, *9*, 848. [CrossRef]
41. Klahr, S.; Levey, A.S.; Beck, G.J.; Caggiula, A.W.; Hunsicker, L.; Kusek, J.W.; Striker, G. The Effects of Dietary Protein Restriction and Blood-Pressure Control on the Progression of Chronic Renal Disease. Modification of Diet in Renal Disease Study Group. *N. Engl. J. Med.* **1994**, *330*, 877–884. [CrossRef] [PubMed]
42. Kasiske, B.L.; Lakatua, J.D.; Ma, J.Z.; Louis, T.A. A Meta-Analysis of the Effects of Dietary Protein Restriction on the Rate of Decline in Renal Function. *Am. J. Kidney Dis.* **1998**, *31*, 954–961. [CrossRef]
43. Shah, B.V.; Patel, Z.M. Role of Low Protein Diet in Management of Different Stages of Chronic Kidney Disease—Practical Aspects. *BMC Nephrol.* **2016**, *17*, 156. [CrossRef]
44. Yan, B.; Su, X.; Xu, B.; Qiao, X.; Wang, L. Effect of Diet Protein Restriction on Progression of Chronic Kidney Disease: A Systematic Review and Meta-Analysis. *PLoS ONE* **2018**, *13*, e0206134. [CrossRef]
45. Lips, P.; Goldsmith, D.; de Jongh, R. Vitamin D and Osteoporosis in Chronic Kidney Disease. *J. Nephrol.* **2017**, *30*, 671–675. [CrossRef]
46. Hou, Y.-C.; Lu, C.-L.; Lu, K.-C. Mineral Bone Disorders in Chronic Kidney Disease. *Nephrology* **2018**, *23*, 88–94. [CrossRef]
47. Kim, S.M.; Choi, H.J.; Lee, J.P.; Kim, D.K.; Oh, Y.K.; Kim, Y.S.; Lim, C.S. Prevalence of Vitamin D Deficiency and Effects of Supplementation with Cholecalciferol in Patients with Chronic Kidney Disease. *J. Ren. Nutr.* **2014**, *24*, 20–25. [CrossRef]
48. Filipov, J.J.; Zlatkov, B.K.; Dimitrov, E.P.; Svinarov, D. Relationship between Vitamin D Status and Immunosuppressive Therapy in Kidney Transplant Recipients. *Biotechnol. Biotechnol. Equip.* **2015**, *29*, 331–335. [CrossRef]
49. Berdanier, C.D.; Berdanier, L.A. *Advanced Nutrition Macronutrients, Micronutrients, and Metabolism*, 2nd ed.; CRC Press: Boca Raton, FL, USA, 2015; ISBN 978-1-4822-0517-6.
50. Garofalo, C.; Provenzano, M.; Andreucci, M.; Pisani, A.; De Nicola, L.; Conte, G.; Borrelli, S. Predictive Effect of Salt Intake on Patient and Kidney Survival in Non-Dialysis CKD: Competing Risk Analysis in Older versus Younger Patients under Nephrology Care. *Nephrol. Dial. Transplant.* **2020**, gfaa252. [CrossRef] [PubMed]

51. Serra-Majem, L.; Román-Viñas, B.; Sanchez-Villegas, A.; Guasch-Ferré, M.; Corella, D.; La Vecchia, C. Benefits of the Mediterranean Diet: Epidemiological and Molecular Aspects. *Mol. Asp. Med.* **2019**, *67*, 1–55. [CrossRef] [PubMed]
52. Kandula, P.; Dobre, M.; Schold, J.D.; Schreiber, M.J.; Mehrotra, R.; Navaneethan, S.D. Vitamin D Supplementation in Chronic Kidney Disease: A Systematic Review and Meta-Analysis of Observational Studies and Randomized Controlled Trials. *Clin. J. Am. Soc. Nephrol.* **2011**, *6*, 50–62. [CrossRef] [PubMed]
53. Jean, G.; Souberbielle, J.C.; Chazot, C. Vitamin D in Chronic Kidney Disease and Dialysis Patients. *Nutrients* **2017**, *9*, 328. [CrossRef]
54. Massart, A.; Debelle, F.D.; Racapé, J.; Gervy, C.; Husson, C.; Dhaene, M.; Wissing, K.M.; Nortier, J.L. Biochemical Parameters After Cholecalciferol Repletion in Hemodialysis: Results From the VitaDial Randomized Trial. *Am. J. Kidney Dis.* **2014**, *64*, 696–705. [CrossRef]
55. Umut, S. Relationship of Dietary Phosphate Intake with Risk of End-Stage Renal Disease and Mortality in Chronic Kidney Disease Stages 3–5: The Modification of Diet in Renal Disease Study | Elsevier Enhanced Reader. Available online: https://reader.elsevier.com/reader/sd/pii/S0085253815000241?token=91355844158B638748824241A0450698282B3693E30793102456034ADB82312BE8F55F454D23184C440C0F6760D9B40D (accessed on 23 February 2021).
56. Malluche, H.; Monier-Faugère, M. Hyperphosphatemia: Pharmacologic Intervention Yesterday, Today and Tomorrow—Abstract—Europe PMC. Available online: https://europepmc.org/article/med/11076107 (accessed on 23 February 2021).
57. Carrero, J.J.; Stenvinkel, P.; Cuppari, L.; Ikizler, T.A.; Kalantar-Zadeh, K.; Kaysen, G.; Mitch, W.E.; Price, S.R.; Wanner, C.; Wang, A.Y.M.; et al. Etiology of the Protein-Energy Wasting Syndrome in Chronic Kidney Disease: A Consensus Statement From the International Society of Renal Nutrition and Metabolism (ISRNM). *J. Ren. Nutr.* **2013**, *23*, 77–90. [CrossRef]
58. Kalantar-Zadeh, K. Patient Education for Phosphorus Management in Chronic Kidney Disease. *PPA* **2013**, *7*, 379. [CrossRef] [PubMed]
59. Melse-Boonstra, A. Bioavailability of Micronutrients From Nutrient-Dense Whole Foods: Zooming in on Dairy, Vegetables, and Fruits. *Front. Nutr.* **2020**, *7*, 101. [CrossRef]
60. Kalantar-Zadeh, K.; Gutekunst, L.; Mehrotra, R.; Kovesdy, C.P.; Bross, R.; Shinaberger, C.S.; Noori, N.; Hirschberg, R.; Benner, D.; Nissenson, A.R.; et al. Understanding Sources of Dietary Phosphorus in the Treatment of Patients with Chronic Kidney Disease. *CJASN* **2010**, *5*, 519–530. [CrossRef] [PubMed]
61. Sigrist, M.K.; Taal, M.W.; Bungay, P.; McIntyre, C.W. Progressive Vascular Calcification over 2 Years Is Associated with Arterial Stiffening and Increased Mortality in Patients with Stages 4 and 5 Chronic Kidney Disease. *Clin. J. Am. Soc. Nephrol.* **2007**, *2*, 1241–1248. [CrossRef]
62. Goraya, N.; Simoni, J.; Jo, C.-H.; Wesson, D.E. Treatment of Metabolic Acidosis in Patients with Stage 3 Chronic Kidney Disease with Fruits and Vegetables or Oral Bicarbonate Reduces Urine Angiotensinogen and Preserves Glomerular Filtration Rate. *Kidney Int.* **2014**, *86*, 1031–1038. [CrossRef] [PubMed]
63. Kawasaki, T.; Maeda, Y.; Matsuki, H.; Matsumoto, Y.; Akazawa, M.; Kuyama, T. Urinary Phosphorus Excretion per Creatinine Clearance as a Prognostic Marker for Progression of Chronic Kidney Disease: A Retrospective Cohort Study. *BMC Nephrol.* **2015**, *16*, 116. [CrossRef] [PubMed]
64. Arnold, R.; Pianta, T.J.; Pussell, B.A.; Kirby, A.; O'Brien, K.; Sullivan, K.; Holyday, M.; Cormack, C.; Kiernan, M.C.; Krishnan, A.V. Randomized, Controlled Trial of the Effect of Dietary Potassium Restriction on Nerve Function in CKD. *Clin. J. Am. Soc. Nephrol.* **2017**, *12*, 1569–1577. [CrossRef] [PubMed]
65. Cupisti, A.; Kovesdy, C.P.; D'Alessandro, C.; Kalantar-Zadeh, K. Dietary Approach to Recurrent or Chronic Hyperkalaemia in Patients with Decreased Kidney Function. *Nutrients* **2018**, *10*, 261. [CrossRef]
66. McMahon, E.J.; Bauer, J.D.; Hawley, C.M.; Isbel, N.M.; Stowasser, M.; Johnson, D.W.; Campbell, K.L. A Randomized Trial of Dietary Sodium Restriction in CKD. *JASN* **2013**, *24*, 2096–2103. [CrossRef]
67. Vogt, L.; Waanders, F.; Boomsma, F.; de Zeeuw, D.; Navis, G. Effects of Dietary Sodium and Hydrochlorothiazide on the Antiproteinuric Efficacy of Losartan. *J. Am. Soc. Nephrol.* **2008**, *19*, 999–1007. [CrossRef]
68. Campbell, K.L.; Johnson, D.W.; Bauer, J.D.; Hawley, C.M.; Isbel, N.M.; Stowasser, M.; Whitehead, J.P.; Dimeski, G.; McMahon, E. A Randomized Trial of Sodium-Restriction on Kidney Function, Fluid Volume and Adipokines in CKD Patients. *BMC Nephrol.* **2014**, *15*, 57. [CrossRef]
69. Watanabe, R.; Watanabe, R. Hyperkalemia in Chronic Kidney Disease. *Rev. Assoc. Médica Bras.* **2020**, *66*, s31–s36. [CrossRef]
70. Hansrivijit, P.; Oli, S.; Khanal, R.; Ghahramani, N.; Thongprayoon, C.; Cheungpasitporn, W. Mediterranean Diet and the Risk of Chronic Kidney Disease: A Systematic Review and Meta-Analysis. *Nephrology* **2020**, *25*, 913–918. [CrossRef]
71. Mirabelli, M.; Chiefari, E.; Arcidiacono, B.; Corigliano, D.M.; Brunetti, F.S.; Maggisano, V.; Russo, D.; Foti, D.P.; Brunetti, A. Mediterranean Diet Nutrients to Turn the Tide against Insulin Resistance and Related Diseases. *Nutrients* **2020**, *12*, 1066. [CrossRef] [PubMed]
72. Raphael, K.L. The Dietary Approaches to Stop Hypertension (DASH) Diet in Chronic Kidney Disease: Should We Embrace It? *Kidney Int.* **2019**, *95*, 1296–1298. [CrossRef] [PubMed]
73. Tyson, C.C.; Lin, P.-H.; Corsino, L.; Batch, B.C.; Allen, J.; Sapp, S.; Barnhart, H.; Nwankwo, C.; Burroughs, J.; Svetkey, L.P. Short-Term Effects of the DASH Diet in Adults with Moderate Chronic Kidney Disease: A Pilot Feeding Study. *Clin. Kidney J.* **2016**, *9*, 592–598. [CrossRef] [PubMed]

74. Willett, W.C.; Sacks, F.; Trichopoulou, A.; Drescher, G.; Ferro-Luzzi, A.; Helsing, E.; Trichopoulos, D. Mediterranean Diet Pyramid: A Cultural Model for Healthy Eating. *Am. J. Clin. Nutr.* **1995**, *61*, 1402S–1406S. [CrossRef] [PubMed]
75. Serra-Majem, L.; Tomaino, L.; Dernini, S.; Berry, E.M.; Lairon, D.; de la Cruz, J.N.; Bach-Faig, A.; Donini, L.M.; Medina, F.-X.; Belahsen, R.; et al. Updating the Mediterranean Diet Pyramid towards Sustainability: Focus on Environmental Concerns. *Int. J. Environ. Res. Public Health* **2020**, *17*, 8758. [CrossRef] [PubMed]
76. Dinu, M.; Pagliai, G.; Casini, A.; Sofi, F. Mediterranean Diet and Multiple Health Outcomes: An Umbrella Review of Meta-Analyses of Observational Studies and Randomised Trials. *Eur. J. Clin. Nutr.* **2018**, *72*, 30–43. [CrossRef]
77. Borrelli, S.; De Nicola, L.; Minutolo, R.; Conte, G.; Chiodini, P.; Cupisti, A.; Santoro, D.; Calabrese, V.; Giannese, D.; Garofalo, C.; et al. Current Management of Hyperkalemia in Non-Dialysis CKD: Longitudinal Study of Patients Receiving Stable Nephrology Care. *Nutrients* **2021**, *13*, 942. [CrossRef] [PubMed]
78. Palmer, B.F. Potassium Binders for Hyperkalemia in Chronic Kidney Disease—Diet, Renin-Angiotensin-Aldosterone System Inhibitor Therapy, and Hemodialysis. *Mayo Clin. Proc.* **2020**, *95*, 339–354. [CrossRef]
79. Murphy, D.; Ster, I.C.; Kaski, J.-C.; Anderson, L.; Banerjee, D. The LIFT Trial: Study Protocol for a Double-Blind, Randomised, Placebo-Controlled Trial of K+-Binder Lokelma for Maximisation of RAAS Inhibition in CKD Patients with Heart Failure. *BMC Nephrol.* **2021**, *22*, 254. [CrossRef]
80. Mattson, D.L. Immune Mechanisms of Salt-Sensitive Hypertension and Renal End-Organ Damage. *Nat. Rev. Nephrol.* **2019**, *15*, 290–300. [CrossRef]
81. Borrelli, S.; Provenzano, M.; Gagliardi, I.; Ashour, M.; Liberti, M.E.; De Nicola, L.; Conte, G.; Garofalo, C.; Andreucci, M. Sodium Intake and Chronic Kidney Disease. *Int. J. Mol. Sci.* **2020**, *21*, 4744. [CrossRef] [PubMed]
82. Garofalo, C.; Borrelli, S.; Provenzano, M.; De Stefano, T.; Vita, C.; Chiodini, P.; Minutolo, R.; Nicola, L.; Conte, G. Dietary Salt Restriction in Chronic Kidney Disease: A Meta-Analysis of Randomized Clinical Trials. *Nutrients* **2018**, *10*, 732. [CrossRef]
83. Chang, A.R.; Lóser, M.; Malhotra, R.; Appel, L.J. Blood Pressure Goals in Patients with CKD. *Clin. J. Am. Soc. Nephrol.* **2019**, *14*, 161–169. [CrossRef] [PubMed]
84. Kotchen, T.A.; Cowley, A.W.; Frohlich, E.D. Salt in Health and Disease—A Delicate Balance. *N. Engl. J. Med.* **2013**, *368*, 1229–1237. [CrossRef] [PubMed]
85. Kempner, W. Some Effects of the Rice Diet Treatment of Kidney Disease and Hypertension. *Bull. N. Y. Acad. Med.* **1946**, *22*, 358–370.
86. Garofalo, C.; Borrelli, S.; Pacilio, M.; Minutolo, R.; Chiodini, P.; De Nicola, L.; Conte, G. Hypertension and Prehypertension and Prediction of Development of Decreased Estimated GFR in the General Population: A Meta-Analysis of Cohort Studies. *Am. J. Kidney Dis.* **2016**, *67*, 89–97. [CrossRef]
87. Tuso, P.J.; Ismail, M.H.; Ha, B.P.; Bartolotto, C. Nutritional Update for Physicians: Plant-Based Diets. *Perm J.* **2013**, *17*, 61–66. [CrossRef]
88. Barnard, N.D.; Cohen, J.; Jenkins, D.J.A.; Turner-McGrievy, G.; Gloede, L.; Green, A.; Ferdowsian, H. A Low-Fat Vegan Diet and a Conventional Diabetes Diet in the Treatment of Type 2 Diabetes: A Randomized, Controlled, 74-Wk Clinical Trial. *Am. J. Clin. Nutr.* **2009**, *89*, 1588S–1596S. [CrossRef]
89. Chiavaroli, L.; Mirrahimi, A.; Sievenpiper, J.L.; Jenkins, D.J.A.; Darling, P.B. Dietary Fiber Effects in Chronic Kidney Disease: A Systematic Review and Meta-Analysis of Controlled Feeding Trials. *Eur. J. Clin. Nutr.* **2015**, *69*, 761–768. [CrossRef]
90. Biörklund, M.; van Rees, A.; Mensink, R.P.; Önning, G. Changes in Serum Lipids and Postprandial Glucose and Insulin Concentrations after Consumption of Beverages with β-Glucans from Oats or Barley: A Randomised Dose-Controlled Trial. *Eur. J. Clin. Nutr.* **2005**, *59*, 1272–1281. [CrossRef] [PubMed]
91. Wu, M.-J.; Chang, C.-S.; Cheng, C.-H.; Chen, C.-H.; Lee, W.-C.; Hsu, Y.-H.; Shu, K.-H.; Tang, M.-J. Colonic Transit Time in Long-Term Dialysis Patients. *Am. J. Kidney Dis.* **2004**, *44*, 322–327. [CrossRef]
92. Rebello, C.J.; O'Neil, C.E.; Greenway, F.L. Dietary Fiber and Satiety: The Effects of Oats on Satiety. *Nutr. Rev.* **2016**, *74*, 131–147. [CrossRef] [PubMed]
93. Bowden, R.G.; Wilson, R.L.; Deike, E.; Gentile, M. Fish Oil Supplementation Lowers C-Reactive Protein Levels Independent of Triglyceride Reduction in Patients with End-Stage Renal Disease. *Nutr. Clin. Pract.* **2009**, *24*, 508–512. [CrossRef]
94. Balk, E.M.; Lichtenstein, A.H.; Chung, M.; Kupelnick, B.; Chew, P.; Lau, J. Effects of Omega-3 Fatty Acids on Serum Markers of Cardiovascular Disease Risk: A Systematic Review. *Atherosclerosis* **2006**, *189*, 19–30. [CrossRef] [PubMed]
95. Kris-Etherton, P.M.; Harris, W.S.; Appel, L.J.; American Heart Association. Nutrition Committee Fish Consumption, Fish Oil, Omega-3 Fatty Acids, and Cardiovascular Disease. *Circulation* **2002**, *106*, 2747–2757. [CrossRef]
96. Sherman, R.A.; Mehta, O. Phosphorus and Potassium Content of Enhanced Meat and Poultry Products: Implications for Patients Who Receive Dialysis. *CJASN* **2009**, *4*, 1370–1373. [CrossRef] [PubMed]
97. Sullivan, C.M.; Leon, J.B.; Sehgal, A.R. Phosphorus-Containing Food Additives and the Accuracy of Nutrient Databases: Implications for Renal Patients. *J. Ren. Nutr.* **2007**, *17*, 350–354. [CrossRef] [PubMed]
98. Siener, R. Dietary Treatment of Metabolic Acidosis in Chronic Kidney Disease. *Nutrients* **2018**, *10*, 512. [CrossRef] [PubMed]
99. Kramer, H. Diet and Chronic Kidney Disease. *Adv. Nutr.* **2019**, *10*, S367–S379. [CrossRef] [PubMed]
100. Joshi, S.; Hashmi, S.; Shah, S.; Kalantar-Zadeh, K. Plant-Based Diets for Prevention and Management of Chronic Kidney Disease. *Curr. Opin. Nephrol. Hypertens.* **2020**, *29*, 16–21. [CrossRef]

101. Levey, A.S.; Adler, S.; Caggiula, A.W.; England, B.K.; Greene, T.; Hunsicker, L.G.; Kusek, J.W.; Rogers, N.L.; Teschan, P.E. Effects of Dietary Protein Restriction on the Progression of Advanced Renal Disease in the Modification of Diet in Renal Disease Study. *Am. J. Kidney Dis.* **1996**, *27*, 652–663. [CrossRef]
102. Zeller, K.; Whittaker, E.; Sullivan, L.; Raskin, P.; Jacobson, H.R. Effect of Restricting Dietary Protein on the Progression of Renal Failure in Patients with Insulin-Dependent Diabetes Mellitus. *N. Engl. J. Med.* **1991**, *324*, 78–84. [CrossRef]
103. Levey, A.S.; Greene, T.; Beck, G.J.; Caggiula, A.W.; Kusek, J.W.; Hunsicker, L.G.; Klahr, S. Dietary Protein Restriction and the Progression of Chronic Renal Disease: What Have All of the Results of the MDRD Study Shown? Modification of Diet in Renal Disease Study Group. *J. Am. Soc. Nephrol.* **1999**, *10*, 2426–2439. [CrossRef]

nutrients

MDPI

Article

Economic Evaluation of Individualized Nutritional Support for Hospitalized Patients with Chronic Heart Failure

Philipp Schuetz [1,2,*], Suela Sulo [3], Stefan Walzer [4,5,6], Sebastian Krenberger [4], Zeno Stagna [7], Filomena Gomes [8], Beat Mueller [1,2] and Cory Brunton [3]

1 Medical University Department, Kantonsspital Aarau, 5001 Aarau, Switzerland; happy.mueller@unibas.ch
2 Medical Faculty, University of Basel, 4001 Basel, Switzerland
3 Abbott Nutrition, Chicago, IL 60045, USA; suela.sulo@abbott.com (S.S.); cory.brunton@abbott.com (C.B.)
4 MArS Market Access & Pricing Strategy GmbH, 79576 Weil am Rhein, Germany; stefan.walzer@marketaccess-pricingstrategy.de (S.W.); sebastian.krenberger@marketaccess-pricingstrategy.de (S.K.)
5 Health Care Management, State University Baden-Wuerttemberg, 70174 Loerrach, Germany
6 Social Work & Health Care, University of Applied Sciences Ravensburg-Weingarten, 88250 Weingarten, Germany
7 Division of Diabetes, Endocrinology, Nutritional Medicine and Metabolism, Inselspital, Bern University Hospital, University of Bern, 4001 Bern, Switzerland; zeno.stanga@insel.ch
8 NOVA Medical School, Universidade NOVA de Lisboa, 1169-056 Lisboa, Portugal; filomenisabel@hotmail.com
* Correspondence: schuetzph@gmail.com; Fax: +41-62-838-4100

Abstract: Background Malnutrition is a highly prevalent risk factor in hospitalized patients with chronic heart failure (CHF). A recent randomized trial found lower mortality and improved health outcomes when CHF patients with nutritional risk received individualized nutritional treatment. Objective To estimate the cost-effectiveness of individualized nutritional support in hospitalized patients with CHF. Methods This analysis used data from CHF patients at risk of malnutrition (N = 645) who were part of the Effect of Early Nutritional Therapy on Frailty, Functional Outcomes and Recovery of Undernourished Medical Inpatients Trial (EFFORT). Study patients with CHF were randomized into (i) an intervention group (individualized nutritional support to reach energy, protein, and micronutrient goals) or (ii) a control group (receiving standard hospital food). We used a Markov model with daily cycles (over a 6-month interval) to estimate hospital costs and health outcomes in the comparator groups, thus modeling cost-effectiveness ratios of nutritional interventions. Results With nutritional support, the modeled total additional cost over the 6-month interval was 15,159 Swiss Francs (SF). With an additional 5.77 life days, the overall incremental cost-effectiveness ratio for nutritional support vs. no nutritional support was 2625 SF per life day gained. In terms of complications, patients receiving nutritional support had a cost savings of 6214 SF and an additional 4.11 life days without complications, yielding an incremental cost-effectiveness ratio for avoided complications of 1513 SF per life day gained. Conclusions On the basis of a Markov model, this economic analysis found that in-hospital nutritional support for CHF patients increased life expectancy at an acceptable incremental cost-effectiveness ratio.

Keywords: economic analysis; chronic heart failure; nutritional support; clinical outcomes; cost savings

Citation: Schuetz, P.; Sulo, S.; Walzer, S.; Krenberger, S.; Stagna, Z.; Gomes, F.; Mueller, B.; Brunton, C. Economic Evaluation of Individualized Nutritional Support for Hospitalized Patients with Chronic Heart Failure. *Nutrients* **2022**, *14*, 1703. https://doi.org/10.3390/nu14091703

Academic Editors: Omorogieva Ojo and Amanda R. Amorim Adegboye

Received: 18 March 2022
Accepted: 18 April 2022
Published: 20 April 2022

1. Highlights

We previously reported a reduced risk for mortality and major cardiovascular events when older hospitalized patients with chronic heart failure and malnutrition received individualized nutritional interventions compared with similar patients who consumed only a usual hospital diet. In this study, we developed a Markov model of healthcare–state transitions and costs to identify the cost-savings and incremental cost-effectiveness ratios

(ICER) of nutritional intervention. With an additional 5.77 life days, the overall ICER for nutritional support vs. no nutritional support was 2625 Swiss francs per life day gained.

2. Introduction

Chronic heart failure (CHF) has high clinical and economic costs worldwide given adverse health outcomes and increased healthcare resource utilization. Globally, HF cases exceed 60 million and account for nearly 10 million life-years lost to disability, with yearly costs estimated at nearly USD 350 billion [1,2]. The annual medical cost for a person with HF was estimated at more than USD 24,000 in the United States, although costs vary widely among individuals and are highest among those who are oldest and have co-morbidities [3]. Since HF imposes the greatest burden on older adults [1], the incidence is increasing as the population grows and ages [4].

Poor nutritional status is common among older people with HF because of multiple negative prognostic factors, such as decreased appetite and weight loss [5], impaired intestinal function [6], the presence of other comorbidities, and catabolic metabolism due to HF-related inflammation [7,8]. Malnutrition with consequent loss of muscle mass and physical functionality has been associated with increased morbidity, poorer quality of life, and worsening of CHF [9]. Nutritional strategies have long been recommended as part of treatment for CHF, but clinical studies often focus on restricting sodium intake and following specific dietary patterns for long-term cardiac health benefits, e.g., the Mediterranean and DASH diets [10,11].

Currently, many HF patients urgently need supportive nutrition care to address nutritional shortfalls and subsequent adverse consequences. Studies have reported improved health outcomes when patients with poor nutritional status receive nutritional interventions. In fact, quality improvement programs can be used across the continuum of care to enhance outcomes for people who have evidence of poor nutritional status in home-care settings, in residential nursing care [12], and during hospital admission [13–17]. An early review by Tappendan et al. found that hospital care with a focus on nutrition can reduce complication rates, length of hospital stays, readmission rates, and mortality [17]. Further, the results of a systematic review and meta-analysis of studies on hospitalized patients with malnutrition showed that nutritional interventions can significantly improve nutritional intake and reduce the risk of mortality [18]. Beyond health benefits, individualized nutritional support during and after hospitalization is also recognized as cost-saving because it spares healthcare resource utilization due to excess hospital lengths of stay, readmissions, and need for intensive care unit (ICU) admission [19–22]. In fact, the added cost of providing nutritional support is considered low, especially relative to the resultant lowered costs of hospitalization and medical treatments [20].

We previously reported results of beneficial health outcomes of nutritional intervention for at-risk patients in Swiss hospitals—a study known as Effect of Early Nutritional Therapy on Frailty, Functional Outcomes and Recovery of Undernourished Medical Inpatients Trial (EFFORT) [23]. In this study of more than 2000 medical inpatients, we found that nutritional interventions helped poorly nourished participants meet calorie and protein goals better than usual hospital food, significantly enhancing survival. When we focused the analysis on a subpopulation of EFFORT patients with CHF, we similarly found better health outcomes for the patients who were given supportive, individualized nutritional care [24]. Specifically, CHF patients at high nutritional risk had significantly reduced risk for mortality and major cardiovascular events when they received individualized nutritional interventions rather than standard hospital food [24]. In our current economic analysis of results from these vulnerable CHF patients in EFFORT, we applied a Markov model of health outcomes to predict how nutritional support would affect costs of healthcare utilization.

3. Methods

3.1. Study Design

This study was a secondary economic analysis of CHF patients who were part of EFFORT—a prospective, noncommercial, multicenter, randomized controlled trial. EFFORT was registered at ClinicalTrials.gov at https://clinicaltrials.gov/ct2/show/NCT02517476 (accessed on 7 August 2015) and conducted in eight Swiss hospitals. The overall objective of the original trial was to compare medical outcomes for patients at risk of malnutrition who were randomized to (i) an intervention group (individualized nutritional support to reach energy, protein, and micronutrient goals) or (ii) a control group (receiving usual hospital food).

Individualized nutritional support included screening patients for malnutrition risk on admission; dietitian-conducted nutritional assessment for patients identified to be at risk for malnutrition; individualized nutritional care plans developed by a dietitian; and implementation of the care plan with monitoring of health outcomes during hospitalization and follow-up post-discharge [23,25].

The rationale for the initial trial, design details, and eligibility features were previously reported [25], and the primary results of the full study were recently published [23,26], as were health outcomes in the CHF patient population [24]. The present study is based on CHF inpatients only, and it represents an analysis of healthcare costs and health outcomes in EFFORT's two comparator groups—patients who were randomized to receive individualized nutritional support (intervention group) and those who received usual hospital food (control group) [24]. EFFORT included a total of 645 patients with CHF, with 234 (36%) acutely decompensated and 411 (64%) with chronic stable HF [24].

3.2. Health Economic Terms Used

Here, we provide definitions of key health economic terms (Appendix A, Table A1) used in our report [20,27,28].

3.3. Description of Markov Simulation Model

We developed a Markov simulation model with daily cycles to analyze the economic impact of nutritional support in malnourished inpatients with CHF; the model reflected the perspective of Swiss health insurers. A modeling timeframe of six months (180 days) with five designated health states was based on findings in a recent systematic review and meta-analysis report [18]. In the present analysis, we assumed that all patients began in a stable health state—hospitalization with HF and evidence of malnutrition risk on admission (Figure 1). During hospitalization, patients could develop complications, such as myocardial infarction or arrhythmia. This complication state was modeled as an autonomous state because the probability of death is higher than for patients not experiencing in-hospital complications. Worsening CHF and complications might require transfer to the ICU. Other modeled states included discharge from the hospital and readmission for a non-elective reason. Notably, patients had different costs for care and risks of death in each state. Transition probabilities between health states were based on the outcome results for CHF patients in our full EFFORT clinical study [24]. Transition values are compiled in Table A2 of Appendix A). Raw data were taken from the original EFFORT study for the CHF population and then put manually into the simulation model via Excel.

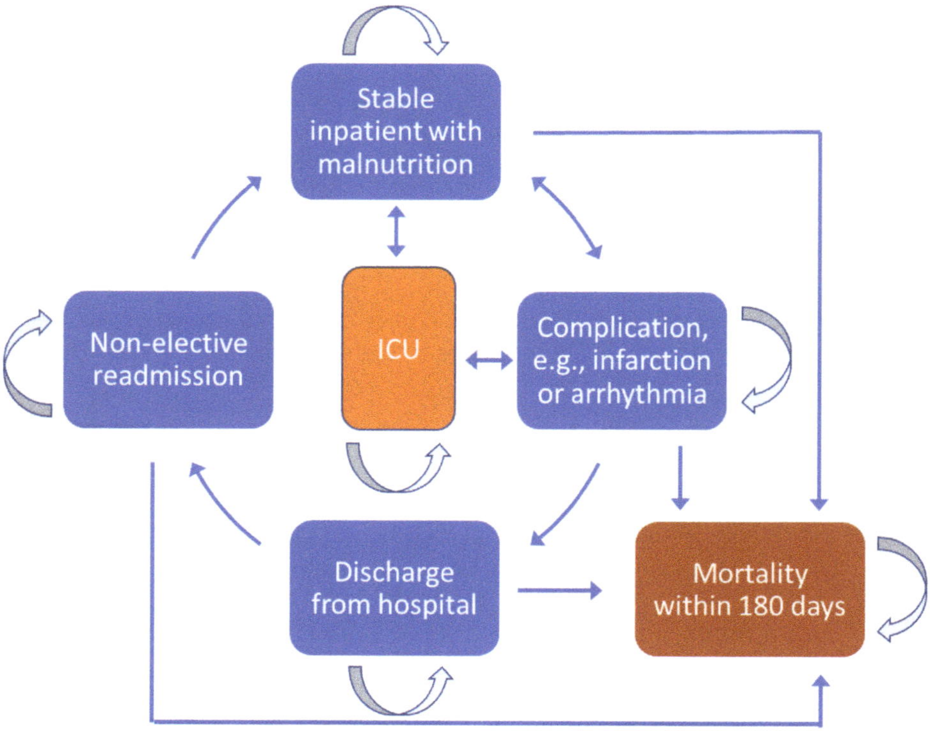

Figure 1. Health states of the Markov model. Light blue arrows represent patients staying within the given health state, while bright blue arrows represent transitions between states. Abbreviation: ICU, intensive care unit.

3.4. Patient Population

For the initial trial, we screened medical patients upon hospital admission for risk of malnutrition using the Nutritional Risk Screening (NRS) 2002 [29]. We included adult patients with a total NRS score ≥ 3 points, an expected length of stay (LOS) > 4 days, and written informed consent. We excluded patients who were treated in the intensive care or surgical units, were unable to have oral intake, or were receiving long-term nutritional support on admission; patients with terminal illnesses, gastric bypass surgery, anorexia nervosa, acute pancreatitis, acute liver failure, cystic fibrosis, stem cell transplantation; and patients previously included in the trial. All patients eligible for this secondary analysis had a documented diagnosis of CHF on hospital admission, which was confirmed and validated by a complete chart review after hospital discharge. In line with the European Society of Cardiology (ESC) guidelines [29], we stratified CHF patients, according to their ejection fraction, into three groups: (1) reduced ejection fraction (HFrEF; rEF < 40%), (2) mid-range ejection fraction (HFmrEF; mr EF 40–49%), and (3) preserved ejection fraction (HFpEF; pEF $\geq$ 50%).

Table A3 of Appendix A gives an overview of the main results from the initial report [24].

3.5. Costs and Utilities

Utility values (cost of gained effectiveness of nutritional support) were derived from a study by Schuetz et al., assuming the utility value for preventing a major cardiovascular event (MACE) was a reasonable proxy for developing a major complication (adverse event) during hospitalization [24,26]. Costs for the different health states were assumed as follows: (1) costs for nutritional inpatient support were based on the publication by Schuetz et al. 2020 [26], assuming a standard deviation of 20% of the input value, for both in- and outpatient nutritional support; (2) costs for 20% of post-discharge patients to continue nutritional supplements were based on cost data from the largest Swiss online pharmacy [30]; (3) costs for a heterogeneous distribution of cardiovascular events were estimated on the basis of the Swiss Disease-related Group (DRG) costs for severe arrhythmia and cardiac arrest [31]; (4) ICU costs were based on the Swiss DRG costs for an intensive care complex treatment [31]; and (5) no costs were assigned for death (Table 1).

Table 1. Cost input values for the health economic model with monetary costs expressed in Swiss francs (SF).

Cost Item	Cost Input, Swiss Francs (SF)	For Probabilistic Analysis		Reference
		Distribution	SD, (SF)	
Nutritional support inpatient	5	Gamma	1	ZRMB [30]
Nutritional support outpatient	5	Gamma	1	ZRMB [30]
Cost per day in normal ward	1650	Gamma	1485	BFS 2020 [32]
Cost per day in ICU	4654	Gamma	3900	DRG [31]
Average cost per complication (per day)	1513	Gamma	1477	DRG [31]

ICU: intensive care unit; SD: standard deviation; SF: Swiss francs. Costs were rounded to the nearest full unit. 1 SFCHF = 0.95 EUR.

3.6. Base-Case and Cost-Effectiveness Analyses

The primary outcomes in our model were *cost-by-health-state* and *total cost*. We calculated days in each health state and calculated utility values as the difference between the total costs of individualized nutritional support compared with no support. Because real-life findings were modeled, we did not apply any discount rates.

3.7. Sensitivity Analyses

Since costs of nutritional supplements may vary in different health states and care sites, we performed a sensitivity analysis to determine whether cost savings would be maintained when the costs of nutritional supplements were 5 SF per day (lower bound), 100 SF per day (medium bound), and 1000 SF per day (upper bound).

Further, we ran sensitivity analyses (1) assuming 50% of discharged patients would continue oral nutritional support in the outpatient setting (5 SF per day, corresponding to one oral supplement per day) and (2) assuming 100% of discharged patients would continue nutritional support in the outpatient setting (5 SF per day). We also analyzed the costs per life-year. Therefore, we extrapolated the data from 180 days to 365 days. Finally, we investigated which costs for nutritional support would still be cost-effective at a threshold of 100,000 SF per life-year.

We followed the international modeling guidelines of the ISPOR SMDM Modeling Good Research Practices Task Force [33,34] and the reporting recommendations of the Consolidated Health Economic Evaluation Reporting Standards (CHEERS) statement [35].

4. Results

4.1. Patient Outcomes

In the original analysis of the EFFORT trial, 645 patients had CHF (321 patients allocated to the intervention group and 324 patients allocated to the control group). Compared with patients in the control group, the 180-day mortality rate for patients who received nutritional support was significantly lower (85 of 321 (26.5%) vs. 102 of 324 (31.5%)) with an adjusted hazard ratio of 0.74 (95% CI: 0.55 to 0.996; $p = 0.047$) [24].

4.2. Base-Case Analyses of Cost-Effectiveness

A base-case analysis summarizes our cost results (Table 2). Here, the term 'Life days' represents the number of patient days in each health state. Utility results are shown as quality-adjusted life days (QALDs), which were calculated in the model. Finally, the calculated costs for each health state are shown. The per-patient costs for in-hospital nutritional support were estimated at 679 SF (EUR 651) per patient across the patient's hospital length of stay. In terms of costs over the 6-month timeframe of the study model, hospital care averaged 229,036 SF (EUR 219,427) per patient in the intervention group versus 213,878 SF (EUR 204,905) in the control group. These totals included costs for days in the normal ward, days in the ICU, and added costs due to complications. Ongoing nutritional support in the outpatient setting amounted to 19 SF (EUR 18) in total since 20% of the patients continued oral nutrition supplements after discharge from the hospital. Sensitivity analysis within a range of 5 SF to 1000 SF per day for nutritional supplements did not overcome the cost benefit for nutritional support at a threshold of 100,000 SF per life-year.

Table 2. Costs and cost differences by nutrition group over 180 days for HF patients in the EFFORT trial.

Cost Item	Life Days		Utilities		Cost (Swiss Francs, CHF)	
	Individualized Nutritional Support	No Nutritional Support	Individualized Nutritional Support	No Nutritional Support	Individualized Nutritional Support	No Nutritional Support
Nutrition (support)					679	–
Days in normal ward	123.84	111.24	0.25	0.23	204,342	183,544
Days in ICU	1.88	1.90	0.00	0.00	8733	8857
Complications	10.09	14.20	0.02	0.03	15,263	21,477
Post-hospital discharge life days	18.77	21.47	0.04	0.04	19	0
Total	154.58	148.81	0.31	0.30	229,036	213,878
Difference	**5.77**		**0.02**		**15,159 SF**	

ICU: intensive care unit; SF: Swiss francs. Costs were rounded to the nearest whole unit. All other data were rounded to two decimal places. 1 SF = EUR 0.95.

Incremental differences in cost, life days, and the incremental cost-effectiveness ratio (ICER) were determined (Table 3). When using nutritional support, the total cost difference over the 6-month modeling interval was 15,159 SF (EUR 14,523), which was mainly driven by increased days in a normal ward (20,798 SF) and by cost savings due to avoided complications (6214 SF). In terms of complications, patients receiving nutritional support had 4.11 more life days without complications. Given the cost savings of 6214 SF (EUR 5953) and the additional 4.11 life days, the ICER per avoided complication was 1513 SF (EUR 1450). The overall ICER for nutritional support vs. no nutritional support was 2625 SF (EUR 2515) per life day saved.

Table 3. Results for incremental differences from base-case analysis of HF patients in EFFORT.

Cost Item	Incremental Changes for Nutritional Support vs. No Nutritional Support		
	Cost, Swiss Francs (SF)	Life Days	ICER LD, SF
Day in normal ward	20,798	12.60	1650
Day in ICU	−123	−0.03	4109
Complication (AE)	−6214	−4.11	1513
Post-hospital stay, life days	19	−2.70	−7
Total	15,159	5.77	2625

AE: adverse event; ICER LD: incremental cost-effectiveness ratio per life day; ICU: intensive care unit; costs were rounded to the nearest full unit, and all other data were rounded to two decimal places. 1 SF = EUR 0.95.

4.3. Sensitivity Analyses

Even when varying input values for sensitivity analyses, findings were consistent with the original analysis (Appendix A, Table A4). When adjusting the proportion of patients continuing nutritional support after being discharged from the hospital, no relevant increases in nutrition costs could be observed. With 50% of patients receiving outpatient nutritional support, 47 SF (EUR 45) would have to be invested for 180 days, and 134 SF (EUR 128) would have to be invested for one year. With 100% of patients, those costs would amount to 94 SF (EUR 90) per 180 days and 269 SF (EUR 258) per year. We also analyzed different cost input values for nutritional support and the maximum cost input to stay under a threshold of 100,000 SF per life-year. The maximum cost input would be 6755 SF (EUR 6472) if 100% of patients continued nutritional support in the outpatient setting; 7497 SF (EUR 7182) if 50% of patients continued nutritional support as outpatients; and 8027 SF (EUR 7690) if only 20% of patients continued nutritional support as outpatients.

5. Discussion

In our prior study of hospitalized CHF patients with malnutrition (or risk of malnutrition) receiving nutritional support, we reported a significantly reduced risk for mortality and major cardiovascular events compared with CHF patients who consumed the usual hospital diet [24]. Importantly, the results of our current modeling study showed that the added cost of providing nutritional support is relatively low, especially when considering the associated reduction in risk for complications and their excess costs (extended hospitalization time and more medical treatments). Altogether, the results from our present Markov healthcare cost modeling for hospitalized CHF patients showed that nutritional care (i.e., in-hospital nutritional support continued post-discharge as needed) is a cost-effective intervention. This finding underscores the benefits of routine and robust nutritional intervention for all patients hospitalized with CHF, i.e., screening patients for malnutrition or its risk when they are admitted to the hospital, then providing nutritional support according to a dietitian-recommended, individualized plan. While the focus of our study and others was on healthcare utilization and cost, we note that such cost savings occur in the context of improved patient outcomes, especially longer survival [23].

Nutrition interventions for hospitalized patients have been established as cost-effective strategies that also yield benefits in terms of better patient outcomes, especially for older adults [36,37]. In terms of health economics, value is determined as outcomes relative to costs; in the value equation, the numerator is the outcome, while the denominator is the cost. Depending on the stakeholder's perspective, high value may be viewed as reduced patient morbidity and mortality, cost containment, or profitability [38]. All stakeholders recognize the value of better patient health outcomes.

Rising healthcare expenditures necessitate the adoption of evidence-based strategies for cost containment, especially for hospital care. The strategy of improving patients' nutritional status to improve health and cost outcomes is well-known and gaining ever-growing supportive evidence. In a recent systematic review, Galekop et al. identified 53 studies that analyzed the cost-effectiveness of personalized nutrition in patient care [39]. Nearly half of the analyses (49%) concluded that nutritional intervention was cost-effective, and 75% of the incremental cost–utility ratios were cost-effective given a willingness-to-pay threshold of USD 50,000 per quality-adjusted life-year [39]. Other researchers performed a specific value analysis on the use of nutritional support therapy to lower the risk of hospital-acquired infections (HAIs), which are life-threatening and expensive to treat [40]. On the basis of decreased HAIs and the shortened length of hospital stay among patients who were critically ill or undergoing major surgery, these researchers reported that nutritional support therapy has the potential to save the United States (US) Centers for Medicare and Medicaid Services approximately USD 104 million annually [40]. A broader Medicare Claims modeling study, the Value Project of the American Society for Enteral and Parenteral Nutrition (ASPEN), projected annual cost savings from nutritional support therapy in five selected therapeutic areas—sepsis, gastrointestinal cancer, hospital-acquired infections, surgical complications, and pancreatitis [41]. The total cost savings was estimated at USD 580 million per year [41]. Another research team conducted an economic evaluation alongside a multicenter randomized controlled clinical trial (the NOURISH Study); the study population was malnourished older patients in US hospitals [42]. Across a 90-day time horizon, nutrition therapy yielded health improvements at a cost of no more than USD 34,000 (EUR 29,800) per quality-adjusted life-year. When extending the time horizon to a patients' entire lifetime, the intervention cost only USD 524 (EUR 460) per life-year saved [42].

However, disease-associated malnutrition often remains undiagnosed and untreated. While medical nutritional support requires multidisciplinary awareness and care, Meehan and colleagues noted that hospital nurses are ideally positioned to play critical roles in nutrition—screening for malnutrition on patient admission to the hospital, monitoring for and addressing conditions that impede nutrition intake, and ensuring that prescribed nutritional interventions are delivered and administered or consumed [14]. Such nursing support in multidisciplinary nutrition care can contribute to better patient outcomes at lower costs [14].

Our economic analysis model has limitations inherent to most modeling analyses. Costs and cost savings were calculated from the perspective of the 27 hospitals included in the Gomes et al. review and meta-analysis [18]; the results may thus not be fully generalizable to other hospitals. Demographics and different levels of need for care could have influenced treatment outcomes and related costs. Populations are becoming increasingly older, and elderly patients are perceived to need more care support. However, only total costs would be influenced by this need for care. Incremental costs would remain the same, as these patients have a need for additional care independent of the nutritional intervention. In addition, concomitant and other diseases could cause additional costs and influence the outcome of CHF treatment. Further, our cost data and reported savings are calculated from the perspective of Swiss hospital payers and their reimbursement system; this model may not be generalizable to other hospitals or to the outpatient setting. The ICER of 100,000 SF used in our sensitivity analysis is hypothetical because in Switzerland, no cost-effectiveness threshold is applied in reimbursement decisions. Finally, our model uses direct costs as the main drivers of economic decision-making from the perspective of hospital administrators and payers; future models could tackle savings in cost terms important to the patients, such as faster recovery with less disability and lower loss of work productivity.

6. Conclusions

This Markov-modeled economic analysis showed that in-hospital nutritional support for chronic HF patients with malnutrition was a cost-effective strategy to improve health outcomes. Compared with other more invasive procedures, nutritional support is easy to implement in hospitals and other care settings and can help protect patients from adverse events that require cost-intensive interventions, such as 21,750 SF (EUR 20,838) for a coronary bypass or 27,818 SF (EUR 26,651) for cardiac defibrillator implants [43].

6.1. Clinical Perspective

Given the high proportion of older people with HF and at risk of malnutrition [9,44], we anticipate that patient-specific nutritional interventions can lead to substantial reductions in healthcare costs in addition to well-recognized health and mortality benefits. The evaluation of other patient-centered outcomes, such as quality of life, should also be explored in future studies.

6.2. Translational Outlook

The significant reduction in hospital complications and the associated costs in the subgroup of HF patients with established malnutrition may be particularly relevant for policymakers. We anticipate that such findings will be confirmed and extended by randomized controlled trials that specifically enroll hospitalized patients with CHF.

Author Contributions: P.S.: conceptualization, investigation, funding acquisition, original draft preparation; S.S. and C.B.: conceptualization, writing—review and editing; S.W. and S.K.: formal analysis, writing—review and editing; Z.S., F.G. and B.M.: conceptualization, investigation, writing—review and editing. All authors have read and agreed to the published version of the manuscript.

Funding: The initial trial was funded by the Swiss National Science Foundation (SNSF) (PP00P3_150531) and the Research Council of the Kantonsspital Aarau (1410.000.058 and 1410.000.044). Abbott provided a grant (HA34) to cover expenses associated with the economic analysis.

Institutional Review Board Statement: The study was conducted in accordance with the Declaration of Helsinki and approved by the Institutional Review Board (or Ethics Committee) of Northwestern part of Switzerland (EKNZ) (protocol code 2014_001 and 15.1.2014).

Informed Consent Statement: Informed consent was obtained from all subjects involved in the study. Not applicable for economic analysis.

Data Availability Statement: The data presented in this study are available on request from the corresponding author. The data are not publicly available due to privacy and ethical reasons.

Acknowledgments: We would like to thank all participating patients and hospital staff for their support of our trial. We thank Cecilia Hofmann for her assistance with manuscript review and editing.

Conflicts of Interest: The initial study was investigator-initiated and supported by a grant from the Swiss National Science Foundation to P. Schuetz (SNSF Professorship, PP00P3_150531) and the Forschungsrat of the Kantonsspital Aarau (1410.000.058 and 1410.000.044). The institution of P. Schuetz previously received unrestricted grant money unrelated to this project from Nestlé Health Science and Abbott Nutrition. The institution of Z. Stanga received speaking honoraria and research support from Nestlé Health Science, Abbott Nutrition, and Fresenius Kabi. S. Sulo and C. Brunton are employees and stockholders of Abbott. S. Walzer and S. Krenberger received funding for the model development by Abbott. S. Walzer has also received funding from Nestlé and Fresenius Kabi for health economic support. All other authors report no conflicts of interest.

Trial registration: ClinicalTrials.gov number NCT02517476.

Appendix A

Table A1. Definition of terms used in health economic analyses.

Term	Definition
Markov model	A Markov model is used to analyze systems that change on a random basis. Applied to healthcare, a Markov model assumes that a patient moves from one discrete health state to another, e.g., inpatient with malnutrition, inpatient with infectious complication, patient discharged from hospital, and patient readmitted to hospital non-electively. In modeling, the patient transitions from one state to another, with death as an irreversible state.
Base-case analysis	A base case analysis refers to the results obtained from running an economic model with the most likely or preferred set of assumptions and input values.
Cost-effectiveness	Cost-effectiveness analysis is a way to examine both the costs and health outcomes of an intervention. It compares an intervention with another intervention (or the status quo) by estimating how much it costs to gain a unit of a health outcome, such as a life-year gained or a death prevented. In healthcare, the goal is to maximize the benefit of treatment for a patient population while using resources efficiently, i.e., obtaining value for the cost.
Incremental cost-effectiveness ratio (ICER)	ICER is used to compare two different interventions in terms of the cost of gained effectiveness. ICER is computed by dividing the difference in cost of two interventions by the difference of their effectiveness, e.g., if treatment A costs 100 per patient and provides 1 quality-adjusted life day (QALD), and treatment B costs 1000 Swiss francs (SF) but provides 4 QALDs, the ICER of treatment B is 100–10 SF/4-1 = 30 SF per QALD. ICER is also called a cost-utility analysis.
Sensitivity analysis (SA)	SA is based on what happens to the dependent variable when other parameters change. It is considered a "what if" evaluation, which is used to determine the robustness of an assessment by examining the extent to which variables are affected by changes in assumptions or methods.

Table A2. Transition probabilities for the health states in the model.

	Transition Probability Per Day *					
Transition Phases	**Individualized Nutritional Support**	**Distribution**	**SD**	**No Nutritional Support**	**Distribution**	**SD**
Stable→stable	0.00418	Beta	0.00258	0.00270	Beta	0.00206
Stable→AE	0.00106	Beta	0.00099	0.00174	Beta	0.00150
Stable→ICU	0.00018	Beta	0.00019	0.00017	Beta	0.00019
Stable→Death	0.00171	Beta	0.00148	0.00210	Beta	0.00173
AE→Stable	0.00000	Beta	0.00000	0.00000	Beta	0.00000
AE→AE	0.00293	Beta	0.00222	0.00206	Beta	0.00174
AE→ICU	0.00000	Beta	0.00000	0.00013	Beta	0.00016
AE→Death	0.00493	Beta	0.00278	0.00608	Beta	0.00285
ICU→Stable	0.00000	Beta	0.00000	0.00000	Beta	0.00000
ICU→AE	0.00000	Beta	0.00000	0.00000	Beta	0.00000
ICU→ICU	0.00508	Beta	0.00270	0.00608	Beta	0.00282
ICU→Death	0.00283	Beta	0.00209	0.00225	Beta	0.00184
Stable→Release	0.00171	Beta	0.00274	0.00210	Beta	0.00279
Release→Stable	0.00233	Beta	0.00187	0.00229	Beta	0.00185
Release→Release	0.00592	Beta	0.00280	0.00601	Beta	0.00280

AE: adverse event; ICU: intensive care unit; SD: standard deviation. * Transition probabilities were calculated from day 180 relative risk. SDs were calculated on the basis of a 95% confidence interval (Clopper–Pearson confidence interval for a binomial proportion, with https://epitools.ausvet.com.au/ciproportion; accessed on 1 Septemeber 2021).

Table A3. Clinical outcomes in patients randomized to the intervention and the control group according to the original report.

Parameters	**Control Group (N = 324)**	**Intervention Group (N = 321)**	***p*-Value**	**Regression Analysis (Adjusted) (95% CI and *p*-Value)**
Outcomes				
All-cause mortality within 30 days	48 (14.8%)	27 (8.4%)	0.013	0.44 (0.26 to 0.75) $p = 0.002$
All-cause mortality within 180 days	102 (31.5%)	85 (26.5%)	0.19	0.74 (0.55 to 0.996) $p = 0.047$
MACE within 30 days	87 (26.9%)	56 (17.4%)	0.005	0.50 (0.34 to 0.75) $p = 0.001$
Admission to the intensive care unit within 30 days	10 (3.1%)	10 (3.1%)	0.96	0.97 (0.39 to 2.40) $p = 0.943$
Non-elective hospital readmission within 180 days	84 (25.9%)	92 (28.7%)	0.38	1.23 (0.86 to 1.76) $p = 0.245$
Non-elective hospital readmission within 30 days	27 (8.3%)	29 (9.0%)	0.72	1.11 (0.64 to 1.94) $p = 0.699$
Mean length of stay (days)	9.8 (6.2)	10.4 (7.1)	0.24	0.53 (−0.46 to 1.57) $p = 0.284$

Data are number of events (%). Models were adjusted for initial nutritional risk screening score and study center. Continuous values are expressed as means and SDs, categorical/binary values as absolute numbers and percentages. MACE: major cardiovascular events, containing myocardial infarction, stroke, and all-cause mortality.

Table A4. Sensitivity analysis results for ICER per life-year.

	20% of Outpatients	50% of Outpatients	100% of Outpatients
Cost Input for Outpatient Nutritional Support in Swiss Francs (SF)			
5 SF	2131 SF	2135 SF	2142 SF
100 SF	3290 SF	3376 SF	3519 SF
1000 SF	14,269 SF	15,131 SF	16,566 SF
Maximum input to remain below SF 100,000/cost-effectiveness threshold	8027 SF	7497 SF	6755 SF

ICER incremental cost-effectiveness ratio; costs were rounded to the full amount. SF 1 = EUR 0.95.

References

1. Lippi, G.; Sanchis-Gomar, F. Global epidemiology and future trends of heart failure. *AME Med. J.* **2020**, *5*, 1–6. [CrossRef]
2. Benjamin, E.J.; Muntner, P.; Alonso, A.; Bittencourt, M.S.; Callaway, C.W.; Carson, A.P.; Chamberlain, A.M.; Chang, A.R.; Cheng, S.; Das, S.R.; et al. Heart disease and stroke statistics-2019 update: A report from the American Heart Association. *Circulation* **2019**, *139*, e56–e528. [CrossRef] [PubMed]
3. Urbich, M.; Globe, G.; Pantiri, K.; Heisen, M.; Bennison, C.; Wirtz, H.S.; Tanna, G.L.D. A systematic review of medical costs associated with heart failure in the USA (2014–2020). *Pharmacoeconomics* **2020**, *38*, 1219–1236. [CrossRef] [PubMed]
4. Groenewegen, A.; Rutten, F.H.; Mosterd, A.; Hoes, A.W. Epidemiology of heart failure. *Eur. J. Heart Fail.* **2020**, *22*, 1342–1356. [CrossRef]
5. Andreae, C.; Stromberg, A.; Arestedt, K. Prevalence and associated factors for decreased appetite among patients with stable heart failure. *J. Clin. Nurs.* **2016**, *25*, 1703–1712. [CrossRef] [PubMed]
6. Rogler, G.; Rosano, G. The heart and the gut. *Eur. Heart J.* **2014**, *35*, 426–430. [CrossRef] [PubMed]
7. Zhou, H.; Qian, H. Relationship between enteral nutrition and serum levels of inflammatory factors and cardiac function in elderly patients with heart failure. *Clin. Interv. Aging* **2018**, *13*, 397–401. [CrossRef]
8. Mosterd, A.; Hoes, A.W. Clinical epidemiology of heart failure. *Heart* **2007**, *93*, 1137–1146. [CrossRef]
9. Lena, A.; Coats, A.J.S.; Anker, M.S. Metabolic disorders in heart failure and cancer. *ESC Heart Fail.* **2018**, *5*, 1092–1098. [CrossRef]
10. Billingsley, H.E.; Hummel, S.L.; Carbone, S. The role of diet and nutrition in heart failure: A state-of-the-art narrative review. *Prog. Cardiovasc. Dis.* **2020**, *63*, 538–551. [CrossRef]
11. Ishikawa, Y.; Sattler, E.L.P. Nutrition as treatment modality in heart failure. *Curr. Atheroscler. Rep.* **2021**, *23*, 13. [CrossRef] [PubMed]
12. Moick, S.; Simon, J.; Hiesmayr, M. Nutrition care quality indicators in hospitals and nursing homes: A systematic literature review and critical appraisal of current evidence. *Clin. Nutr.* **2020**, *39*, 1667–1680. [CrossRef] [PubMed]
13. Meehan, A.; Loose, C.; Bell, J.; Partridge, J.; Nelson, J.; Goates, S. Health system quality improvement: Impact of prompt nutrition care on patient outcomes and health care costs. *J. Nurs. Care Qual.* **2016**, *31*, 217–223. [CrossRef] [PubMed]
14. Meehan, A.; Partridge, J.; Jonnalagadda, S.S. Clinical and economic value of nutrition in healthcare: A nurse's perspective. *Nutr. Clin. Pract.* **2019**, *34*, 832–838. [CrossRef] [PubMed]
15. Sriram, K.; Sulo, S.; VanDerBosch, G.; Partridge, J.; Feldstein, J.; Hegazi, R.A.; Summerfelt, W.T. A comprehensive nutrition-focused Quality Improvement Program reduces 30-day readmissions and length of stay in hospitalized patients. *J. Parenter. Enter. Nutr.* **2017**, *41*, 384–391. [CrossRef] [PubMed]
16. Riley, K.; Sulo, S.; Dabbous, F.; Partridge, J.; Kozmic, S.; Landow, W.; VanDerBosch, G.; Falson, M.K.; Sriram, S. Reducing hospitalizations and costs: A home health nutrition-focused Quality Improvement Program. *J. Parenter. Enter. Nutr.* **2020**, *44*, 58–68. [CrossRef]
17. Tappenden, K.A.; Quatrara, B.; Parkhurst, M.L.; Malone, A.M.; Fanjiang, G.; Ziegler, T.R. Critical role of nutrition in improving quality of care: An interdisciplinary call to action to address adult hospital malnutrition. *J. Parenter. Enter. Nutr.* **2013**, *37*, 482–497. [CrossRef]
18. Gomes, F.; Baumgartner, A.; Bounoure, L.; Bally, M.; Deutz, N.E.; Greenwald, J.L.; Stanga, Z.; Mueller, B.; Schuetz, P. Association of nutritional support with clinical outcomes among medical inpatients who are malnourished or at nutritional risk: An updated systematic review and meta-analysis. *JAMA Netw. Open* **2019**, *2*, e1915138. [CrossRef]
19. Deutz, N.E.; Matheson, E.M.; Matarese, L.E.; Luo, M.; Baggs, G.E.; Nelson, J.L.; Hegazi, R.A.; Tappenden, K.A.; Ziegler, T.R.; NOURISH Study Group. Readmission and mortality in malnourished, older; hospitalized adults treated with a specialized oral nutritional supplement: A randomized clinical trial. *Clin. Nutr.* **2016**, *35*, 18–26. [CrossRef]
20. Schuetz, P.; Sulo, S.; Walzer, S.; Vollmer, L.; Brunton, C.; Kaegi-Braun, N.; Stanga, Z.; Mueller, B.; Gomes, F. Cost savings associated with nutritional support in medical inpatients: An economic model based on data from a systematic review of randomised trials. *BMJ Open* **2021**, *11*, e046402. [CrossRef]

21. Sulo, S.; Feldstein, J.; Partridge, J.; Schwander, B.; Sriram, K.; Summerfelt, W.T. Budget impact of a comprehensive nutrition-focused Quality Improvement Program for malnourished hospitalized patients. *Am. Health Drug Benefits* **2017**, *10*, 262–270. [PubMed]
22. Sulo, S.; Vargas, J.; Gomez, G.; Misas, J.D.; Serralde-Zuniga, A.E.; Correia, M. Hospital nutrition care informs potential cost-savings for healthcare: A budget impact analysis. *Clin. Nutr. ESPEN* **2021**, *42*, 195–200. [CrossRef] [PubMed]
23. Schuetz, P.; Fehr, R.; Baechli, V.; Geiser, M.; Deiss, M.; Gomes, F.; Kutz, A.; Tribolet, P.; Bregenzer, T.; Braun, N.; et al. Individualised nutritional support in medical inpatients at nutritional risk: A randomised clinical trial. *Lancet* **2019**, *393*, 2312–2321. [CrossRef]
24. Hersberger, L.; Dietz, A.; Burgler, H.; Bargetzi, A.; Bargetzi, L.; Kägi-Braun, N.; Tribolet, P.; Gomes, F.; Hoess, C.; Pavlicek, V.; et al. Individualized nutritional support for hospitalized patients with chronic heart failure. *J. Am. Coll. Cardiol.* **2021**, *77*, 2307–2319. [CrossRef]
25. Schuetz, P.; Fehr, R.; Baechli, V.; Geiser, M.; Gomes, F.; Kutz, A.; Tribolet, P.; Bregenzer, T.; Hoess, C.; Pavlicek, V.; et al. Design and rationale of the effect of early nutritional therapy on frailty, functional outcomes and recovery of malnourished medical inpatients trial (EFFORT): A pragmatic, multicenter, randomized-controlled trial. *Int. J. Clin. Trials* **2018**, *5*, 142–150. [CrossRef]
26. Schuetz, P.; Sulo, S.; Walzer, S.; Vollmer, Z.; Stanga, Z.; Gomes, F.; Rueda, R.; Mueller, B.; Partridge, J.; EFFORT trial collaborators. Economic evaluation of individualized nutritional support in medical inpatients: Secondary analysis of the EFFORT trial. *Clin. Nutr.* **2020**, *39*, 3361–3368. [CrossRef] [PubMed]
27. Briggs, A.; Sculpher, M. An introduction to Markov modelling for economic evaluation. *Pharmacoeconomics* **1998**, *13*, 397–409. [CrossRef]
28. Komorowski, M.; Raffa, J. Markov models and cost effectiveness analysis: Applications in medical research. In *MIT Critical Data, Editor Secondary Analysis of Electronic Health Records*; Springer International Publishing: Cham, Switzerland, 2016; pp. 351–368.
29. Ponikowski, P.; Voors, A.A.; Anker, S.D.; Bueno, H.; Cleland, J.G.F.; Coats, A.J.S.; Falk, V.; González-Juanatey, J.R.; Harjola, V.-P.; Jankowska, E.A.; et al. 2016 ESC Guidelines for the diagnosis and treatment of acute and chronic heart failure: The Task Force for the diagnosis and treatment of acute and chronic heart failure of the European Society of Cardiology (ESC). Developed with the special contribution of the Heart Failure Association (HFA) of the ESC. *Eur. J. Heart Fail.* **2016**, *18*, 891–975.
30. ZRMB MARKETPLACE AG. Zur Rose Online-Apotheke [Online]. 2021. Available online: https://www.zurrose-shop.ch/de/lp/about-us? (accessed on 18 March 2022).
31. SWISS DRG. SwissDRG-Version 10.0 SwissDRG-Version 10.0. Abrechnungsversion (2021/2021). Stand: 09.12.2020 [Online]. 2020. Available online: https://www.swissdrg.org/de/akutsomatik/swissdrg-system-1002021/fallpauschalenkatalog (accessed on 18 March 2022).
32. Bundesamt für Statistik (BFS). BFS Krankenhäuser 2019 [Online]. 2020. Available online: Extension://bfdogplmndidlpjfhoijckpakkdjkkil/pdf/viewer.html?file=https%3A%2F%2Fspitalstatistik.bagapps.ch%2Fdata%2Fdownload%2Fkzp19_publication.pdf%3Fv%3D1616491353 (accessed on 18 March 2022).
33. Caro, J.J.; Briggs, A.H.; Siebert, U.; Kuntz, K.M.; ISPOR-SMDM Modeling Good Research Practices Task Force. Modeling good research practices—Overview: A report of the ISPOR-SMDM Modeling Good Research Practices Task Force-1. *Med. Decis. Mak.* **2012**, *32*, 667–677. [CrossRef]
34. Pitman, R.; Fisman, D.; Zaric, G.S.; Postma, M.; Kretzschmar, M.; Edmunds, J.; Brisson, M.; ISPOR-SMDM Modeling Good Research Practices Task Force. Dynamic transmission modeling: A report of the ISPOR-SMDM Modeling Good Research Practices Task Force Working Group-5. *Med. Decis. Mak.* **2012**, *32*, 712–721. [CrossRef]
35. Husereau, D.; Drummond, M.; Petrou, S.; Carswell, C.; Moher, D.; Greenberg, D.; Augustovski, F.; Briggs, A.H.; Mauskopf, J.; Loder, E.; et al. Consolidated Health Economic Evaluation Reporting Standards (CHEERS) statement. *Eur. J. Health Econ.* **2013**, *14*, 367–372. [CrossRef] [PubMed]
36. Elia, M.; Normand, C.; Norman, K.; Laviano, A. A systematic review of the cost and cost effectiveness of using standard oral nutritional supplements in the hospital setting. *Clin. Nutr.* **2016**, *35*, 370–380. [CrossRef] [PubMed]
37. Stratton, R.J.; Hebuterne, X.; Elia, M. A systematic review and meta-analysis of the impact of oral nutritional supplements on hospital readmissions. *Ageing Res. Rev.* **2013**, *12*, 884–897. [CrossRef] [PubMed]
38. Sulo, S.; Gramlich, L.; Benjamin, J.; McCauley, S.; Powers, J.; Sriram, K.; Mitchell, K. Nutrition interventions deliver value in healthcare: Real-world evidence. *Nutr. Diet. Suppl.* **2020**, *12*, 139–146. [CrossRef]
39. Galekop, M.M.J.; Uyl-de Groot, C.A.; Ken Redekop, W. A systematic review of cost-effectiveness studies of interventions with a personalized nutrition component in adults. *Value Health* **2021**, *24*, 325–335. [CrossRef]
40. Bechtold, M.L.; Regunath, H.; Tyler, R.; Guenter, P.; Barrocas, A.; Collins, N.A. Impact of a nutrition support therapy on hospital-acquired infections: A value analysis. *Nutr. Clin. Pract.* **2021**, *36*, 1034–1040. [CrossRef]
41. Tyler, R.; Barrocas, A.; Guenter, P.; Torres, K.A.; Bechtold, M.L.; Chan, L.-N.; Collier, B.; Collins, N.A.; Evans, D.C.; Godamunne, K.; et al. Value of nutrition support therapy: Impact on clinical and economic outcomes in the United States. *J. Parenter. Enter. Nutr.* **2020**, *44*, 395–406. [CrossRef]
42. Zhong, Y.; Cohen, J.T.; Goates, S.; Luo, M.; Nelson, J.; Neumann, P.J. The cost-effectiveness of oral nutrition supplementation for malnourished older hospital patients. *Appl. Health Econ. Health Policy* **2017**, *15*, 75–83. [CrossRef]

43. Reisman, A.; Farrell, K.; Leitman, I. Value analysis of the costliest elective lifesaving procedures at an academic medical center. *J. Sci. Innov. Med.* **2018**, *1*, 1–10. [CrossRef]
44. Vest, A.R.; Chan, M.; Deswal, A.; Givertz, M.M.; Lekavich, C.; Lennie, T.; Litwin, S.E.; Parsly, L.; Rodgers, J.E.; Rich, M.W.; et al. Nutrition, obesity, and cachexia in patients with heart failure: A consensus statement from the Heart Failure Society of America Scientific Statements Committee. *J. Card. Fail.* **2019**, *25*, 380–400. [CrossRef]

nutrients MDPI

Article

Handgrip Strength Values Depend on Tumor Entity and Predict 180-Day Mortality in Malnourished Cancer Patients

Pascal Tribolet [1,2,3], Nina Kaegi-Braun [3], Carla Gressies [3], Annic Baumgartner [3], Karl-Heinz Wagner [2], Zeno Stanga [4] and Philipp Schuetz [3,5,*]

1 Department of Health Professions, Bern University of Applied Sciences, 3008 Bern, Switzerland; pascal.tribolet@bfh.ch
2 Department of Nutritional Sciences and Research Platform Active Ageing, University of Vienna, 1090 Vienna, Austria; karl-heinz.wagner@univie.ac.at
3 Division of General Internal and Emergency Medicine, Medical University Department, Kantonsspital Aarau, 5001 Aarau, Switzerland; nina.kaegi@ksa.ch (N.K.-B.); carla.gressies@ksa.ch (C.G.); annic.baumgartner@ksa.ch (A.B.)
4 Division of Diabetes, Endocrinology, Nutritional Medicine and Metabolism, Inselspital, Bern University Hospital, University of Bern, 4001 Bern, Switzerland; zeno.stanga@insel.ch
5 Medical Faculty, University of Basel, 4056 Basel, Switzerland
* Correspondence: schuetzph@gmail.com; Tel.: +41-62-838-95-24

Abstract: Background: Cancer-related malnutrition is a prevalent condition associated with a loss of muscle mass and impaired functional status, leading to immunodeficiency, impaired quality of life and adverse clinical outcomes. Handgrip strength (HGS) is a practical measure to assess muscle strength in individual patients during clinical practice. However, HGS reference values refer to populations of healthy people, and population-specific values, such as those in the population of cancer patients, still need to be defined. Methods: Within a secondary analysis of a previous randomized controlled nutritional trial focusing on hospitalized cancer patients at risk for malnutrition, we investigated sex-specific HGS values stratified by age and tumor entity. Additionally, we examined the association between HGS and 180-day all-cause mortality. Results: We included data from 628 cancer patients, which were collected from eight hospitals in Switzerland. Depending on the age of patients, HGS varied among female patients from 7 kg to 26 kg and among male patients from 20.5 kg to 44 kg. An incremental decrease in handgrip strength by 10 kg resulted in a 50% increase in 180-day all-cause mortality (odds ratio 1.52 (95%CI 1.19 to 1.94), $p = 0.001$). Conclusion: Our data provide evidence of the prognostic implications of HGS measurement in cancer patients and validate the prognostic value of handgrip strength in regard to long-term mortality. In addition, our results provide expected HGS values in the population of hospitalized malnourished cancer patients, which may allow better interpretation of values in individual patients.

Keywords: handgrip strength; malnutrition; cancer; nutritional support; clinical outcomes

Citation: Tribolet, P.; Kaegi-Braun, N.; Gressies, C.; Baumgartner, A.; Wagner, K.-H.; Stanga, Z.; Schuetz, P. Handgrip Strength Values Depend on Tumor Entity and Predict 180-Day Mortality in Malnourished Cancer Patients. *Nutrients* **2022**, *14*, 2173. https://doi.org/10.3390/nu14102173

Academic Editors: Omorogieva Ojo and Amanda R Amorim Adegboye

Received: 21 April 2022
Accepted: 20 May 2022
Published: 23 May 2022

1. Introduction

Malnutrition is a highly prevalent condition among oncology patients [1]. Up to 70% of cancer patients are at increased risk for malnutrition [2,3], a condition that is strongly associated with higher mortality and morbidity, functional decline, prolonged hospital stays and increased health care costs [4–9]. The pathophysiology of malnutrition in cancer patients is complex and involves different direct and indirect mechanisms, including inflammation, direct tumor effects, chemotherapy-induced effects and a decrease in appetite [10–12]. In addition, a reduced nutrient intake leads to protein and energy deficits, which in turn lead to muscle wasting and impairment of muscle strength [13,14].

Handgrip strength (HGS) measured through dynamometry is an important tool for the assessment of sarcopenia [15]. HGS has been proposed as an easy-to-use, noninvasive, objective and inexpensive tool to detect and monitor changes in nutritional status, and to predict functional decline during hospitalization and post-discharge [16–19]. HGS correlates with nutritional status and may detect changes in functional capacity in an early stage, before changes in the body composition are manifest [20]. Therefore, the use of HGS has been advocated by different international guidelines as an important adjunct in the assessment of malnutrition [21–24]. Moreover, for the population of cancer patients, studies have suggested that lower HGS is associated with higher risks for mortality and sarcopenia, as well as a decrease in quality of life (Qol) [25].

Based on published data of cancer patients, HGS reference values for this population are expected to be lower compared to healthy people [26–30], but there is a lack of data on reference values for this specific patient population. Herein, we investigated sex-specific HGS levels by tumor entity and additionally studied the prognostic information regarding 180-day all-cause mortality and other adverse outcomes from cancer patients included in the Effect of Early Nutritional Support on Frailty, Functional Outcome, and Recovery of Malnourished Medical Inpatients Trial (EFFORT) [31]. Knowledge of such data may provide health care workers with information about expected results when managing cancer patients in their clinical routine.

2. Material and Methods

2.1. Study Design and Setting

This is a secondary analysis of the EFFORT trial, which was a pragmatic, investigator-initiated, open-label, randomized controlled trial conducted in eight Swiss hospitals between April 2014 and February 2018. The original study investigated the effects of a protocol-guided individualized nutritional treatment algorithm on medical outcomes in patients at nutritional risk. The protocol and the main results [31,32], as well as several predefined secondary analyses [33–42], have been previously published. The EFFORT trial was registered with ClinicalTrials.gov, number NCT02517476.

The Ethics Committee of Northwest/Central Switzerland (EKNZ) approved the study protocol in January 2014 (EKNZ; 2014_001). The eight participating sites were secondary and tertiary care hospitals in Switzerland and included the University Clinic in Aarau, the University Hospital in Bern, the cantonal hospitals in Solothurn, Lucerne, St. Gallen, Baselland, Muensterlingen and the hospital in Lachen.

2.2. Patient Population

The methods of the trial have been previously published in detail [31,32]. In brief, for the current analysis, all patients with a cancer diagnosis (wither as the main diagnosis or a comorbidity, or both) from the original trial with available HGS measurements at the time of hospital admission were eligible. EFFORT enrolled consecutive adult patients with a Nutritional Risk Screening 2002 (NRS 2002) [43] total score ≥ 3 points, an expected length of hospital stay (LOS) ≥ 5 days and a willingness to provide informed consent. Patients initially admitted to a surgical unit or intensive care unit were excluded. Other exclusion criteria related to some diseases, including anorexia nervosa, acute pancreatitis, acute liver failure or cystic fibrosis, terminal condition, stem cell transplantation, history of gastric bypass surgery, contraindications for nutritional support, nutritional support at the time of admission and previous inclusion in the trial.

2.3. Assessment and Classification of Handgrip Strength

Grip strength data were collected at the time of admission by trained dieticians with a dynamometer (North Coast Medical Exacta™ Hydraulic Hand Dynamometer, North Coast Medical, Inc., 780 Jarvis Drive, Suite 100, Morgan Hill, CA 95037, USA [44]). The unit of measurement was kg, and measurements were performed in a seated position at the edge of the bed using the dominant hand at a 90° angle position without contacting any surface [45]. The patients performed three attempts, interrupted by a one-minute break, and the highest result was collected.

2.4. Outcomes

For this analysis, the primary endpoint was defined as 180-day all-cause mortality. We prespecified additional short-term and long-term secondary endpoints, including adverse clinical outcomes within 30 days (composite endpoint of the original trial including all-cause mortality, admission to intensive care unit (ICU), 30-day readmission rate, functional decline, length of hospital stay, non-elective hospital readmission and major complications (including nosocomial infection, respiratory failure, a major cardiovascular event and acute renal failure or gastrointestinal failure during hospitalization)) and activities of daily living assessed by Barthel Index. Further long-term secondary outcomes included QoL and incidence of falls during the 180-days follow-up period. QoL was assessed using: (a) the EuroQol Group 5-Dimension Self-Report Questionnaire (EQ-5D), which ranges from 0 to 1, with higher scores indicating better life quality, and (b) the EQ-5D visual analogue scale (VAS) from 0 to 100, with higher scores indicating better health status.

2.5. Statistical Analysis

Categorical variables are expressed as counts and percentages, and continuous variables as means and standard deviations. We performed descriptive statistics by calculating mean HGS according to tumor entities (hematological tumors, lung cancer, gastrointestinal tumors, prostate carcinoma, breast carcinoma and others (gynecological cancers, kidney and urothelial cancers, ear, nose and throat carcinoma, genital cancer, skin cancer, pleural mesothelioma and cancer of unknown primary and similar)) and age (10-year intervals) stratified by sex.

The association between sex-specific HGS and clinical outcome was investigated using logistic regression analyses for categorical variables with reporting of odd ratios (ORs) and linear regression for continuous variables with reporting of coefficients (Coef) and 95% confidence intervals (CI). We adjusted the results for important confounders (sex, age, weight, height, NRS 2002 score, center), several main diagnoses (cardiovascular, infectious, renal, frailty), various comorbidities (hypertension, chronic kidney disease, chronic heart failure, diabetes mellitus) and for randomization group.

All statistical analyses were performed with STATA 15.1 (Stata Corp, College Station, TX, USA). A p value < 0.05 (for a 2-sided test) was considered to indicate statistical significance.

3. Results

3.1. Patient Cohort

Of the initial population of 2028 patients included in the original trial recruiting patients from April 2014 to February 2018, we had complete data for 628 (368 male and 260 female) cancer patients from eight hospitals in Switzerland. The baseline characteristics for all patients included in this analysis, stratified according to sex, are shown in Table 1. Patients had a mean age of 72 years (±12.5), and 41.4% were females. The mean (SD) BMI was 24.6 (±4.8) with a similar distribution in both sexes, and the most common admission diagnosis was cancer (50.8%), followed by infection (21.0%). Patients had a high burden of comorbidities, including hypertension (49.8%), chronic renal disease (30.3%), coronary heart disease (24.8%), diabetes mellitus (19.9%) and chronic heart failure (11.6%). The most frequent types of cancer were hematological tumors (19.7%), lung cancer (16.4%), and gastrointestinal tumors (12.4%).

Table 1. Baseline characteristics of malnourished cancer patients.

	Overall	Female	Male
n	628	260	368
Sociodemographic			
Age (years), mean (SD)	72.0 (12.5)	72.3 (11.5)	71.9 (13.2)
Nutritional status			

Table 1. *Cont.*

	Overall	Female	Male
BMI (kg/m^2), mean (SD)	24.6 (4.8)	24.6 (5.4)	24.6 (4.3)
Weight (kg), mean (SD)	70.7 (14.9)	65.3 (14.0)	74.4 (14.5)
Height (cm), mean (SD)	168.6 (8.8)	162.2 (6.5)	173.1 (7.2)
NRS 3	175 (27.9%)	72 (27.7%)	103 (28.0%)
NRS 4	200 (31.8%)	85 (32.7%)	115 (31.3%)
NRS 5	253 (40.3%)	103 (39.6%)	150 (40.8%)
Main diagnosis			
Cancer	319 (50.8%)	141 (54.2%)	178 (48.4%)
Infection	132 (21.0%)	44 (16.9%)	88 (23.9%)
Cardiovascular	34 (5.4%)	16 (6.2%)	18 (4.9%)
Frailty	45 (7.2%)	21 (8.1%)	24 (6.5%)
Lung	22 (3.5%)	7 (2.7%)	15 (4.1%)
Gastrointestinal	29 (4.6%)	15 (5.8%)	14 (3.8%)
Neurological/psychiatric	13 (2.1%)	5 (1.9%)	8 (2.2%)
Renal	11 (1.8%)	3 (1.2%)	8 (2.2%)
Metabolic	6 (1.0%)	3 (1.2%)	3 (0.8%)
Other	12 (1.9%)	4 (1.5%)	8 (2.2%)
Comorbidities			
Tumor	580 (92.4%)	237 (91.2%)	343 (93.2%)
Hypertension	313 (49.8%)	137 (52.7%)	176 (47.8%)
Chronic kidney disease (without kidney replacement therapy)	190 (30.3%)	71 (27.3%)	119 (32.3%)
Coronary heart disease	156 (24.8%)	47 (18.1%)	109 (29.6%)
Diabetes mellitus	125 (19.9%)	49 (18.8%)	76 (20.7%)
Chronic heart failure	73 (11.6%)	23 (8.8%)	50 (13.6%)
Chronic obstructive pneumopathypulmonary disease	70 (11.1%)	24 (9.2%)	46 (12.5%)
Peripheral arterial vascular disease	43 (6.8%)	13 (5.0%)	30 (8.2%)
Stroke	39 (6.2%)	10 (3.8%)	29 (7.9%)
Dementia	14 (2.2%)	6 (2.3%)	8 (2.2%)
Tumor entity			
Hematological tumors	124 (19.7%)	53 (20.4%)	71 (19.3%)
Lung cancer	103 (16.4%)	29 (11.2%)	74 (20.1%)
Gastrointestinal tumors	78 (12.4%)	30 (11.5%)	48 (13.0%)
Prostate carcinoma	62 (9.9%)		62 (16.8%)
Breast carcinoma	56 (8.9%)	55 (21.2%)	1 (0.3%)
Other *	205 (32.6%)	93 (35.8%)	112 (30.4%)
Handgrip strength (kg), mean (SD)			
Overall HGS	23.6 (10.7)	17.3 (6.3)	28.0 (10.8)

* Gynecological cancers, kidney and urothelial cancers, ear, nose and throat carcinoma, genital cancer, skin cancer, pleural mesothelioma and cancer of unknown primary.

3.2. Handgrip Measurement in the Study Population

The overall mean (SD) HGS was 23.6 (±10.7 kg) with lower values in females (17.3 ± 6.3) compared to males (28.0 ± 10.8). Age, tumor entity and sex-specific HGS data are presented in Table 2. With higher age, the mean (SD) HGS decreased. In younger male cancer patients (<50 years), the mean (SD) HGS was 45.1 kg (±12.7 kg), while in patients ≥ 90 years, there was a mean (SD) HGS of 19.5 kg (±8.0 kg). For female cancer patients, the mean (SD) HGS values were lower, ranging from 23.1 kg (±8.9 kg) in young patients to 8.8 kg (±4.8 kg) in the oldest age group (≥90 years). Stratified by tumor entity, lung cancer patients had the highest mean HGS with 27.4 kg (±10 kg), which was consistent in both sexes (male: mean HGS of 30.8 kg (±9.6 kg), female: mean HGS of 18.9 kg (±4.5 kg)). In the female population, the lowest mean HGS was found in gastrointestinal tumor patients: 16.4 kg (±6.0 kg), whereas male patients had the lowest HGS with prostate carcinoma: 23.6 (±7.4 kg).

Table 2. Handgrip strength according to tumor entity and age.

	Overall (n = 628)			Female (n = 260)			Male (n = 368)		
Age (year)	n	**HGS Mean (kg) (SD)**	p	n	**HGS Mean (kg) (SD)**	p	n	**HGS Mean (kg) (SD)**	p
<50	30	38.5 (15.4)	<0.001	21	23.1 (8.9)	<0.001	9	45.1 (12.7)	<0.001
50–59	66	29.6 (10.3)		40	23.2 (6.0)		26	33.7 (10.5)	
60–69	119	24.4 (9.9)		66	19.0 (6.6)		53	28.8 (10.0)	
70–79	233	23.3 (8.9)		135	17.3 (6.3)		98	27.6 (8.0)	
80–89	146	19.6 (9.2)		84	14.5 (5.3)		62	23.4 (9.6)	
≥90	34	15.7 (8.7)		22	8.8 (4.8)		12	19.5 (8.0)	
Tumor entity									
Hematological tumors	124	23.1 (11.0)	<0.001	53	18.3 (7.0)	0.48	71	26.7 (12.1)	0.002
Lung cancer	103	27.4 (10)		29	18.9 (4.5)		74	30.8 (9.6)	
Gastrointestinal tumors	78	24.1 (11.3)		30	16.4 (6.0)		48	28.9 (11.2)	
Prostate carcinoma	62	23.6 (7.4)		-	-		62	23.6 (7.4)	
Breast carcinoma	56	17 (17.4)		55	16.9 (7.4)		1	18	
Other *	205	23.7 (11.5)		93	16.9 (7.4)		112	29.3 (11.3)	

Abbreviations: HGS, handgrip strength; SD, standard deviation. * Gynecological cancers, kidney and urothelial cancers, ear, nose and throat carcinoma, genital cancer, skin cancer, pleural mesothelioma and cancer of unknown primary.

3.3. Association of Handgrip with Adverse Outcomes

In a second step, we investigated the prognostic value of HGS in this population of cancer patients stratified by sex (Table 3). In our overall adjusted statistical model, a 10 kg decrease in HGS was associated with a 50% increase in the risk of 180-day all-cause mortality (adjusted OR 1.52 (95% CI 1.19 to 1.94), p = 0.001). The effect was similar among male and female patients, but the association was only significant in male patients (adjusted OR 1.59 (95% CI 1.19 to 2.12), p = 0.002 vs. adjusted OR 1.54 (95% CI 0.89 to 2.65), p = 0.122 for female patients).

Table 3. Association of handgrip strength with short- and long-term outcomes stratified by sex.

	HGS Mean (SD), Patients with No Event	HGS Mean (SD), Patients with Event	HGS Decrease Cont (−10 kg)	HGS Decrease Cont (−10 kg)
All patients			Unadjusted OR or Coef (95% CI), *p*-value	* Adjusted OR or Coef (95% CI), *p*-value
Primary endpoint				
180-day all-cause mortality	24.42 (11.13)	22.35 (9.99)	1.2 (1.03 to 1.41) $p = 0.019$	1.52 (1.19 to 1.94), $p = 0.001$
Short-term endpoints (30 days)				
All-cause mortality	23.81 (10.65)	22.15 (11.37)	1.16 (0.92 to 1.48) $p = 0.211$	1.59 (1.13 to 2.22), $p = 0.007$
Adverse outcome	23.82 (10.73)	23.17 (10.76)	1.06 (0.9 to 1.24) $p = 0.481$	1.23 (0.98 to 1.54), $p = 0.077$
Admission to the intensive care unit	23.71 (10.8)	19 (6.1)	1.64 (0.89 to 3.01) $p = 0.114$	2.58 (1.08 to 6.16), $p = 0.033$
Non-elective hospital readmission	23.42 (10.71)	25.1 (10.92)	0.87 (0.7 to 1.08) $p = 0.211$	0.84 (0.61 to 1.15), $p = 0.283$
Any major complication	23.87 (10.85)	20.53 (8.84)	1.39 (1.02 to 1.89) $p = 0.038$	1.65 (1.09 to 2.51), $p = 0.018$
Decline in functional status of ≥10% *	23.75 (10.5)	22.93 (11.9)	1.08 (0.88 to 1.31) $p = 0.475$	1.18 (0.89 to 1.58), $p = 0.254$
Mean length of stay (days)	-	-	0.22 (−0.29 to 0.73) $p = 0.398$	0.65 (−0.08 to 1.37), $p = 0.081$
Mean Barthel Index score (points)	-	-	−1.69 (−2.48 to −0.9) $p < 0.001$	−1.44 (−2.56 to −0.33), $p = 0.011$
Long-term endpoints (180 days)				
Mean EQ-5D VAS (points)	-	-	−0.81 (−2.87 to 1.25) $p = 0.442$	−1.2 (−4.14 to 1.75), $p = 0.425$
Mean EQ-5D index (points)	-	-	−0.02 (−0.03 to 0) $p = 0.027$	−0.01 (−0.03 to 0.01), $p = 0.363$
Incidence of one or more falls	23.84 (10.69)	20.37 (10.53)	1.41 (1.04 to 1.91) $p = 0.027$	1.58 (1.02 to 2.46), $p = 0.04$
Female patients				
Primary endpoint				
180-day all-cause mortality	18.14 (7.08)	15.9 (6.31)	1.62 (1.11 to 2.37) $p = 0.013$	1.54 (0.89 to 2.65), $p = 0.122$
Short-term endpoints (30 days)				
All-cause mortality	17.68 (6.79)	14.33 (7.17)	2.05 (1.12 to 3.74) $p = 0.02$	2.26 (1.03 to 4.95), $p = 0.041$
Adverse outcome	17.39 (6.77)	17.24 (7.22)	1.03 (0.7 to 1.52) $p = 0.876$	1.31 (0.8 to 2.15), $p = 0.275$
Admission to the intensive care unit	17.32 (6.94)	18.43 (4.83)	0.79 (0.26 to 2.37) $p = 0.673$	1.33 (0.3 to 5.83), $p = 0.704$
Non-elective hospital readmission	17.05 (6.89)	19.63 (6.55)	0.57 (0.32 to 1.01) $p = 0.055$	0.75 (0.37 to 1.55), $p = 0.444$
Any major complication	17.4 (6.84)	16.63 (7.59)	1.18 (0.6 to 2.32) $p = 0.638$	1.55 (0.67 to 3.57), $p = 0.304$

Table 3. *Cont.*

	HGS Mean (SD), Patients with No Event	HGS Mean (SD), Patients with Event	HGS Decrease Cont (−10 kg)	HGS Decrease Cont (−10 kg)
Decline in functional status of ≥10%	17.66 (6.84)	15.51 (6.98)	1.58 (0.95 to 2.62) $p = 0.076$	1.23 (0.64 to 2.39), $p = 0.532$
Mean length of stay (days)	-	-	0.33 (−0.88 to 1.53) $p = 0.596$	0.43 (−1.06 to 1.92), $p = 0.569$
Mean Barthel Index score (points)	-	-	−2.89 (−4.92 to −0.86) $p = 0.005$	−2.44 (−4.94 to 0.06), $p = 0.056$
Long-term endpoints (180 days)				
Mean EQ-5D VAS (points)	-	-	−2.91 (−7.42 to 1.6) $p = 0.204$	−1.47 (−6.89 to 3.95), $p = 0.592$
Mean EQ-5D index (points)	-	-	−0.05 (−0.09 to −0.01) $p = 0.013$	−0.04 (−0.09 to 0.01), $p = 0.084$
Incidence of one or more falls	17.59 (6.84)	13.56 (6.13)	2.38 (1.14 to 4.95) $p = 0.021$	3.57 (1.36 to 9.41), $p = 0.01$
Male patients				
Primary endpoint				
180-day all-cause mortality	29.33 (11.26)	26.23 (9.79)	1.32 (0.01 to 1.63) $p = 2.69$	1.59 (1.19 to 2.12), $p = 0.002$
Short-term endpoints (30 days)				
All-cause mortality	28.29 (10.73)	26.38 (11.01)	1.19 (0.25 to 1.61) $p = 1.14$	1.61 (1.09 to 2.38), $p = 0.016$
Adverse outcome	28.54 (10.66)	27 (10.96)	1.15 (0.2 to 1.41) $p = 1.28$	1.18 (0.91 to 1.55), $p = 0.218$
Admission to the intensive care unit	28.18 (10.77)	19.67 (7.76)	2.77 (0.05 to 7.73) $p = 1.95$	4.28 (0.83 to 22.16), $p = 0.083$
Non-elective hospital readmission	27.92 (10.65)	29.02 (11.78)	0.91 (0.53 to 1.22) $p = -0.62$	0.79 (0.54 to 1.15), $p = 0.217$
Any major complication	28.47 (10.83)	23.09 (8.77)	1.76 (0.01 to 2.71) $p = 2.58$	1.61 (0.95 to 2.73), $p = 0.08$
Decline in functional status of ≥10% *	28.24 (10.47)	27.14 (12.09)	1.1 (0.45 to 1.42) $p = 0.76$	1.19 (0.86 to 1.66), $p = 0.297$
Mean length of stay (days)	-	-	0.47 (0.17 to 1.15) $p = 1.38$	0.59 (−0.28 to 1.46), $p = 0.182$
Mean Barthel Index score (points)	-	-	−1.67 (0 to −0.69) $p = -3.35$	−0.96 (−2.2 to 0.29), $p = 0.132$
Long-term endpoints (180 days)				
Mean EQ-5D VAS (points)	-	-	−1.27 (0.39 to 1.63) $p = -0.86$	−0.45 (−4.18 to 3.28), $p = 0.813$
Mean EQ-5D index (points)	-	-	−0.01 (0.47 to 0.01) $p = -0.73$	0 (−0.03 to 0.02), $p = 0.810$
Incidence of one or more falls	28.42 (10.69)	23.78 (10.67)	1.61 (0.02 to 2.37) $p = 2.4$	1.29 (0.78 to 2.11), $p = 0.32$

Abbreviations: OR, odds ratio; Coef, coefficient; SD, standard deviation; EQ-5D, EuroQol Group 5-Dimension Self-Report Questionnaire; HGS, handgrip strength; VAS, visual analogue scale. * Adjusted for randomization, age, weight, height, NRS 2002, center, main diagnosis (cardiovascular, infection, renal disease, failure to thrive) and comorbidities (hypertension, chronic kidney failure, chronic heart failure, diabetes mellitus) and for randomization group.

We also found significant associations between HGS and other clinical endpoints, namely 30-day all-cause mortality (adjusted OR 1.59 (95% CI 1.13 to 2.22), p = 0.007), admission to ICU (adjusted OR 2.58 (95% CI 1.08 to 6.16), p = 0.033), major complications (adjusted OR 1.65 (95% CI 1.09 to 2.51), p = 0.018), mean Barthel Index score (points) (adjusted Coef −1.44 (95% CI-2.56 to −0.33), p = 0.011) and incidence of one or more falls within 180 days (adjusted OR 1.58 (1.02 to 2.46), p = 0.04).

4. Discussion

This secondary analysis of a large, randomized trial found HGS to be highly predictive of long- and short-term mortality and other adverse outcomes among cancer patients. In line with previous research, HGS values depended on sex and patient age [26,27,36], but we additionally found important differences among different types of cancers. Our data provide important HGS reference values for the specific population of cancer patients, which may help future counseling of patients and interpretation of HGS results.

Through the additional stratification by tumor entity, a more precise assessment of functional status and muscle strength via HGS measurements is possible within the population of malnourished cancer patients. It may help to better understand the value of a single HGS measurement in an individual cancer patient and puts this measurement in the perspective of what is expected of the specific population. For this reason, our HGS data from a large, randomized controlled trial may contribute to a better classification of the functional status of malnourished cancer patients by HGS values. Nevertheless, our analysis is still limited by sample size within the different tumor entities, and larger studies would be useful to provide better estimates. Further investigations should also focus on the predictive value of HGS in different tumor entities stratified by age and sex.

Our analysis also showed a significant association of HGS in cancer patients and different clinical outcomes, such as all-cause mortality within 180 days, which is consistent with findings from our research group including other patient populations [36]. In fact, an incremental decrease in HGS by 10 kg resulted in more than doubling the risk for 180-d all-cause mortality among all tumor entities and sexes. These results persisted after adjustment for important cofounders, such as randomization, age, weight, height, NRS 2002, main diagnosis and comorbidities. These findings underline the prognostic value of HGS in malnourished cancer patients and are consistent with results from studies that also included patients with different diseases [36,41,46,47]. Additionally, our results show that, in the overall population, there is a significant association between the Barthel Index score and a decrease in HGS. As both are instruments for assessing functional status, this is an expected result, which was not stable in sex-specific subgroups. A decrease in HGS by 10 kg was also associated with other short-term endpoints, as shown in Table 3.

While the role of nutrition in cancer patients has received little attention, several studies observed a strong increase in mortality in patients with higher nutritional risk [41,48–50]. Indeed, patients with an NRS of ≥5 points had a 19% higher risk of long-term mortality compared to those with 3 points in a previous analysis [41]. Our data now suggest that, in addition to clinical information about weight and low appetite included in the NRS score, HGS is an additional parameter that helps providers understand the risk of a patient and may also help with decisions regarding the start of nutritional interventions. Importantly, we recently found that HGS was also predictive for treatment response, with patients in the lowest HGS ranges showing the best response rates [36].

The present analysis has several strengths worth mentioning, including the rather large population of patients with different types of cancer and the prospective gathering of data as part of the EFFORT trial [31,32]. High adherence to the study protocol in the main trial further increases the value of data collected. Limitations include the secondary analysis with limited power and the exploratory nature of our analyses, with the risk for model overfitting and type I error. Validation of our results is thus necessary. Further, we did not include critically ill and surgical cancer patients, which makes our findings only applicable to medical cancer patients. Since we had only limited information about the CKD stages, we did not consider a further stratification.

5. Conclusions

Our data provide evidence about the prognostic implications of HGS measurement in cancer patients and validate the prognostic value of HGS in regard to long-term mortality. In addition, our results provide expected HGS values in the population of hospitalized malnourished cancer patients, which may allow better interpretation of values in individual patients.

Author Contributions: Conceptualization, P.S., P.T. and N.K.-B.; Methodology, P.S., P.T. and N.K.-B.; Formal Analysis, P.S., P.T. and N.K.-B.; Investigation, P.S., P.T. and N.K.-B.; Data Curation, P.S., P.T. and N.K.-B.; Writing—Original Draft Preparation, P.T., C.G., N.K.-B., A.B., K.-H.W., Z.S. and P.S.; Writing—Review and Editing, P.T., N.K.-B. and C.G.; Visualization, P.T.; Supervision, P.S. and K.-H.W.; Project Administration, P.S.; Funding Acquisition, P.S. and Z.S. All authors have read and agreed to the published version of the manuscript.

Funding: This research was funded by the Swiss National Science Foundation (SNSF professorship, PP00P3_150531 and PP00P3_176972) and the Research Council of the Kantonsspital Aarau, Switzerland (1410.000.058 and 1410.000.044).

Institutional Review Board Statement: The study was conducted according to the guidelines of the Declaration of Helsinki and approved by the Ethics Committee Northwest/Central Switzerland (EKNZ) in January 2014 (EKNZ; 2014_001).

Informed Consent Statement: Informed consent was obtained from all subjects involved in the study.

Data Availability Statement: The datasets used and analyzed during the current study are available from the corresponding author on reasonable request.

Acknowledgments: We thank all patients and hospital staff for support of our trial.

Conflicts of Interest: Unrelated to this trial, Nestlé Health Science and Abbott Nutrition previously provided unrestricted grant money to the institution of P.S. The institution of Z.S. received research support from Nestlé Health Science, Abbott Nutrition, B. Braun and Fresenius Kabi. All other authors report no conflict of interest.

References

1. Schuetz, P.; Seres, D.; Lobo, D.N.; Gomes, F.; Kaegi-Braun, N.; Stanga, Z. Management of disease-related malnutrition for patients being treated in hospital. *Lancet* **2021**, *398*, 1927–1938. [CrossRef]
2. Ryan, A.M.; Power, D.G.; Daly, L.; Cushen, S.J.; Ni Bhuachalla, E.; Prado, C.M. Cancer-associated malnutrition, cachexia and sarcopenia: The skeleton in the hospital closet 40 years later. *Proc. Nutr. Soc.* **2016**, *75*, 199–211. [CrossRef] [PubMed]
3. Muscaritoli, M.; Lucia, S.; Farcomeni, A.; Lorusso, V.; Saracino, V.; Barone, C.; Plastino, F.; Gori, S.; Magarotto, R.; Carteni, G.; et al. Prevalence of malnutrition in patients at first medical oncology visit: The PreMiO study. *Oncotarget* **2017**, *8*, 79884–79896. [CrossRef] [PubMed]
4. Imoberdorf, R.; Meier, R.; Krebs, P.; Hangartner, P.J.; Hess, B.; Staubli, M.; Wegmann, D.; Ruhlin, M.; Ballmer, P.E. Prevalence of undernutrition on admission to Swiss hospitals. *Clin. Nutr.* **2010**, *29*, 38–41. [CrossRef] [PubMed]
5. Felder, S.; Lechtenboehmer, C.; Bally, M.; Fehr, R.; Deiss, M.; Faessler, L.; Kutz, A.; Steiner, D.; Rast, A.C.; Laukemann, S.; et al. Association of nutritional risk and adverse medical outcomes across different medical inpatient populations. *Nutrition* **2015**, *31*, 1385–1393. [CrossRef] [PubMed]
6. Khalatbari-Soltani, S.; Marques-Vidal, P. The economic cost of hospital malnutrition in Europe; a narrative review. *Clin. Nutr. ESPEN* **2015**, *10*, e89–e94. [CrossRef] [PubMed]
7. Planas, M.; Alvarez-Hernandez, J.; Leon-Sanz, M.; Celaya-Perez, S.; Araujo, K.; Garcia de Lorenzo, A.; PREDyCES®Researchers. Prevalence of hospital malnutrition in cancer patients: A sub-analysis of the PREDyCES(R) study. *Support. Care Cancer* **2016**, *24*, 429–435. [CrossRef]
8. Lis, C.G.; Gupta, D.; Lammersfeld, C.A.; Markman, M.; Vashi, P.G. Role of nutritional status in predicting quality of life outcomes in cancer–a systematic review of the epidemiological literature. *Nutr. J.* **2012**, *11*, 27. [CrossRef]
9. Hebuterne, X.; Lemarie, E.; Michallet, M.; de Montreuil, C.B.; Schneider, S.M.; Goldwasser, F. Prevalence of malnutrition and current use of nutrition support in patients with cancer. *J. Parenter. Enter. Nutr.* **2014**, *38*, 196–204. [CrossRef]
10. Anker, S.D.; Laviano, A.; Filippatos, G.; John, M.; Paccagnella, A.; Ponikowski, P.; Schols, A.M.; Espen. ESPEN Guidelines on Parenteral Nutrition: On cardiology and pneumology. *Clin. Nutr.* **2009**, *28*, 455–460. [CrossRef]

11. Prado, C.M.; Anker, S.D.; Coats, A.J.S.; Laviano, A.; von Haehling, S. Nutrition in the spotlight in cachexia, sarcopenia and muscle: Avoiding the wildfire. *J. Cachexia Sarcopenia Muscle* **2021**, *12*, 3–8. [CrossRef] [PubMed]
12. Bargetzi, L.; Bargetzi, M.; Laviano, A.; Stanga, Z.; Schuetz, P. Inflammation reduces the effect of nutritional therapy on clinical outcomes in cancer patients. *Ann. Oncol.* **2021**, *32*, 1451–1452. [CrossRef] [PubMed]
13. Schutz, P.; Bally, M.; Stanga, Z.; Keller, U. Loss of appetite in acutely ill medical inpatients: Physiological response or therapeutic target? *Swiss Med. Wkly.* **2014**, *144*, w13957. [CrossRef] [PubMed]
14. Casaer, M.P.; Van den Berghe, G. Nutrition in the acute phase of critical illness. *N. Engl. J. Med.* **2014**, *370*, 1227–1236. [CrossRef]
15. Steemburgo, T.; Averbuch, N.C.; Belin, C.H.S.; Behling, E.B. Hand Grip Strength and nutritional status in hospitalized oncological patients. *Rev. Nutr.* **2018**, *31*, 489–499.
16. Humphreys, J.; de la Maza, P.; Hirsch, S.; Barrera, G.; Gattas, V.; Bunout, D. Muscle strength as a predictor of loss of functional status in hospitalized patients. *Nutrition* **2002**, *18*, 616–620. [CrossRef]
17. Norman, K.; Stobaus, N.; Gonzalez, M.C.; Schulzke, J.D.; Pirlich, M. Hand grip strength: Outcome predictor and marker of nutritional status. *Clin. Nutr.* **2011**, *30*, 135–142. [CrossRef]
18. Veronese, N.; Stubbs, B.; Punzi, L.; Soysal, P.; Incalzi, R.A.; Saller, A.; Maggi, S. Effect of nutritional supplementations on physical performance and muscle strength parameters in older people: A systematic review and meta-analysis. *Ageing Res. Rev.* **2019**, *51*, 48–54. [CrossRef]
19. Olguin, T.; Bunout, D.; de la Maza, M.P.; Barrera, G.; Hirsch, S. Admission handgrip strength predicts functional decline in hospitalized patients. *Clin. Nutr. ESPEN* **2017**, *17*, 28–32. [CrossRef]
20. Pereira, A.A.C.; Zaia, R.D.; Souza, G.H.G.; Luizeti, B.O.; Andreola, R.; Junior, A.O.V.; Ferrari, A. The Correlation between Hand Grip Strength and Nutritional Variables in Ambulatory Cancer Patients. *Nutr. Cancer* **2021**, *73*, 221–229. [CrossRef]
21. Dent, E.; Morley, J.E.; Cruz-Jentoft, A.J.; Arai, H.; Kritchevsky, S.B.; Guralnik, J.; Bauer, J.M.; Pahor, M.; Clark, B.C.; Cesari, M.; et al. International Clinical Practice Guidelines for Sarcopenia (ICFSR): Screening, Diagnosis and Management. *J. Nutr. Health Aging* **2018**, *22*, 1148–1161. [CrossRef] [PubMed]
22. Cruz-Jentoft, A.J.; Bahat, G.; Bauer, J.; Boirie, Y.; Bruyere, O.; Cederholm, T.; Cooper, C.; Landi, F.; Rolland, Y.; Sayer, A.A.; et al. Sarcopenia: Revised European consensus on definition and diagnosis. *Age Ageing* **2019**, *48*, 16–31. [CrossRef] [PubMed]
23. Cederholm, T.; Jensen, G.L.; Correia, M.; Gonzalez, M.C.; Fukushima, R.; Higashiguchi, T.; Baptista, G.; Barazzoni, R.; Blaauw, R.; Coats, A.; et al. GLIM criteria for the diagnosis of malnutrition—A consensus report from the global clinical nutrition community. *Clin. Nutr.* **2019**, *38*, 1–9. [CrossRef] [PubMed]
24. Gomes, F.; Schuetz, P.; Bounoure, L.; Austin, P.; Ballesteros-Pomar, M.; Cederholm, T.; Fletcher, J.; Laviano, A.; Norman, K.; Poulia, K.A.; et al. ESPEN guidelines on nutritional support for polymorbid internal medicine patients. *Clin. Nutr.* **2018**, *37*, 336–353. [CrossRef]
25. Kilgour, R.D.; Vigano, A.; Trutschnigg, B.; Lucar, E.; Borod, M.; Morais, J.A. Handgrip strength predicts survival and is associated with markers of clinical and functional outcomes in advanced cancer patients. *Support. Care Cancer* **2013**, *21*, 3261–3270. [CrossRef]
26. Steiber, N. Strong or Weak Handgrip? Normative Reference Values for the German Population across the Life Course Stratified by Sex, Age, and Body Height. *PLoS ONE* **2016**, *11*, e0163917. [CrossRef]
27. Werle, S.; Goldhahn, J.; Drerup, S.; Simmen, B.R.; Sprott, H.; Herren, D.B. Age- and gender-specific normative data of grip and pinch strength in a healthy adult Swiss population. *J. Hand Surg. Eur. Vol.* **2009**, *34*, 76–84. [CrossRef]
28. Wong, S.L. Grip strength reference values for Canadians aged 6 to 79: Canadian Health Measures Survey, 2007 to 2013. *Health Rep.* **2016**, *27*, 3–10.
29. Amaral, C.A.; Amaral, T.L.M.; Monteiro, G.T.R.; Vasconcellos, M.T.L.; Portela, M.C. Hand grip strength: Reference values for adults and elderly people of Rio Branco, Acre, Brazil. *PLoS ONE* **2019**, *14*, e0211452. [CrossRef]
30. Leong, D.P.; Teo, K.K.; Rangarajan, S.; Kutty, V.R.; Lanas, F.; Hui, C.; Quanyong, X.; Zhenzhen, Q.; Jinhua, T.; Noorhassim, I.; et al. Reference ranges of handgrip strength from 125,462 healthy adults in 21 countries: A prospective urban rural epidemiologic (PURE) study. *J. Cachexia Sarcopenia Muscle* **2016**, *7*, 535–546. [CrossRef]
31. Schuetz, P.; Fehr, R.; Baechli, V.; Geiser, M.; Deiss, M.; Gomes, F.; Kutz, A.; Tribolet, P.; Bregenzer, T.; Braun, N.; et al. Individualised nutritional support in medical inpatients at nutritional risk: A randomised clinical trial. *Lancet* **2019**, *393*, 2312–2321. [CrossRef]
32. Schuetz, P.; Fehr, R.; Baechli, V.; Geiser, M.; Gomes, F.; Kutz, A.; Tribolet, P.; Bregenzer, T.; Hoess, C.; Pavlicek, V.; et al. Design and rationale of the effect of early nutritional therapy on frailty, functional outcomes and recovery of malnourished medical inpatients trial (EFFORT): A pragmatic, multicenter, randomized-controlled trial. *Int. J. Clin. Trials* **2018**, *5*, 1–9. [CrossRef]
33. Merker, M.; Felder, M.; Gueissaz, L.; Bolliger, R.; Tribolet, P.; Kagi-Braun, N.; Gomes, F.; Hoess, C.; Pavlicek, V.; Bilz, S.; et al. Association of Baseline Inflammation With Effectiveness of Nutritional Support Among Patients With Disease-Related Malnutrition: A Secondary Analysis of a Randomized Clinical Trial. *JAMA Netw. Open* **2020**, *3*, e200663. [CrossRef] [PubMed]
34. Bretschera, C.; Boesiger, F.; Kaegi-Braun, N.; Hersberger, L.; Lobo, D.N.; Evans, D.C.; Tribolet, P.; Gomes, F.; Hoess, C.; Pavlicek, V.; et al. Admission serum albumin concentrations and response to nutritional therapy in hospitalised patients at malnutrition risk: Secondary analysis of a randomised clinical trial. *EClinicalMedicine* **2022**, *45*, 101301. [CrossRef]
35. Kaegi-Braun, N.; Tribolet, P.; Gomes, F.; Fehr, R.; Baechli, V.; Geiser, M.; Deiss, M.; Kutz, A.; Bregenzer, T.; Hoess, C.; et al. Six-month outcomes after individualized nutritional support during the hospital stay in medical patients at nutritional risk: Secondary analysis of a prospective randomized trial. *Clin. Nutr.* **2021**, *40*, 812–819. [CrossRef] [PubMed]

36. Kaegi-Braun, N.; Tribolet, P.; Baumgartner, A.; Fehr, R.; Baechli, V.; Geiser, M.; Deiss, M.; Gomes, F.; Kutz, A.; Hoess, C.; et al. Value of handgrip strength to predict clinical outcomes and therapeutic response in malnourished medical inpatients: Secondary analysis of a randomized controlled trial. *Am. J. Clin. Nutr.* **2021**, *114*, 731–740. [CrossRef] [PubMed]
37. Hersberger, L.; Stanga, Z.; Schuetz, P. Reply: The Importance of Objective Nutritional Indexes in Heart Failure Patients. *J. Am. Coll. Cardiol.* **2021**, *78*, 856–857. [CrossRef]
38. Efthymiou, A.; Hersberger, L.; Reber, E.; Schonenberger, K.A.; Kagi-Braun, N.; Tribolet, P.; Mueller, B.; Schuetz, P.; Stanga, Z.; EFFORT Study Group. Nutritional risk is a predictor for long-term mortality: 5-Year follow-up of the EFFORT trial. *Clin. Nutr.* **2021**, *40*, 1546–1554. [CrossRef]
39. Baumgartner, A.; Pachnis, D.; Parra, L.; Hersberger, L.; Bargetzi, A.; Bargetzi, L.; Kaegi-Braun, N.; Tribolet, P.; Gomes, F.; Hoess, C.; et al. The impact of nutritional support on malnourished inpatients with aging-related vulnerability. *Nutrition* **2021**, *89*, 111279. [CrossRef]
40. Baumgartner, A.; Hasenboehler, F.; Cantone, J.; Hersberger, L.; Bargetzi, A.; Bargetzi, L.; Kaegi-Braun, N.; Tribolet, P.; Gomes, F.; Hoess, C.; et al. Effect of nutritional support in patients with lower respiratory tract infection: Secondary analysis of a randomized clinical trial. *Clin. Nutr.* **2021**, *40*, 1843–1850. [CrossRef]
41. Bargetzi, L.; Brack, C.; Herrmann, J.; Bargetzi, A.; Hersberger, L.; Bargetzi, M.; Kaegi-Braun, N.; Tribolet, P.; Gomes, F.; Hoess, C.; et al. Nutritional support during the hospital stay reduces mortality in patients with different types of cancers: Secondary analysis of a prospective randomized trial. *Ann. Oncol.* **2021**, *32*, 1025–1033. [CrossRef] [PubMed]
42. Bargetzi, A.; Emmenegger, N.; Wildisen, S.; Nickler, M.; Bargetzi, L.; Hersberger, L.; Segerer, S.; Kaegi-Braun, N.; Tribolet, P.; Gomes, F.; et al. Admission kidney function is a strong predictor for the response to nutritional support in patients at nutritional risk. *Clin. Nutr.* **2021**, *40*, 2762–2771. [CrossRef] [PubMed]
43. Kondrup, J.; Rasmussen, H.H.; Hamberg, O.; Stanga, Z.; An Ad Hoc Espen Working Group. Nutritional risk screening (NRS 2002): A new method based on an analysis of controlled clinical trials. *Clin. Nutr.* **2003**, *22*, 321–336. [CrossRef]
44. Available online: https://www.ncmedical.com/item_699.html (accessed on 29 April 2020).
45. Hillman, T.E.; Nunes, Q.M.; Hornby, S.T.; Stanga, Z.; Neal, K.R.; Rowlands, B.J.; Allison, S.P.; Lobo, D.N. A practical posture for hand grip dynamometry in the clinical setting. *Clin. Nutr.* **2005**, *24*, 224–228. [CrossRef] [PubMed]
46. Bohannon, R.W.; Maljanian, R.; Ferullo, J. Mortality and readmission of the elderly one year after hospitalization for pneumonia. *Aging Clin. Exp. Res.* **2004**, *16*, 22–25. [CrossRef] [PubMed]
47. Bohannon, R.W. Muscle strength: Clinical and prognostic value of hand-grip dynamometry. *Curr. Opin. Clin. Nutr. Metab. Care* **2015**, *18*, 465–470. [CrossRef]
48. Sanson, G.; Sadiraj, M.; Barbin, I.; Confezione, C.; De Matteis, D.; Boscutti, G.; Zaccari, M.; Zanetti, M. Prediction of early- and long-term mortality in adult patients acutely admitted to internal medicine: NRS-2002 and beyond. *Clin. Nutr.* **2020**, *39*, 1092–1100. [CrossRef] [PubMed]
49. Hersberger, L.; Bargetzi, L.; Bargetzi, A.; Tribolet, P.; Fehr, R.; Baechli, V.; Geiser, M.; Deiss, M.; Gomes, F.; Kutz, A.; et al. Nutritional risk screening (NRS 2002) is a strong and modifiable predictor risk score for short-term and long-term clinical outcomes: Secondary analysis of a prospective randomised trial. *Clin. Nutr.* **2020**, *39*, 2720–2729. [CrossRef]
50. Ravasco, P. Nutrition in Cancer. *Nestle Nutr. Inst. Workshop Ser.* **2015**, *82*, 91–102. [CrossRef]

nutrients

MDPI

Article

Association of Dietary Fiber Intake with All-Cause Mortality and Cardiovascular Disease Mortality: A 10-Year Prospective Cohort Study

Yu-Jin Kwon [1,†], Hye-Sun Lee [2,†], Goeun Park [3], Hyung-Mi Kim [4] and Ji-Won Lee [5,*]

1 Department of Family Medicine, Yongin Severance Hospital, Yonsei University College of Medicine, Seoul 16995, Korea; digda3@yuhs.ac
2 Biostatistics Collaboration Unit, Department of Research Affairs, Yonsei University College of Medicine, Seoul 03277, Korea; hslee1@yuhs.ac
3 Biomedical Statistics Center, Research Institute for Future Medicine, Samsung Medical Center, Seoul 06351, Korea; say.goeun@gmail.com
4 Department of Food and Nutrition, Dongduck Women's University, Seoul 02748, Korea; veronykim@naver.com
5 Department of Family Medicine, Severance Hospital, Yonsei University College of Medicine, Seoul 03722, Korea
* Correspondence: indi5645@yuhs.ac
† These authors contributed equally to this work.

Abstract: Although previous studies have established that dietary fiber (DF) intake reduces the total cardiovascular disease (CVD) mortality in general populations, limited studies have been conducted in individuals with pre-existing chronic conditions, especially in Asian countries. We aimed to investigate the association of DF intake with all-cause and CVD mortality in the general population and in the subpopulation with hypertension, diabetes, and dyslipidemia. We examined the relationship between DF intake and all-cause and CVD mortality using the Korean genome and epidemiology study. Diet was assessed using a food-frequency questionnaire at baseline. Cox proportional hazard models were used to estimate the hazard ratio (HR) and 95% confidence intervals (CIs) after adjusting for confounders. During the mean 10.1 years of follow-up, higher DF intake was significantly associated with a lower risk of all-cause mortality after adjusting for confounders (HR and 95% CIs for Q5 vs. Q1: 0.84 (0.76–0.93); $p < 0.001$). DF intake was inversely associated with a lower risk of CVD mortality after adjusting for the same confounders (HR and 95% CIs for Q5 vs. Q1: 0.61 (0.47–0.78); $p < 0.001$). Total DF intake was inversely associated with all-cause and CVD mortality in middle-aged and older adults.

Keywords: dietary fiber; mortality; cardiovascular diseases; cohort study

Citation: Kwon, Y.-J.; Lee, H.-S.; Park, G.; Kim, H.-M.; Lee, J.-W. Association of Dietary Fiber Intake with All-Cause Mortality and Cardiovascular Disease Mortality: A 10-Year Prospective Cohort Study. *Nutrients* **2022**, *14*, 3089. https://doi.org/10.3390/nu14153089

Academic Editors: Omorogieva Ojo and Amanda R Amorim Adegboye

Received: 2 July 2022
Accepted: 25 July 2022
Published: 27 July 2022

1. Introduction

At the global level, non-communicable diseases (NCDs) contribute to 73.4% of total deaths, whereas cardiovascular disease (CVD) is the leading cause of death and disability [1]. In Korea, CVD mortality and hospitalization have steadily increased over the last decade [2]. In addition, a considerable proportion of adults in Korea have multiple CVD risk factors, such as hypertension, dyslipidemia, and type 2 diabetes. In 2018, approximately 12.1 million adults had hypertension, 4.3 million had diabetes, and 8.7 million had hypercholesterolemia [2].

Since a significant portion of CVD is preventable, the importance of adequate prevention strategies has long been emphasized [3]. Dietary intervention is the first-line approach for preventing CVD, and dietary fiber (DF) has been established as a nutritionally important and health-promoting food component [4]. DF is composed of plant substances that include non-digestible carbohydrates and lignin, which resist digestion by human endogenous enzymes [5]. DF is primarily derived from plant-based foods, such as whole grain, seeds, vegetables, and fruits [6]. Various beneficial effects of fibers on serum low-density lipoprotein (LDL) cholesterol, blood pressure, insulin sensitivity, controlling body weight, and chronic inflammation may exert a protective effect on the cardiovascular system [7–10]. Indeed, previous studies have established that DF intake reduced the total mortality and CVD mortality in healthy populations [8,11] Accumulating observational studies indicates that DF is inversely associated with risk of hypertension, diabetes, dyslipidemia, peripheral vascular disease, and coronary heart disease [12–14].

However, most studies have been conducted in Western countries, and only a limited number of prospective cohort studies on the Asian population have been published [11,15]. Dietary habits and cultures differ by ethnicity, regions, and countries; hence, the main sources of DF could also be different for various countries and subpopulations [11,15,16]. Moreover, few studies have analyzed the effect of DF intake on total and CVD mortality in individuals with pre-existing chronic conditions, which included hypertension, type 2 diabetes, and dyslipidemia [12]. Therefore, we aimed to investigate the association of DF intake with all-cause and CVD mortality in middle-aged and older Korean adults. Further, we analyzed the effect of DF intake on all-cause and CVD mortality in the subpopulation with hypertension, diabetes, and dyslipidemia.

2. Materials and Methods

2.1. Study Population

We analyzed baseline survey data from the Korean genome and epidemiology study (KoGES),_Ansan–Ansung study (2001–2002), KoGES_health examinee study (2004–2013), and the KoGES_cardiovascular disease association study (2005–2011), which were large-scale, longitudinal, and prospective cohort studies that investigated the risk factors for NCDs [17]. This study included 211,571 adults aged 40 years and older, who had lived in urban and rural areas for $\geq$6 months. The survey began in the year 2001 and is presently ongoing.

Figure 1 shows the participant selection process. From the 211,571 participants included in the baseline survey, participants were excluded for further study based on the following criteria: (1) lack of data on age and other lifestyle factors (n = 2231); (2) missing data on laboratory tests (n = 5853); (3) missing dietary information and total calorie intake, <500 kcal/day or >6000 kcal/day (n = 14,007); (3) missing data regarding death information (n = 54,530); and (4) death during the enrolled year (n = 63). Finally, we included a total of 143,050 participants. All participants provided written informed consent. The study protocol conformed to the ethical guidelines of the 1975 Declaration of Helsinki. This study was approved by the IRB of Yongin Severance Hospital.

2.2. Dietary Assessment

A food frequency questionnaire (FFQ) was used to assess the dietary intake. The FFQ of the KoGES-Ansan-Ansung baseline study contains 103 food items, while those of the KoGES-HEXA and KoGES_CAVAS contain 106 food items.

The FFQ evaluated how often the participants consumed each food item (never or seldom, once a month, two to three times per month, one to two times per week, three to four times per week, five to six times per week, once a day, twice per day, or thrice per day) and the amount of a particular food that they consumed in each meal (a half serving, one serving, or two or more servings) during the past 1 year. DF intake and other nutrients were calculated using the FFQ. We recorded the DF intake as g/day and divided it into quintiles (Q1 to Q5). The FFQ used in this study was available on the following website: http://www.cdc.go.kr/contents.es?mid=a40504100100 (accessed on 27 December 2021).

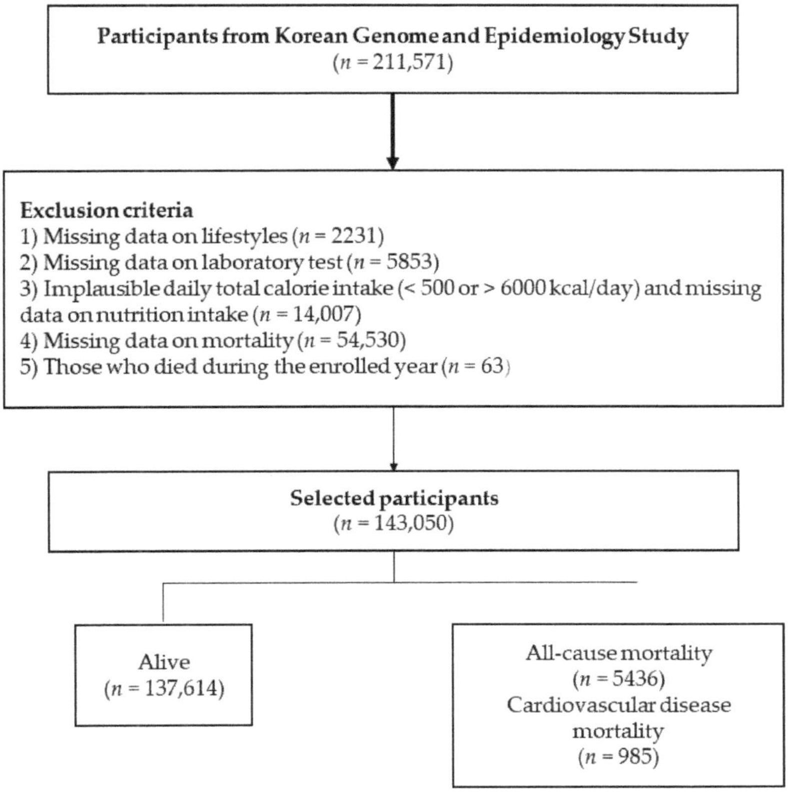

Figure 1. Selection process of study population.

2.3. Covariates

Trained medical staff performed all health examination procedures according to the standardized protocols published by KoGES. Measurements for height were obtained to the nearest 0.1 cm, and body mass index (BMI) was calculated using the height and weight. Blood pressure was measured after 5minutes resting in a sitting position. Blood tests were conducted after 8 h fasting. Serum total cholesterol, high-density lipoprotein (HDL), triglyceride, and glucose levels were enzymatically analyzed using a Chemistry Analyzer (Hitachi 7600, Tokyo, Japan from August 2002 and ADVIA 1650, Siemens, Tarrytown, NY from September 2002). A self-questionnaire was used to assess the status of smoking, alcohol drinking and exercise. Smokers were classified as current smokers, former smokers, and non-smokers. People consuming alcohol were classified as current drinkers, former drinkers, and non-drinkers. A person who regularly exercised enough to sweat was defined as a regular exerciser. A person with a systolic blood pressure $\geq$140 mmHg, diastolic blood pressure $\geq$90 mmHg, or diagnosis by physician was corresponded to hypertension. A person with fasting plasma glucose level $\geq$126 mg/dL, glycosylated hemoglobin $\geq$6.5%, or diagnosis by physician was corresponded to diabetes. A person with total cholesterol level $\geq$200 mg/dL, triglyceride $\geq$150 mg/dL, or diagnosis by physician was corresponded to dyslipidemia.

2.4. Study Outcomes

The primary outcome was all-cause mortality. Mortality outcomes were ascertained by the death records provided by Korea National Statistical Office. Participant deaths were tracked from the initial assessment up to 2019. All-cause mortality included all the specified and unknown causes of death. CVD mortality included deaths from International Classification of Diseases-10 codes I00-I99.

2.5. Statistical Analyses

Continuous data are presented as mean ± standard deviation (SD) and categorical data are presented as number (%). Among-group comparisons of continuous and categorical variables were performed using analysis of variance and the chi-square test, respectively. Cox proportional hazard regression model was used to calculate the hazard ratio (HR) and determine the association between DF intake and all-cause mortality and CVD mortality. Kaplan–Meier curves with the log-rank test were used to calculated the cumulative all-cause mortality and CVD mortality, according to the DF intake quintiles. The warranty period was defined as the time taken to reach the cumulative mortality rate of >0.5% for each group. If the 0.5% threshold was not met, the value was expressed as the time of the last follow-up. Incidence per 1000 person-years was calculated for each group. We calculated the HR and 95% confidence interval (CI) for all-cause mortality and CVD mortality, based on the DF intake quintile groups. In model 1, we adjusted for age, sex, body mass index, smoking, alcohol intake, exercise, and total calorie. In model 2, we further adjusted for hypertension, diabetes, and dyslipidemia. We conducted subgroup analysis according to the presence of hypertension, diabetes, and dyslipidemia. SAS statistical software (version 9.4; SAS Institute Inc., Cary, NC, USA) and R (Version 4.0.3; R Foundation for Statistical Computing, Vienna, Austria) were used for statistical analyses. Statistical significance was set at $p \leq 0.05$.

3. Results

During the median (min, max) 10.1 years (0.2, 15.9) follow-up period, 5436 cases of all-cause mortality and 985 cases of CVD mortality were noted. Among the total 143,050 participants, the mean age ± SD was 53.9 ± 8.7 years, and the proportion of men was 35.6%.

Table 1 shows the baseline characteristics of study population, according to the DF intake quintiles. Men ($p < 0.001$), older adults ($p < 0.001$), and current smokers ($p < 0.001$) were less likely to have a higher DF intake. Participants with the highest DF intake had lower serum glucose level ($p < 0.001$) and triglyceride level ($p = 0.022$). Moreover, participants with the highest DF intake had higher BMI ($p < 0.001$), waist circumference ($p < 0.001$), and serum HDL-C level ($p = 0.001$) and were more likely to regularly exercise ($p < 0.001$) and live in urban areas ($p < 0.001$). Regarding nutritional intake, participants with the highest DF intake consumed higher total energy and other nutrients, while they consumed less carbohydrate proportion.

Table 1. Baseline characteristics of the cohort according to fiber intake (g/day).

Variables	Q1 (0.37, 3.51)	Q2 (3.51, 4.64)	Q3 (4.64, 5.79)	Q4 (5.79, 7.44)	Q5 (7.44, 52.65)	*p*-Value
N	28,610	28,610	28,610	28,610	28,610	
Sex, men, n (%)	10,033 (35.1)	10,308 (36.0)	10,438 (36.5)	10,315 (36.1)	9850 (34.4)	<0.001
Age, years	54.7 ± 9.4	53.9 ± 8.8	53.8 ± 8.6	53.6 ± 8.4	53.2 ± 8.2	<0.001
BMI, kg/m^2	23.8 ± 3.0	23.9 ± 2.9	24.0 ± 2.9	24.0 ± 2.9	24.1 ± 2.9	<0.001
WC, cm	81.0 ± 8.9	81.3 ± 8.7	81.4 ± 8.8	81.5 ± 8.7	81.4 ± 8.8	<0.001
SBP, mmHg	122.5 ± 15.7	122.7 ± 15.5	122.8 ± 15.3	122.7 ± 15.3	122.5 ± 15.2	<0.001
DBP, mmHg	76.1 ± 10.2	76.1 ± 10.1	76.2 ± 10.0	76.2 ± 10.0	76.4 ± 10.0	<0.001
Glucose, mg/dL	96.0 ± 21.9	95.9 ± 21.9	95.7 ± 20.6	95.4 ± 21.0	95.2 ± 21.3	<0.001
HbA1c, %	5.72 ± 0.77	5.71 ± 0.75	5.72 ± 0.73	5.72 ± 0.75	5.72 ± 0.74	0.901
TC, mg/dL	197.5 ± 36.1	197.2 ± 35.6	197.3 ± 35.7	197.7 ± 35.6	197.2 ± 35.5	0.392

Table 1. *Cont.*

Variables	Q1 (0.37, 3.51)	Q2 (3.51, 4.64)	Q3 (4.64, 5.79)	Q4 (5.79, 7.44)	Q5 (7.44, 52.65)	*p*-Value
HDL-C, mg/dL	52.4 ± 13.3	52.6 ± 13.1	52.6 ± 13.0	52.7 ± 13.0	52.9 ± 12.9	0.001
LDL-C (mg/dL)	119.3 ± 32.9	118.9 ± 32.7	119.0 ± 32.8	119.2 ± 32.6	119.0 ± 32.4	0.469
TG, mg/dL	129.9 ± 90.3	130.0 ± 92.4	129.3 ± 91.3	130.1 ± 91.1	127.9 ± 89.4	0.022
Smoking status, n (%)						<0.001
Never smoker	20,428 (71.4)	20,424 (71.4)	20,406 (71.3)	20,525 (71.7)	21,004 (73.4)	
Former smoker	4177 (14.6)	4421 (15.5)	4602 (16.1)	4467 (15.6)	4218 (14.7)	
Current smoker	4005 (14.0)	3765 (13.2)	3602 (12.6)	3618 (12.7)	3388 (11.8)	
Alcohol intake, n (%)						<0.001
Never drinker	14,618 (51.1)	14,355 (50.2)	14,348 (50.2)	14,418 (50.4)	14,633 (51.2)	
Former drinker	1271 (4.4)	1155 (4.0)	1100 (3.8)	1049 (3.7)	1210 (4.2)	
Current drinker	12,721 (44.5)	13,100 (45.8)	13,162 (46.0)	13,143 (45.9)	12,767 (44.6)	
Regular exercise (No)	12,359 (43.2)	13,410 (46.9)	14,468 (50.6)	15,139 (52.9)	16,353 (57.2)	<0.001
HTN, n (%)	4997 (17.5)	4861 (17.0)	4842 (16.9)	4809 (16.8)	4898 (17.1)	0.272
DM	2165 (7.6)	2157 (7.5)	2064 (7.2)	2019 (7.1)	1959 (6.9)	0.003
Dyslipidemia	16,216 (56.7)	16,348 (57.1)	16,228 (56.7)	16,344 (57.1)	16,083 (56.2)	0.147
Residential area, n (%)						<0.001
Urban	23,776 (83.1)	24,648 (86.2)	24,993 (87.4)	25,467 (89.0)	25,472 (89.0)	
Rural	4834 (16.9)	3962 (13.9)	3617 (12.6)	3143 (11.0)	3138 (11.0)	
Total energy, kcal/day	1310.3 ± 344.0	1541.9 ± 356.2	1695.2 ± 377.3	1874.8 ± 415.3	2256.4 ± 630.7	<0.001
Carbohydrate intake, g/day	242.8 ± 64.9	279.7 ± 64.8	304.1 ± 66.3	332.3 ± 71.3	389.4 ± 100.7	<0.001
Carbohydrate (%)	74.2 ± 7.1	72.8 ± 6.7	72.1 ± 6.5	71.2 ± 6.5	69.70± 7.4	<0.001
Fat, g/day	17.4 ± 10.5	22.6 ± 12.0	26.1 ± 13.1	30.3 ± 15.0	40.2 ± 22.9	<0.001
Fat (%)	11.8 ± 5.7	13.01 ± 5.3	13.6 ± 5.1	14.2 ± 5.0	15.4 ± 5.4	<0.001
Protein, g/day	38.9 ± 12.7	49.0 ± 14.1	55.9 ± 15.6	64.4 ± 17.9	84.2 ± 31.3	<0.001
Protein (%)	11.9 ± 2.3	12.8 ± 2.2	13.2 ± 2.2	13.8 ± 2.3	14.9 ± 2.8	<0.001
Sodium, mg/day	1135.9 ± 492.0	1837.7 ± 562.3	2355.9 ± 665.6	2919.9 ± 783.2	4306.2 ± 1666.1	<0.001
Potassium, mg/day	1183.5 ± 392.1	1684.8 ± 377.2	2059.3 ± 410.6	2512.0 ± 474.2	3642.7 ± 1151.1	<0.001
Ca, mg	229.8 ± 125.2	328.7 ± 135.5	402.5 ± 150.7	491.9 ± 168.0	736.1 ± 321.0	<0.001
P, mg	571.8 ± 168.2	731.7 ± 177.5	843.6 ± 197.1	975.3 ± 222.5	1286.9 ± 415.7	<0.001
Fe, mg	5.4 ± 1.6	7.5 ± 1.7	9.1 ± 1.9	11.0 ± 2.3	16.3 ± 6.0	<0.001
Vit. A, R.E	201.4 ± 98.0	316.4 ± 117.5	405.8 ± 139.1	524.9 ± 173.3	920.7 ± 474.9	<0.001
Vit. B1, mg	0.64 ± 0.23	0.82 ± 0.25	0.94 ± 0.26	1.10 ± 0.30	1.47 ± 0.51	<0.001
Vit. B2, mg	0.53 ± 0.22	0.71 ± 0.25	0.84 ± 0.27	0.99 ± 0.30	1.38 ± 0.54	<0.001
Niacin, mg	9.38 ± 3.09	11.92 ± 3.36	13.66 ± 3.68	15.77 ± 4.22	20.80 ± 7.42	<0.001
Vit. C, mg	42.45 ± 20.01	70.18 ± 22.86	92.71 ± 27.07	121.30 ± 33.19	196.14 ± 82.27	<0.001
Zinc, μg	5.38 ± 1.76	6.65 ± 2.05	7.52 ± 2.30	8.59 ± 2.70	11.15 ± 4.90	<0.001
Vit. B6, mg	0.93 ± 0.25	1.24 ± 0.25	1.46 ± 0.28	1.74 ± 0.33	2.47 ± 0.80	<0.001
Folate, μg	98.94 ± 30.53	152.45 ± 28.69	193.63 ± 33.93	243.46 ± 44.54	385.96 ± 146.10	<0.001
Retinol, μg	41.20 ± 38.49	54.77 ± 46.34	63.83 ± 48.57	74.72 ± 54.54	102.65 ± 87.49	<0.001
Carotene, μg	925.6 ± 469.8	1520.1 ± 585.1	1991.6 ± 720.7	2621.8 ± 924.6	4779.4 ± 2648.2	<0.001
Vit. E, mg	4.49 ± 1.75	6.09 ± 2.04	7.31 ± 2.37	8.86 ± 2.77	13.24 ± 5.48	<0.001
Fiber, g	2.7 ± 0.6	4.1 ± 0.3	5.2 ± 0.3	6.5 ± 0.5	10.1 ± 3.0	<0.001

BMI, body mass index; DBP, diastolic blood pressure; DM, diabetes mellitus; HTN, hypertension; HbA1c, hemoglobin A1c; HDL-C, high density lipoprotein cholesterol; LDL-C, low density lipoprotein cholesterol; R.E, retinol equivalents; SBP, systolic blood pressure; TC, total cholesterol; TG, triglycerides; WC, waist circumference.

Figure 2 presents the cumulative all-cause mortality and CVD mortality of DF intake quintiles as Kaplan–Meier curves. The lowest DF intake group showed significantly the highest cumulative all-cause mortality, followed by Q2, Q3, Q4, and Q5 (log-rank test, $p < 0.001$). There was a similar trend in the cumulative CVD mortality event curves among the DF intake quintile groups (log-rank test, $p < 0.001$).

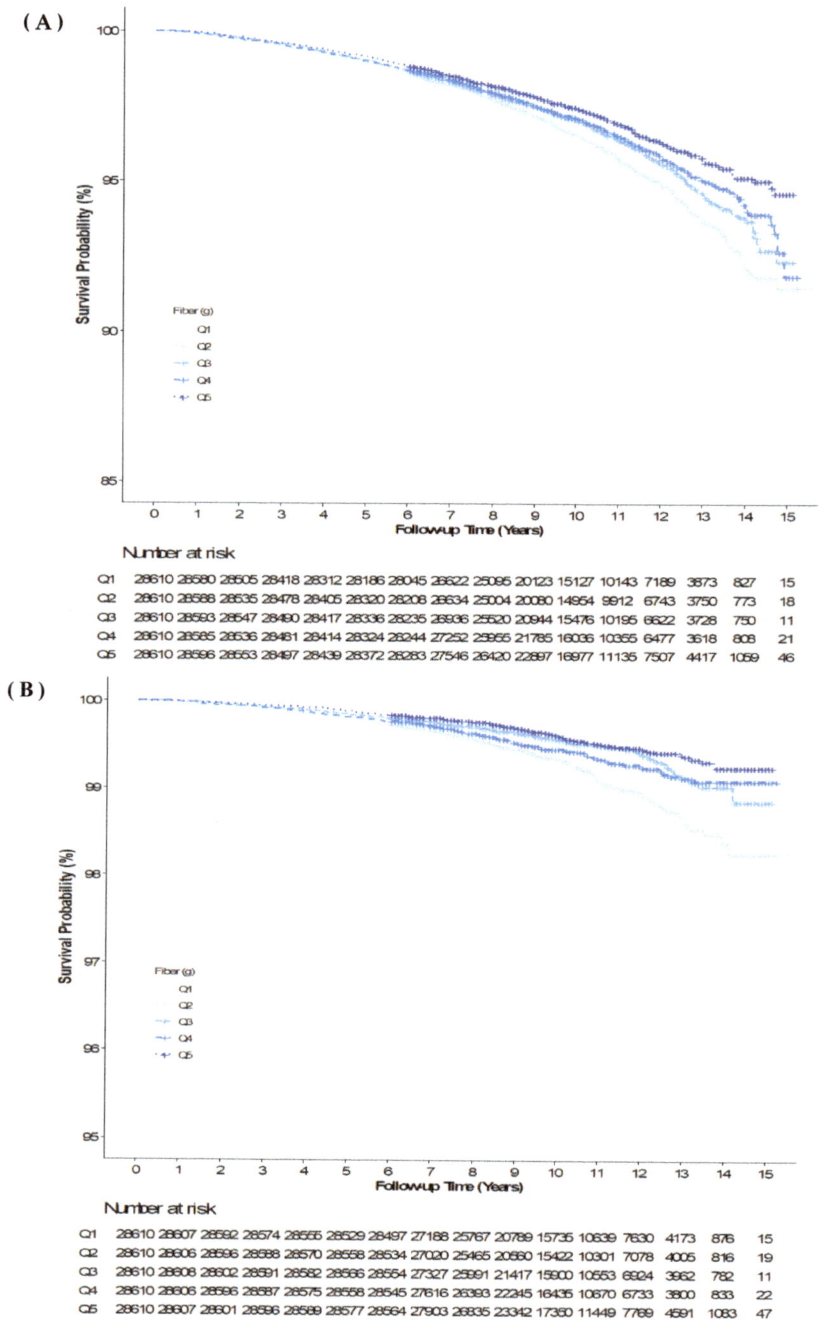

Figure 2. Kaplan–Meier curves displaying the estimated survival probability for fiber intake quintile groups. (**A**) All-cause mortality across the dietary fiber quintiles. (**B**) Cardiovascular disease mortality across the dietary fiber quintiles.

Table 2 shows the warranty period of all-cause mortality and CVD mortality for all the DF intake quintiles. All-cause mortality was associated with the longest 0.5% warranty period of 3.50 years for participants in the highest DF intake group. CVD mortality was associated with the longest 0.5% warranty period of 11.17 years for those in the highest DF intake group. The observed durations of the warranty period for both all-cause mortality and CVD mortality was the shortest in Q1 of DF intake. The incidence rate for all-cause mortality was lowest for participants with the highest DF intake (2.87 per 1000 person-years; 95% CI, 2.25–3.49). The incidence rate for CVD mortality was also the lowest for participants with the highest DF intake (0.42 per 1000 person-years; 95% CI, 0.18–0.66). The absolute numbers of all-cause mortality were 1511 in Q1, 1126 in Q2, 990 in Q3, 951 in Q4, and 858 in Q5. The number of CVD mortality was the lowest in Q5 (n = 127).

Table 2. Warranty periods for mortality according to fiber intake (g/day).

All-Cause Mortality					
Variables	**Warranty Period (0.5%)**	**n**	**Person-Time (Years)**	**Events, n**	**Incidence per 1000 Person-Years (95% CI)**
Fiber (g)					
Q1 (0.37, 3.51)	2.42	28,610	289,394.9	1511	5.22 (4.39–6.06)
Q2 (3.51, 4.64)	3.09	28,610	288,648.6	1126	3.90 (3.18–4.62)
Q3 (4.64, 5.79)	3.34	28,610	291,149.0	990	3.40 (2.73–4.08)
Q4 (5.79, 7.44)	3.09	28,610	293,349.0	951	3.24 (2.58–3.90)
Q5 (7.44, 52.65)	3.50	28,610	299,095.9	858	2.87 (2.25–3.49)
CVD Mortality					
Variables	**Warranty Period (0.5%)**	**n**	**Person-Time (Years)**	**Events, n (%)**	**Incidence per 1000 Person-Years (95% CI)**
Fiber (g)					
Q1 (0.37, 3.51)	6.84	28,610	294,427.0	311	1.06 (0.68–1.43)
Q2 (3.51, 4.64)	8.50	28,610	292,325.6	231	0.79 (0.47–1.12)
Q3 (4.64, 5.79)	10.83	28,610	294,656.7	145	0.49 (0.24–0.75)
Q4 (5.79, 7.44)	9.33	28,610	296,603.7	171	0.58 (0.30–0.86)
Q5 (7.44, 52.65)	11.17	28,610	302,202.9	127	0.42 (0.18–0.66)

CVD, cardiovascular disease; CI, confidence interval.

Table 3 shows the multiple Cox proportional hazard regression analysis for all-cause mortality and CVD mortality. The HR (95% CI) for all-cause mortality in Q5 with reference to Q1 was 0.84 (0.76–0.93, $p < 0.001$) after adjusting for age, sex, BMI, smoking, alcohol intake, exercise, total calories, hypertension, diabetes, and dyslipidemia. The HR (95% CI) for CVD mortality in Q5 with reference to Q1 was 0.61 (0.47–0.78, $p < 0.001$), after adjusting for the same confounders.

Figure 3 depicts the HR and 95% CI for all-cause mortality and CVD mortality, according to DF intake by participants with hypertension, diabetes, and dyslipidemia. In the baseline survey, there were 24,407 hypertension cases, 10,364 diabetes cases, and 81,219 dyslipidemia cases. During the median follow-up period of 10.3 years, 9.4 years, and 10.2 years, there were 1411 deaths in patients with hypertension, 838 deaths in patients with diabetes, and 3008 deaths in patients with dyslipidemia, respectively. Further, the numbers of CVD deaths were 324 in patients with hypertension, 153 in patients with diabetes, and 591 in patients with dyslipidemia, respectively. The HR (95% CI) for all-cause mortality in Q5 compared with Q1 was 0.55 (0.47–0.65) in patients with hypertension. Similar trends remained after adjusting for confounders. The HR (95% CI) for CVD mortality in Q5 compared with Q1 was 0.55 (0.39–0.77) in patients with hypertension. However, the significant association was attenuated after adjusting for confounders. In patients with diabetes, the HRs (95% CI) for all-cause mortality and CVD mortality in Q5 compared with Q1 were 0.53 (0.43, 0.66) and 0.33 (0.19, 0.59), respectively. However, this significant association between DF intake and total mortality disappeared after adjusting for confounders. Regarding

CVD mortality, a significant inverse association between DF intake and CVD mortality was observed after adjusting for confounders. The HRs (95% CIs) for all-cause mortality and CVD mortality were 0.50 (0.49, 0.56) and 0.35 (0.27, 0.45), respectively, in patients with dyslipidemia, and significant association remained after adjusting for confounders. The detailed figures were described in Supplementary Table S1.

Table 3. Multiple Cox proportional hazard regression analysis for all-cause mortality and cardiovascular disease (CVD) mortality of fiber intake quintiles.

	Q1 (0.37, 3.51)	Q2 (3.51, 4.64)	Q3 (4.64, 5.79)	Q4 (5.79, 7.44)	Q5 (7.44, 52.65)
			Hazard ratios (95% Confidence intervals)		
All-cause mortality					
Unadjusted	1.00 (ref)	0.75 (0.70, 0.81)	0.65 (0.60, 0.71)	0.62 (0.57, 0.67)	0.54 (0.50, 0.59)
Model 1	1.00 (ref)	0.89 (0.82, 0.96)	0.83 (0.76, 0.90)	0.85 (0.78, 0.93)	0.84 (0.76, 0.94)
Model 2	1.00 (ref)	0.88 (0.82, 0.96)	0.82 (0.76, 0.90)	0.85 (0.78, 0.93)	0.84 (0.76, 0.93)
CVD mortality					
Unadjusted	1.00 (ref)	0.75 (0.64–0.89)	0.47 (0.38–0.57)	0.55 (0.45–0.66)	0.39 (0.32–0.48)
Model 1	1.00 (ref)	0.94 (0.79–1.12)	0.62 (0.51–0.77)	0.79 (0.64–0.97)	0.62 (0.48–0.79)
Model 2	1.00 (ref)	0.93 (0.78–1.10)	0.62 (0.50–0.76)	0.78 (0.64–0.96)	0.61 (0.47–0.78)

Model 1: adjusted for age, sex, body mass index, smoking, alcohol intake, exercise, and total calories. Model 2: adjusted for age, sex, body mass index, smoking, alcohol intake, exercise, total calories, hypertension, diabetes, and dyslipidemia

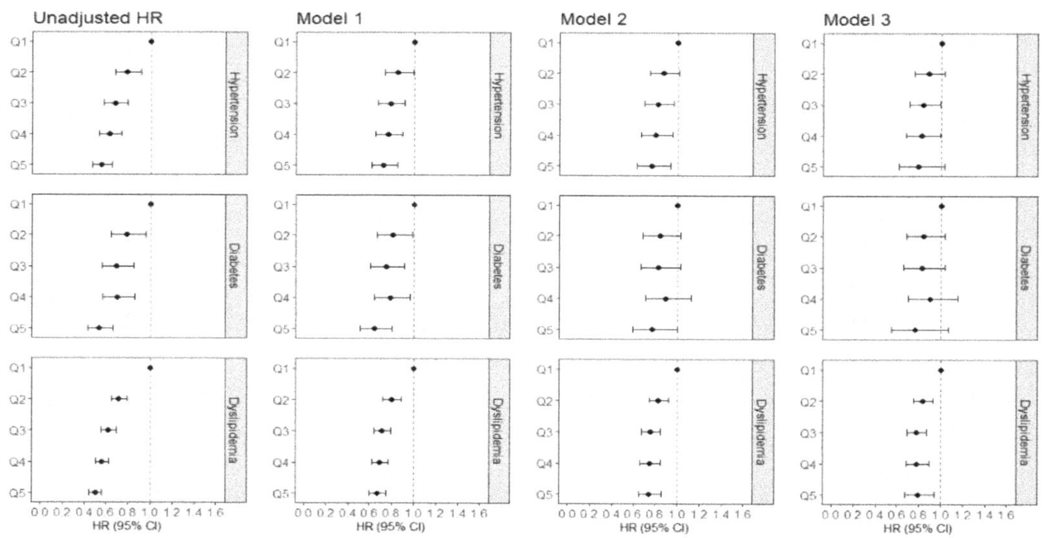

(A)

Figure 3. *Cont.*

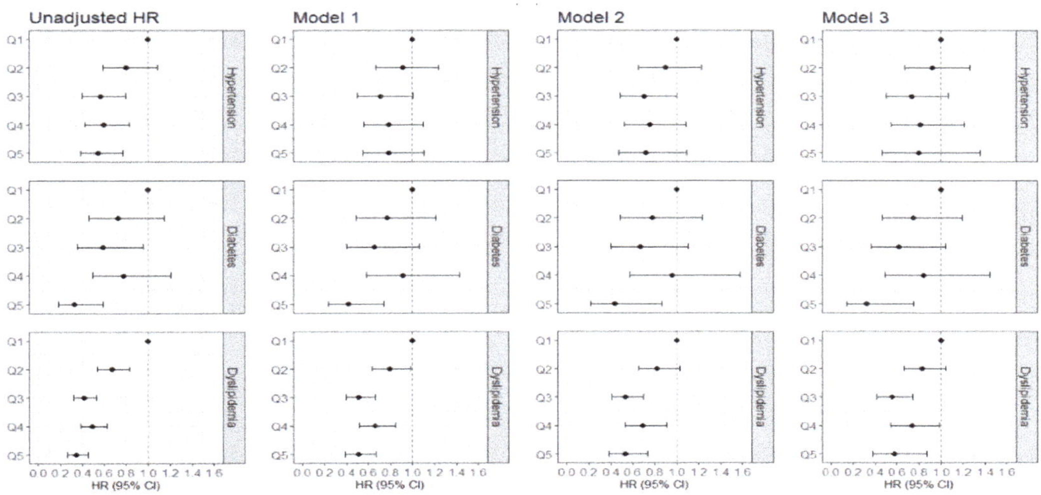

(B)

Figure 3. Association between dietary fiber intake and all-cause mortality and cardiovascular disease mortality according to hypertension, diabetes, and dyslipidemia. (**A**) All-cause mortality, (**B**) Cardiovascular disease mortality. HR: hazard ratio; CI: confidence interval.

4. Discussion

In this study, we found that DF intake was significantly inversely associated with all-cause mortality and with CVD mortality in middle-aged and older Korean adults.

Poor dietary habits are associated with many chronic diseases and can be a major contributor to mortality [18]. In contrast, a healthy diet is one of the most promising factors in primary and secondary prevention of NCD [19]. DF, "the seventh nutrient" of the body, is a component of fruits, vegetables, and whole grains, and is non-replaceable to maintain health [20]. DF has anti-inflammatory [10] and antioxidative properties [21] and has been shown to lower blood pressure [7], reduce serum cholesterol [22] and glucose levels [9], improve endothelial function [23], reduce body weight loss, aid favorable changes of gut microbial composition, and reduce the contact time between carcinogens and intestinal mucosal cells by increasing fecal bulking and viscosity [24]. All of these biochemical effects may be potentially related to a lower risk of chronic diseases and mortality.

Prospective studies have shown an inverse association between DF intake and all-cause and CVD mortality [11,15,25–30]. The NIH-AARP Diet and Health Study that had an average 9 years follow-up (567,169 participants aged 50–71 years from six states of the U.S.) found that 10 g/day increment of DF consumption reduced total mortality by 12% and 15% in men and women, respectively [25]. The Zutphen Study reported that 10 g/day of fiber increment reduced all-cause mortality by 9% and CVD mortality by 17% [29]. A meta-analysis from the prospective cohort studies reported that DF intake was inversely associated with the risk of CVD and coronary heart diseases [14]. Recently, the Japan Public Health Center-based prospective study with 92,924 participants found that total mortality was reduced by 23% in men and 18% in women for the highest quintile of total fiber intake [11]. Our data were consistent with the results of the previous studies.

Current recommendations for DF intake for adults in most European countries and in the U.S. are 30–35 g/day for men and 25–32 g/day for women [31]. Similarly, the Korean Nutrition Society suggested that sufficient amount of daily fiber intake for Korean adults is 25 g/day for men and 20 g/day for women regardless of age [32]. However, the global consumption of fiber falls below the recommended levels, regardless of the country [31]. In our study, the amounts of DF intake were far below the recommended daily intake in Korea.

Nonetheless, we found that for the highest quintile of total fiber intake, all-cause mortality was reduced by 16% and CVD mortality by 39%. Our data suggested that a relatively small increase in fiber intake may offer a public health benefit in reducing all-cause and CVD mortality.

DF reportedly exerts its influence on CVD mortality by decreasing the risk of hypertension, diabetes, and dyslipidemia in different patterns [7,9]. Epidemiologic studies consistently show that increased fiber intake is significantly associated with reduced hypertension risk. Meta-analyses of intervention trials showed that increased fiber intake is more effective in lowering blood pressure in hypertensive individuals than in normotensive individuals [33]. Therefore, increased fiber intake may adjunctively further lower blood pressure in hypertensive individuals. Further, systematic reviews conducted in 2010–2018 reported that diet high in soluble fiber caused at least moderate (i.e., 0.20–0.40 mmol/L) reductions in LDL cholesterol [34]. The bulking effect, viscosity, fermentation, and increased production of short-chain fatty acids are believed to be responsible for the antihyperlipidemic benefits of DFs [24]. Although the blood glucose-lowering effects of DF are inconsistent, three meta-analyses presented a 15–19% reduction in the incidence of developing type 2 diabetes, when participants with the highest intake of total DF to those with the lowest intake were compared [35–37]. A few studies have examined the effect of DF intake on mortality in patients who had hypertension, diabetes, or dyslipidemia and have reported inconsistent results [25,30,38]. Patients with hypertension, diabetes, and dyslipidemia have a 2–3 times higher risk of CVD and premature mortality than the general population [39,40]. We analyzed the effect of DF intake on total and CVD mortality in these patients and found that DF intake lowered the risk of all-cause mortality in patients with hypertension and dyslipidemia, and the risk of CVD mortality in patients with diabetes and dyslipidemia. Our findings suggest that low DF may be considered an important modifiable risk factor for decreasing mortality in patients with hypertension, diabetes, and dyslipidemia. However, the significant association between DF intake in patients with hypertension and CVD mortality, and that between DF intake in patients with diabetes and all-cause mortality were attenuated after adjusting for confounders. Although the exact mechanism is not known, different effects of DF subtypes and sources may explain the results to some extent. Vegetables, cereals, and fruits were three major sources of DF for Koreans who obtained approximately 75% of DF from those sources. Kimchi, a traditional fermented food made by salting and fermenting vegetables, was the first major source of DF for Koreans aged over 12 years [41,42]. A negative impression of kimchi is that as it is high in sodium and could be the cause of high blood pressure and metabolic syndrome [43]. Moreover, the sugar content in fruits is generally higher than in vegetables, leading to concerns about its potentially harmful impacts on patients with diabetes [44]. There are inconsistent results about the effects of fruit consumption on risks of death and major complications among people with established diabetes [45–47]. Further studies are needed to identify the association between DF and mortality considering the specific dietary type and sources.

Our study had several limitations. First, the FFQ method used in this study has the potential for recall bias, and DF could be underestimated. However, KoGES data are reliable and widely used. Moreover, it is the only available large dataset that contains both information about nutrition assessed by FFQ and mortality status in Korea. Second, we could not measure the overall diet quality due to restricted data, and there could be potential and unmeasured confounders that were not fully considered. However, the sample size used in this study was sufficiently large to detect significant HRs over 90% power (Q1 vs. Q5). Therefore, our study population was large enough to detect meaningful associations. Third, our study could not differentiate subtypes and sources of DF. Since we obtained an integrated dataset using the three KoGES data and not the original FFQ dataset, we could only use the total DF amount calculated by FFQ. Fourth, since our analysis only included middle-aged and older Korean adults, the current findings cannot be generalized to other countries and ethnic groups. Fifth, we could not consider the information about anti-hypertensive, anti-diabetic, and anti-dyslipidemia medications. Finally, we assessed the baseline DF intake, which was not updated during the follow-up period. Thus, baseline exposures might not reflect changes in DF intake over time.

To the best of our knowledge, this is the first study that investigated the association between DF intake and mortality in Korea using the largest population-based dataset over a long duration of follow-up. Furthermore, we comprehensively analyzed the effect of DF intake on total and CVD mortality in individuals with pre-existing hypertension, type 2 diabetes, and dyslipidemia.

5. Conclusions

In the current study, the total fiber intake was inversely associated with all-cause and CVD mortality. DF intake lowered the risk of all-cause mortality in patients with hypertension and dyslipidemia and the risk of CVD mortality in patients with diabetes and dyslipidemia. In the view of public health policies, adequate fiber consumption with healthy diet habits should be emphasized for reducing mortality and premature NCDs mortality.

Supplementary Materials: The following are available online at https://www.mdpi.com/article/10.3390/nu14153089/s1. Table S1: Multiple Cox proportional hazard regression analysis for all-cause mortality and cardiovascular disease (CVD) mortality of fiber intake quintiles according to the presence of hypertension, diabetes, and dyslipidemia.

Author Contributions: J.-W.L. had full access to all the data in the study and takes responsibility for the integrity of the data and the accuracy of the data analysis; Concept and design: Y.-J.K., H.-S.L. and J.-W.L.; Acquisition, analysis, or interpretation of data: Y.-J.K., H.-S.L., G.P., H.-M.K. and J.-W.L.; Drafting of the manuscript: Y.-J.K., H.-S.L. and J.-W.L.; Critical revision of the manuscript for important intellectual content: Y.-J.K., H.-S.L. and J.-W.L.; Statistical analysis: H.-S.L. and G.P.; Obtained funding: Y.-J.K. and J.-W.L.; Supervision: J.-W.L. All authors have read and agreed to the published version of the manuscript.

Funding: This work was supported by the Korea Institute of Planning and Evaluation for Technology in Food, Agriculture and Forestry (IPET) through the High Value-added Food Technology Development Program funded by the Ministry of Agriculture, Food and Rural Affairs (MAFRA) (321030051HD030) to J.-W.L. and Y.-J.K.

Institutional Review Board Statement: The study protocol conformed to the ethical guidelines of the 1975 Declaration of Helsinki. This study was approved by the IRB of Yongin Severance Hospital (IRB number: 3-2020-0043).

Informed Consent Statement: All participants provided written informed consent.

Data Availability Statement: The data used in this study were available on the following website: http://www.cdc.go.kr/contents.es?mid=a40504100100 (accessed on 27 December 2021).

Acknowledgments: Data used in this study were from the Korean Genome and Epidemiology Study (KoGES; 4851-302), National Research Institute of Health, Centers for Disease Control and Prevention, Ministry for Health and Welfare, Republic of Korea.

Conflicts of Interest: The authors declare no conflict of interest.

References

1. GBD 2017 Causes of Death Collaborators. Global, regional, and national age-sex-specific mortality for 282 causes of death in 195 countries and territories, 1980–2017: A systematic analysis for the global burden of disease study 2017. *Lancet* **2018**, *392*, 1736–1788. [CrossRef]
2. Lee, H.H.; Cho, S.M.J.; Lee, H.; Baek, J.; Bae, J.H.; Chung, W.J.; Kim, H.C. Korea heart disease fact sheet 2020: Analysis of nationwide data. *Korean Circ. J.* **2021**, *51*, 495–503. [CrossRef] [PubMed]
3. Mensah, G.A.; Wei, G.S.; Sorlie, P.D.; Fine, L.J.; Rosenberg, Y.; Kaufmann, P.G.; Mussolino, M.E.; Hsu, L.L.; Addou, E.; Engelgau, M.M.; et al. Decline in cardiovascular mortality: Possible causes and implications. *Circ. Res.* **2017**, *120*, 366–380. [CrossRef] [PubMed]

4. Cronin, P.; Joyce, S.A.; O'Toole, P.W.; O'Connor, E.M. Dietary fibre modulates the gut microbiota. *Nutrients* **2021**, *13*, 1655. [CrossRef]
5. Joye, I.J. Dietary fibre from whole grains and their benefits on metabolic health. *Nutrients* **2020**, *12*, 3045.
6. Dahl, W.J.; Stewart, M.L. Position of the academy of nutrition and dietetics: Health implications of dietary fiber. *J. Acad. Nutr. Diet.* **2015**, *115*, 1861–1870. [CrossRef]
7. Gibbs, J.; Gaskin, E.; Ji, C.; Miller, M.A.; Cappuccio, F.P. The effect of plant-based dietary patterns on blood pressure: A systematic review and meta-analysis of controlled intervention trials. *J. Hypertens* **2021**, *39*, 23–37. [CrossRef]
8. Zeng, X.; Li, X.; Zhang, Z.; Li, H.; Wang, Y.; Zhu, Y.; Hu, A.; Zhao, Q.; Tang, M.; Zhang, X.; et al. A prospective study of carbohydrate intake and risk of all-cause and specific-cause mortality. *Eur. J. Nutr.* **2022**. [CrossRef]
9. Xie, Y.; Gou, L.; Peng, M.; Zheng, J.; Chen, L. Effects of soluble fiber supplementation on glycemic control in adults with type 2 diabetes mellitus: A systematic review and meta-analysis of randomized controlled trials. *Clin. Nutr.* **2021**, *40*, 1800–1810. [CrossRef]
10. Milesi, G.; Rangan, A.; Grafenauer, S. Whole grain consumption and inflammatory markers: A systematic literature review of randomized control trials. *Nutrients* **2022**, *14*, 374. [CrossRef]
11. Katagiri, R.; Goto, A.; Sawada, N.; Yamaji, T.; Iwasaki, M.; Noda, M.; Iso, H.; Tsugane, S. Dietary fiber intake and total and cause-specific mortality: The japan public health center-based prospective study. *Am. J. Clin. Nutr.* **2020**, *111*, 1027–1035. [CrossRef]
12. Reynolds, A.N.; Akerman, A.P.; Mann, J. Dietary fibre and whole grains in diabetes management: Systematic review and meta-analyses. *PLoS Med.* **2020**, *17*, e1003053. [CrossRef]
13. Kulezic, A.; Bergwall, S.; Fatemi, S.; Sonestedt, E.; Zarrouk, M.; Gottsäter, A.; Acosta, S. Healthy diet and fiber intake are associated with decreased risk of incident symptomatic peripheral artery disease—A prospective cohort study. *Vasc. Med.* **2019**, *24*, 511–518. [CrossRef] [PubMed]
14. Threapleton, D.E.; Greenwood, D.C.; Evans, C.E.; Cleghorn, C.L.; Nykjaer, C.; Woodhead, C.; Cade, J.E.; Gale, C.P.; Burley, V.J. Dietary fibre intake and risk of cardiovascular disease: Systematic review and meta-analysis. *BMJ* **2013**, *347*, f6879. [CrossRef] [PubMed]
15. Rebello, S.A.; Koh, H.; Chen, C.; Naidoo, N.; Odegaard, A.O.; Koh, W.P.; Butler, L.M.; Yuan, J.M.; van Dam, R.M. Amount, type, and sources of carbohydrates in relation to ischemic heart disease mortality in a chinese population: A prospective cohort study. *Am. J. Clin. Nutr.* **2014**, *100*, 53–64. [CrossRef]
16. McGill, C.R.; Fulgoni, V.L., 3rd; Devareddy, L. Ten-year trends in fiber and whole grain intakes and food sources for the united states population: National health and nutrition examination survey 2001–2010. *Nutrients* **2015**, *7*, 1119–1130. [CrossRef] [PubMed]
17. Kim, Y.; Han, B.-G.; the KoGES Group. Cohort profile: The korean genome and epidemiology study (koges) consortium. *Int. J. Epidemiol.* **2017**, *46*, e20. [CrossRef]
18. GBD 2017 Diet Collaborators. Health effects of dietary risks in 195 countries, 1990–2017: A systematic analysis for the global burden of disease study 2017. *Lancet* **2019**, *393*, 1958–1972. [CrossRef]
19. Hu, E.A.; Steffen, L.M.; Coresh, J.; Appel, L.J.; Rebholz, C.M. Adherence to the healthy eating index-2015 and other dietary patterns may reduce risk of cardiovascular disease, cardiovascular mortality, and all-cause mortality. *J. Nutr.* **2020**, *150*, 312–321. [CrossRef]
20. Nie, Y.; Luo, F. Dietary fiber: An opportunity for a global control of hyperlipidemia. *Oxid. Med. Cell. Longev.* **2021**, *2021*, 5542342. [CrossRef]
21. Barber, T.M.; Kabisch, S.; Pfeiffer, A.F.H.; Weickert, M.O. The health benefits of dietary fibre. *Nutrients* **2020**, *12*, 3209. [CrossRef]
22. Tani, S.; Matsuo, R.; Imatake, K.; Suzuki, Y.; Takahashi, A.; Matsumoto, N. Association of daily fish intake with serum non-high-density lipoprotein cholesterol levels and healthy lifestyle behaviours in apparently healthy males over the age of 50 years in japanese: Implication for the anti-atherosclerotic effect of fish consumption. *Nutr. Metab. Cardiovasc. Dis.* **2020**, *30*, 190–200.
23. Yubero-Serrano, E.M.; Fernandez-Gandara, C.; Garcia-Rios, A.; Rangel-Zuñiga, O.A.; Gutierrez-Mariscal, F.M.; Torres-Peña, J.D.; Marin, C.; Lopez-Moreno, J.; Castaño, J.P.; Delgado-Lista, J.; et al. Mediterranean diet and endothelial function in patients with coronary heart disease: An analysis of the cordioprev randomized controlled trial. *PLoS Med.* **2020**, *17*, e1003282. [CrossRef] [PubMed]
24. Gill, S.K.; Rossi, M.; Bajka, B.; Whelan, K. Dietary fibre in gastrointestinal health and disease. *Nat. Rev. Gastroenterol. Hepatol.* **2021**, *18*, 101–116. [CrossRef] [PubMed]
25. Park, Y.; Subar, A.F.; Hollenbeck, A.; Schatzkin, A. Dietary fiber intake and mortality in the nih-aarp diet and health study. *Arch. Intern. Med.* **2011**, *171*, 1061–1068. [CrossRef]
26. Huang, T.; Xu, M.; Lee, A.; Cho, S.; Qi, L. Consumption of whole grains and cereal fiber and total and cause-specific mortality: Prospective analysis of 367,442 individuals. *BMC Med.* **2015**, *13*, 59.
27. Kim, Y.; Je, Y. Dietary fiber intake and total mortality: A meta-analysis of prospective cohort studies. *Am. J. Epidemiol.* **2014**, *180*, 565–573. [CrossRef]
28. Buil-Cosiales, P.; Zazpe, I.; Toledo, E.; Corella, D.; Salas-Salvadó, J.; Diez-Espino, J.; Ros, E.; Fernandez-Creuet Navajas, J.; Santos-Lozano, J.M.; Arós, F.; et al. Fiber intake and all-cause mortality in the prevención con dieta mediterránea (predimed) study. *Am. J. Clin. Nutr.* **2014**, *100*, 1498–1507. [CrossRef]

29. Streppel, M.T.; Ocké, M.C.; Boshuizen, H.C.; Kok, F.J.; Kromhout, D. Dietary fiber intake in relation to coronary heart disease and all-cause mortality over 40 y: The zutphen study. *Am. J. Clin. Nutr.* **2008**, *88*, 1119–1125. [CrossRef]
30. Chuang, S.C.; Norat, T.; Murphy, N.; Olsen, A.; Tjønneland, A.; Overvad, K.; Boutron-Ruault, M.C.; Perquier, F.; Dartois, L.; Kaaks, R.; et al. Fiber intake and total and cause-specific mortality in the european prospective investigation into cancer and nutrition cohort. *Am. J. Clin. Nutr.* **2012**, *96*, 164–174. [CrossRef]
31. Stephen, A.M.; Champ, M.M.; Cloran, S.J.; Fleith, M.; van Lieshout, L.; Mejborn, H.; Burley, V.J. Dietary fibre in europe: Current state of knowledge on definitions, sources, recommendations, intakes and relationships to health. *Nutr Res. Rev.* **2017**, *30*, 149–190. [CrossRef] [PubMed]
32. The Korean Nutrition Society. *2020 Dietary Reference Intakes for Koreans*; Ministry of Health and Welfare: Sejong-si, Korea, 2020.
33. Dreher, M.L. Fiber and hypertension. In *Dietary Fiber in Health and Disease*; Springer: Berlin/Heidelberg, Germany, 2018; pp. 291–303.
34. Schoeneck, M.; Iggman, D. The effects of foods on ldl cholesterol levels: A systematic review of the accumulated evidence from systematic reviews and meta-analyses of randomized controlled trials. *Nutr. Metab. Cardiovasc. Dis.* **2021**, *31*, 1325–1338. [CrossRef] [PubMed]
35. Yao, B.; Fang, H.; Xu, W.; Yan, Y.; Xu, H.; Liu, Y.; Mo, M.; Zhang, H.; Zhao, Y. Dietary fiber intake and risk of type 2 diabetes: A dose-response analysis of prospective studies. *Eur. J. Epidemiol.* **2014**, *29*, 79–88. [CrossRef]
36. Ye, E.Q.; Chacko, S.A.; Chou, E.L.; Kugizaki, M.; Liu, S. Greater whole-grain intake is associated with lower risk of type 2 diabetes, cardiovascular disease, and weight gain. *J. Nutr.* **2012**, *142*, 1304–1313. [CrossRef]
37. Dietary fibre and incidence of type 2 diabetes in eight european countries: The epic-interact study and a meta-analysis of prospective studies. *Diabetologia* **2015**, *58*, 1394–1408. [CrossRef] [PubMed]
38. Bazzano, L.A.; He, J.; Ogden, L.G.; Loria, C.M.; Whelton, P.K. Dietary fiber intake and reduced risk of coronary heart disease in us men and women: The national health and nutrition examination survey i epidemiologic follow-up study. *Arch. Intern. Med.* **2003**, *163*, 1897–1904. [CrossRef] [PubMed]
39. Fuchs, F.D.; Whelton, P.K. High blood pressure and cardiovascular disease. *Hypertension* **2020**, *75*, 285–292. [CrossRef]
40. Welty, F.K. Cardiovascular disease and dyslipidemia in women. *Arch. Intern. Med.* **2001**, *161*, 514–522. [CrossRef]
41. Park, S.; Na, W.; Kim, M.; Kim, E.; Sohn, C. Correlation between intake of dietary fiber and adherence to the korean national dietary guidelines in adolescents from jeonju. *Prev. Nutr. Food Sci.* **2012**, *17*, 254–260. [CrossRef]
42. Lee, H.-J.; Kim, Y.-A.; Lee, H.-S. The estimated dietary fiber intake of korean by age and sex. *J. Korean Soc. Food Sci. Nutr.* **2006**, *35*, 1207–1214.
43. Ma, Y.; He, F.J.; MacGregor, G.A. High salt intake: Independent risk factor for obesity? *Hypertension* **2015**, *66*, 843–849. [CrossRef]
44. Christensen, A.S.; Viggers, L.; Hasselström, K.; Gregersen, S. Effect of fruit restriction on glycemic control in patients with type 2 diabetes—A randomized trial. *Nutr. J.* **2013**, *12*, 29. [CrossRef] [PubMed]
45. Tanaka, S.; Yoshimura, Y.; Kawasaki, R.; Kamada, C.; Tanaka, S.; Horikawa, C.; Ohashi, Y.; Araki, A.; Ito, H.; Akanuma, Y.; et al. Fruit intake and incident diabetic retinopathy with type 2 diabetes. *Epidemiology* **2013**, *24*, 204–211. [CrossRef] [PubMed]
46. Nöthlings, U.; Schulze, M.B.; Weikert, C.; Boeing, H.; van der Schouw, Y.T.; Bamia, C.; Benetou, V.; Lagiou, P.; Krogh, V.; Beulens, J.W.; et al. Intake of vegetables, legumes, and fruit, and risk for all-cause, cardiovascular, and cancer mortality in a european diabetic population. *J. Nutr.* **2008**, *138*, 775–781. [CrossRef] [PubMed]
47. Sluik, D.; Boeing, H.; Li, K.; Kaaks, R.; Johnsen, N.F.; Tjønneland, A.; Arriola, L.; Barricarte, A.; Masala, G.; Grioni, S.; et al. Lifestyle factors and mortality risk in individuals with diabetes mellitus: Are the associations different from those in individuals without diabetes? *Diabetologia* **2014**, *57*, 63–72. [CrossRef] [PubMed]

nutrients

Article

Influence of Obesity on Bone Turnover Markers and Fracture Risk in Postmenopausal Women

Juan J. López-Gómez [1,2,*], José L. Pérez-Castrillón [2,3], Isabel García de Santos [4], María Pérez-Alonso [2], Olatz Izaola-Jauregui [1,2], David Primo-Martín [1,2] and Daniel A. De Luis-Román [1,2]

1 Department of Endocrinology and Nutrición, Hospital Clínico Universitario de Valladolid, 47003 Valladolid, Spain; olatzizaola@yahoo.es (O.I.-J.); dprimoma@saludcastillayleon.es (D.P.-M.); dadluis@yahoo.es (D.A.D.L.-R.)
2 Centro de Investigación Endocrinología y Nutrición (IENVA), University of Valladolid, 47002 Valladolid, Spain; uvacastrv@gmail.com (J.L.P.-C.); joseluis.perez@uva.es (M.P.-A.)
3 Department of Internal Medicine, Hospital Universitario Rio Hortega, 47012 Valladolid, Spain
4 School of Medicine, University of Valladolid, 47002 Valladolid, Spain; isa.garcia.desantos@gmail.com
* Correspondence: jjlopez161282@hotmail.com

Abstract: Background and aims: The relationship between obesity and bone metabolism is controversial. In recent decades, the protective role of obesity in the development of osteoporosis is questioned. The aims of this study are the following: to evaluate the differences in bone turnover markers between postmenopausal women with and without obesity and to compare the risk of fracture at five years between these groups. Methods: An observational longitudinal prospective cohort study of postmenopausal women with obesity (O) (body mass index (BMI) > 30 kg/m^2) and non-obesity (NoO) (BMI < 30 kg/m^2) is designed. 250 postmenopausal women are included in the study (NoO: 124 (49.6%) and O: 126 (50.4%)). It measures epidemiological variables, dietary variables (calcium intake, vitamin D intake, smoking, alcohol consumption, and physical activity), biochemicals (β-crosslap, type I procollagen amino-terminal peptide (P1NP), 25OH-vitamin D, and parathyroid hormone (PTH)), anthropometric variables, and fracture data five years after the start of the study. The mean age is 56.17 (3.91) years. Women with obesity showed lower levels of vitamin D (O: 17.27 (7.85) ng/mL, NoO: 24.51 (9.60) ng/mL; $p < 0.01$), and higher levels of PTH (O: 53.24 (38.44–65.96) pg/mL, NoO: 35.24 (25.36–42.40) pg/mL; $p < 0.01$). Regarding the bone formation marker (P1NP), it was found to be high in women without obesity, O: 45.46 (34.39–55.16) ng/mL, NoO: 56.74 (45.34–70.74) ng/mL; $p < 0.01$; the bone resorption marker (β-crosslap) was found to be high in women with obesity, being significant in those older than 59 years (O: 0.39 (0.14) ng/mL, NoO 0.24 (0.09) ng/mL; $p < 0.05$). No differences are observed in the risk of fracture at 5 years based on BMI (OR = 0.90 (95%CI 0.30–2.72); $p = 0.85$). Conclusions: Postmenopausal women with obesity showed lower levels of bone formation markers; older women with obesity showed higher markers of bone resorption.

Keywords: bone metabolism; obesity; osteoporosis; fracture; bone turnover markers

Citation: López-Gómez, J.J.; Pérez-Castrillón, J.L.; García de Santos, I.; Pérez-Alonso, M.; Izaola-Jauregui, O.; Primo-Martín, D.; De Luis-Román, D.A. Influence of Obesity on Bone Turnover Markers and Fracture Risk in Postmenopausal Women. *Nutrients* **2022**, *14*, 1617. https://doi.org/10.3390/nu14081617

Academic Editors: Omorogieva Ojo and Amanda R. Amorim Adegboye

Received: 18 March 2022
Accepted: 11 April 2022
Published: 13 April 2022

1. Introduction

Bone is an active organ on which many factors act. Osteoporosis and the risk of osteoporotic fracture are pathologies that affect the bone. In recent years, these entities are acquiring great relevance due to the progressive aging of the population and the impact on quality of life and the economy that they have on society [1,2].

Many epidemiological studies have shown that low weight and body mass index (BMI) are indicators of high risk of fracture, just as high weight and BMI are protective factors. Although, recently, this relationship does not seem to be so clear, and it is observed that obesity can be related to certain types of fractures [3,4].

Obesity is the most prevalent metabolic disease in the developed world and is one of the main causes of morbidity and mortality. The prevalence of obesity has tripled between 1975 and 2016 according to WHO data. In 2016, more than 1.9 billion adults were overweight, and more than 650 million people were obese [5].

During the past decades, obesity and osteoporosis have become major health problems, and the belief that obesity protects against osteoporosis has been questioned. In fact, some clinical and epidemiological studies have shown that excess fat mass could be a risk factor for osteoporosis and fragility fractures [6].

The factors related to obesity that negatively influence bone mass are mainly associated with an increase in the percentage of fat mass since obesity is a proinflammatory state that is associated with the secretion of a series of cytokines (IL-6, TNF-α) and adipokines (adiponectin, leptin ...). Although the cytokines have been observed to have a negative influence on bone, the role of the adipokines is still partially unknown in humans [7]. On the other hand, in patients with obesity, there is a decrease in the levels of circulating 25OHvitamin D, mostly due to its sequestration by adipose tissue. This situation can produce an alteration in the formation of bone, altering its quantity as well as its quality (architecture). In relation to this last point, the decrease in 25OHvitamin D can be associated in some cases, with an increase in PTH that can independently influence 25OHvitamin D in bone metabolism [8,9].

Increased lean mass or fat-free mass is associated with increased bone mass due to an increased mechanical load on the bone relative to weight and muscle hypertrophy. The positive effect of increased lean mass is attributed to lifestyle factors such as exercise and diet, estrogenic sufficiency, genetic influences, or a combination of these factors. On the other hand, increased muscle mass has an independent effect on fracture risk by reducing frailty and falls related to osteoporotic fracture [10]. In obesity, the concept of sarcopenic obesity should be considered, which entails a relative decrease in muscle mass in the situation of obesity in some individuals. This sarcopenia would be associated with a worse influence of muscle mass on bone, in addition to an increased risk of fracture due to frailty (increased falls). The proinflammatory situation related to this entity could also have a negative influence on the bone [11].

According to the described situation, the effect of obesity on bone health and fracture risk has yet to be determined. For this reason, this study has been proposed to evaluate the influence of obesity on bone metabolism. Given that bone metabolism, osteoporosis, and the risk of fracture are related to many risk factors, it is necessary to use highly selected populations to control possible confounding factors. For this reason, it was decided to evaluate the differences in markers of bone metabolism and the risk of fracture in postmenopausal female patients with and without obesity.

2. Materials and Methods

2.1. Study Design

An observational prospective longitudinal cohort study has been designed. All patients included in the study were informed and gave their consent.

This study has been carried out according to the ethical principles of the Declaration of Helsinki and has been approved by the Clinical Research Ethics Committee (CEIC) of the East Area of Valladolid with code PI 19-1517.

2.2. Population and Study Period

The study has been carried out from the following two cohorts of postmenopausal women: one cohort was women with obesity defined as BMI > 30 kg/m^2; the other cohort was women without obesity defined as BMI < 30 kg/m^2. The patients belonged to the East and West health areas of Valladolid, Spain.

A total of 250 patients were included in the study, including 124 postmenopausal women without obesity and 126 postmenopausal women with obesity, from whom data on anthropometry, food intake, and biochemical parameters related to bone metabolism in the initial assessment were taken; we also took fracture data five years after the start of study (January 2014).

The selection was made based on the following inclusion criteria: being a postmenopausal woman and being under 65 years of age. Exclusion criteria were being older than 65 years; criteria of premature or early menopause; have severe chronic kidney or liver disease and have the following toxic habits: active alcoholism and/or drug abuse.

2.3. *Variables*

2.3.1. Epidemiological Variables

Age, physical activity, and toxic habits such as alcohol consumption and smoking were recorded as follows:

- Age: it was calculated based on the date of birth and the of entry into the study;
- Physical activity: habitual physical activity is defined as that with a minimum duration of 30 min of exercise per day or 60 min on two days;
- Alcohol consumption: It is considered with the intake of more than 5 g per day. We have considered alcohol abuse with an intake of more than 20 g per day;
- Smoking: This variable was considered with a smoking habit of more than 6 months;
- Calcium intake in the diet: Calcium intake was considered based on the dairy rations consumed per day. In total, 200 mg of calcium per dairy ration consumed per day were considered. It was considered a dairy ration (1 glass of milk of 200 mL, 2 yogurts, or 1 portion of 100 g cheese);
- Consumption of vitamin D in the diet: The consumption of vitamin D was considered based on the rations of dairy products consumed per day. In total, 0.2 μg of vitamin D per dairy ration was considered. It was considered a dairy ration (1 glass of milk of 200 mL, 2 yogurts, or 1 portion of 100 g cheese).

2.3.2. Biochemical Variables

Vitamin D, plasma calcium, and bone turnover markers were recorded as follows:

- Vitamin D: This was determined by electrochemiluminescence immunoassay. Cobas 6000 e-601 (Roche Diagnostics, Basel, Switzerland) with a measurement range between 3.00–70.0 ng/mL;
- Plasma calcium: Total calcium was determined by the ocresolphthalein Schearzenbach method;
- Bone turnover markers: Three parameters were collected as bone turnover markers. These were beta-crosslap, type I procollagen amino-terminal propeptide (P1NP) and bone non-specific alkaline phosphatase (FA);
    - Beta-Crosslap: This is a marker of bone resorption. The measurement range was 0.010–6.00 ng/mL with a functional sensitivity of 0.07 ng/mL. Normal values are different depending on the stage of life. In the case of a postmenopausal woman, the reference value is 0.556–1.008 ng/mL;
    - P1NP: This bone formation marker is called the amino-terminal propeptide of type I procollagen. The measurement range was 5–1200 μg/L. The reference value in a postmenopausal woman is <76.3 ng/mL;
    - Alkaline phosphatase: Total alkaline phosphatase (not bone-specific) was considered. This is a marker of bone formation.

2.3.3. Anthropometric Variables

The anthropometric assessment of the subjects was performed by determining weight, height, and body mass index (BMI).

Weight was measured without clothing with an accuracy of ±0.5 kg using a manual scale to the nearest 0.1 kg (SECA, Birmingham, UK). Height was measured with the patient in an upright position to the nearest centimeter using a stadiometer (SECA, Birmingham, UK). The Body Mass Index (BMI) was calculated using the following formula:

$$BMI = weight\ (kg)/height\ (m) \times height\ (m) \quad (1)$$

In this case, the obesity cut-off with a BMI > 30 kg/m^2 was used for comparison between groups.

2.3.4. Fracture Variables

- Osteoporotic-type fracture: For this type of fracture, classic osteoporotic locations were considered (vertebral compression, femoral neck fracture, and distal radius fracture (Colles fracture));
- Non-osteoporotic fracture: Those that were not found in the typical osteoporotic locations already described (fractures due to trauma, or low impact fractures in non-classical locations of osteoporosis);
- Incidental fracture: To detect this type of fracture, the available imaging tests were reviewed in search of fractures that had gone unnoticed.

2.4. Statistical Analysis

Data were processed using the SPSS statistical package (SPSS for windows version 15.0, 2008 SPSS INC, Chicago, IL, USA).

Quantitative variables with normal distribution were described as mean and standard deviation (Mean (SD)), quantitative variables with non-normal distribution were described as Median and interquartile range (Median (p25–p75)) and, finally, qualitative variables as total number and percentages (Total number (%)).

The inferential analysis tests used were Student's *t*-test to compare means of normal quantitative variables; the Mann-Whitney U test to compare means of non-normal variables. Chi-square test to compare qualitative variables. Linear regression analysis to compare continuous variables. Multivariate logistic regression analysis to assess the causal relationships between qualitative variables. The significance level was conventionally set at $p < 0.05$.

3. Results

3.1. Description of the Sample

A total of 250 postmenopausal women with a mean age of 56.17 (3.91) years and a median body mass index (BMI) of 32.27 (24.14–39.59) kg/m^2 were analyzed.

The presence or absence of obesity was established as a study parameter, so the comparison will be established based on BMI (patients with a BMI > 30 kg/m^2 (O) vs. patients with a BMI < 30 kg/m^2 (NoO)). When we made this division, we had 124 patients (49.6%) with a BMI < 30 kg/m^2; 126 patients (50.4%) with a BMI of > 30 kg/m^2.

The differences between the different variables related to bone metabolism between the two cohorts are shown in Table 1. An increase in tobacco consumption was observed in patients with a BMI of less than 30 kg/m^2 and a decrease in the amount of physical activity among patients with a BMI greater or equal to 30 kg/m^2. On the other hand, there was an increased dietary intake of calcium and vitamin D in the group of patients with a BMI greater than 30 kg/m^2 (Table 1).

Table 1. Differences between the presence or absence of obesity in variables than may affect in bone metabolism.

	TOTAL	BMI > 30	BMI < 30	*p*-Value
BMI [1] (kg/m^2)	32.27 (24.14–39.59)	39.49 (5.13)	24.14 (2.87)	$p < 0.01$
Age (years)	56.17 (3.91)	56.99 (4.48)	55.33 (3.01)	$p < 0.01$
Alcohol (%)	18 (7.2%)	13 (10.3%)	5 (4%)	0.055
Smoking (%)	59 (23.6%) Exfum 27 (10.8%)	18 (14.4%) 4 (3.2%)	41 (33.1%) 23 (18.5%)	<0.01
Physical activity (%)	144 (57.6%)	38 (30.2%)	106 (85.5%)	<0.01
Dietary calcium (mg)	742.50 (330.18)	857.50 (302.16)	633.52 (322.53)	<0.01
Dietary vitamin D3 (μg)	1.05 (1.39)	1.58 (1.95)	0.63 (0.32)	<0.01

[1] BMI: Body Mass Index.

3.2. Differences in Bone Turnover Biochemical Parameters

In obese patients (BMI > 30 kg/m^2) there is a lower 25OHvitamin D levels (O:17.27 (7.85) ng/mL; NoO: 24.51 (9.60); $p < 0.01$), and a higher in intact PTH levels (O:53.24 (38.44–65.96) ng/mL; NoO: 35.24 (25.36–42.40) ng/mL $p < 0.01$). It was also observed to have lower P1NP levels (O: 45.46 (34.39) ng/mL; NoO: 56.74 (45.34–70.74) ng/mL; $p < 0.01$), a marker of bone formation, and we did not observe an effect on beta-crosslaps (O:0.34 (0.14) ng/mL; NoO: 0.33 (0.14) ng/mL; $p > 0.05$).

A stratified analysis by age quartiles was performed (Q1: under 53 years: 89 (35.6%); Q2: 53–56 years: 69 (27.6%); Q3: 56–59 years: 65 (26%); Q4: older than 60 years: 27 (10.8%) patients). The differences in the parameters according to the quartiles are shown in Table 2. A lower vitamin D value and a higher PTH maintained in the four quartiles were observed in patients with BMI > 30 kg/m^2. In these patients, P1NP was observed to be lower in the lowest age quartiles, while crosslaps were found to be high in the highest age quartile Table 2.

Table 2. Differences in bone metabolism biochemical markers between BMI agruped by age Quartiles.

	Q1 (<53 Years)		Q2 (53–56 Years)		Q3 (56–59 Years)		Q4 (>60 Years)	
Variable	**BMI [1] > 30**	**BMI [1] < 30**	**BMI [1] > 30**	**BMI [1] < 30**	**BMI [1] > 30**	**BMI [1] < 30**	**BMI [1] > 30**	**BMI [1] < 30**
Calcium (mg/dL)	9.4 (9.15–9.59)	9.48 (9.27–9.66)	9.5 (9.28–9.77)	9.43 (9.24–9.67)	9.60 (9.50–9.90)	9.45 (9.29–9.67)	9.40 (9.30–9.67)	9.71 (9.62–9.81)
25OHvitamin D (ng/mL)	***17.30 (7.46)***	***23.56 (10.35)***	***16.59 (8.86)***	***25.66 (8.95)***	***18.24 (8.66)***	***24.39 (9.54)***	***16.48 (6.01)***	***26.42 (7.21)***
PTH (pg/mL)	***54.36 (40.77–64.44)***	***36.23 (28.74–42.01)***	***51.89 (37.34–69.52)***	***29.9 (24.72–39.90)***	***51.27 (38.68–68.96)***	***36.39 (24.04–46.04)***	54 (38.06–60.55)	48.13 (31.18–48.18)
P1NP (ng/mL)	***46.11 (29.91–55.81)***	***65.49 (50.64–85.55)***	***43.40 (32.86–46.40)***	***56.04 (44.78–70.01)***	48.35 (37.89–62.08)	47.84 (40.55–56.83)	43.49 (34.91–59.63)	61.16 (30.42–77.01)
CROSSLAPS (ng/mL)	0.32 (0.16)	0.38 (0.17)	0.32 (0.11)	0.30 (0.98)	0.35 (0.13)	0.29 (0.12)	***0.39 (0.14)***	***0.24 (0.09)***
Alcaline Phosphatase (mg/dL)	***77.10 (23.66)***	***92.65 (20.96)***	75.92 (17.80)	81.21 (25.79)	83.61 (20.97)	82.39 (17.19)	75.83 (21.21)	85.66 (29.29)

[1] BMI: Body Mass Index; PTH: Parathyroid Hormone; P1NP: amino-terminal propeptide of type I procollagen. ***p-value < 0.05 (bold and italics).***

3.3. Correlation between BMI and Bone Turnover Biochemical Markers

A negative correlation was observed between the vitamin D levels and P1NP, while a positive correlation was observed between PTH and body mass index (Table 3).

When stratifying according to age quartiles, a negative correlation was observed that was maintained in all quartiles of vitamin D and in the youngest age quartiles of P1NP. PTH maintained the correlation in all quartiles except in the highest age quartile, and the resorption parameter showed a correlation in the third age quartile (Table 3).

Table 3. Correlation analysis of body mass index (BMI) with bone metabolism parameters in total sample and grouped by age quartiles.

	Calcium (mg/dL)	25OHvitamin D (ng/mL)	PTH [1] Intact (pg/mL)	P1NP1 [2] (ng/mL)	Crosslaps (ng/mL)	Alcaline Phosphatase (mg/dL)
TOTAL	r = 0.09 *p* = 0.17	***r* = −0.39** ***p* < 0.01**	***r* = 0.52** ***p* < 0.01**	***r* = −0.29** ***p* < 0.01**	r = 0.04 *p* = 0.24	***r* = −0.13** ***p* = 0.04**
Q1 (<53 years)	r = −0.22 *p* = 0.04	***r* = −0.34** ***p* < 0.01**	***r* = 0.55** ***p* < 0.01**	***r* = −0.37** ***p* < 0.01**	r = −0.128 *p* = 0.24	***r* = −0.27** ***p* = 0.01**
Q2 (53–56 years)	r = 0.03 *p* = 0.81	***r* = −0.43** ***p* < 0.01**	***r* = 0.53** ***p* < 0.01**	***r* = −0.39** ***p* < 0.01**	r = 0.04 *p* = 0.77	r = −0.06 *p* = 0.63
Q3 (56–59 years)	r = 0.23 *p* = 0.08	***r* = −0.35** ***p* < 0.01**	***r* = 0.51** ***p* < 0.01**	r = 0.08 *p* = 0.54	***r* = 0.27** ***p* = 0.03**	r = 0.06 *p* = 0.65
Q4 (>59 years)	r = −0.14 *p* = 0.49	***r* = −0.47** ***p* = 0.01**	r = 0.32 *p* = 0.23	r = −0.20 *p* = 0.31	r = 0.11 *p* = 0.59	r = −0.01 *p* = 0.96

[1] PTH: Parathyroid Hormone; [2] P1NP: amino-terminal propeptide of type I procollagen. ***p-value* < 0.05 (bold and italics).**

3.4. Multivariate Analysis

No significant differences were observed in fractures at five years of any type according to BMI (Figure 1).

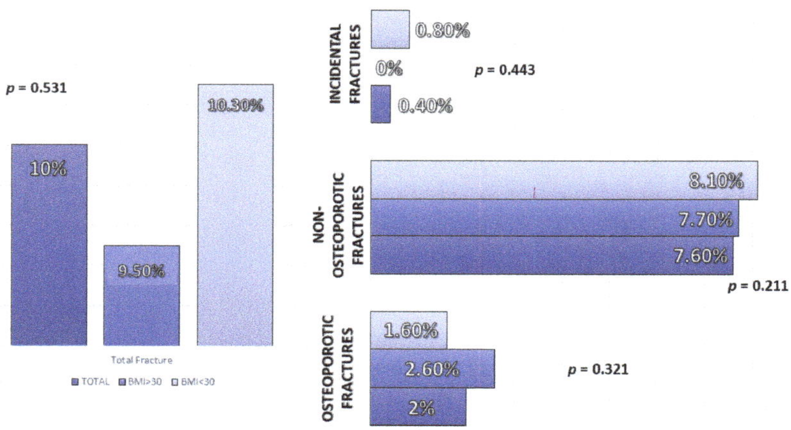

Figure 1. Differences in type fracture depending on body mass index (BMI).

A multivariate analysis was performed based on age, body mass index, and risk factors for fracture (age, smoking, physical activity, and alcohol consumption) to assess the risk factors for fracture without detecting data on the relationship between BMI and risk of total fracture risk (OR = 0.90 (95%CI 0.30–2.72); *p* = 0.85) (Osteoporotic Fracture: OR = 1.61 (95%CI 0.17–15.06); *p* = 0.68. Non-osteoporotic Fracture: OR = 0.77 (95%CI 0.23–2.65), *p* = 0.68).

4. Discussion

In the study carried out, lower levels of 25OHvitamin D were observed in obese women regardless of age. When evaluating bone metabolism parameters, a higher bone formation marker was observed in younger non-obese postmenopausal women and a higher bone resorption marker (β-crosslap) in older obese postmenopausal women.

An inverse relationship was found between BMI and vitamin D levels, such that obese women had less circulating vitamin D, especially in 59-year-old women, together with high intact PTH values compared to women who were not obese. This effect may be related to vitamin D, which has an inhibiting influence on PTH [8].

The most likely mechanism may be the dilution of vitamin D (as it is a fat-soluble vitamin) in the large fat deposits, decreasing its concentration in the blood [8]. The basis of this dilution is found in the sequestration of vitamin D by adipose tissue [12]. Other mechanisms may be low sun exposure, poor nutrition, or a decrease in the 25-hydroxylation of vitamin D in the liver, but they can also play a role, along with other factors such as inflammation or insulin resistance. However, there is no evidence that these low levels of vitamin D that occur in people with obesity have consequences for bone tissue, although they may have effects on other organs [9].

In obese postmenopausal women, higher resorption markers and lower bone formation markers were observed compared to non-obese postmenopausal women. Regarding the bone resorption marker (β-crosslap) in non-obese women, lower values were found compared to obese women, although it was only significant in older postmenopausal women. The role of age is very important in this sample since they are patients under 65 years of age and this stratification allows us to properly categorize risk. In fact, according to the NHANES study in its 2011 to 2018 cutoff, from the age of 60, the increase in fat mass and the decrease in lean mass are associated with a loss of bone mineral density [13].

The cause could be that in women who were not obese and who were in an early stage of menopause, the estrogen level had not dropped enough to reduce bone formation, since as Cui et al. demonstrated in their study, the significant decrease in bone mass occurs from 45–49 to 55–59 years of age [14]. This evidence contradicts obesity protects bone tissue [6], although obese people maintain circulating estrogens due to peripheral aromatization of androgens in relation to increased fat mass [10,15]. However, estrogen levels are not the only regulators of bone mass and, in fact, in studies such as the one carried out by Corina et al., the estrogenic action of adipose tissue was not observed to have a significant effect on bone, especially in the case of postmenopausal patients [16]. In addition, the adipose tissue also secretes proinflammatory cytokines that could interfere with the balance between bone resorption and formation [3].

No differences were found in the risk of fracture at five years (with a total of 10.30% in the obese versus 10.50% in the non-obese). Regarding the osteoporotic type, it was 2.6% in obese women versus 1.60% in non-obese women. The reason for not finding differences in the risk of fracture between obese and non-obese women could be that there are already alterations at the metabolic level in the bone tissue but that there is not yet a sufficient degree of disease to increase the risk of fracture in obese women [17].

Along these lines, a meta-analysis of 25 cohorts of women aged 63 years and older showed that osteoporotic fractures are less frequent in obese women, but, however, wrist fractures are more frequent compared to non-obese women. No association was found with respect to the most distal part of the lower extremity, but it is thought that these differences in location are related to the pattern of falls, the mechanical force induced by the fall, and that the low BMI of the controls could have masked the fracture risk associated with obesity [18]. However, Adachi et al. showed that obesity does not protect against osteoporotic fractures and even increases the risk of ankle and femur fractures [19].

On the other hand, it has been seen that the distribution of fat mass may be important for bone health, since in two meta-analyses it was observed that abdominal obesity, related to visceral adiposity, was associated with a higher rate of hip fractures [20,21].

These findings can have some implications for the management of these patients. In the first place, it is important to know all the parameters that may affect bone health to plan a therapeutic approach and prevent this bone loss. It is necessary to know how the supplementation of vitamin D and the consumption of calcium and phosphorus may affect this [5]. On the other hand, this bone turnover environment could be affected by weight loss strategies in patients with obesity, and it could increase bone resorption parameters, as it was shown in a study by our group where significant weight loss was associated with an increase in beta-crosslaps [22].

Regarding the main limitations of the study, having chosen two cohorts of women under 65 years of age with recent menopause interferes with the adequate assessment of the risk of fracture, since the negative influence of the decrease in estrogen has not been sufficient. Another associated limitation could be having chosen a short follow-up time (5 years) since a greater number of fractures could have been found in a longer-term follow-up. Another limitation is the lack of measures of adiposity in these patients with body composition, waist and hip, or diabetes presence or absence in all patients. These points could influence over the bone metabolism, and they can be parameters to measure in other studies. Lastly, not having imaging studies that indicate bone mineral density limits the adequate evaluation of the influence on bone metabolism.

In view of the results, there are several possible lines of research, such as the effect of the circulating estrogens in obese postmenopausal women on bone tissue and their functionality, since there are not many studies that identify this functionality.

Likewise, it would also be interesting to carry out a long-term prospective study on the net detrimental or beneficial effect of obesity on bone mass (using imaging tests) and the adequate categorization of fracture risk. It is necessary for the treatment strategies to prevent this bone loss and the possible fracture risk.

5. Conclusions

Obese postmenopausal women showed lower marker levels of bone formation, especially at younger ages. Older, obese women showed higher markers of bone resorption. This situation may be related to the fact that obese patients showed a decrease in vitamin D levels regardless of age, which is associated with a high PTH. However, an increased risk of fracture at five years was not found among obese patients.

The interaction of obesity and bone metabolism is complex due to the multitude of factors that interfere with it. The belief that obesity protects against osteoporosis is not a completely clear concept, and more studies are required to evaluate its relationship, determine its incidence, and be able to propose measures to prevent its negative influence on bone.

Author Contributions: Conceptualization, J.J.L.-G., J.L.P.-C. and D.A.D.L.-R.; Data curation, J.J.L.-G., I.G.d.S., M.P.-A., O.I.-J. and D.P.-M.; Formal analysis, J.J.L.-G.; Funding acquisition, D.A.D.L.-R.; Investigation, J.J.L.-G., I.G.d.S., M.P.-A., O.I.-J. and D.P.-M.; Methodology, J.J.L.-G.; Project administration, J.J.L.-G. and D.A.D.L.-R.; Resources, J.J.L.-G.; Software, J.J.L.-G.; Supervision, J.J.L.-G., J.L.P.-C. and D.A.D.L.-R.; Validation, J.J.L.-G. and D.A.D.L.-R.; Visualization, J.J.L.-G.; Writing—original draft, J.J.L.-G. and D.A.D.L.-R.; Writing—review and editing, J.J.L.-G., J.L.P.-C. and D.A.D.L.-R. All authors have read and agreed to the published version of the manuscript.

Funding: This research received no external funding.

Institutional Review Board Statement: The study was conducted in accordance with the Declaration of Helsinki and approved by the Institutional Review Board (or Ethics Committee) of "Comité de Ética en Investigación con Medicamentos (CEIm) of East Valladolid Area (protocol PI 19-1517, septiembre 2019).

Informed Consent Statement: Informed consent was obtained from all subjects involved in the study.

Data Availability Statement: Not applicable.

Conflicts of Interest: The authors declare no conflict of interest.

References

1. Ström, O.; Borgström, F.; Kanis, J.A.; Compston, J.; Cooper, C.; McCloskey, E.V.; Jönsson, B. Osteoporosis: Burden, health care provision and opportunities in the EU: A report prepared in collaboration with the International Osteoporosis Foundation (IOF) and the European Federation of Pharmaceutical Industry Associations (EFPIA). *Arch. Osteoporos.* **2011**, *6*, 59–155. [CrossRef] [PubMed]
2. Hernlund, E.; Svedbom, A.; Ivergård, M.; Compston, J.; Cooper, C.; Stenmark, J.; McCloskey, E.V.; Jönsson, B.; Kanis, J.A. Osteoporosis in the European Union: Medical management, epidemiology and economic burden. A report prepared in collaboration with the International Osteoporosis Foundation (IOF) and the European Federation of Pharmaceutical Industry Associations (EFPIA). *Arch. Osteoporos.* **2013**, *8*, 136. [CrossRef] [PubMed]

3. Fassio, A.; Idolazzi, L.; Rossini, M.; Gatti, D.; Adami, G.; Giollo, A.; Viapiana, O. The obesity paradox and osteoporosis. *Eat. Weight Disord. EWD* **2018**, *23*, 293–302. [CrossRef] [PubMed]
4. Gonnelli, S.; Caffarelli, C.; Nuti, R. Obesity and fracture risk. *Clin. Cases Miner. Bone Metab.* **2014**, *11*, 9–14. [CrossRef]
5. López-Gómez, J.J.; Pérez Castrillón, J.L.; de Luis Román, D.A. Impact of obesity on bone metabolism. *Endocrinol. Nutr.* **2016**, *63*, 551–559. [CrossRef] [PubMed]
6. Greco, E.A.; Lenzi, A.; Migliaccio, S. The obesity of bone. *Ther. Adv. Endocrinol. Metab.* **2015**, *6*, 273–286. [CrossRef] [PubMed]
7. Holecki, M.; Wiecek, A. Relationship between body fat mass and bone metabolism. *Pol. Arch. Intern. Med.* **2010**, *120*, 361–367. [CrossRef]
8. Migliaccio, S.; Di Nisio, A.; Mele, C.; Scappaticcio, L.; Savastano, S.; Colao, A.; Obesity Programs of Nutrition, Education, Research and Assessment (OPERA) Group. Obesity and hypovitaminosis D: Causality or casualty? *Int. J. Obes. Suppl.* **2019**, *9*, 20–31. [CrossRef]
9. Walsh, J.S.; Evans, A.L.; Bowles, S.; Naylor, K.E.; Jones, K.S.; Schoenmakers, I.; Jacques, R.M.; Eastell, R. Free 25-hydroxyvitamin D is low in obesity, but there are no adverse associations with bone health. *Am. J. Clin. Nutr.* **2016**, *103*, 1465–1471. [CrossRef]
10. Reid, I.R. Relationships among body mass, its components, and bone. *Bone* **2002**, *31*, 547–555. [CrossRef]
11. Gower, B.A.; Casazza, K. Divergent effects of obesity on bone health. *J. Clin. Densitom.* **2013**, *16*, 450–454. [CrossRef]
12. Collins, K.H.; Herzog, W.; Macdonald, G.Z.; Reimer, R.A.; Rios, J.L.; Smith, I.C.; Zernicke, R.F.; Hart, D.A. Obesity, Metabolic Syndrome, and Musculoskeletal Disease: Common Inflammatory Pathways Suggest a Central Role for Loss of Muscle Integrity. *Front. Physiol.* **2018**, *9*, 112. [CrossRef]
13. Jain, R.K.; Vokes, T. Fat Mass Has Negative Effects on Bone, Especially in Men: A Cross-Sectional Analysis of NHANES 2011–2018. *J. Clin. Endocrinol. Metab.* **2022**, dgac040. [CrossRef]
14. Yao, W.J.; Wu, C.H.; Wang, S.T.; Chang, C.J.; Chiu, N.T.; Yu, C.Y. Differential changes in regional bone mineral density in healthy Chinese: Age-related and sex-dependent. *Calcif. Tissue Int.* **2001**, *68*, 330–336. [CrossRef]
15. Zhao, L.-J.; Jiang, H.; Papasian, C.J.; Maulik, D.; Drees, B.; Hamilton, J.; Deng, H.-W. Correlation of obesity and osteoporosis: Effect of fat mass on the determination of osteoporosis. *J. Bone Miner. Res.* **2008**, *23*, 17–29. [CrossRef]
16. Corina, M.; Vulpoi, C.; Brănişteanu, D. Relationship between bone mineral density, weight, and estrogen levels in pre and postmenopausal women. *Med.-Surg. J.* **2012**, *116*, 946–950.
17. Savvidis, C.; Tournis, S.; Dede, A.D. Obesity and bone metabolism. *Hormones* **2018**, *17*, 205–217. [CrossRef]
18. Johansson, H.; Kanis, J.A.; Odén, A.; McCloskey, E.; Chapurlat, R.D.; Christiansen, C.; Cummings, S.R.; Diez-Perez, A.; Eisman, J.A.; Fujiwara, S.; et al. A meta-analysis of the association of fracture risk and body mass index in women. *J. Bone Miner. Res.* **2014**, *29*, 223–233. [CrossRef]
19. Compston, J.E.; Watts, N.B.; Chapurlat, R.; Cooper, C.; Boonen, S.; Greenspan, S.; Pfeilschifter, J.; Silverman, S.; Díez-Pérez, A.; Lindsay, R.; et al. Obesity is not protective against fracture in postmenopausal women: GLOW. *Am. J. Med.* **2011**, *124*, 1043–1050. [CrossRef]
20. Li, X.; Gong, X.; Jiang, W. Abdominal obesity and risk of hip fracture: A meta-analysis of prospective studies. *Osteoporos. Int.* **2017**, *28*, 2747–2757. [CrossRef]
21. Sadeghi, O.; Saneei, P.; Nasiri, M.; Larijani, B.; Esmaillzadeh, A. Abdominal Obesity and Risk of Hip Fracture: A Systematic Review and Meta-Analysis of Prospective Studies. *Adv. Nutr.* **2017**, *8*, 728–738. [CrossRef] [PubMed]
22. López-Gómez, J.J.; Izaola-Jauregui, O.; Primo-Martín, D.; Torres-Torres, B.; Gómez-Hoyos, E.; Ortolá-Buigues, A.; Martín-Ferrero, M.A.; Pérez-Castrillón, J.L.; De Luis-Román, D.A. Effect of weight loss on bone metabolism in postmenopausal obese women with osteoarthritis. *Obes. Res. Clin. Pract.* **2019**, *13*, 378–384. [CrossRef] [PubMed]

nutrients

MDPI

Article

Coffee Restores Expression of lncRNAs Involved in Steatosis and Fibrosis in a Mouse Model of NAFLD

Stefania Di Mauro [1,†], Federico Salomone [2,†], Alessandra Scamporrino [1,†], Agnese Filippello [1], Filomena Morisco [3], Maria Guido [4], Vincenzo Lembo [3], Valentina Cossiga [3], Rosaria Maria Pipitone [5], Stefania Grimaudo [5], Roberta Malaguarnera [6], Francesco Purrello [1,*] and Salvatore Piro [1]

1 Department of Clinical and Experimental Medicine, University of Catania, 95122 Catania, Italy; 8stefaniadimauro6@gmail.com (S.D.M.); alessandraska@hotmail.com (A.S.); agnese.filippello@gmail.com (A.F.); salvatore.piro@unict.it (S.P.)
2 Division of Gastroenterology, Ospedale di Acireale, Azienda Sanitaria Provinciale di Catania, 95024 Catania, Italy; federicosalomone@rocketmail.com
3 Department of Clinical Medicine and Surgery, University of Naples "Federico II", 80125 Naples, Italy; filomena.morisco@unina.it (F.M.); vincenzo.lembo@unina.it (V.L.); valentina.cossiga@unina.it (V.C.)
4 Department of Medicine, University of Padua, 35121 Padua, Italy; mguido@unipd.it
5 Department PROMISE, University of Palermo, 90128 Palermo, Italy; rosariamaria.pipitone@unina.it (R.M.P.); stefania.grimaudo@unipa.it (S.G.)
6 Faculty of Medicine and Surgery, "Kore" University of Enna, 94100 Enna, Italy; roberta.malaguarnera@unikore.it
* Correspondence: fpurrell@unict.it; Tel.: +39-095-759-8401
† These authors contributed equally to this work.

Citation: Di Mauro, S.; Salomone, F.; Scamporrino, A.; Filippello, A.; Morisco, F.; Guido, M.; Lembo, V.; Cossiga, V.; Pipitone, R.M.; Grimaudo, S.; et al. Coffee Restores Expression of lncRNAs Involved in Steatosis and Fibrosis in a Mouse Model of NAFLD. *Nutrients* **2021**, *13*, 2952. https://doi.org/10.3390/nu13092952

Academic Editor: Marilyn Cornelis

Received: 8 June 2021
Accepted: 22 August 2021
Published: 25 August 2021

Abstract: Background and aim: Coffee intake exerts protective effects against non-alcoholic fatty liver disease (NAFLD), although without fully cleared mechanisms. In this study we aimed to assess whether coffee consumption may influence the expression of long non-coding RNAs (lncRNAs) in the liver. Methods: C57BL/6J mice were fed a 12-week standard diet (SD), high-fat diet (HFD) or HFD plus decaffeinated coffee solution (HFD + coffee). Expression of specific lncRNAs involved in NAFLD was analyzed by real-time PCR. For the most differentially expressed lncRNAs, the analysis was also extended to their mRNA targets. Results: Decaffeinated coffee intake reduced body weight gain, prevented NAFLD, lowered hyperglycemia and hypercholesterolemia. NAFLD was associated with lower hepatic expression of Gm16551, a lncRNA inhibiting *de novo* lipogenesis, and higher expression of H19, a lncRNA promoting fibrogenesis. Coffee intake restored Gm16551 to levels observed in lean mice and downregulated gene expression of its targets acetyl coenzyme A carboxylase 1 and stearoyl coenzyme A desaturase 1. Furthermore, coffee consumption markedly decreased hepatic expression of H19 and of its target gene collagen alpha-1(I) chain; consistently, in mice fed HFD + coffee liver expression of αSMA protein returned to levels of mice fed SD. Expression of lncRNA involved in circadian clock such as fatty liver-related lncRNA 1 (FLRL1) and fatty liver-related lncRNA 2 (FLRL2) were upregulated by HFD and were also modulated by coffee intake. Conclusion. Hepatoprotective effects of coffee may be depending on the modulation of lncRNAs involved in key pathways of NAFLD onset and progression.

Keywords: NAFLD; coffee; lncRNA; Gm16551; H19

1. Introduction

Coffee is the most consumed beverage worldwide and coffee production plays a relevant role in several countries [www.fao.org]. Coffee is consumed for its taste, flavor, and its psychoactive properties. In the last decade, it has been also shown that coffee consumption associates with beneficial effects on several health outcomes [1].

Epidemiological studies including prospective cohorts have shown that coffee consumption may prevent type 2 diabetes mellitus (T2DM) and may confer protection against the metabolic syndrome in general [2]. Coffee consumption is inversely associated with the degree of fibrosis in subjects with non-alcoholic fatty liver disease (NAFLD) [3,4]. Among the thousand molecules contained in coffee, pre-clinical studies have showed that the main components exerting beneficial metabolic effects are those of the polyphenolic fraction, i.e., chlorogenic acids [5], whereas the molecular mechanisms by which coffee exerts hepatoprotective effects have been only in part elucidated [6].

Long noncoding RNAs (lncRNAs) are transcripts longer than 200 nucleotides lacking a long protein-coding open reading frame (ORFs) [7,8]. LncRNAs are involved in a myriad of cellular processes through the regulation of gene expression at epigenetic, transcriptional, post-transcriptional, translational, and post-translational levels [9]. Furthermore, they are also involved in the regulation of protein localization and activity [10]. All these functions are probably determined by the ability of lncRNAs to bind DNA, other RNAs and proteins [11]. Several lncRNAs have been associated with metabolic homeostasis and disorders related to insulin resistance [7,12,13].

In recent years, the involvement of specific lncRNA in metabolic pathways relevant to NAFLD including lipid metabolism, fibrosis, clock gene regulation, apoptosis and inflammation has been reported [14].

The aim of this study was to establish whether the intake of coffee might influence the liver expression of lncRNAs in a diet-induced murine model of NAFLD.

2. Materials and Methods

2.1. Animals and Treatments

This animal study was reviewed by the Ethics Committee of the University of Naples and approved by the Italian Minister of Scientific Research (Code 2014/0013808). Twenty-four 4-week-old male C57BL/6J mice were purchased from Harlan (San Pietro al Natisone, Italy). Animals were housed randomly in wire-bottomed cages and were maintained under controlled temperature conditions of 22 ± 1 °C, with a 12 h light–dark cycle and free access to water. After 1-week's acclimation, the mice were divided into three groups and were randomly assigned to one of the following 12-week diets: (1) standard diet (SD) $n = 8$; (2) high-fat diet (HFD) $n = 8$; (3) HFD plus decaffeinated coffee solution $n = 8$ (HFD + coffee). A detailed composition of the diets is reported in Supplementary Table S1.

Coffee-containing beverages were prepared by filtering on a filter paper (Whatman grade 113; Merck KGaA), a mix of boiling water and decaffeinated coffee powder (4:1, v/w) (Illy Caffè). Filtered coffee was portioned and stored at −20 °C until used. In a preliminary experiment, we found that the average daily consumption of solution (water or the coffee solution) was about 3.5 mL/mouse/d. The coffee-based beverage was prepared by diluting 1.5 mL of coffee in 100 mL of water. The dose administered coffee corresponded to six cups of espresso coffee or two cups of filtered coffee for a person weighing 70 kg [15]. Food and energy intake as well as body weight were recorded weekly. Food intake was calculated based on the amount of food remaining from a known amount administered weekly. After 12 weeks of the experimental diet, the mice were fasted overnight, anesthetized by Tribromoethanol 250 mg/kg intraperitoneally and sacrificed. Blood and liver samples were harvested, processed and snap-frozen until analyses.

2.2. Liver Histology

A portion of the liver was fixed in 4% formaldehyde and embedded in paraffin. Sections (5 μm thick) were obtained and stained with haematoxylin and eosin. A pathologist (M.G.) blindly evaluated liver sections. Macrovesicular steatosis was assessed at low magnification (4×) and scored as Grade 0 (between 0 and 5%), Grade 1 (between 6 and 33%), Grade 2 (between 34 and 66%) and Grade 3 (>66%). Microvesicular steatosis was evaluated at higher magnification (20×) and expressed as percentage of affected cells. Necro-inflammatory foci were scored as present or absent.

2.3. Biochemical Analysis and Real Time PCR

Serum ALT, total cholesterol and glucose were measured on frozen sera using automated assays following the manufacturer's instructions (Reflotron Plus, Roche Diagnostic).

RNA extraction was performed from murine liver tissue (20 mg) of 23 samples: 8 SD, 8 HFD, 7 HFD + coffee (one sample was lost). Total RNA was extracted by using miRNeasy mini kit (Qiagen, Milan, Italy) according to the manufacturer's instructions [16,17]. The quantity and quality of total RNA were measured with *NanoDrop* (Thermo Fisher Scientific, Monza, Italy). Specific qPCR primers for NAFLD-associated lncRNAs and their mRNA targets were generated through Primer Blast [18] and are reported in Supplementary Table S2. Transcript expression was analyzed through real-time PCR assays by using *Power SYBR Green RNA-to-CT 1-Step kit* (Thermo Fisher Scientific) in *QuantStudio 5 Real-time PCR System* (Thermo Fisher Scientific). Gene expression fold changes (FC) were determined by applying the $2^{-\Delta\Delta Ct}$ method and analyzing GAPDH as endogenous control [19,20].

Protein was extracted from 80 mg of liver tissue with RIPA lysis buffer and Western blot was made as previously reported [21]. All the immunoblot signals were detected using the Odyssey Fc System Infrared Scanner (LI-COR Biosciences, Lincoln, NE, USA) and densitometric analyses were performed by using Odyssey software Image Studio Lite Ver 5.2. We used the following antibodies: anti α-smooth muscle actin (Cell Signaling Technology, Danvers, MA, USA) and anti β-actin (Sigma Aldrich, St. Louis, MO, USA).

2.4. Statistical Analysis

Continuous variables are presented as mean ± SD or median IQR (interquartile range), based on data distribution assessed by D'Agostino and Pearson test. Statistical significance was evaluated applying the ordinary one-way ANOVA with Tukey's multiple comparisons test. Statistical significance was established at a two-tailed p-value < 0.05. GraphPad Prism 8 (GraphPad Software, Inc., San Diego, CA, USA) was employed for statistical analysis and graph-figure design.

3. Results

3.1. Metabolic Parameters and Liver Histology

At the beginning of the study, the three groups of mice had similar body weight (Table 1). At the end of the 12-week study period, mice of both HFD-fed groups, with or without coffee, had higher body weight compared to mice fed SD (Table 1).

Table 1. Metabolic parameters.

Parameter (Units)	Standard Diet (SD)	High Fat Diet (HFD)	High Fat Diet (HFD) + Coffee
Initial body weight (g)	20.5 ± 1.7	20.7 ± 1.3	20.9 ± 0.7
Final body weight (g)	30.2 ± 2.5	38.4 ± 3.0 *	34.8 ± 2.5 *†
Food intake (Kcal/day)	4.52 ± 0.32	6.59 ± 0.38 *	6.90 ± 0.78 *
Glucose (mg/dl)	365 ± 33.6	450 ± 49.7 *	178 ± 57.6 *†
Total cholesterol (mg/dl)	107 ± 11.3	239 ± 42.3 *	161 ± 23.8 *†
ALT (IU/L)	53.3 ± 39.8	56.9 ± 16.1	45.9 ± 36.2

All variables are presented as mean ± SD because of normal distribution assessed by D'Agostino and Pearson test. Statistical significance was assessed by ordinary one-way ANOVA with Tukey's multiple comparisons test. * $p < 0.5$ vs. SD, † $p < 0.5$ vs. HFD.

Mice fed HFD + coffee had lower body weight compared to HFD + vehicle despite similar food intake (Table 1). In agreement with body weight reduction, mice fed HFD + coffee displayed lower serum levels of total cholesterol and fasting glucose compared to mice fed HFD alone, whereas ALT were not significantly different among groups (Table 1).

Figure 1 shows representative pictures of liver hematoxylin-eosin staining in the three groups. All HFD animals showed some degree of steatosis, which was predominantly microvesicular in most cases (Supplementary Table S3). In coffee treated mice, macrovesicular steatosis disappeared and the degree of microvesicular steatosis was less severe, with most cases showing only Grade 1. Rare inflammatory foci were seen in four HFD mice and in none of the coffee treated animals.

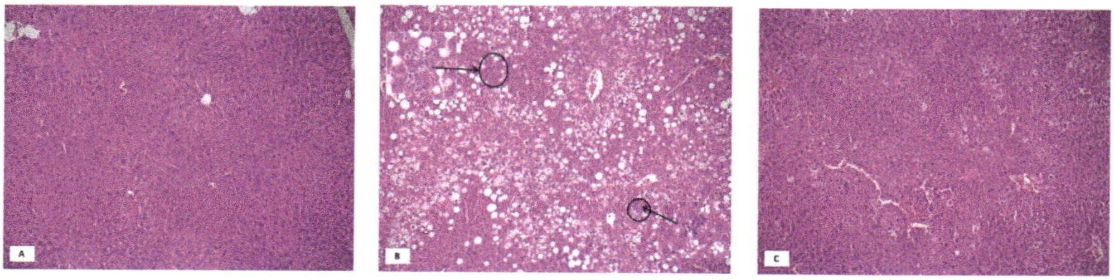

Figure 1. Representative pictures of liver hematoxylin-eosin staining in the three groups. Normal liver histology in mice fed standard diet (Panel (**A**), original magnification 10×). Mice on HFD for 12 weeks showed severe mixed, micro- and macrovesicular steatosis (Panel (**B**), original magnification 10×). Two necro-inflammatory foci are visible in this field (Original magnification in the inserts 40×). Amelioration of liver histology in mice fed high fat diet + decaffeinated coffee for 12 weeks: absence of macrovesicular steatosis and inflammatory foci, reduction of microvescicular steatosis (Panel (**C**), original magnification 10×).

3.2. Liver Expression of Long Non-Coding RNAs

Based on literature data, we chose 14 specific lncRNAs involved in pathways related to NAFLD onset and progression including lipid metabolism, oxidative stress, inflammation, fibrosis, circadian rhythm regulation and apoptosis. For significantly ($p < 0.05$) deregulated lncRNAs with fold change values ≤ -2 or ≥ 2 in HFD + coffee versus HFD, qPCR analysis was extended also to their known validated direct or indirect targets.

3.3. Coffee Inhibits De Novo Lipogenesis via lncRNA Gm16551/Srebf1 Pathway

Figure 2 shows expression levels of Gm16551, a liver-specific lncRNA that regulates de novo lipogenesis through its interaction with the transcription factor Sterol regulatory element-binding protein isoform 1c (SREBP-1c) [22] (UniProtKB-Q9WTN3, encoded by the gene Srebf1). HFD caused a 2-fold downregulation of Gm16551 while administration of decaffeinated coffee solution determined a 3-fold upregulation of Gm16551 compared to HFD alone, restoring its expression to levels similar to SD condition (Figure 2, Panel A).

Surprisingly, mRNA for Srebf1 displayed an increasing trend of expression from SD towards HFD to HFD + coffee conditions (Figure 2, Panel B). This may depend on the fact that we evaluated the transcript for Srebf1 instead of measuring this factor at the translational level. However, although Srebf1 mRNA was upregulated by coffee, mRNA expression of its downstream targets acetyl coenzyme A carboxylase 1 (UniProtKB-Q5SWU9 encoded by Acaca) and stearoyl coenzyme A desaturase 1 (UniProtKB-P13516 encoded by Scd1) was downregulated. In detail, the administration of coffee in HFD mice induced a 3-fold down-regulation of mRNA for Acaca in comparison both to mice fed SD and HFD + vehicle (Figure 2, Panel C). mRNA for Scd1 had a similar expression trend, with a six-fold downregulation in HFD + coffee vs. HFD alone and respect to SD (Figure 2, Panel D).

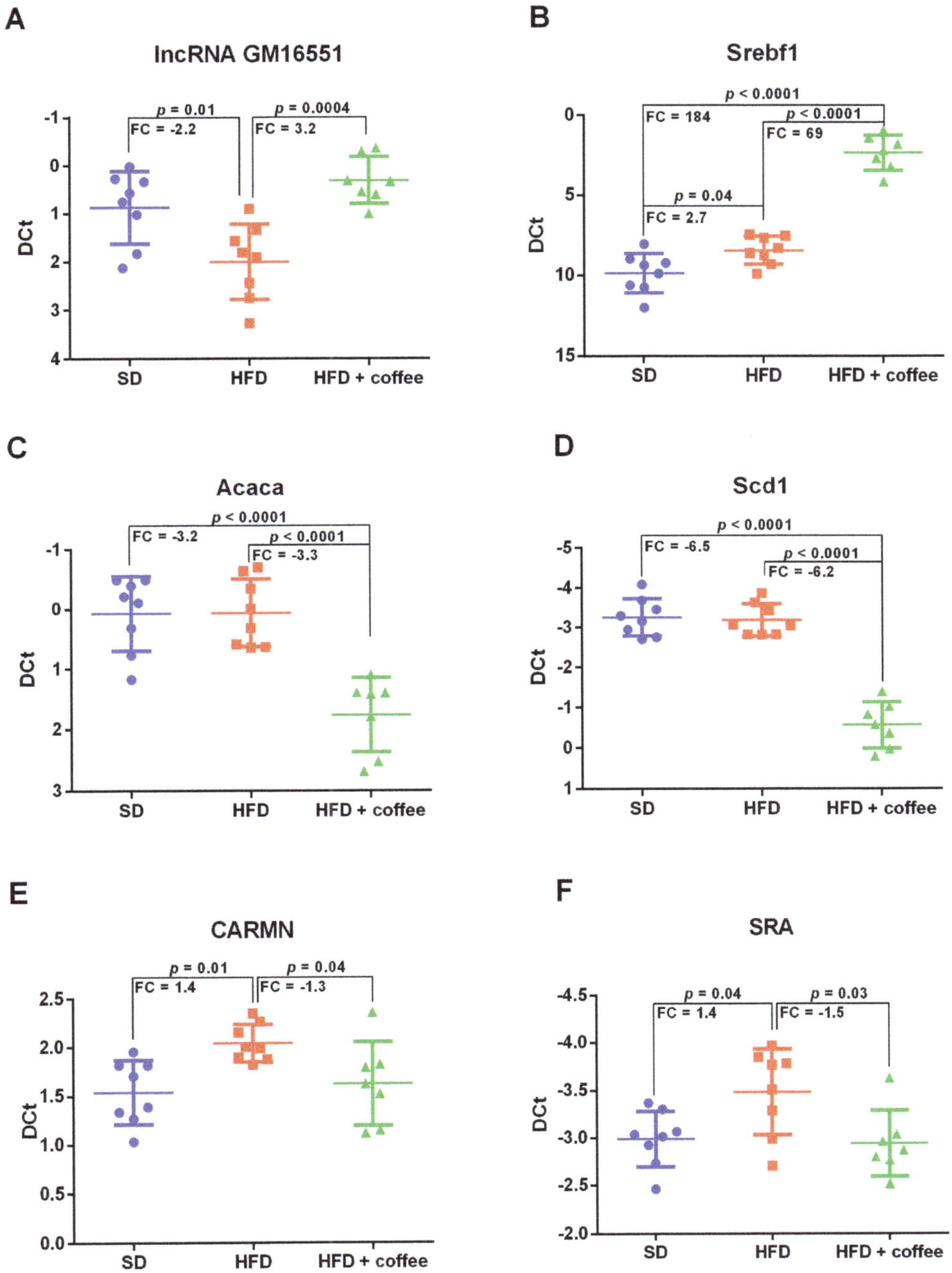

Figure 2. Dot plots of the hepatic expression of *Gm16551* lncRNA (Panel **A**), *sterol regulatory element-binding protein factor 1* (Srebf1) mRNA (Panel **B**), *acetyl coenzyme A carboxylase alpha* (Acaca) mRNA (Panel **C**), *stearoyl coenzyme A desaturase 1* (Scd1) mRNA (Panel **D**), *cardiac mesoderm enhancer-associated* (CARMN) lncRNA (Panel **E**) and *steroid receptor RNA activator* (SRA) lncRNA (Panel **F**), analyzed through qPCR, in mice fed standard diet (SD), high fat diet (HFD) and HFD plus decaffeinated coffee; $n = 23$: 8 SD, 8 HFD, 7 HFD + coffee. Transcript statistical significance of DE transcripts was evaluated with one-way ANOVA with Tukey post-hoc test for multiple comparisons (two-tailed p-value < 0.05); FC = fold change.

Two other lncRNAs, also involved in lipid metabolism, were slightly modified by HFD and coffee intake, the lncRNA *cardiac mesoderm enhancer-associated* (CARMN) [23] and the *steroid receptor RNA activator* (SRA) [24]. HFD induced a slight increase in the expression of CARMN and SRA versus SD, while coffee supplementation significantly decreased their expression and restored them to levels of mice fed SD (Figure 2, Panel E,F).

3.4. Coffee Inhibits Expression of the Fibrosis-Associated lncRNA H19

Figure 3 shows the expression levels of *H19*, a lncRNA that is involved in liver fibrogenesis [25]. We found a 2.6 up-regulation of H19 in mice fed HFD compared to SD, whereas decaffeinated coffee reduced the expression of this lncRNA to levels lower than those observed in mice fed HFD alone and even SD (Figure 3, Panel A). We observed that mRNA for *Collagen alpha-1(I) chain* (UniProtKB-P11087 encoded by Col1a1) was downregulated in HFD + coffee in comparison to HFD and SD (Figure 3, Panel B). Although HFD is a model of early NAFLD without histological fibrosis, we also found an up-regulation of α-SMA protein expression evaluated by Western blot analysis (Figure 3, Panel C) suggesting the activation of hepatic stellate cells. In agreement with H19 down-regulation by coffee, the expression of α-SMA was restored by coffee intake to levels observed in mice fed SD (Figure 3, Panel D).

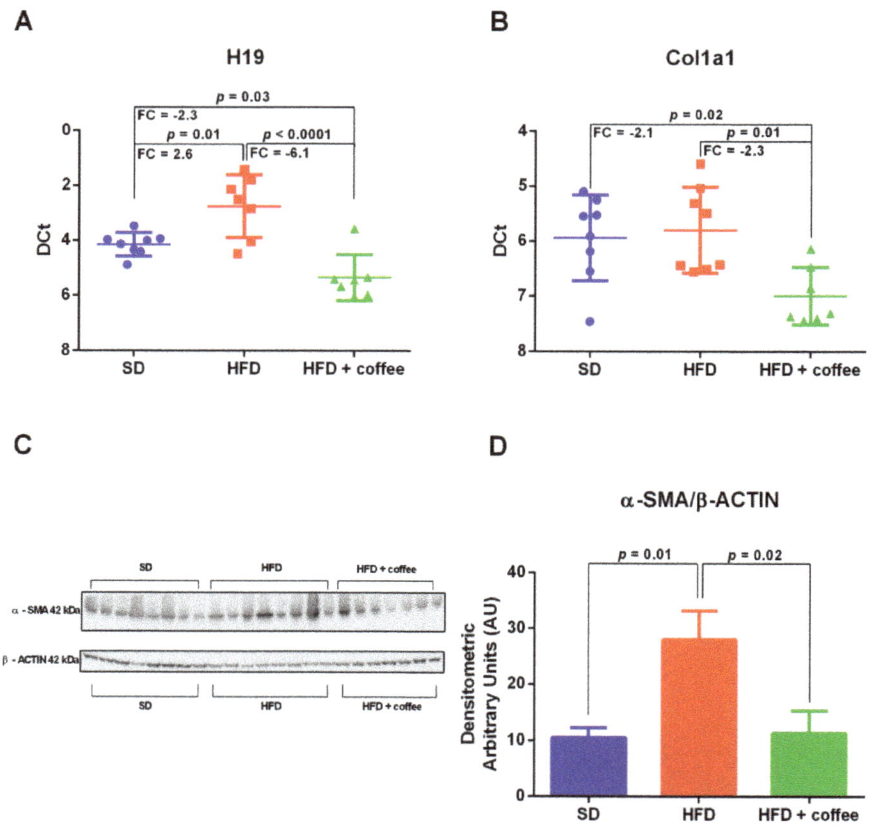

Figure 3. Hepatic expression of *H19* lncRNA (Panel **A**) and *Collagen alpha-1(I) chain* (Col1a1) mRNA (Panel **B**), analyzed through qPCR, in mice fed standard diet (SD), high fat diet (HFD) and HFD plus decaffeinated coffee. Transcript statistical significance was evaluated with one-way ANOVA with Tukey post-hoc test for multiple comparisons (two-tailed p-value < 0.05); FC = fold change. Liver expression of *alpha-smooth muscle actin* (α-SMA) protein, analyzed by Western blot (Panel **C**), and relative densitometry normalized for the housekeeping β-actin (Panel **D**). $n = 23$: 8 SD, 8 HFD, 7 HFD+ coffee.

3.5. Coffee Modulates Expression of lncRNAs Associated with Circadian Clock Regulation

Figure 4 shows the expression of fatty liver-related lncRNA 1 (FLRL1) and fatty liver-related lncRNA 2 (FLRL2), two lncRNAs that are involved in circadian clock regulation and whose liver expression is changed by HFD in mice [26]. Overall, we found an up-regulation of both lncRNAs in the liver of mice fed HFD versus SD, whereas their expression was differently modulated by coffee intake. In detail, coffee consumption further increased FLRL1 (Figure 4, Panel A), while FLRL2 was decreased by coffee intake to levels lower than those of mice fed SD (Figure 4, Panel B). It has been reported that FLRL1 and FLRL2 target *period circadian protein homolog 3* (UniProtKB-O70361 encoded by Per3) and *aryl hydrocarbon receptor nuclear translocator-like protein 1* (UniProtKB-Q9WTL8 encoded by Arntl), respectively [27]. Therefore, we extended qPCR analysis also to these target genes. mRNA for Per3 showed the same expression trend observed for its regulator lncRNA FLRL1; thus, it was upregulated by HFD and further increased by coffee intake (Figure 4, Panel C). Arntl mRNA expression, which was downregulated by HFD, was unaffected by coffee administration (Figure 4, Panel D).

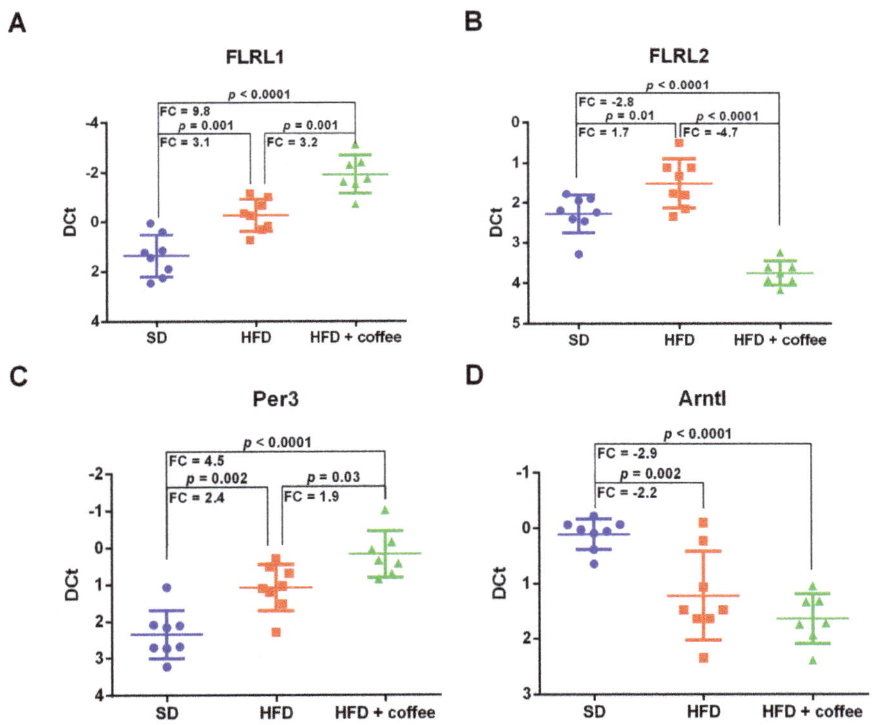

Figure 4. Hepatic expression of fatty liver-related lncRNA 1 (FLRL1) (Panel **A**), Fatty liver-related lncRNA 2 (FLRL2) (Panel **B**), period circadian protein homolog 3 (Per3) mRNA (Panel **C**) and Aryl hydrocarbon receptor nuclear translocator-like protein 1 (Arntl) mRNA (Panel **D**), analyzed through qPCR, in mice fed standard diet (SD), high fat diet (HFD) and HFD plus decaffeinated coffee; $n = 23$: 8 SD, 8 HFD, 7 HFD+ coffee. Transcript statistical significance was evaluated with one-way ANOVA with Tukey post-hoc test for multiple comparisons (two-tailed p-value < 0.05); FC = fold change.

3.6. lncRNAs Not Modified by Coffee Consumption

Another lncRNA involved in the regulation of metabolic processes is *colorectal neoplasia differentially expressed* (CRNDE) [28]. Although CRNDE was upregulated about three-fold by HFD, its expression was not modified by coffee consumption (Supplementary Figure S1A). Similarly, *nuclear enriched abundant transcript 1* (NEAT1), that plays a role in LDL uptake [29], was downregulated by HFD but its expression was unchanged by coffee intake (Supplementary Figure S1B). A summary of lncRNAs modified by coffee consumption and relative targets is shown in Supplementary Table S4.

4. Discussion

Epidemiological studies indicate that coffee intake favourably impacts on NAFLD prevalence and severity [30], although without fully clarified mechanisms. In this study we provide the first evidence that hepatoprotection induced by coffee in a mouse model is associated with the modulation of selected lncRNAs known to be involved in mechanisms related to NAFLD onset and progression such as impairment of lipid metabolism and circadian clock, pro-inflammatory state and activation of hepatic stellate cells.

Among the mechanisms connected to lipid metabolism and steatogenesis, Gm16551 has recently reported as a liver specific lncRNA downregulated in mice subjected to 24-h or a 12-week HFD that, through a negative feedback loop, reduces SREBP-1c functional activity thus inhibiting *de novo* lipogenesis [22]. In our study, Gm16551 was downregulated by a 12-week HFD, whereas coffee administration induced its expression. In agreement with histological improvement of steatosis, the induction of Gm16551 reduced the transcript for acetyl-CoA carboxylase 1 (Acaca), the enzyme that catalyzes the carboxylation of acetyl-CoA to malonyl-CoA, the first and rate-limiting step of *de novo* fatty acid biosynthesis [31]. Coffee intake reduced the mRNA level of Scd1, an enzyme that also contributes to steatogenesis [32]. Therefore, according to our data, a potential mechanism by which coffee reduces steatosis could be represented by Gm16551 expression induction.

It is known that NAFLD is associated with a chronic inflammatory state as evidenced in the liver of animal models and patients [33]. Although the 12-week HFD is a model of early NAFLD, as showed by histology, we found a slight increase of the lncRNA CARMN that is a pro-inflammatory mediator that is upregulated in macrophages treated in vitro with high glucose and palmitic acid and in macrophages isolated from diabetic mice and whose transient overexpression stimulates the expression of inflammatory genes and of CD36 [23]. This last aspect is relevant because in HepG2 treated with palmitate, lipid overload is exacerbated by the upregulation of the receptor involved in the uptake of lipids such as CD36 [16]. Thus, in our model, the downregulation of CARMN induced by coffee administration could contribute to the observed inflammation reduction, and to the reduction of lipid uptake and consequent steatosis grade. However, the downregulation of CARMN by coffee administration could explain the complete absence of inflammatory foci in coffee treated mice. Further studies are needed to confirm this hypothesis.

Another possible contribution in this direction may rely on lncRNA SRA. In fact, it has been reported that SRA genetic knockout protects against high fat diet-induced obesity [34] and hepatic steatosis [24]. In accordance with this evidence, HFD induces the expression of lncRNA SRA, while coffee co-administration decreases its expression level with respect to HFD. Thus, SRA downregulation could contribute to the observed reduced steatosis levels.

Since fibrosis is the main predictor of mortality in patients with NAFLD [35], it is relevant to identify molecular determinants of fibrogenesis. In this respect, experimental studies have reported the important role of the lncRNA H19. Zhu J et al. showed that H19 is overexpressed in the liver and primary hepatic stellate cells (HSCs) of mice with CCl4-induced liver fibrosis and demonstrated that the stable H19 overexpression induces the upregulation of α-SMA and Col1a1 both in vitro and in vivo [36]. Cholangiocyte-derived exosomal H19 stimulates trans-differentiation of mouse primary HSCs and induces proliferation and collagen production in HSC-derived fibroblasts [25]. In our study, we showed an up-regulation of H19 by HFD and a downregulation of H19, Col1a1 and αSMA by coffee intake. A main limitation in the interpretation of these results lies in the fact that we studied a model of early NAFLD that does not display fibrosis at H&E staining, although we cannot exclude the presence of small amount of pericellular or perisinusoidal fibrosis that could have been detected by Sirius Red. However, it is reliable to consider the upregulation of α-SMA as a marker of onset of the fibrogenesis process since in mice fed steatogenic diets the increase of α-SMA expression is confined at hepatic stellate cell level as showed by immunohistochemical analysis [37,38].

As concerns the circadian clock lncRNAs, Yi Chen et al., after performing a whole transcriptome analysis in an eight-week HFD mouse model, identified 266 differentially expressed lncRNA, among which they validated the expression of eight lncRNA through real time PCR [26]. To gain further insights into the molecular mechanisms regulated by such lncRNAs they performed a computational analysis that led to the identification of two fatty liver related lncRNAs associated with clock gene regulation, FLRL1 and FLRL2. They identified Per3 as a molecular target of FLRL1 computationally. The role of FLRL2 was investigated through transient inhibition in a cellular model of NAFLD; the authors demonstrated that FLRL2 downregulation is associated with Arnt downregulation at protein level [26]. However, physiological and pathophysiological functions of FLRL1 and FLRL2 and of their targets have not been elucidated so far and thus we cannot speculate on this aspect, although it deserves further exploration.

5. Conclusions

In this study we observed that decaffeinated coffee modulates expression of lncRNAs involved in key pathways of NAFLD onset and progression. Our data extend the knowledge concerning the molecular mechanism underlying beneficial effects exerted by coffee consumption against NAFLD.

Supplementary Materials: The following are available online at https://www.mdpi.com/article/10.3390/nu13092952/s1, Table S1: Diet composition, Table S2: Primer sequences of selected mouse lncRNAs and relative targets, Table S3: Liver histology scores, Table S4: List of analyzed lncRNAs modulated or not modulated by coffee supplementation and relative targets, Figure S1: Dot plots of CRNDE lncRNA and NEAT1 lncRNA, analyzed through qPCR in mice fed with Standard Diet (SD), High Fat Diet (HFD) and HFD plus decaffeinate coffee $n = 23$: 8 SD, 8 HFD, 7 HFD+ coffee. Transcript statistical significance of DE transcripts was evaluated with one-way ANOVA with Tukey post-hoc test for multiple comparisons (two-tailed p-value < 0.05); FC= Fold Change.

Author Contributions: S.D.M. Investigation, formal analysis, methodology, writing and visualization. F.S. Conceptualization, investigation, formal analysis, methodology writing and visualization. A.S. Investigation, formal analysis, methodology and visualization. A.F. Investigation and methodology. F.M. Investigation and resources. M.G. Investigation, methodology and formal analysis. V.L. Investigation. V.C. Investigation. R.M.P. Resources. S.G. Resources. R.M. Supervision. F.P. Resources and supervision. S.P. Project administration, supervision, writing—review. All authors have read and agreed to the published version of the manuscript.

Funding: This research received no external funding.

Institutional Review Board Statement: All experiments were performed according to the Ethics Committee of the University of Naples and approved by the Italian Minister of Scientific Research (Code 2014/0013808).

Informed Consent Statement: Not applicable.

Data Availability Statement: Data are contained within the article or Supplementary Material.

Acknowledgments: We wish to thank the Scientific Bureau of the University of Catania for language support. This study was in keeping with the objectives of the project "DEGENER-action", Department of Clinical and Experimental Medicine University of Catania.

Conflicts of Interest: The authors declare no conflict of interest.

References

1. Poole, R.; Kennedy, O.J.; Roderick, P.; Fallowfield, J.A.; Hayes, P.C.; Parkes, J. Coffee consumption and health: Umbrella review of meta-analyses of multiple health outcomes. *BMJ* **2017**, *359*, j5024. [CrossRef]
2. Marventano, S.; Salomone, F.; Godos, J.; Pluchinotta, F.; Del Rio, D.; Mistretta, A.; Grosso, G. Coffee and tea consumption in relation with non-alcoholic fatty liver and metabolic syndrome: A systematic review and meta-analysis of observational studies. *Clin. Nutr.* **2016**, *35*, 1269–1281. [CrossRef]

3. Zelber-Sagi, S.; Salomone, F.; Webb, M.; Lotan, R.; Yeshua, H.; Halpern, Z.; Santo, E.; Oren, R.; Shibolet, O. Coffee consumption and nonalcoholic fatty liver onset: A prospective study in the general population. *Transl. Res. J. Lab. Clin. Med.* **2015**, *165*, 428–436. [CrossRef] [PubMed]
4. Alferink, L.J.M.; Fittipaldi, J.; Kiefte-de Jong, J.C.; Taimr, P.; Hansen, B.E.; Metselaar, H.J.; Schoufour, J.D.; Ikram, M.A.; Janssen, H.L.A.; Franco, O.H.; et al. Coffee and herbal tea consumption is associated with lower liver stiffness in the general population: The Rotterdam study. *J. Hepatol.* **2017**, *67*, 339–348. [CrossRef] [PubMed]
5. Salomone, F.; Godos, J.; Zelber-Sagi, S. Natural antioxidants for non-alcoholic fatty liver disease: Molecular targets and clinical perspectives. *Liver Int. Off. J. Int. Assoc. Study Liver* **2016**, *36*, 5–20. [CrossRef] [PubMed]
6. Salomone, F.; Galvano, F.; Li Volti, G. Molecular Bases Underlying the Hepatoprotective Effects of Coffee. *Nutrients* **2017**, *9*, 85. [CrossRef]
7. Zhao, X.Y.; Lin, J.D. Long Noncoding RNAs: A New Regulatory Code in Metabolic Control. *Trends Biochem. Sci.* **2015**, *40*, 586–596. [CrossRef] [PubMed]
8. Wang, S.; Mao, C.; Liu, S. Peptides encoded by noncoding genes: Challenges and perspectives. *Signal Transduct. Target. Ther.* **2019**, *4*, 57. [CrossRef] [PubMed]
9. Zhang, X.; Wang, W.; Zhu, W.; Dong, J.; Cheng, Y.; Yin, Z.; Shen, F. Mechanisms and Functions of Long Non-Coding RNAs at Multiple Regulatory Levels. *Int. J. Mol. Sci.* **2019**, *20*, 5573. [CrossRef] [PubMed]
10. Noh, J.H.; Kim, K.M.; McClusky, W.G.; Abdelmohsen, K.; Gorospe, M. Cytoplasmic functions of long noncoding RNAs. *Wiley Interdiscip. Rev. RNA* **2018**, *9*, e1471. [CrossRef]
11. Marchese, F.P.; Raimondi, I.; Huarte, M. The multidimensional mechanisms of long noncoding RNA function. *Genome Biol.* **2017**, *18*, 206. [CrossRef] [PubMed]
12. Losko, M.; Kotlinowski, J.; Jura, J. Long Noncoding RNAs in Metabolic Syndrome Related Disorders. *Mediat. Inflamm.* **2016**, *2016*, 5365209. [CrossRef]
13. Giroud, M.; Scheideler, M. Long Non-Coding RNAs in Metabolic Organs and Energy Homeostasis. *Int. J. Mol. Sci.* **2017**, *18*, 2578. [CrossRef] [PubMed]
14. Shabgah, A.G.; Norouzi, F.; Hedayati-Moghadam, M.; Soleimani, D.; Pahlavani, N.; Navashenaq, J.G. A comprehensive review of long non-coding RNAs in the pathogenesis and development of non-alcoholic fatty liver disease. *Nutr. Metab.* **2021**, *18*, 22. [CrossRef] [PubMed]
15. Vitaglione, P.; Mazzone, G.; Lembo, V.; D'Argenio, G.; Rossi, A.; Guido, M.; Savoia, M.; Salomone, F.; Mennella, I.; De Filippis, F.; et al. Coffee prevents fatty liver disease induced by a high-fat diet by modulating pathways of the gut-liver axis. *J. Nutr. Sci.* **2019**, *8*, e15. [CrossRef]
16. Di Mauro, S.; Ragusa, M.; Urbano, F.; Filippello, A.; Di Pino, A.; Scamporrino, A.; Pulvirenti, A.; Ferro, A.; Rabuazzo, A.M.; Purrello, M.; et al. Intracellular and extracellular miRNome deregulation in cellular models of NAFLD or NASH: Clinical implications. *Nutr. Metab. Cardiovasc. Dis. NMCD* **2016**, *26*, 1129–1139. [CrossRef]
17. Scicali, R.; Di Pino, A.; Pavanello, C.; Ossoli, A.; Strazzella, A.; Alberti, A.; Di Mauro, S.; Scamporrino, A.; Urbano, F.; Filippello, A.; et al. Analysis of HDL-microRNA panel in heterozygous familial hypercholesterolemia subjects with LDL receptor null or defective mutation. *Sci. Rep.* **2019**, *9*, 20354. [CrossRef] [PubMed]
18. Di Mauro, S.; Scamporrino, A.; Fruciano, M.; Filippello, A.; Fagone, E.; Gili, E.; Scionti, F.; Purrazzo, G.; Di Pino, A.; Scicali, R.; et al. Circulating Coding and Long Non-Coding RNAs as Potential Biomarkers of Idiopathic Pulmonary Fibrosis. *Int. J. Mol. Sci.* **2020**, *21*, 8812. [CrossRef]
19. Di Mauro, S.; Scamporrino, A.; Petta, S.; Urbano, F.; Filippello, A.; Ragusa, M.; Di Martino, M.T.; Scionti, F.; Grimaudo, S.; Pipitone, R.M.; et al. Serum coding and non-coding RNAs as biomarkers of NAFLD and fibrosis severity. *Liver Int. Off. J. Int. Assoc. Study Liver* **2019**, *39*, 1742–1754. [CrossRef]
20. Filippello, A.; Urbano, F.; Di Mauro, S.; Scamporrino, A.; Di Pino, A.; Scicali, R.; Rabuazzo, A.M.; Purrello, F.; Piro, S. Chronic Exposure to Palmitate Impairs Insulin Signaling in an Intestinal L-cell Line: A Possible Shift from GLP-1 to Glucagon Production. *Int. J. Mol. Sci.* **2018**, *19*, 3791. [CrossRef] [PubMed]
21. Filippello, A.; Scamporrino, A.; Di Mauro, S.; Malaguarnera, R.; Di Pino, A.; Scicali, R.; Purrello, F.; Piro, S. Direct Effects of D-Chiro-Inositol on Insulin Signaling and Glucagon Secretion of Pancreatic Alpha Cells. *Biomolecules* **2020**, *10*, 1404. [CrossRef] [PubMed]
22. Yang, L.; Li, P.; Yang, W.; Ruan, X.; Kiesewetter, K.; Zhu, J.; Cao, H. Integrative Transcriptome Analyses of Metabolic Responses in Mice Define Pivotal LncRNA Metabolic Regulators. *Cell Metab.* **2016**, *24*, 627–639. [CrossRef] [PubMed]
23. Reddy, M.A.; Chen, Z.; Park, J.T.; Wang, M.; Lanting, L.; Zhang, Q.; Bhatt, K.; Leung, A.; Wu, X.; Putta, S.; et al. Regulation of inflammatory phenotype in macrophages by a diabetes-induced long noncoding RNA. *Diabetes* **2014**, *63*, 4249–4261. [CrossRef]
24. Liu, S.; Sheng, L.; Miao, H.; Saunders, T.L.; MacDougald, O.A.; Koenig, R.J.; Xu, B. SRA gene knockout protects against diet-induced obesity and improves glucose tolerance. *J. Biol. Chem.* **2014**, *289*, 13000–13009. [CrossRef] [PubMed]
25. Liu, R.; Li, X.; Zhu, W.; Wang, Y.; Zhao, D.; Wang, X.; Gurley, E.C.; Liang, G.; Chen, W.; Lai, G.; et al. Cholangiocyte-Derived Exosomal Long Noncoding RNA H19 Promotes Hepatic Stellate Cell Activation and Cholestatic Liver Fibrosis. *Hepatology* **2019**, *70*, 1317–1335. [CrossRef] [PubMed]
26. Chen, Y.; Huang, H.; Xu, C.; Yu, C.; Li, Y. Long Non-Coding RNA Profiling in a Non-Alcoholic Fatty Liver Disease Rodent Model: New Insight into Pathogenesis. *Int. J. Mol. Sci.* **2017**, *18*, 21. [CrossRef] [PubMed]

27. Chen, Y.; Chen, X.; Gao, J.; Xu, C.; Xu, P.; Li, Y.; Zhu, Y.; Yu, C. Long noncoding RNA FLRL2 alleviated nonalcoholic fatty liver disease through Arntl-Sirt1 pathway. *FASEB J. Off. Publ. Fed. Am. Soc. Exp. Biol.* **2019**, *33*, 11411–11419. [CrossRef] [PubMed]
28. Ellis, B.C.; Graham, L.D.; Molloy, P.L. CRNDE, a long non-coding RNA responsive to insulin/IGF signaling, regulates genes involved in central metabolism. *Biochim. Biophys. Acta* **2014**, *1843*, 372–386. [CrossRef] [PubMed]
29. Huang-Fu, N.; Cheng, J.S.; Wang, Y.; Li, Z.W.; Wang, S.H. Neat1 regulates oxidized low-density lipoprotein-induced inflammation and lipid uptake in macrophages via paraspeckle formation. *Mol. Med. Rep.* **2018**, *17*, 3092–3098. [CrossRef] [PubMed]
30. Chen, Y.P.; Lu, F.B.; Hu, Y.B.; Xu, L.M.; Zheng, M.H.; Hu, E.D. A systematic review and a dose-response meta-analysis of coffee dose and nonalcoholic fatty liver disease. *Clin. Nutr.* **2019**, *38*, 2552–2557. [CrossRef]
31. Colbert, C.L.; Kim, C.W.; Moon, Y.A.; Henry, L.; Palnitkar, M.; McKean, W.B.; Fitzgerald, K.; Deisenhofer, J.; Horton, J.D.; Kwon, H.J. Crystal structure of Spot 14, a modulator of fatty acid synthesis. *Proc. Natl. Acad. Sci. USA* **2010**, *107*, 18820–18825. [CrossRef]
32. Sampath, H.; Miyazaki, M.; Dobrzyn, A.; Ntambi, J.M. Stearoyl-CoA desaturase-1 mediates the pro-lipogenic effects of dietary saturated fat. *J. Biol. Chem.* **2007**, *282*, 2483–2493. [CrossRef] [PubMed]
33. Farrell, G.C.; van Rooyen, D.; Gan, L.; Chitturi, S. NASH is an Inflammatory Disorder: Pathogenic, Prognostic and Therapeutic Implications. *Gut Liver* **2012**, *6*, 149–171. [CrossRef]
34. Chen, G.; Yu, D.; Nian, X.; Liu, J.; Koenig, R.J.; Xu, B.; Sheng, L. LncRNA SRA promotes hepatic steatosis through repressing the expression of adipose triglyceride lipase (ATGL). *Sci. Rep.* **2016**, *6*, 35531. [CrossRef] [PubMed]
35. Taylor, R.S.; Taylor, R.J.; Bayliss, S.; Hagstrom, H.; Nasr, P.; Schattenberg, J.M.; Ishigami, M.; Toyoda, H.; Wai-Sun Wong, V.; Peleg, N.; et al. Association Between Fibrosis Stage and Outcomes of Patients With Nonalcoholic Fatty Liver Disease: A Systematic Review and Meta-Analysis. *Gastroenterology* **2020**, *158*, 1611–1625. [CrossRef] [PubMed]
36. Zhu, J.; Luo, Z.; Pan, Y.; Zheng, W.; Li, W.; Zhang, Z.; Xiong, P.; Xu, D.; Du, M.; Wang, B.; et al. H19/miR-148a/USP4 axis facilitates liver fibrosis by enhancing TGF-beta signaling in both hepatic stellate cells and hepatocytes. *J. Cell. Physiol.* **2019**, *234*, 9698–9710. [CrossRef] [PubMed]
37. Dungubat, E.; Watabe, S.; Togashi-Kumagai, A.; Watanabe, M.; Kobayashi, Y.; Harada, N.; Yamaji, R.; Fukusato, T.; Lodon, G.; Sevjid, B.; et al. Effects of Caffeine and Chlorogenic Acid on Nonalcoholic Steatohepatitis in Mice Induced by Choline-Deficient, L-Amino Acid-Defined, High-Fat Diet. *Nutrients* **2020**, *12*, 3886. [CrossRef] [PubMed]
38. Wei, G.; An, P.; Vaid, K.A.; Nasser, I.; Huang, P.; Tan, L.; Zhao, S.; Schuppan, D.; Popov, Y.V. Comparison of murine steatohepatitis models identifies a dietary intervention with robust fibrosis, ductular reaction, and rapid progression to cirrhosis and cancer. *Am. J. Physiol. Gastrointest. Liver Physiol.* **2020**, *318*, G174–G188. [CrossRef] [PubMed]

nutrients

MDPI

Article

Prevalence of Micronutrient Deficiencies in Patients Hospitalized with COVID-19: An Observational Cohort Study

Manyola Voelkle [1,2], Claudia Gregoriano [1], Peter Neyer [3], Daniel Koch [1], Alexander Kutz [1], Luca Bernasconi [3], Anna Conen [2,4], Beat Mueller [1,2] and Philipp Schuetz [1,2,*]

1 Medical University Department of Medicine, Kantonsspital Aarau, 5001 Aarau, Switzerland; manyola.voelkle@ksa.ch (M.V.); claudia.gregoriano@ksa.ch (C.G.); daniel.koch@ksa.ch (D.K.); alexander.kutz@ksa.ch (A.K.); beat.mueller@ksa.ch (B.M.)
2 Faculty of Medicine, University of Basel, 4056 Basel, Switzerland; anna.conen@ksa.ch
3 Institute of Laboratory Medicine, Kantonsspital Aarau, 5001 Aarau, Switzerland; peter.neyer@ksa.ch (P.N.); luca.bernasconi@ksa.ch (L.B.)
4 Department of Infectious Diseases and Infection Prevention, Kantonsspital Aarau, 5001 Aarau, Switzerland
* Correspondence: schuetzph@gmail.com; Tel.: +41-62-838-9524

Abstract: Background: A higher risk for severe clinical courses of coronavirus disease 2019 (COVID-19) has been linked to deficiencies of several micronutrients. We therefore studied the prevalence of deficiencies of eight different micronutrients in a cohort of hospitalized COVID-19-patients. Methods: We measured admission serum/plasma levels of vitamins A, B12, D, and E, as well as folic acid, zinc, selenium, and copper in 57 consecutively admitted adult patients with confirmed COVID-19 and analyzed prevalence of micronutrient deficiencies and correlations among micronutrient levels. Further, we studied associations of micronutrient levels with severe disease progression, a composite endpoint consisting of in-hospital mortality and/or need for intensive care unit (ICU) treatment with logistic regression. Results: Median age was 67.0 years (IQR 60.0, 74.2) and 60% (n = 34) were male. Overall, 79% (n = 45) of patients had at least one deficient micronutrient level and 33% (n = 19) had ≥3 deficiencies. Most prevalent deficiencies were found for selenium, vitamin D, vitamin A, and zinc (51%, 40%, 39%, and 39%, respectively). We found several correlations among micronutrients with correlation coefficients ranging from r = 0.27 to r = 0.42. The strongest associations with lower risk for severe COVID-19 disease progression (adjusted odds ratios) were found for higher levels of vitamin A (0.18, 95% CI 0.05–0.69, p = 0.01), zinc (0.73, 95% CI 0.55–0.98, p = 0.03), and folic acid (0.88, 95% CI 0.78–0.98, p = 0.02). Conclusions: We found a high prevalence of micronutrient deficiencies in mostly older patients hospitalized for COVID-19, particularly regarding selenium, vitamin D, vitamin A, and zinc. Several deficiencies were associated with a higher risk for more severe COVID-19 courses. Whether supplementation of micronutrients is useful for prevention of severe clinical courses or treatment of COVID-19 warrants further research.

Keywords: COVID-19; micronutrients; deficiency; SARS-CoV-2; hospital outcomes

Citation: Voelkle, M.; Gregoriano, C.; Neyer, P.; Koch, D.; Kutz, A.; Bernasconi, L.; Conen, A.; Mueller, B.; Schuetz, P. Prevalence of Micronutrient Deficiencies in Patients Hospitalized with COVID-19: An Observational Cohort Study. *Nutrients* **2022**, *14*, 1862. https://doi.org/10.3390/nu14091862

Academic Editors: Omorogieva Ojo and Amanda R Amorim Adegboye

Received: 30 March 2022
Accepted: 26 April 2022
Published: 29 April 2022

1. Introduction

Over the last two years, severe acute respiratory syndrome coronavirus 2 (SARS-CoV-2) led to a global pandemic with high morbidity and mortality, causing over 426 million infections and almost 6 million deaths up to February 2022 [1]. Clinical courses range from asymptomatic infection to severe disease with need for intensive care unit (ICU) stay and death [2]. Different risk factors for severe COVID-19 have been described and include older age, frailty, and higher burden of comorbidities [3,4]. However, because older and frail patients are often malnourished and have a higher likelihood for low levels in different specific micronutrients [5,6], micronutrient deficiencies could additionally contribute to more severe courses in patients infected with SARS-CoV-2. From a preclinical perspective, the importance of various micronutrients for a functioning immune system has been well documented [7–9]. For example, it has been shown that vitamin A plays an important role in maintaining mucosal integrity [10], while zinc plays an essential role in protecting against reactive oxygen and nitrogen species [11,12]. Furthermore, vitamin D increases the excretion of antimicrobial peptides in epithelial lining cells in the respiratory tract [13] and is involved in the modulation of pro- and anti-inflammatory cytokine production [14]. Consequently, deficiencies in micronutrients may lead to higher susceptibility for infections.

With respect to COVID-19, deficiencies in different micronutrients, including vitamin D [15,16], zinc [17,18], and selenium [19,20], have been discussed as risk factors for a more severe disease course, with need for ICU admission and mechanical ventilation, or higher incidence of death. Based on these observations, there has been a call for more wide-spread supplementation of the above-mentioned micronutrients during the COVID-19 pandemic to prevent and improve courses of infected patients [7,21]. This call is particularly timely, since a high prevalence of deficiencies has been reported from different countries, including Switzerland [22,23]. Still, until now, there is insufficient research concerning the association of micronutrients levels with clinical courses of COVID-19.

Herein, we analyzed different micronutrient levels in a COVID-19 cohort and described the distribution of deficiencies, as well as the correlation among levels of different micronutrients. Further, the association between deficiencies in micronutrients and severe progression of COVID-19 disease was investigated.

2. Subjects and Methods

2.1. Patient Population

This prospective observational study involved adult patients (≥18 years) hospitalized with a confirmed SARS-CoV-2 infection between 17 March 2020 and 30 April 2020 at the Cantonal Hospital Aarau (Switzerland), a tertiary care hospital. Baseline characteristics of this cohort have been published elsewhere [24]. In brief, patients were included if they had typical clinical symptoms of a SARS-CoV-2 infection (e.g., respiratory symptoms with or without fever and/or pulmonary infiltrates) and a positive real-time reverse transcription polymerase chain reaction test (RT-PCR) taken from nasopharyngeal swabs or lower respiratory tract specimens, according to the World Health Organization (WHO) guidance [25]. Written general informed consent was obtained from all analyzed patients. The study was approved by the local ethics committee (EKZN, 2020-01306) and performed in conformance with the Declaration of Helsinki ethical guidelines. For the present analysis, only patients with a complete micronutrient status were included, whereas patients receiving either an oral or intravenous substitution of the analyzed micronutrients at the hospital before obtaining the blood samples were excluded. Patients taking micronutrient supplements at home were included, as we aimed to display micronutrient status at the time of hospital admission.

2.2. Data Collection

Clinical data were collected by chart abstraction and automatic export from the electronic health records and included socio-demographics, comorbidities, and pre-existing risk factors for a severe COVID-19 course. Comorbidities were classified according to International Statistical Classification of Diseases and Related Health Problems codes (ICD10). For all patients, age-adjusted Charlson Comorbidity Index [26] and Clinical Frailty Scale [27] were calculated. Patient outcomes, including in-hospital mortality, admission to ICU, need for invasive ventilation, and length of hospital stay (LOS), were collected by chart review. To assess nutritional status, we calculated nutritional risk screening (NRS) 2002 score [28] and body mass index (BMI).

2.3. Laboratory Analysis

Laboratory values correspond to blood samples obtained within the first four days of hospitalization. Serum or plasma levels for vitamins A, B12, D, and E, as well as folic acid, zinc, selenium, and copper, were measured. As a surrogate for vitamin A and for vitamin

E we measured total retinol and total α-tocopherol concentrations, respectively, by high performance liquid chromatography (HPLC). The method was modified for small sample volume and based on best practice guideline ([29], modified). Vitamin B12 and folates were measured by chemiluminescence microparticle immunoassays on an Abbott Alinity i system. Total vitamin D (cholecalciferol) concentrations were measured by chemiluminescence immunoassay on a DiaSorin LIAISON XL system. Trace elements (copper, selenium, zinc) were measured by inductively coupled plasma mass spectrometry in collision mode (helium). The method was set up without digestion and simple dilution with an alkaline solution containing an internal standard element (rhodium). Table S1 shows cut-off values for deficiencies.

2.4. Outcomes

The primary outcome included the assessment of different micronutrient levels and the prevalence of deficiencies in patients hospitalized with COVID-19. Secondary outcomes included the association of micronutrient levels and a composite adverse outcome, defined as ICU admission and/or all-cause in-hospital mortality.

2.5. Statistical Analyses

Discrete variables are expressed as frequency (percentage) and continuous variables as medians (interquartile range (IQR)) or means (standard deviation (SD)). Values for vitamin D, vitamin B12, and folic acid were left-censored and values for vitamin B12 and folic acid were right-censored. For statistical analyses, we replaced these values with the corresponding limit values. To test for normal distribution of the analyzed variables, the Shapiro–Wilk test was used. The correlation of different micronutrients was investigated by using a Spearman's rank correlation analysis and reported as Spearman's rank coefficient rho with the corresponding *p*-value. Further, we investigated the association of initial micronutrient levels with the composite endpoint of transfer to the ICU and/or all-cause in-hospital mortality with a logistic regression analysis. Odds ratios (OR) including the corresponding 95% confidence intervals (CI) were reported as a measure of association for both micronutrients as continuous and binary (deficient vs. non-deficient) variables. We adjusted the analyses only for age, since a multivariable regression was not possible due to the small sample size, to avoid over-fitting of the model. A two-sided *p*-value of < 0.05 was considered significant. Statistical analyses were performed using Stata, version 15.1 (StataCorp LLC, College Station, TX, USA).

3. Results

Overall, 74 patients with confirmed COVID-19 were eligible. Ten patients had to be excluded because of micronutrient substitution before blood draw and seven patients because of incomplete micronutrient values. Therefore, 57 patients were included for the final analysis. Figure 1 provides an overview of the study flow.

3.1. Baseline Characteristics

Table 1 shows patient demographics and comorbidities, as well as micronutrient levels and deficiencies, outcomes, and nutritional status stratified by the number of micronutrient deficiencies. Patients were divided into groups with "no", "one", "two", or "multiple" micronutrient deficiencies. In total, a third (n = 19, 33%) had three or more (multiple) micronutrient deficiencies and 79% (n = 45) at least one deficiency. Among all patients, selenium, vitamin D, vitamin A, and zinc deficiencies were most prevalent (51%, 40%, 39%, and 39%, respectively). Selenium, vitamin D, vitamin A, and zinc levels were lower when patients had more micronutrient deficiencies ($p < 0.01$). In patients with a single micronutrient deficiency, selenium deficiency was most prevalent (n = 5, 50%). No patient had a vitamin E deficiency and vitamin B12, folic acid, and copper deficiencies were rare with 7%, 5%, and 2%, respectively.

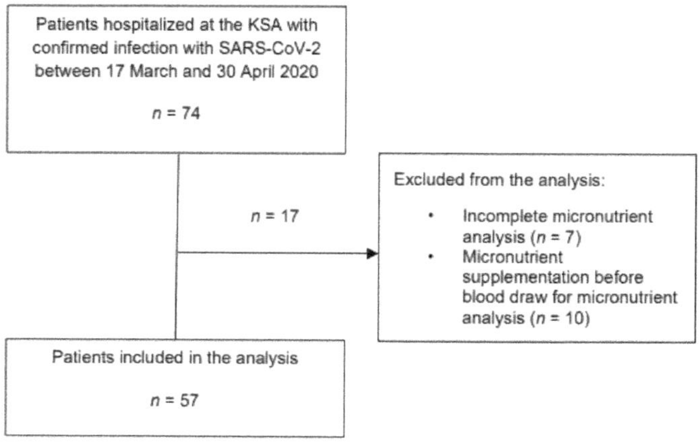

Figure 1. Overview of the study flow.

Table 1. Baseline characteristics and micronutrient levels stratified by number of micronutrient deficiencies.

	All	No Deficiency	One Deficiency	Two Deficiencies	Multiple Deficiencies	*p*-Value °
	n = 57	*n* = 12	*n* = 10	*n* = 16	*n* = 19	
Socio-demographics						
Age (years), median (IQR)	67.0 (60.0, 74.2)	64.5 (57.8, 74.3)	71.0 (65.1, 72.4)	63.6 (58.9, 73.4)	67.5 (55.9, 76.6)	0.87
Gender (male), *n* (%)	34 (60)	6 (50)	7 (70)	10 (63)	11 (58)	0.80
Nationality						
Swiss, *n* (%)	37 (65)	9 (75)	8 (80)	11 (69)	9 (47)	0.46
Others, *n* (%)	13 (23)	3 (25)	1 (10)	3 (19)	6 (32)	
Unknown, *n* (%)	7 (12)	0 (0)	1 (10)	2 (13)	4 (21)	
Pre-existing risk-factors						
Active smoker, *n* (%)	5 (12)	1 (11)	1 (14)	1 (7)	2 (15)	0.92
Immunosuppressant, *n* (%)	1 (2)	0 (0)	1 (10)	0 (0)	0 (0)	0.19
Pre-admission history						
Transfer from another hospital, *n* (%)	14 (25)	0 (0)	1 (10)	6 (38)	7 (37)	0.05
Symptom onset before admission (days), median (IQR)	7.0 (5.0, 11.0)	8.0 (6.0, 11.0)	9.0 (6.0, 14.0)	7.5 (4.5, 12.5)	5.5 (3.0, 9.0)	0.38
Comorbidities						
Cancer, *n* (%)	5 (9)	0 (0)	1 (10)	2 (13)	2 (11)	0.68
Hypertension, *n* (%)	35 (61)	6 (50)	7 (70)	11 (69)	11 (58)	0.70
Coronary artery disease, *n* (%)	16 (28)	3 (25)	6 (60)	3 (19)	4 (21)	0.10
Chronic heart failure, *n* (%)	3 (5)	1 (8)	2 (20)	0 (0)	0 (0)	0.09
Asthma, *n* (%)	11 (19)	2 (17)	3 (30)	3 (19)	3 (16)	0.82

Table 1. *Cont.*

	All	No Deficiency	One Deficiency	Two Deficiencies	Multiple Deficiencies	*p*-Value °
Chronic obstructive pulmonary disease, *n* (%)	3 (5)	0 (0)	2 (20)	0 (0)	1 (5)	0.12
Obstructive sleep apnea syndrome, *n* (%)	10 (18)	2 (17)	3 (30)	2 (13)	3 (16)	0.71
Active rheumatic disease, *n* (%)	1 (2)	0 (0)	1 (10)	0 (0)	0 (0)	0.19
Chronic kidney disease, *n* (%)	17 (30)	3 (25)	5 (50)	5 (31)	4 (21)	0.42
Diabetes, *n* (%)	18 (32)	4 (33)	5 (50)	3 (19)	6 (32)	0.42
Age-adjusted Charlson comorbidity index, median (IQR)	3.0 (2.0, 6.0)	2.0 (1.0, 6.5)	6.5 (2.0, 9.0)	3.0 (2.0, 5.0)	3.0 (2.0, 5.0)	0.21
Clinical frailty score, median (IQR)	3.0 (2.0, 4.0)	2.5 (2.0, 3.5)	3.0 (3.0, 5.0)	3.0 (3.0, 5.0)	3.0 (2.0, 4.0)	0.51
Outcomes						
Length of hospital stay (days), median (IQR)	9.0 (5.0, 14.0)	5.0 (4.0, 8.5)	5.0 (4.0, 10.0)	10.0 (8.0, 11.5)	19.0 (6.0, 22.0)	0.02
ICU admission, *n* (%)	12 (21)	0 (0)	0 (0)	3 (19)	9 (47)	<0.01
Need for mechanical ventilation, *n* (%)	10 (18)	0 (0)	0 (0)	2 (13)	8 (42)	<0.01
In-hospital death, *n* (%)	6 (11)	0 (0)	3 (30)	1 (6)	2 (11)	0.12
Micronutrients						
Vitamin A						
Median (µmol/L), (IQR)	1.2 (0.8, 1.7)	1.8 (1.6, 2.5)	1.9 (1.4, 2.4)	1.0 (0.7, 1.4)	0.8 (0.6, 1.1)	<0.01
Deficiency, *n* (%)	22 (39)	0 (0)	1 (10)	8 (50)	13 (68)	<0.01
Vitamin B12						
Median (pmol/L), (IQR)	292.0 (207.0, 548.0)	255.0 (213.0, 560.0)	268.0 (171.0, 493.0)	545.0 (335.5, 706.0)	237.0 (182.0, 310.0)	0.04
Deficiency, *n* (%)	4 (7)	0 (0)	0 (0)	1 (6)	3 (16)	0.27
Vitamin D						
Median (nmol/L), (IQR)	34.6 (24.3, 62.2)	61.0 (51.2, 80.2)	65.0 (36.4, 70.0)	32.4 (26.8, 52.0)	23.6 (15.2, 31.5)	<0.01
Deficiency, *n* (%)	23 (40)	0 (0)	2 (20)	7 (44)	14 (74)	<0.01
Vitamin E						
Mean (µmol/L), (SD)	32.9 (8.6)	35.7 (7.5)	31.6 (8.3)	35.2 (10.1)	30.0 (7.7)	0.19
Deficiency, *n* (%)	0 (0)	0 (0)	0 (0)	0 (0)	0 (0)	n.a.
Folic acid						
Median (nmol/L), (IQR)	15.0 (10.2, 23.3)	18.4 (13.6, 23.9)	19.6 (11.7, 27.6)	15.3 (9.7, 21.6)	11.8 (8.2, 17.9)	0.07
Deficiency, *n* (%)	3 (5)	0 (0)	0 (0)	0 (0)	3 (16)	0.10
Zinc						
Median (µmol/L), (IQR)	10.9 (8.8, 12.8)	13.9 (12.4, 16.0)	12.6 (11.4, 14.4)	10.7 (8.5, 11.7)	9.3 (8.3, 10.7)	<0.01
Deficiency, *n* (%)	22 (39)	0 (0)	2 (20)	7 (44)	13 (68)	<0.01
Selenium						
Mean (µmol/L), (SD)	0.96 (0.29)	1.20 (0.18)	1.06 (0.28)	0.92 (0.26)	0.78 (0.24)	<0.01

Table 1. *Cont.*

	All	No Deficiency	One Deficiency	Two Deficiencies	Multiple Deficiencies	*p*-Value °
Deficiency, *n* (%)	29 (51)	0 (0)	5 (50)	9 (56)	15 (79)	<0.01
Copper						
Mean (μmol/L), (SD)	21.2 (4.0)	21.3 (3.3)	22.1 (3.6)	21.3 (4.4)	20.7 (4.4)	0.85
Deficiency, *n* (%)	1 (2)	0 (0)	0 (0)	0 (0)	1 (5)	0.57
Nutritional assessment						
NRS $\geq$ 3, *n* (%)	8 (19)	0 (0)	3 (38)	2 (20)	3 (19)	0.27
BMI						
18.5–24.9 kg/m^2, *n* (%)	15 (33)	3 (33)	0 (0)	6 (46)	6 (38)	0.36
25–29.9 kg/m^2, *n* (%)	17 (37)	4 (44)	5 (63)	4 (31)	4 (25)	
$\geq$30 kg/m^2, *n* (%)	14 (30)	2 (22)	3 (38)	3 (23)	6 (38)	

° ANOVA for normally distributed variables, Kruskal–Wallis for continuous variables, Pearson's chi squared test for binary and categorical variables. Abbreviations: BMI, body mass index; IQR, interquartile range; NRS, nutritional risk score; SD, standard deviation.

Median age was 67.0 years (IQR 60.0, 74.2) and 60% (*n* = 34) were male. Age, gender and most comorbidities were relatively equally distributed within the different categories of micronutrient deficiencies. Regarding clinical outcomes of COVID-19, we found higher risks for longer LOS, ICU admissions, and mechanical ventilation with more deficiencies.

3.2. Correlation of Different Micronutrient Values

The correlations of different micronutrient levels are shown in Table 2. Significant correlations for vitamin D and folic acid (r = 0.39) were found, as well as for vitamin D and vitamin A (r = 0.27) and vitamin D and selenium (r = 0.32). Further, there was a significant positive correlation for folic acid and selenium (r = 0.3) and vitamin A and zinc (r = 0.42), as well as copper and zinc and copper and selenium (r = 0.37 and 0.29, respectively). All these correlations were moderate, with correlation coefficients in the range of 0.28 to 0.42. A negative correlation was observed between vitamin A and B12 (r = −0.28). No significant correlations were found between vitamin E and other micronutrients.

Table 2. Spearman's rank correlation among micronutrient levels.

	Vitamin A	Vitamin B12	Vitamin D	Vitamin E	Folic Acid	Zinc	Selenium	Copper
Vitamin A	1							
Vitamin B12	**−0.28, *p* = 0.04**	1						
Vitamin D	**0.27, *p* = 0.04**	−0.04, *p* = 0.75	1					
Vitamin E	0.24, *p* = 0.08	0.19, *p* = 0.15	0.08, *p* = 0.56	1				
Folic acid	0.21, *p* = 0.11	−0.008, *p* = 0.95	**0.39, *p* = 0.002**	0.04, *p* = 0.79	1			
Zinc	**0.42, *p* = 0.001**	−0,17, *p* = 0.22	0.19, *p* = 0.16	0.18, *p* = 0.18	−0.03, *p* = 0.83	1		
Selenium	0.20, *p* = 0.13	0.07, *p* = 0.59	**0.32, *p* = 0.02**	0.25, *p* = 0.06	**0.30, *p* = 0.02**	0.26, *p* = 0.05	1	
Copper	−0.07, *p* = 0.58	−0.04, *p* = 0.75	−0.14, *p* = 0.28	0.26, *p* = 0.05	0.05, *p* = 0.72	**0.37, *p* = 0.004**	**0.29, *p* = 0.03**	1

Bold if statistically significant.

3.3. Association of Micronutrient Levels with ICU Admission and In-Hospital Mortality

In Table 3, micronutrient levels in association with the composite outcome of ICU admission and/or in-hospital mortality in patients with SARS-CoV-2 infection are shown. Median admission folic acid values were over 1.5-fold higher in patients with a mild course of COVID-19 compared to those with a severe course (16.6 nmol/L (IQR 11.4, 24.0) vs. 10.2 nmol/L (IQR 8.2, 14.4), $p < 0.01$). Median vitamin A levels were over two times higher in patients with mild compared to severe courses of COVID-19 (1.5 µmol/L (IQR 1.0, 2.0) vs. 0.7 µmol/L (IQR 0.4, 1.1), $p < 0.01$). Vitamin A and zinc deficiencies were almost threefold more prevalent in patients with severe COVID-19 (73% vs. 26%, $p < 0.01$, 67% vs. 29%, $p < 0.01$, respectively).

Table 3. Micronutrient levels stratified by composite endpoint and crude and adjusted association of micronutrient levels and composite endpoint of ICU admission and/or in-hospital mortality.

	Mild Disease	Severe Disease	p-Value °	Univariable OR (95% CI), p-Value	Adjusted OR * (95% CI), p-Value
	n = 42	n = 15			
Vitamin A					
Median (µmol/L), (IQR)	1.5 (1.0, 2.0)	0.7 (0.4, 1.1)	<0.01	0.17 (0.05–0.66), **p = 0.01**	0.18 (0.05–0.69), **p = 0.01**
Deficiency, n (%)	11 (26)	11 (73)	<0.01	7.75 (2.04–29.46), **p = 0.003**	7.41 (1.91–28.68), **p = 0.004**
Vitamin B12					
Median (pmol/L), (IQR)	290.0 (200.0, 597.0)	310.0 (220.0, 497.0)	0.82	1.00 (0.99–1.00), p = 0.91	1.00 (0.99–1.00), p = 0.91
Deficiency, n (%)	4 (10)	0 (0)	0.22	n.a.	n.a.
Vitamin D					
Median (nmol/L), (IQR)	34.4 (24.3, 65.0)	34.6 (16.2, 46.8)	0.31	0.99 (0.96–1.01), p = 0.32	0.99 (0.96–1.01), p = 0.38
Deficiency, n (%)	16 (38)	7 (47)	0.56	1.42 (0.43–4.68), p = 0.56	1.44 (0.43–4.79), p = 0.55
Vitamin E					
Mean (µmol/L), (SD)	33.1 (9.0)	32.5 (7.7)	0.81	0.99 (0.92–1.06), p = 0.81	0.99 (0.93–1.07), p = 0.88
Deficiency, n (%)	0 (0)	0 (0)	n.a.	n.a.	n.a.
Folic acid					
Median (nmol/L), (IQR)	16.6 (11.4, 24.0)	10.2 (8.2, 14.4)	<0.01	0.88 (0.79–0.98), **p = 0.02**	0.88 (0.78–0.98), **p = 0.02**
Deficiency, n (%)	0 (0)	3 (20)	<0.01	n.a.	n.a.
Zinc					
Median (µmol/L), (IQR)	11.7 (9.8, 13.5)	9.3 (8.3, 11.4)	0.03	0.77 (0.60–0.99), **p = 0.04**	0.73 (0.55–0.98), **p = 0.03**
Deficiency, n (%)	12 (29)	10 (67)	<0.01	5 (1.41–17.72), **p = 0.01**	7.18 (1.73–29.76), **p = 0.007**
Selenium					
Mean (µmol/L), (SD)	0.9 (0.3)	1.0 (0.3)	0.66	1.62 (0.20–13.11), p = 0.65	1.39 (0.16–12.27), p = 0.77
Deficiency, n (%)	21 (50)	8 (53)	0.82	1.14 (0.35–3.72), p = 0.83	1.19 (0.36–3.93), p = 0.77

Table 3. *Cont.*

	Mild Disease	Severe Disease	*p*-Value °	Univariable OR (95% CI), *p*-Value	Adjusted OR * (95% CI), *p*-Value
Copper					
Mean (µmol/L), (SD)	21.2 (4.4)	21.3 (2.9)	0.93	1.00 (0.87–1.17), $p = 0.92$	1.00 (0.86–1.16), $p = 0.99$
Deficiency, *n* (%)	1 (2)	0 (0)	0.55	n.a.	n.a.

° ANOVA for normally distributed variables, Wilcoxon rank-sum test for continuous variables, * adjusted for age. Bold if statistically significant. Abbreviations: CI, confidence interval; ICU, intensive care unit; IQR, interquartile range; SD, standard deviation; n.a., not applicable; OR, odds ratio.

Higher levels of folic acid, vitamin A, or zinc were associated with a lower risk for a severe course of COVID-19 (adjusted OR 0.88 (95% CI 0.78–0.98, $p = 0.02$), adjusted OR 0.18 (95% CI 0.05–0.69, $p = 0.01$), adjusted OR 0.73 (95% CI 0.55–0.98, $p = 0.03$), respectively). Accordingly, both vitamin A and zinc deficiencies were associated with a more than sevenfold higher risk for the composite endpoint of ICU admission and/or in-hospital mortality (adjusted OR 7.41 (95% CI 1.91–29.68, $p = 0.004$, adjusted OR 7.18 (95% CI 1.73–29.76, $p = 0.007$, respectively) (Figure 2).

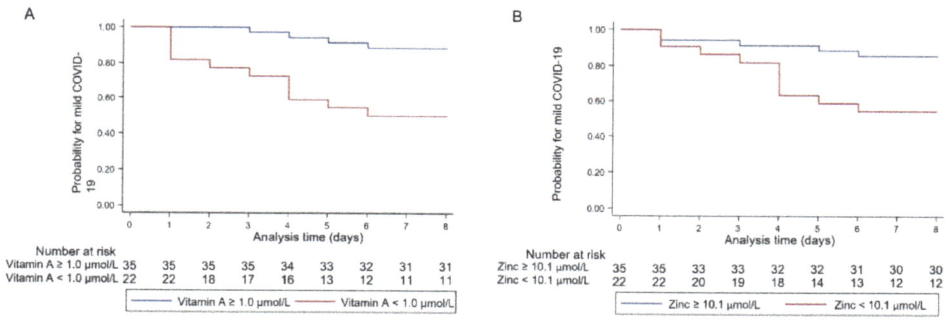

Figure 2. Kaplan–Meier estimates for time to composite endpoint stratified by (**A**) vitamin A deficiency and (**B**) zinc deficiency.

4. Discussion

In this prospective cohort of mostly older patients hospitalized with COVID-19, micronutrient deficiencies were highly prevalent, mainly for selenium, vitamin D, vitamin A, and zinc. There was a high proportion of patients with multiple deficiencies and there were correlations among the different micronutrient levels. Importantly, our results indicate that micronutrient deficiencies are associated with more severe clinical courses of COVID-19 and worse outcomes, especially low levels of folic acid, vitamin A, and zinc.

In our study population, a high prevalence of vitamin D deficiency was found with 40%. This may be partly explained by the seasonal period in which the patients were analyzed (March and April). Data from the Swiss Federal Office of Public Health on vitamin D status show that during the winter period, more than 60% of the population have insufficient vitamin D levels with values < 50 nmol/L [22]. Yet, unlike other studies, we found no difference between vitamin D levels in patients with mild or severe COVID-19 [30–32]. Both patients with mild and severe courses of the disease had levels that were only slightly above the cut-off for deficiency (i.e., 30 nmol/L), similar to insufficient vitamin D levels in both patients with COVID-19 and healthy controls in a study by Elham et al. [33]. Compared to our study, studies that found an association of vitamin D deficiency and adverse outcomes in COVID-19 had different outcomes defined, including the need for non-invasive ventilation in one study [15] and a composite of invasive mechanical ventilation and/or death in another study [34]. The lack of an association in our study may be explained by the small sample size with risk for a type II error and the lack of a healthier control group with higher vitamin D baseline levels. Currently, there are several vitamin D treatment trials in COVID-19 ongoing, and it will be interesting to learn whether treatment improves clinical outcomes [35].

Further, our study found a high prevalence of selenium deficiency—in 51% of cases—which is in agreement with data from South Korea, where 42% of COVID-19 patients were selenium deficient [36]. In our study, no association between low selenium levels and adverse clinical outcomes was found. These findings are not consistent with data from China [37] and Belgium [19], but are in agreement with a study from Iran, where an association of selenium deficiency and COVID-19 severity disappeared after adjusting for confounders [20]. Again, the small sample size in our study did not allow us to draw strong conclusions.

Tomasa-Irriguible et al. showed an association of low vitamin A and zinc levels with the need for ICU treatment, but not with mortality in patients with COVID-19 [18]. In a study by Tepasse et al. [38], lower vitamin A levels were associated with acute respiratory distress syndrome and mortality in patients with COVID-19. In our study, the prevalence of vitamin A deficiency was high and vitamin A levels were significantly lower in patients with a severe course of COVID-19 compared to patients with mild courses. Stephensen et al. [10] described that vitamin A levels decrease in the state of inflammation, which could be a reason for low vitamin A levels in COVID-19. However, we found decreased levels only in patients with severe COVID-19. Sarohan et al. [39] hypothesized a possible association of retinoid signaling with the COVID-19 pathogenesis, which may explain the fact that in COVID-19, many different organ systems are affected in a similar way to patients with defects in retinoid signaling. It is well known that vitamin A, through its metabolite retinoid acid, is a regulator in balancing anti- and pro-inflammatory processes, for example regulatory T-cells vs. T-helper-17-cells [40]. There is evidence that this balance is disturbed in severe COVID-19 [41,42], which, therefore, could be partially explained by low vitamin A levels. Further, a study from Bangladesh [43] showed higher interleukin 6 (IL-6) levels in men with low vitamin A stores. Given that high IL-6 levels were observed in patients with severe COVID-19 [44,45], IL-6 receptors are a target in COVID-19 treatment. Two meta-analyses showed that tocilizumab treatment was associated with lower mortality [46,47].

Zinc deficiency showed an over sevenfold higher risk for an adverse outcome in COVID-19 in our study. In a French study, low zinc levels were an independent risk factor for hospitalization due to respiratory deterioration in COVID-19 [48] and Du Laing et al. [19] found significantly lower zinc levels in deceased COVID-19 patients in comparison with survivors. However, studies showed that during inflammation, zinc redistribution into hepatocytes is upregulated by cytokines like IL-6, resulting in lower plasma levels [49,50]. Therefore, low zinc levels are possibly a consequence of the infection and may not reflect a pre-existing zinc deficiency. Folic acid levels were almost 1.5 times higher in mild versus severe courses of COVID-19. However, levels in patients with a clinical progression to severe COVID-19 were only slightly below the cut-off values, so the clinical relevance may be marginal.

In our study, we found a stepwise increase in LOS and the need for ICU care in patients with a higher number of deficiencies. However, the causality of these observations is unclear. It is known that infections may lead to micronutrient deficiencies via various mechanisms, for example lower intake, malabsorption, and redistribution in the body [51]. Therefore, it remains unclear, if deficiencies in micronutrients are a risk factor for infections and a more severe disease course, or if the infection itself leads to the deficiencies due to higher need.

In addition, the question of whether deficiencies occur in a specific pattern was analyzed. Indeed, we found multiple positive correlations among different micronutrient levels with moderate correlation coefficients between r = 0.27 and 0.42, suggesting that the risk for additional deficiencies is increased if one such deficiency is found. This finding is in line with a study from China that included pediatric patients with respiratory infections [52] and showed correlations among vitamin A, vitamin D, and vitamin E. As a consequence, it may be advised to supplement different micronutrients at the same time using a multivitamin drug instead of only focusing on single micronutrients.

There are several limitations for this study. Firstly, it was a single-center study with a relatively small number of patients. Due to the small sample size, adjusting for multiple confounders was not possible. Secondly, the analyzed blood samples were not obtained on the day of admission but during the first four days of hospitalization. This may have led to slight differences compared to admission levels. Therefore, we excluded patients who already received micronutrient supplementation before the blood draw. In two patients, the beginning of enteral nutrition by stomach tube was not clearly documented. Given that we did not consider food intake as a supplementation and recent data showed no association between standard enteral nutrition and micronutrients levels in patients in the ICU [53], we decided to include these two patients in the analysis. A further limitation is that blood collection was independent of fasting state. Additionally, some patients received micronutrient supplements after the blood draw, especially if hospitalized in ICU, which made the interpretation of data more difficult. Lastly, our study did not include a healthy control group in order to better understand the significance of micronutrient deficiency in patients with COVID-19.

In conclusion, micronutrient deficiencies are common in patients with COVID-19. This study is in line with previous data and shows an association of micronutrient deficiencies and adverse outcomes in COVID-19. As the link between a well-functioning immune system and sufficient levels of micronutrients is well described, further research is warranted to assess the benefits of micronutrient supplementation for either prevention or treatment of COVID-19.

Supplementary Materials: The following are available online at https://www.mdpi.com/article/10.3390/nu14091862/s1, Table S1: Cut-off values for different micronutrients [54].

Author Contributions: Conceptualization, M.V., C.G. and P.S.; data curation, C.G., P.N. and L.B.; statistical analysis, M.V., C.G. and P.S.; writing—original draft preparation: M.V., C.G. and P.S.; writing—review and editing, P.N., D.K., A.K., L.B., A.C. and B.M. All authors have read and agreed to the published version of the manuscript.

Funding: This research was funded by Research Council KSA (Kantonsspital Aarau), (1410.000.131).

Institutional Review Board Statement: The study was conducted according to the guidelines of the Declaration of Helsinki and approved by the local ethics committee (EKZN, 2020-01306, 11 June 2020).

Informed Consent Statement: Informed consent was obtained from all subjects involved in the study.

Data Availability Statement: The data presented in this study are available on request from the corresponding author.

Conflicts of Interest: P.S. and B.M. received research support paid to the Institution from Thermofisher, bioMerieux, Roche Diagnostics, Nestle Health Science, and Abbott Nutrition. All other authors reported no conflict of interest.

References

1. World Health Organization. WHO Coronavirus Disease (COVID-19) Dashboard. Available online: https://covid19.who.int/ (accessed on 24 February 2022).
2. Wu, Z.; McGoogan, J.M. Characteristics of and Important Lessons From the Coronavirus Disease 2019 (COVID-19) Outbreak in China: Summary of a Report of 72314 Cases From the Chinese Center for Disease Control and Prevention. *JAMA* **2020**, *323*, 1239–1242. [CrossRef] [PubMed]
3. Gao, Y.D.; Ding, M.; Dong, X.; Zhang, J.J.; Kursat Azkur, A.; Azkur, D.; Gan, H.; Sun, Y.L.; Fu, W.; Li, W.; et al. Risk factors for severe and critically ill COVID-19 patients: A review. *Allergy* **2021**, *76*, 428–455. [CrossRef]
4. Fang, X.; Li, S.; Yu, H.; Wang, P.; Zhang, Y.; Chen, Z.; Li, Y.; Cheng, L.; Li, W.; Jia, H.; et al. Epidemiological, comorbidity factors with severity and prognosis of COVID-19: A systematic review and meta-analysis. *Aging* **2020**, *12*, 12493–12503. [CrossRef] [PubMed]

5. Conzade, R.; Koenig, W.; Heier, M.; Schneider, A.; Grill, E.; Peters, A.; Thorand, B. Prevalence and Predictors of Subclinical Micronutrient Deficiency in German Older Adults: Results from the Population-Based KORA-Age Study. *Nutrients* **2017**, *9*, 1276. [CrossRef]
6. Chandra, R.K. Nutrition and the immune system from birth to old age. *Eur. J. Clin. Nutr.* **2002**, *56* (Suppl. S3), S73–S76. [CrossRef] [PubMed]
7. Berger, M.M.; Herter-Aeberli, I.; Zimmermann, M.B.; Spieldenner, J.; Eggersdorfer, M. Strengthening the immunity of the Swiss population with micronutrients: A narrative review and call for action. *Clin. Nutr. ESPEN* **2021**, *43*, 39–48. [CrossRef]
8. Gombart, A.F.; Pierre, A.; Maggini, S. A Review of Micronutrients and the Immune System-Working in Harmony to Reduce the Risk of Infection. *Nutrients* **2020**, *12*, 236. [CrossRef]
9. Calder, P.C.; Carr, A.C.; Gombart, A.F.; Eggersdorfer, M. Optimal Nutritional Status for a Well-Functioning Immune System Is an Important Factor to Protect against Viral Infections. *Nutrients* **2020**, *12*, 1181. [CrossRef]
10. Stephensen, C.B. Vitamin A, infection, and immune function. *Annu. Rev. Nutr.* **2001**, *21*, 167–192. [CrossRef]
11. Wintergerst, E.S.; Maggini, S.; Hornig, D.H. Immune-enhancing role of vitamin C and zinc and effect on clinical conditions. *Ann. Nutr. Metab.* **2006**, *50*, 85–94. [CrossRef]
12. Gammoh, N.Z.; Rink, L. Zinc in Infection and Inflammation. *Nutrients* **2017**, *9*, 624. [CrossRef]
13. Gombart, A.F. The vitamin D-antimicrobial peptide pathway and its role in protection against infection. *Future Microbiol.* **2009**, *4*, 1151–1165. [CrossRef]
14. Lin, Z.; Li, W. The Roles of Vitamin D and Its Analogs in Inflammatory Diseases. *Curr. Top. Med. Chem.* **2016**, *16*, 1242–1261. [CrossRef]
15. Baktash, V.; Hosack, T.; Patel, N.; Shah, S.; Kandiah, P.; Van den Abbeele, K.; Mandal, A.K.J.; Missouris, C.G. Vitamin D status and outcomes for hospitalised older patients with COVID-19. *Postgrad. Med. J.* **2021**, *97*, 442–447. [CrossRef] [PubMed]
16. Chen, J.; Mei, K.; Xie, L.; Yuan, P.; Ma, J.; Yu, P.; Zhu, W.; Zheng, C.; Liu, X. Low vitamin D levels do not aggravate COVID-19 risk or death, and vitamin D supplementation does not improve outcomes in hospitalized patients with COVID-19: A meta-analysis and GRADE assessment of cohort studies and RCTs. *Nutr. J.* **2021**, *20*, 89. [CrossRef] [PubMed]
17. Jothimani, D.; Kailasam, E.; Danielraj, S.; Nallathambi, B.; Ramachandran, H.; Sekar, P.; Manoharan, S.; Ramani, V.; Narasimhan, G.; Kaliamoorthy, I.; et al. COVID-19: Poor outcomes in patients with zinc deficiency. *Int. J. Infect. Dis.* **2020**, *100*, 343–349. [CrossRef]
18. Tomasa-Irriguible, T.M.; Bielsa-Berrocal, L.; Bordeje-Laguna, L.; Tural-Llacher, C.; Barallat, J.; Manresa-Dominguez, J.M.; Toran-Monserrat, P. Low Levels of Few Micronutrients May Impact COVID-19 Disease Progression: An Observational Study on the First Wave. *Metabolites* **2021**, *11*, 565. [CrossRef]
19. Du Laing, G.; Petrovic, M.; Lachat, C.; De Boevre, M.; Klingenberg, G.J.; Sun, Q.; De Saeger, S.; De Clercq, J.; Ide, L.; Vandekerckhove, L.; et al. Course and Survival of COVID-19 Patients with Comorbidities in Relation to the Trace Element Status at Hospital Admission. *Nutrients* **2021**, *13*, 3304. [CrossRef]
20. Razeghi Jahromi, S.; Moradi Tabriz, H.; Togha, M.; Ariyanfar, S.; Ghorbani, Z.; Naeeni, S.; Haghighi, S.; Jazayeri, A.; Montazeri, M.; Talebpour, M.; et al. The correlation between serum selenium, zinc, and COVID-19 severity: An observational study. *BMC Infect. Dis.* **2021**, *21*, 899. [CrossRef] [PubMed]
21. Schuetz, P.; Gregoriano, C.; Keller, U. Supplementation of the population during the COVID-19 pandemic with vitamins and micronutrients—How much evidence is needed? *Swiss Med Wkly* **2021**, *151*, w20522. [CrossRef]
22. Federal Commission for Nutrition. *Vitamin D Deficiency: Evidence, Safety, and Recommendations for the Swiss Population*; Expert Report of the FCN; Federal Office for Public Health: Zurich, Switzerland, 2012.
23. Schupbach, R.; Wegmuller, R.; Berguerand, C.; Bui, M.; Herter-Aeberli, I. Micronutrient status and intake in omnivores, vegetarians and vegans in Switzerland. *Eur. J. Nutr.* **2017**, *56*, 283–293. [CrossRef] [PubMed]
24. Gregoriano, C.; Koch, D.; Haubitz, S.; Conen, A.; Fux, C.A.; Mueller, B.; Bernasconi, L.; Hammerer-Lercher, A.; Oberle, M.; Burgermeister, S.; et al. Characteristics, predictors and outcomes among 99 patients hospitalised with COVID-19 in a tertiary care centre in Switzerland: An observational analysis. *Swiss Med. Wkly.* **2020**, *150*, w20316. [CrossRef] [PubMed]
25. World Health Organization. Clinical Management of Severe Acute Respiratory Infection When Novel Coronovirus (nCov) Infection Is Suspected; Interim Guidance. 2020. Available online: https://apps.who.int/iris/handle/10665/330893 (accessed on 2 February 2022).
26. Charlson, M.; Szatrowski, T.P.; Peterson, J.; Gold, J. Validation of a combined comorbidity index. *J. Clin. Epidemiol.* **1994**, *47*, 1245–1251. [CrossRef]
27. Juma, S.; Taabazuing, M.M.; Montero-Odasso, M. Clinical Frailty Scale in an Acute Medicine Unit: A Simple Tool That Predicts Length of Stay. *Can. Geriatr. J.* **2016**, *19*, 34–39. [CrossRef] [PubMed]
28. Kondrup, J.; Rasmussen, H.H.; Hamberg, O.; Stanga, Z.; Ad Hoc, E.W.G. Nutritional risk screening (NRS 2002): A new method based on an analysis of controlled clinical trials. *Clin. Nutr.* **2003**, *22*, 321–336. [CrossRef]
29. Greaves, R.F.; Woollard, G.A.; Hoad, K.E.; Walmsley, T.A.; Johnson, L.A.; Briscoe, S.; Koetsier, S.; Harrower, T.; Gill, J.P. Laboratory medicine best practice guideline: Vitamins a, e and the carotenoids in blood. *Clin. Biochem. Rev.* **2014**, *35*, 81–113.
30. Jain, A.; Chaurasia, R.; Sengar, N.S.; Singh, M.; Mahor, S.; Narain, S. Analysis of vitamin D level among asymptomatic and critically ill COVID-19 patients and its correlation with inflammatory markers. *Sci. Rep.* **2020**, *10*, 20191. [CrossRef]

31. Munshi, R.; Hussein, M.H.; Toraih, E.A.; Elshazli, R.M.; Jardak, C.; Sultana, N.; Youssef, M.R.; Omar, M.; Attia, A.S.; Fawzy, M.S.; et al. Vitamin D insufficiency as a potential culprit in critical COVID-19 patients. *J. Med. Virol.* **2021**, *93*, 733–740. [CrossRef]
32. Beigmohammadi, M.T.; Bitarafan, S.; Abdollahi, A.; Amoozadeh, L.; Salahshour, F.; Mahmoodi Ali Abadi, M.; Soltani, D.; Motallebnejad, Z.A. The association between serum levels of micronutrients and the severity of disease in patients with COVID-19. *Nutrition* **2021**, *91–92*, 111400. [CrossRef]
33. Elham, A.S.; Azam, K.; Azam, J.; Mostafa, L.; Nasrin, B.; Marzieh, N. Serum vitamin D, calcium, and zinc levels in patients with COVID-19. *Clin. Nutr. ESPEN* **2021**, *43*, 276–282. [CrossRef]
34. Radujkovic, A.; Hippchen, T.; Tiwari-Heckler, S.; Dreher, S.; Boxberger, M.; Merle, U. Vitamin D Deficiency and Outcome of COVID-19 Patients. *Nutrients* **2020**, *12*, 2757. [CrossRef] [PubMed]
35. Jaun, F.; Boesing, M.; Luthi-Corridori, G.; Abig, K.; Makhdoomi, A.; Bloch, N.; Lins, C.; Raess, A.; Grillmayr, V.; Haas, P.; et al. High-dose vitamin D substitution in patients with COVID-19: Study protocol for a randomized, double-blind, placebo-controlled, multi-center study-VitCov Trial. *Trials* **2022**, *23*, 114. [CrossRef] [PubMed]
36. Im, J.H.; Je, Y.S.; Baek, J.; Chung, M.H.; Kwon, H.Y.; Lee, J.S. Nutritional status of patients with COVID-19. *Int. J. Infect. Dis.* **2020**, *100*, 390–393. [CrossRef]
37. Zhang, J.; Taylor, E.W.; Bennett, K.; Saad, R.; Rayman, M.P. Association between regional selenium status and reported outcome of COVID-19 cases in China. *Am. J. Clin. Nutr.* **2020**, *111*, 1297–1299. [CrossRef] [PubMed]
38. Tepasse, P.R.; Vollenberg, R.; Fobker, M.; Kabar, I.; Schmidt, H.; Meier, J.A.; Nowacki, T.; Husing-Kabar, A. Vitamin A Plasma Levels in COVID-19 Patients: A Prospective Multicenter Study and Hypothesis. *Nutrients* **2021**, *13*, 2173. [CrossRef] [PubMed]
39. Sarohan, A.R.; Kizil, M.; Inkaya, A.C.; Mahmud, S.; Akram, M.; Cen, O. A novel hypothesis for COVID-19 pathogenesis: Retinol depletion and retinoid signaling disorder. *Cell Signal.* **2021**, *87*, 110121. [CrossRef]
40. Mucida, D.; Park, Y.; Kim, G.; Turovskaya, O.; Scott, I.; Kronenberg, M.; Cheroutre, H. Reciprocal TH17 and regulatory T cell differentiation mediated by retinoic acid. *Science* **2007**, *317*, 256–260. [CrossRef] [PubMed]
41. Sadeghi, A.; Tahmasebi, S.; Mahmood, A.; Kuznetsova, M.; Valizadeh, H.; Taghizadieh, A.; Nazemiyeh, M.; Aghebati-Maleki, L.; Jadidi-Niaragh, F.; Abbaspour-Aghdam, S.; et al. Th17 and Treg cells function in SARS-CoV2 patients compared with healthy controls. *J. Cell Physiol.* **2021**, *236*, 2829–2839. [CrossRef]
42. Chen, G.; Wu, D.; Guo, W.; Cao, Y.; Huang, D.; Wang, H.; Wang, T.; Zhang, X.; Chen, H.; Yu, H.; et al. Clinical and immunological features of severe and moderate coronavirus disease 2019. *J. Clin. Investig.* **2020**, *130*, 2620–2629. [CrossRef]
43. Ahmad, S.M.; Haskell, M.J.; Raqib, R.; Stephensen, C.B. Markers of innate immune function are associated with vitamin a stores in men. *J. Nutr.* **2009**, *139*, 377–385. [CrossRef]
44. Coomes, E.A.; Haghbayan, H. Interleukin-6 in COVID-19: A systematic review and meta-analysis. *Rev. Med. Virol.* **2020**, *30*, 1–9. [CrossRef] [PubMed]
45. Broman, N.; Rantasarkka, K.; Feuth, T.; Valtonen, M.; Waris, M.; Hohenthal, U.; Rintala, E.; Karlsson, A.; Marttila, H.; Peltola, V.; et al. IL-6 and other biomarkers as predictors of severity in COVID-19. *Ann. Med.* **2021**, *53*, 410–412. [CrossRef] [PubMed]
46. Berardicurti, O.; Ruscitti, P.; Ursini, F.; D'Andrea, S.; Ciaffi, J.; Meliconi, R.; Iagnocco, A.; Cipriani, P.; Giacomelli, R. Mortality in tocilizumab-treated patients with COVID-19: A systematic review and meta-analysis. *Clin. Exp. Rheumatol.* **2020**, *38*, 1247–1254. [PubMed]
47. The WHO Rapid Evidence Appraisal for COVID-19 Therapies (REACT) Working Group. Association Between Administration of IL-6 Antagonists and Mortality Among Patients Hospitalized for COVID-19: A Meta-analysis. *JAMA* **2021**, *326*, 499–518. [CrossRef]
48. Fromonot, J.; Gette, M.; Ben Lassoued, A.; Gueant, J.L.; Gueant-Rodriguez, R.M.; Guieu, R. Hypozincemia in the early stage of COVID-19 is associated with an increased risk of severe COVID-19. *Clin. Nutr.* **2021**. [CrossRef]
49. Liuzzi, J.P.; Lichten, L.A.; Rivera, S.; Blanchard, R.K.; Aydemir, T.B.; Knutson, M.D.; Ganz, T.; Cousins, R.J. Interleukin-6 regulates the zinc transporter Zip14 in liver and contributes to the hypozincemia of the acute-phase response. *Proc. Natl. Acad. Sci. USA* **2005**, *102*, 6843–6848. [CrossRef]
50. Schroeder, J.J.; Cousins, R.J. Interleukin 6 regulates metallothionein gene expression and zinc metabolism in hepatocyte monolayer cultures. *Proc. Natl. Acad. Sci. USA* **1990**, *87*, 3137–3141. [CrossRef]
51. Calder, P.C. Feeding the immune system. *Proc. Nutr. Soc.* **2013**, *72*, 299–309. [CrossRef]
52. Zhang, J.; Sun, R.R.; Yan, Z.X.; Yi, W.X.; Yue, B. Correlation of serum vitamin A, D, and E with recurrent respiratory infection in children. *Eur. Rev. Med. Pharmacol. Sci.* **2019**, *23*, 8133–8138. [CrossRef]
53. Koekkoek, W.A.C.; Hettinga, K.; de Vries, J.H.M.; van Zanten, A.R.H. Micronutrient deficiencies in critical illness. *Clin. Nutr.* **2021**, *40*, 3780–3786. [CrossRef]
54. Munns, C.F.; Shaw, N.; Kiely, M.; Specker, B.L.; Thacher, T.D.; Ozono, K.; Michigami, T.; Tiosano, D.; Mughal, M.Z.; Mäkitie, O.; et al. Global Consensus Recommendations on Prevention and Management of Nutritional Rickets. *J. Clin. Endocrinol. Metab.* **2016**, *101*, 394–415. [CrossRef]

nutrients

MDPI

Article

The Side-Effects of the COVID-19 Pandemic: Increased BMI z-Score in Children with Overweight and Obesity in a Personalised Lifestyle Intervention One Year after the Start of the Pandemic in The Netherlands

Lisanne Arayess [1,2,3], Nienke Knockaert [1,3,4], Bjorn Winkens [5,6], Judith W. Lubrecht [1,2,3], Marjoke Verweij [4] and Anita C. E. Vreugdenhil [1,2,3,*]

1 Centre for Overweight Adolescent and Children's Healthcare (COACH), Maastricht University Medical Centre+, 6229 HX Maastricht, The Netherlands; lisanne.arayess@mumc.nl (L.A.); n.knockaert@student.maastrichtuniversity.nl (N.K.); judith.lubrecht@mumc.nl (J.W.L.)
2 School of Nutrition and Translational Research (NUTRIM), Maastricht University Medical Centre+, 6229 ER Maastricht, The Netherlands
3 Department of Paediatrics, Maastricht University Medical Centre+, 6229 HX Maastricht, The Netherlands
4 Department of Paediatrics, VieCuri Hospital, 5912 BL Venlo, The Netherlands; mverweij@viecuri.nl
5 Department of Methodology and Statistics, Maastricht University Medical Centre+, 6229 HX Maastricht, The Netherlands; bjorn.winkens@maastrichtuniversity.nl
6 Care and Public Health Research Institute (CAPHRI), Maastricht University Medical Centre+, 6229 ER Maastricht, The Netherlands
* Correspondence: a.vreugdenhil@mumc.nl; Tel.: +31-(0)433875284

Citation: Arayess, L.; Knockaert, N.; Winkens, B.; Lubrecht, J.W.; Verweij, M.; Vreugdenhil, A.C.E. The Side-Effects of the COVID-19 Pandemic: Increased BMI z-Score in Children with Overweight and Obesity in a Personalised Lifestyle Intervention One Year after the Start of the Pandemic in The Netherlands. *Nutrients* **2022**, *14*, 1942. https://doi.org/10.3390/nu14091942

Academic Editors: Omorogieva Ojo and Amanda R Amorim Adegboye

Received: 5 April 2022
Accepted: 3 May 2022
Published: 5 May 2022

Abstract: Background: Early research showed weight gain in children during the COVID-19 pandemic. Objective: To compare changes in BMI z-score of children with overweight and obesity in a personalised lifestyle intervention before and during the pandemic. Methods: Changes in BMI z-score half a year (T6) and twelve months (T12) after the first lockdown were included for 71 children in the '2020 during COVID' group and compared to 48 children in the '2019 before COVID' group, using a marginal model for repeated measures (model 1). Model 2 corrected for lifestyle intervention characteristics, and model 3 corrected additionally for family characteristics. Results: The mean difference in BMI z-score change was significantly different at T12 (+0.07 in 2020 versus −0.09 in 2019, $p = 0.022$). Model 3 showed significant differences in BMI z-score change at both T6 (+0.15, $p = 0.024$) and T12 (+0.18, $p = 0.016$). This model also defined 'having a mother with obesity' (+0.13, $p = 0.019$) and the frequency of no-show consultations (+0.41 per missed consultation per month, $p = 0.025$) as related factors. Conclusions: Lifestyle intervention in children with overweight and obesity is less successful in decreasing BMI z-score during the COVID-pandemic. Identified risk factors for less success could contribute to identifying children with higher risks for, and possibly prevent, BMI z-score increase.

Keywords: childhood obesity; COVID-19; pandemic; personalised lifestyle intervention; overweight; obesity; BMI

1. Introduction

In 2020, the world was confronted with a pandemic caused by the SARS-CoV-2 virus. Several countries had to take governmental measures to cope with the consequences of the virus, such as national lockdowns. In the Netherlands, the first national lockdown started on 15 March 2020 and included several important measures for children, including school and sports club closures and the advice to stay at home as much as possible [1].

Although the SARS-CoV-2 virus itself seems to have a less severe pattern in children, it is assumed that the changing circumstances in daily life have led to drastic changes in the lifestyle of children [2]. Previous research showed lifestyle changes towards more unhealthy behaviour during the pandemic in both children and adults [3–5]. In the early phase of the pandemic, research on short term weight development was published, showing weight gain in children [6–8].

Children with overweight especially seem to be a risk subgroup for weight gain. Before the pandemic, studies showed the vulnerability of these children in comparable periods in terms of changing lifestyle patterns, such as school closure during summer holidays [9–11]. This is worrisome since childhood overweight and obesity is associated with serious health consequences, physical as well psychological, in both childhood and adulthood [12–16]. Childhood obesity also leads to problematic social and economic consequences, such as increasing health costs in the future [17–19].

Weight gain is likely to remain a persistent problem during this pandemic for children, especially for those with overweight, since the pandemic is still a threat for many countries, with corresponding governmental measures such as lockdowns. However, little is known about the long-term consequences of these lifestyle behaviours and weight changes early in the pandemic and which subgroups are at risk. Additionally, the impact of the changed circumstances on existing lifestyle interventions for children due to the pandemic is unclear.

Therefore, the aim of this study is to determine the change in BMI z-score of children with overweight and obesity in a personalised lifestyle intervention six and twelve months after the start of the first lockdown due to the COVID-19 pandemic and compare this to the same period one year earlier.

2. Methods

2.1. Setting

Data of participants were collected from the Centre for Overweight Adolescent and Children's Healthcare (COACH) at the Maastricht University Medical Centre (MUMC+) and VieCuri Medical Centre in Venlo, the Netherlands. COACH is an expertise centre for children with overweight and (severe) obesity for both clinical evaluation and treatment, by providing a family-based, interdisciplinary, tailored lifestyle intervention. The setting and design of this intervention pre-pandemic are described in detail elsewhere [20].

Data of Maastricht were obtained within local regulations of the hospital and registered at the Ethics Committee of the Maastricht University Medical Centre (METC 2022-3105). The Ethics Committee stated that this research did not fall under the scope of the Medical Research Involving Human Subjects Act (WMO) and therefore no ethical approval and informed consent were needed. Data of Venlo was collected within the "Kijk op Overgewicht study" (METC 13-4-130, Clinicaltrial.gov (NCT02091544)). All parents and/or children in Venlo gave written informed consent for this study.

2.2. Participants

Participants were included in either the 2020 group, with measurements during the COVID-19 pandemic, or in the 2019 group, used as a control group.

Participants were included in the 2020 group if they had a measurement at baseline (T0) and at least one measurement after about six months (T6) and twelve months (T12). The baseline (T0) measurements were obtained between 1 January 2020 and 15 March 2020. For T6, this period was from 15 August 2020 to 15 October 2020, and for T12, from 1 January 2021 to 15 April 2021.

For 2019, these three periods ran from 1 January 2019 to 15 March 2019 (baseline, T0), from 15 August 2019 to 15 October 2019 (T6) and from 1 January 2020 to 15 March 2020 (T12). Since the lockdown started on 15 March 2020, no measurements were obtained between 15 March 2020 and 15 April 2020.

Children that participated in the long-term intervention during both 2019 and 2020, and therefore had anthropometric measurements in both years ($N = 20$), were randomly distributed between the 2019 and 2020 groups to avoid overlap and with the intention to create two independent groups.

To have a representative cohort for children with overweight in the school age, children younger than 4 years and older than 18 years at T0 or with a normal weight at T0, were excluded.

Since COACH is an ongoing, long-term family intervention, children could be in different phases of the intervention (waiting list, intake phase, diagnostic phase, active intervention or relapse prevention phase) at T0. All children with anthropometric measurements in the corresponding periods were included, regardless of the length or intensity of the lifestyle intervention.

2.3. Lockdown Due to the COVID-19 Pandemic in the Netherlands

The first lockdown in the Netherlands was characterised by several measures, including but not limited to school closures of both primary and high school (online education), closure of restaurants and sports clubs, cancellation of large gatherings and advice to stay at home if you have COVID-19-related symptoms, to work from home and to avoid large gatherings [21]. Schools were fully re-opened in August 2020 [22]. In December 2020, a second lockdown was announced, including school closure (online education) until February 2021 for primary schools and March 2021 for high schools [23].

2.4. Study Measurements

Anthropometric data (height and body weight) and data on child- and parental characteristics at the start of the intervention (sex, age, ethnicity, BMI and IOTF status of mother and father) were extracted from the medical record.

Measurements for the height and weight of the children were obtained by a healthcare professional following the Dutch guidelines for measuring weight and height of children [24]. Most of the measurements were performed in the COACH outpatient clinic using an electric scale (Seca© 877, Seca, Hamburg, Germany) to the nearest 0.1 kg. Standing height was measured using a portable stadiometer (Seca© 213 stadiometer, Seca, Hamburg, Germany). A minority of the measurements were collected through the medical record via referrals of general youth health doctors or appointments at other disciplines of the Paediatric department of the Medical University Hospital Maastricht.

Self-reported home measurements during remote visits (by phone or video conference) were excluded.

BMI score (weight [kg]/height $[m]^2$) and the BMI z-score were calculated in the Growth Viewer of the patient file, according to the reference data of the TNO Growth Calculator [25,26]. Weight classification was based on the International Obesity Task Force (IOTF criteria) classification system [27]. The change from baseline in BMI z-score was calculated for T6 and T12.

2.5. Lifestyle Intervention Determinants

Due to the national governmental measures because of the COVID-19 pandemic, outpatient hospital visits were not possible during the first lockdown in the Netherlands. Remote consultations via phone or video conference were scheduled to stay in contact with the patients and their families. Additionally, since 15 March 2020, appointments needed to be cancelled or rescheduled when the patient or somebody in the household has symptoms related to COVID-19. The number of the different types of consultations, namely physical visit at the outpatient clinic, remote consultation (video conference or phone) or no-show (consultations that were cancelled by the patient or the family, or when the patient did not show up at the pre-arranged consultation), were retrieved from the patient files. The total number of consultations per month was calculated as the sum of the physical visits and remote consultations, divided by the time in months between T0 and the measurement moment.

2.6. Statistical Analysis

Data are presented as mean (SD) or the number of children (%). Independent-samples *t*-test or chi-square tests were used to examine differences in numerical and categorical characteristics between 2019 and 2020, respectively.

Three models were created and applied to the study population. To assess the group effect (2020 versus 2019) on change in BMI z-score from baseline after 6 and 12 months, we used a marginal model for repeated measures with group, time (6 or 12 months), the interaction between group and time, and centre as fixed factors and an unstructured covariance structure for repeated measures (model 1). To get insight in the effects for the separate subgroups of children that had an increase in their BMI z-score and children that had a stabilisation or decrease in BMI z-score, this model was re-applied to those subgroups separately as a post hoc analysis.

To account for the potential effects of lifestyle intervention factors (number of both physical visits to the outpatient clinic and offline consultations per month, as well the number of no-show consultations) on the change in BMI z-score, we added these factors to the fixed part of model 1 (model 2). As a sensitivity analysis, baseline family characteristics (ethnicity, IOTF classification at T0, length in intervention, educational level of parents, age or having a mother or father with obesity) were separately added to model 2 to see which characteristics contributed significantly to the model. All characteristics that were significantly contributing to the model were included in the final model (model 3). Estimates of fixed effects together with their 95% confidence intervals and *p*-values are presented. Two-sided *p*-values ≤ 0.05 were considered statistically significant.

Statistical analyses were performed using IBM SPSS Statistics for Windows version 25 (Armonk, NY, USA).

3. Results

3.1. Baseline Characteristics

Seventy-one children had a measurement of their height and weight in the three months prior to the start of the lockdown in the Netherlands (15 March 2020) and were included in the 2020 group, while 48 children were included in the control group (2019). (Figure 1) Baseline characteristics regarding weight status, family factors and length in the intervention were similar between the groups (Table 1).

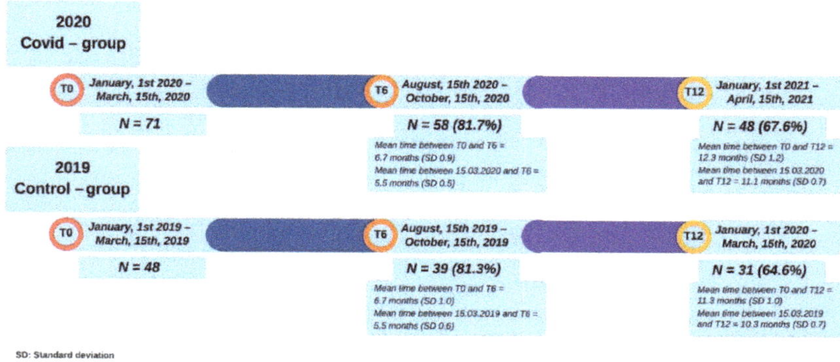

Figure 1. Timeline and inclusion at the several measurement moments for the 2020 COVID-19 group and the 2019 control group.

Table 1. Baseline characteristics.

	2020 (COVID-19 Group) N = 71	2019 (Control Group) N = 48	*p*-Value
Age, mean (SD); years	12.6 (3.1)	11.7 (2.5)	0.094
Gender, % female (*N*)	49.3 (*N* = 35)	52.1 (*N* = 25)	0.765
BMI score (SD); kg/m^8	28.59 (5.95)	27.56 (3.46)	0.237
BMI z-score (SD)	3.09 (0.70)	3.11 (0.60)	0.888
IOTF at T0			
Overweight, % (N)	33.8 (*N* = 24)	29.2 (*N* = 14)	0.744
Obesity, % (N)	40.8 (*N* = 29)	47.9 (*N* = 23)	
Severe obesity, % (N)	25.4 (*N* = 18)	22.9 (*N* = 11)	
Months in intervention at T0 (SD) mean	13.1 (18.3)	15.0 (18.7)	0.578
<1 year since start intervention at T0, % (*N*)	59.2 (*N* = 42)	64.6 (*N* = 31)	0.551
>1 year since start intervention at T0, % (*N*)	40.8 (*N* = 29)	35.4 (*N* = 17)	
Ethnicity			
Dutch, % (N)	62.9 (*N* = 39) #	64.6 (*N* = 31)	0.856
Migration background, % (N)	37.1 (*N* = 23) #	35.4 (*N* = 17)	
Parental factors			
BMI mother, mean (SD); kg/m^8	28.41 (4.99) ^	30.31 (6.08)^	0.068
BMI father, mean (SD); kg/m^8	29.67 (4.91) ^^	29.48 (4.82) ^^	0.845
Having a mother with obesity, % (N)	40.6 (*N* = 28) ^	46.8 (*N* = 22) ^	0.506
Having a father with obesity, % (N)	36.7 (*N* = 22) ^^	42.5% (*N* = 17) ^^	0.558
Having≥1 parent with obesity, % (N)	63.5 (*N* = 40) ^^^	65.1 (*N* = 28) ^^^	0.864
Educational level mother			
Low, % (N)	38.9 (*N* = 21) ◊	34.9 (*N* = 15) ◊	0.16
Medium, % (N)	40.7 (*N* = 22) ◊	27.9 (*N* = 12) ◊	
High, % (N)	20.4 (*N* = 11) ◊	37.2 (*N* = 16) ◊	
Educational level father			
Low, % (N)	32.0 (*N* = 16) ◊◊	34.1 (*N* = 14) ◊◊	0.89
Medium, % (N)	34.0 (*N* = 17) ◊◊	36.6 (*N* = 15) ◊◊	
High, % (N)	34.0 (*N* = 17) ◊◊	29.3 (*N* = 12) ◊◊	

Data available for *N* = 62 in 2020. ^ Data available for *N* = 69 in 2020, *N* = 47 in 2019. ^^ Data available for *N* = 60 in 2020, *N* = 40 in 2019. ^^^ Data available for *N* = 63 in 2020, *N* = 43 in 2019. ◊ Data available for *N* = 54 in 2020, *N* = 43 in 2019. ◊◊ Data available for *N* = 50 in 2020, *N* = 41 in 2019.

3.2. BMI z-Score Change for Children with Overweight and (Severe) Obesity on the Mid-Long Term after the Start of the COVID-19 Pandemic

In 2020, 30 (51.7%) children had an increase in BMI z-score at T6 when compared to T0, while there were 19 children (48.7%) with an increase in the control group of 2019.

At T12, 29 (60.4%) and 18 (58.1%) children had an increase of the BMI z-score in 2020 and 2019, respectively. The change in BMI z-score at T6 and T12 in the 2020 group shows an increasing trend, compared to a decreasing trend in 2019 (Figure 2).

Based on model 1, the mean difference in BMI z-score change was not significant at T6 (+0.05 in 2020 versus −0.05 in 2019, difference = +0.10, 95% CI −0.01, +0.21, $p = 0.061$), while it was significant at T12 (+0.07 in 2020 versus −0.09 in 2019, difference = +0.16, 95% CI 0.02, 0.30, $p = 0.022$).

The same model was re-applied for the different subgroups of children that had an increase in their BMI z-score and children that had a decrease in BMI z-score. This showed significant differences at both T6 and T12 when 2020 was compared to 2019 for children with a BMI z-score increase (mean difference at T6 = 0.10, 95% CI 0.00, 0.21, $p = 0.047$ and mean difference at T12 0.22, 95% CI 0.11, 0.34, $p < 0.001$). It also showed a significant difference at T6 for children with a BMI z-score decrease when 2020 was compared to 2019 (mean difference 0.11, 95% CI 0.03, 0.20, *p*-value = 0.010) (Table 2).

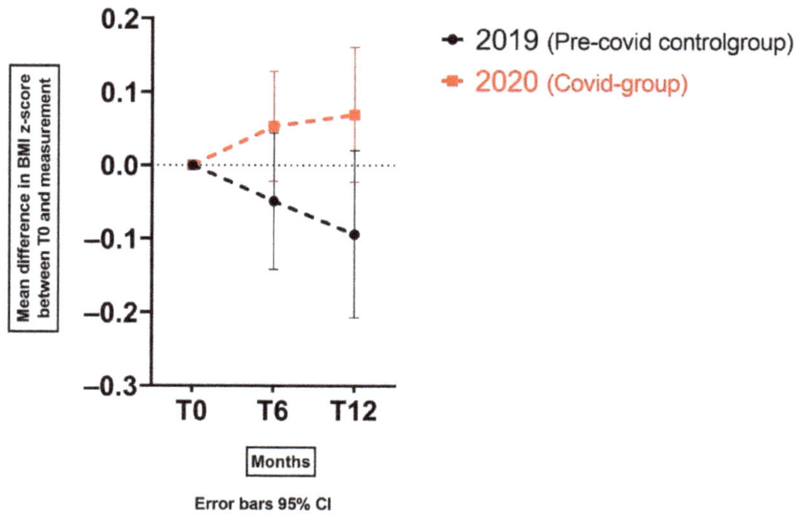

Figure 2. Change in BMI z-score for children with overweight and (severe) obesity at T6 and T12 in 2019 and 2020.

Table 2. Mean change for subgroups that had an increase or decrease/stabilisation in the BMI z-score at T6 and T12 in 2020, when compared to 2019.

	T6		*p*-Value ^	T12		*p*-Value ^
	2020	2019		2020	2019	
BMI z-score increase, mean (SD) change for subgroup	0.24 (0.03) *N* = 30	0.14 (0.04) *N* = 19	0.047 *	0.32 (0.04) *N* = 29	0.10 (0.05) *N* = 18	<0.001 *
BMI z-score decrease or stabilisation, mean (SD) change for subgroup	−0.16 (0.03) *N* = 28	−0.27 (0.04) *N* = 20	0.010 *	−0.23 (0.05) *N* = 19	−0.34 (0.06) *N* = 13	0.178

* Statistically significant, $p \leq 0.05$. ^ based on model 1, which means that the differences in mean change from baseline scores (at T6 and T12) are corrected for centre, measurement moment, year*measurement moment.

3.3. Lifestyle Intervention Changes during the COVID-19 Pandemic

When characteristics of the lifestyle intervention between the group in 2020 and 2019 were compared, no significant differences in total consultations per month or no-show consultations were observed. However, significant differences were observed in the number of outpatient clinic visits and remote consultations when 2020 and 2019 were compared (Figure 3).

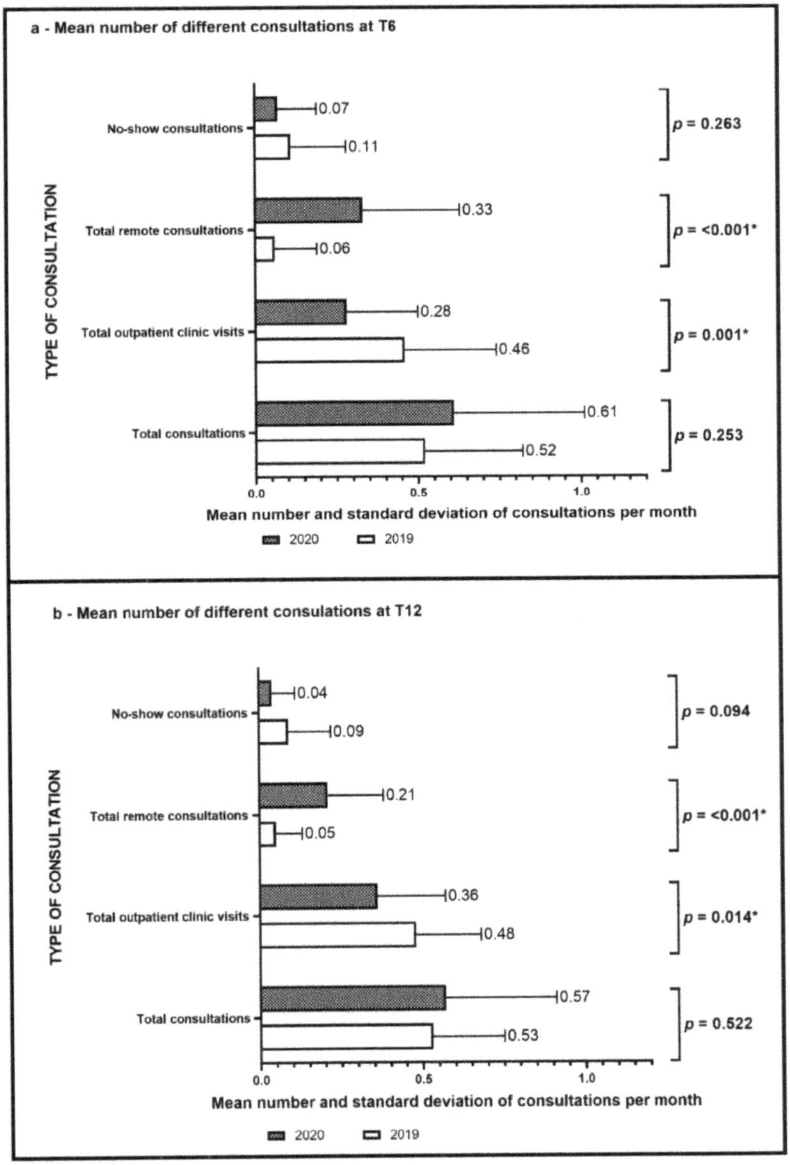

*significant $p \leq 0.05$

Figure 3. Characteristics of the lifestyle intervention: different consultations at T6 (**a**) and T12 (**b**) in 2020 and 2019.

According to model 2, i.e., after correction for the lifestyle intervention characteristics, the mean change in BMI z-score was significantly higher in 2020 when compared to 2019 at both T6 and T12 (see Table 3).

Table 3. Estimated fixed effects in the model for BMI z-score difference for children with overweight and (severe) obesity.

Parameter	Model 2				Model 3			
	Estimate	95% CI		*p*-Value	Estimate	95% CI		*p*-Value
		Lower Bound	Upper Bound			Lower Bound	Upper Bound	
Year [2020 versus 2019] at T6	+0.14	0.01	0.27	0.033 *	+0.15	0.02	0.27	0.024 *
Year [2020 versus 2019] at T12	+0.18	0.04	0.33	0.014 *	+0.18	0.03	0.32	0.016 *
Centre [Maastricht]	+0.04	−0.10	0.18	0.602	+0.05	−0.09	0.19	0.472
Contact moments outpatient clinic/month	+0.05	−0.15	0.25	0.635	+0.05	−0.15	0.25	0.609
Remote contact moments/month	−0.03	−0.24	0.17	0.752	−0.00	−0.21	0.20	0.993
No-show appointments per month	+0.43	0.07	0.80	0.021 *	+0.41	0.05	0.77	0.025 *
>=1 Year in intervention at baseline	−0.02	−0.14	0.09	0.680	−0.02	−0.13	0.10	0.772
Having a mother with obesity					+0.13	0.02	0.23	0.019 *

* Significant, $p \leq 0.05$.

A significant contributing lifestyle factor in this model was the frequency of no-show consultations per month. With every missed consultation per month, the change in BMI z-score at both time points increased by +0.43 (95% CI 0.07, 0.80, $p = 0.021$).

3.4. Model 3: Creating a Model to Identify Family Characteristics Influencing BMI z-Score Change

The only family characteristic that added significantly to model 2, and was therefore included in model 3, was having a mother with obesity ($p = 0.019$).

This model also showed that the corrected mean differences in change of BMI z-score were significant, both at T6 (+0.07 in 2020 versus −0.08 in 2019, difference = 0.15, 95% CI 0.02, 0.27, $p = 0.024$) and T12 (+0.08 in 2020 versus −0.10 in 2019, difference = 0.18, 95% CI 0.03, 0.32, $p = 0.016$). Additionally, the frequency of no-show consultations remained a significant contributor ($p = 0.025$).

4. Discussion

To the best of our knowledge, this is the first study that shows the weight gain in children with overweight and (severe) obesity on the mid-long term, approximately half a year and one year after the start of the COVID-19 pandemic. This study adds knowledge to previous studies on children of the general population by showing the drastic changes in BMI z-score in children with overweight and (severe) obesity that are already participating in a lifestyle intervention. Under pre-pandemic circumstances in 2019, the mean BMI z-score change over the total cohort in the lifestyle intervention is decreasing at both time points, in line with previously described positive effects of the COACH personalised lifestyle intervention [20,28]. Unfortunately, during the lifestyle intervention in 2020, the mean BMI z-score change for children increased. Even after correction for several determinants related to the lifestyle intervention and the family, such as frequency of consultations, our models showed that the difference in change of BMI z-score was significantly higher when 2020 was compared to 2019.

Furthermore, the mean BMI z-score increase for children with weight gain in the lifestyle intervention was significantly larger in 2020 when compared to 2019. Additionally, the success of the lifestyle intervention six months after the lockdown for the subgroup of children that had a decrease or stabilisation of the BMI z-score was less, since our study found a smaller mean BMI z-score decrease at T6 in 2020 when compared to 2019. These results should be considered alarming.

It should be noted that the children in this study were participating in a personalised lifestyle intervention. In general, the main goal of this lifestyle intervention is a decrease in BMI z-score of 0.15 SDS, based on the relationship with cardiovascular health outcomes, although this may vary depending on individual characteristics such as age, severity of the overweight and phase in the intervention [29]. Since overweight is known to be associated with several weight-related comorbidities, even at a younger age, and an increasing BMI z-score is continuously associated with cardiovascular complications such as high blood pressure and dyslipidaemia, it is advisable to keep track of the weight status of children with overweight and obesity [13,30]. However, our study results suggest this is even more important during a pandemic, since even the help of a lifestyle intervention in an expertise centre was not sufficient for stabilising or decreasing the mean BMI z-score of the cohort. Furthermore, we do know that lifestyle interventions for children with overweight in general are efficacious in the treatment of paediatric overweight when compared to children who do not receive guidance [31]. A previous study from our research group showed that parents of children with overweight or obesity more frequently reported perceived weight gain during the lockdown, when compared to parents of children with normal weight [7]. Therefore, the increase in BMI z-score in our 2020 group could be even larger for the youth with overweight that did not receive guidance, since the lifestyle intervention aimed to create guidance and an incentive for healthy behaviour in this study group during the pandemic. To cope with the changed daily lives of children with overweight during the COVID-19 pandemic and possibly other crisis periods, lifestyle interventions should be attentive to the impact of the COVID-19 pandemic and its effects and consider adaptations. Therefore, this study forms the basis for further studies to obtain more insight in contributing determinants.

Our results highlight the importance to take family-related determinants into account in a lifestyle intervention, as children of mothers with obesity were especially at risk for an increase in BMI z-score. It is known that parental obesity and lifestyle influence the weight status of children [32,33]. However, it remains unclear to what extent a pandemic influences these family-related associations. A qualitative study found that parents were concerned that their child with overweight has more access to unhealthy food and that the child is more frequently in surroundings where overweight is normalised [34]. Additionally, other mental health stressors were mentioned in the literature, possibly influencing lifestyle behaviour [35]. To obtain more in-depth data on how the lifestyle in families exactly changed for children with overweight during COVID-19, and what changes in lifestyle behaviours are long-lasting and influencing lifestyle interventions, more long term and qualitative data is needed.

Besides the previously mentioned influence of the weight status of the mother, most of the family determinants in this study did not seem to influence the BMI z-score change of children. Possibly, data on parental variables (weight status or educational level) could be outdated or less accurate for children that are in the lifestyle intervention for a longer period, since data of parents is self-reported at the intake of the lifestyle intervention. Specifically, anthropometric data of fathers were not complete for all children, leading to higher inaccuracy for these variables. This is considered a limitation of this study.

The study provides new insights into the characteristics of a lifestyle intervention during a pandemic, showing the total number of consultations per child did not differ significantly between 2019 and 2020 although a shift to more remote consultations was observed. In addition, it indicates that the number of physical and remote consultations did not have a significant influence on the BMI z-score change. These results are in line with the goal of the personalised intervention, that determines the frequency of contact that is adapted by individual and family needs and possibilities. However, the number of no-show appointments was significantly, positively associated with the change in BMI z-score. Thus, missing appointments most likely will lead to an increase in BMI z-score. A previous Dutch study assessing barriers and facilitators in the adherence to a lifestyle group intervention for children with obesity, describes motivation, satisfaction and means (such as accessibility and time) as barriers [36]. Specifically for 2020, it is hypothesised that the families with high no-shows had less access to online possibilities, such as remote consultations and online lifestyle activities, or are less motivated to make changes in the lifestyle of the child, because of other priorities or less awareness of overweight. Since anxiety during the pandemic in children with overweight is described previously, it is also possible that families with high no-shows were more afraid of COVID-19 and were also avoiding other contacts and activities besides the intervention, and therefore changed their daily lifestyle more [37]. Regardless of the reasons for missing the appointments, this study illustrates the importance of adherence to a tailored intervention. Since personalisation of a lifestyle intervention is seen as a facilitator in adherence, it is advisable to consider children with high no-show rates after the start of the pandemic as a risk group for weight gain and to adapt the interventions to the needs of these families [36,38].

A previous study of our group, conducted directly after the end of the first lockdown, showed that in children with obesity that participated longer than 1 year in the lifestyle intervention, no significant increase in BMI z-score was shown [7]. In contradiction, the current study in this cohort did not see any influence of the length of the intervention on the BMI z-score change. This difference in results is an important finding, that requires deeper analysis. An explanation could be that the length of the pandemic, and therefore the duration of the changed circumstances, could potentially have a stronger negative impact on the lifestyle and weight, than the preparation that was learned at the lifestyle intervention before the pandemic. Given the threat of new COVID-19 virus mutations with possible new corresponding governmental measures influencing daily lifestyle, it is of great importance to keep track of the weight status of children with overweight and obesity in the upcoming period [39,40].

5. Conclusions

Overall, the current study shows the alarming results of weight gain one year after the start of the COVID-19 pandemic in a unique cohort of children within a lifestyle intervention that already suffer from overweight or (severe) obesity. This should be a concern for not only children, parents and health care professionals, but also governments, as it threatens both the current and the future public health, with potential risks for the health of individual children, and financial and social long-term effects. It also shows that, although lifestyle interventions were forced to make changes in their programs due to regulations, such as offering (more) remote consultations, possibly more adaptations should be made to previously successful lifestyle interventions to cope with the changed circumstances of a pandemic. The outcomes of this study clearly show the subgroups that are more at risk for weight gain within a lifestyle intervention, such as children with a mother with obesity and children with high no-show rates to the lifestyle intervention. Thus, health care professionals working with children with overweight and obesity should especially focus on these two subgroups during and post-pandemic. These results offer the first opportunities to tailor lifestyle interventions for these risk groups during and after the current pandemic and future crisis periods.

Author Contributions: L.A. and A.C.E.V. designed the study, N.K. performed data collection (under supervision of J.W.L. and L.A. in Maastricht and M.V. in Venlo), L.A. performed data cleaning, and L.A. and B.W. performed data analysis. L.A. wrote the manuscript, and A.C.E.V. had the primary responsibility for the final content. All authors have read and agreed to the published version of the manuscript.

Funding: This research received no external funding.

Institutional Review Board Statement: Data of Maastricht were obtained within local regulations of the hospital and registered at the Ethics Committee of the Maastricht University Medical Centre (METC 2022-3105). The Ethics Committee stated that this research did not fall under the scope of the Medical Research Involving Human Subjects Act (WMO) and therefore no ethical approval and informed consent were needed. Data of Venlo was collected within the "Kijk op Overgewicht study"

(METC 13-4-130, Clinicaltrial.gov (NCT02091544)). All parents and/or children in Venlo gave written informed consent for this study.

Data Availability Statement: The data presented in this study are available on request from the corresponding author. The data are not publicly available due to privacy and ethical rules.

Acknowledgments: The authors would like to acknowledge all the children and parents involved.

Conflicts of Interest: The authors declare no conflict of interest.

Abbreviations

COVID-19	**Coronavirus Disease 2019**
BMI	Body mass index
BMI z-score	Body mass index z-score
COACH	Centre for Overweight Adolescent and Children's Healthcare

References

1. Coronavirus Measures in Brief. Coronavirus COVID-19. Government.nl Ministerie van Algemene Zaken. Published 20 July 2020. Available online: https://www.government.nl/topics/c/coronavirus-covid-19/tackling-new-coronavirus-in-the-netherlands/coronavirus-measures-in-brief (accessed on 16 December 2021).
2. Liguoro, I.; Pilotto, C.; Bonanni, M.; Ferrari, M.E.; Pusiol, A.; Nocerino, A.; Vidal, E.; Cogo, P. SARS-CoV-2 infection in children and newborns: A systematic review. *Eur. J. Pediatr.* **2020**, *179*, 1029–1046. [CrossRef]
3. Ten Velde, G.; Lubrecht, J.; Arayess, L.; Van Loo, C.; Hesselink, M.; Reijnders, D.; Vreugdenhil, A. Physical activity behavior and screen time in Dutch children during the COVID-19 pandemic: Pre-, during- and post-school closures. *Pediatr. Obes.* **2021**, *16*, e12779.
4. di Renzo, L.; Gualtieri, P.; Pivari, F.; Soldati, L.; Attinà, A.; Cinelli, G.; Leggeri, C.; Caparello, G.; Barrea, L.; Scerbo, F.; et al. Eating habits and lifestyle changes during COVID-19 lockdown: An Italian survey. *J. Transl. Med.* **2020**, *18*, 229. [CrossRef]
5. Adams, E.L.; Caccavale, L.J.; Smith, D.; Bean, M.K. Food Insecurity, the Home Food Environment, and Parent Feeding Practices in the Era of COVID-19. *Obesity (Silver Spring)* **2020**, *28*, 2056–2063. [CrossRef]
6. Baysun, Ş.; Ş, M. Weight gain in children during the COVID-19 quarantine period. *J. Paediatr. Child Health* **2020**, *56*, 1487–1488. [CrossRef]
7. Arayess, L.; Lubrecht, J.W.; Reijnders, R.; Hesselink, M.; Ten Velde, G.; Janse, A.; Von Rosenstiel, I.A.; van Mil, E.G.A.H.; Verweij, M.; Vreugdenhil, A.C.E. Weight changes during the COVID-19 pandemic in children with and without overweight and obesity, and the effects of prior lifestyle intervention. *Obes. Facts.* (in press)..
8. Pietrobelli, A.; Pecoraro, L.; Ferruzzi, A.; Heo, M.; Faith, M.; Zoller, T.; Antoniazzi, F.; Piacentini, G.; Fearnbach, S.N.; Heymsfield, S.B. Effects of COVID-19 Lockdown on Lifestyle Behaviors in Children with Obesity Living in Verona, Italy: A Longitudinal Study. *Obesity (Silver Spring)* **2020**, *28*, 1382–1385. [CrossRef]
9. Rundle, A.G.; Park, Y.; Herbstman, J.B.; Kinsey, E.W.; Wang, Y.C. COVID-19-Related School Closings and Risk of Weight Gain Among Children. *Obesity (Silver Spring)* **2020**, *28*, 1008–1009. [CrossRef]
10. Franckle, R.; Adler, R.; Davison, K. Accelerated Weight Gain among Children during Summer Versus School Year and Related Racial/Ethnic Disparities: A Systematic Review. *Prev. Chronic Dis.* **2014**, *11*, E101. [CrossRef]
11. Browne, N.T.; Snethen, J.A.; Greenberg, C.S.; Frenn, M.; Kilanowski, J.F.; Gance-Cleveland, B.; Burke, P.J.; Lewandowski, L. When Pandemics Collide: The Impact of COVID-19 on Childhood Obesity. *J. Pediatr. Nurs.* **2020**, *56*, 90–98. [CrossRef]
12. Umer, A.; Kelley, G.A.; Cottrell, L.E.; Giacobbi, P., Jr.; Innes, K.E.; Lilly, C.L. Childhood obesity and adult cardiovascular disease risk factors: A systematic review with meta-analysis. *BMC Public Health* **2017**, *17*, 683. [CrossRef]
13. Styne, D.M.; Arslanian, S.A.; Connor, E.L.; Farooqi, I.S.; Murad, M.H.; Silverstein, J.H.; Yanovski, J.A. Pediatric Obesity-Assessment, Treatment, and Prevention: An Endocrine Society Clinical Practice Guideline. *J. Clin. Endocrinol. Metab.* **2017**, *102*, 709–757. [CrossRef]
14. Rijks, J.M.; Plat, J.; Dorenbos, E.; Penders, B.; Gerver, W.-J.M.; Vreugdenhil, A.C.; Gerver, W. Association of TSH With Cardiovascular Disease Risk in Overweight and Obese Children During Lifestyle Intervention. *J. Clin. Endocrinol. Metab.* **2017**, *102*, 2051–2058. [CrossRef]
15. Rijks, J.; Karnebeek, K.; Van Dijk, J.-W.; Dorenbos, E.; Gerver, W.-J.; Stouthart, P.; Plat, J.; Vreugdenhil, A. Glycaemic Profiles of Children With Overweight and Obesity in Free-living Conditions in Association With Cardiometabolic Risk. *Sci. Rep.* **2016**, *6*, 31892. [CrossRef]
16. Karnebeek, K.; Thapar, S.; Willeboordse, M.; Van Schayck, O.C.P.; E Vreugdenhil, A.C. Comorbidities in Primary vs Secondary School Children With Obesity and Responsiveness to Lifestyle Intervention. *J. Clin. Endocrinol. Metab.* **2019**, *104*, 3803–3811. [CrossRef]
17. Gortmaker, S.L.; Must, A.; Perrin, J.M.; Sobol, A.M.; Dietz, W.H. Social and Economic Consequences of Overweight in Adolescence and Young Adulthood. *N. Engl. J. Med.* **1993**, *329*, 1008–1012. [CrossRef]

18. Pelone, F.; Specchia, M.L.; Veneziano, M.A.; Capizzi, S.; Bucci, S.; Mancuso, A.; Ricciardi, W.; De Belvis, A.G. Economic impact of childhood obesity on health systems: A systematic review. *Obes. Rev.* **2011**, *13*, 431–440. [CrossRef]
19. Seidell, J.C.; Halberstadt, J. The Global Burden of Obesity and the Challenges of Prevention. *Ann. Nutr. Metab.* **2015**, *66* (Suppl. 2), 7–12. [CrossRef]
20. Rijks, J.; Plat, J.; Mensink, R.; Dorenbos, E.; Buurman, W.; Vreugdenhil, A. Children with morbid obesity benefit equally as children with overweight and obesity from an on-going care program. *J. Clin. Endocrinol. Metab.* **2015**, *100*, 3572–3580. [CrossRef]
21. Ministerie van Algemene Zaken, COVID-19: Additional Measures in Schools, the Hospitality Sector and Sport, News Item | Government.nl. 2020. Available online: https://www.government.nl/latest/news/2020/03/15/additional-measures-in-schools-the-hospitality-sector-and-sport (accessed on 20 December 2021).
22. Ministerie van Onderwijs, Cultuur en Wetenschap, Alle Scholen Weer Open, Nieuwsbericht | Rijksoverheid.nl. 2020. Available online: https://www.rijksoverheid.nl/actueel/nieuws/2020/08/31/alle-scholen-weer-open (accessed on 20 December 2021). (In Dutch).
23. Government.nl. Lockdown in Order to Minimise Contact between People. 2021. Available online: https://www.government.nl/latest/news/2020/12/14/lockdown-in-order-to-minimise-contact-between-people (accessed on 20 December 2021).
24. Talma, H.; Schonbeck, Y.; Bakker, B.; Hirasing, R.A.; van Buuren, S. Groeidiagrammen 2010. Handleiding bij het Meten en Wegen van Kinderen en het Invullen van Groeidiagrammen. TNO: Leiden, the Netherlands, 2011. Available online: https://repository.tno.nl/islandora/object/uuid%3A5460503b-1519-4db0-8309-d8a63ecca01e (accessed on 2 May 2022).
25. TNO. Growth Calculator for Youth Healthcare in the Netherlands 2020. Available online: https://tnochildhealthstatistics.shinyapps.io/JGZRichtlijnLengtegroei (accessed on 16 December 2020).
26. Cole, T.J.; Roede, M.J. Centiles of body mass index for Dutch children aged 0-20 years in 1980 - a baseline to assess recent trends inobesity. *Ann. Hum. Biol.* **1999**, *26*, 303–308. [CrossRef]
27. Cole, T.J.; Lobstein, T. Extended international (IOTF) body mass index cut-offs for thinness, overweight and obesity. *Pediatr. Obes.* **2012**, *7*, 284–294. [CrossRef]
28. Karnebeek, K.; Rijks, J.M.; Dorenbos, E.; Gerver, W.-J.M.; Plat, J.; Vreugdenhil, A.C.E. Changes in Free-Living Glycemic Profiles after 12 Months of Lifestyle Intervention in Children with Overweight and with Obesity. *Nutrients* **2020**, *12*, 1228. [CrossRef]
29. Kirk, S.; Zeller, M.; Claytor, R.; Santangelo, M.; Khoury, P.R.; Daniels, S.R. The Relationship of Health Outcomes to Improvement in BMI in Children and Adolescents. *Obes. Res.* **2005**, *13*, 876–882. [CrossRef]
30. Bell, L.M.; Byrne, S.; Thompson, A.; Ratnam, N.; Blair, E.; Bulsara, M.; Jones, T.W.; Davis, E.A. Increasing Body Mass Index z-Score Is Continuously Associated with Complications of Overweight in Children, Even in the Healthy Weight Range. *J. Clin. Endocrinol. Metab.* **2006**, *92*, 517–522. [CrossRef]
31. Wilfley, D.E.; Tibbs, T.L.; Van Buren, D.; Reach, K.P.; Walker, M.S.; Epstein, L.H. Lifestyle interventions in the treatment of childhood overweight: A meta-analytic review of randomized controlled trials. *Health Psychol.* **2007**, *26*, 521–532. [CrossRef]
32. Fuemmeler, B.F.; Lovelady, C.A.; Zucker, N.L.; Østbye, T. Parental obesity moderates the relationship between childhood appetitive traits and weight. *Obesity* **2012**, *21*, 815–823. [CrossRef]
33. Gray, L.A.; Alava, M.H.; Kelly, M.P.; Campbell, M.J. Family lifestyle dynamics and childhood obesity: Evidence from the millennium cohort study. *BMC Public Health* **2018**, *18*, 1–15. [CrossRef]
34. Razi, M.; Nasiri, A. Concerns of parents about children's overweight and obesity during the COVID-19 pandemic: A qualitative study. *J. Pediatr. Nurs.* **2021**, *63*, 111–116. [CrossRef]
35. Fegert, J.M.; Vitiello, B.; Plener, P.L.; Clemens, V. Challenges and burden of the Coronavirus 2019 (COVID-19) pandemic for child and adolescent mental health: A narrative review to highlight clinical and research needs in the acute phase and the long return to normality. *Child Adolesc. Psychiatry Ment. Health* **2020**, *14*, 20. [CrossRef]
36. Grootens-Wiegers, P.; van den Eynde, E.; Halberstadt, J.; Seidell, J.C.; Dedding, C. The "Stages towards Completion Model": What helps and hinders children with overweight or obesity and their parents to be guided towards, adhere to and complete a group lifestyle intervention. *Int. J. Qual. Stud. Health Well-Being* **2020**, *15*, 1735093. [CrossRef]
37. Alves, J.M.; Yunker, A.G.; DeFendis, A.; Xiang, A.H.; Page, K.A. BMI status and associations between affect, physical activity and anxiety among U.S. children during COVID-19. *Pediatr. Obes.* **2021**, *16*, e12786. [CrossRef]
38. Zolotarjova, J.; Ten Velde, G.; Vreugdenhil, A.C.E. Effects of multidisciplinary interventions on weight loss and health outcomes in children and adolescents with morbid obesity. *Obes. Rev.* **2018**, *19*, 931–946. [CrossRef]
39. Karim, S.S.A.; Karim, Q.A. Omicron SARS-CoV-2 variant: A new chapter in the COVID-19 pandemic. *Lancet* **2021**, *398*, 2126–2128. [CrossRef]
40. Ministerie van Algemene Zaken. (14 December 2021); Evening Closures Prolonged, Primary Schools to Shut a Week Early before Christmas. News Item | Government.nl. Available online: https://www.government.nl/latest/news/2021/12/14/evening-closures-prolonged-primary-schools-to-shut-a-week-early-before-christmas (accessed on 16 December 2021).

MDPI
St. Alban-Anlage 66
4052 Basel
Switzerland
Tel. +41 61 683 77 34
Fax +41 61 302 89 18
www.mdpi.com

Nutrients Editorial Office
E-mail: nutrients@mdpi.com
www.mdpi.com/journal/nutrients